Troubleshooting
Electronic Circuits

About the Author

Ronald Quan has a BSEE degree from the University of California at Berkeley and is a member of SMPTE, IEEE, and the AES. With a General Radiotelephone license, he has worked as a broadcast and maintenance engineer for FM and AM radio stations, and he also holds an Extra Class amateur radio license. Before becoming a design engineer, he worked in TV-Radio-HiFi repair shops.

He is the author of *Build Your Own Transistor Radios* (McGraw-Hill/TAB Electronics, 2012) and *Electronics from the Ground Up* (McGraw-Hill/TAB Electronics, 2014).

For over 30 years he has worked for companies related to video and audio equipment (Ampex, Sony, Macrovision, Monster Cable, and Portal Player). At Ampex, he designed CRT (cathode ray tube) TV monitors and low-noise preamplifiers for audio and video circuits. Other designs included wide-band FM detectors for an HDTV tape recorder at Sony Corporation, and a twice color subcarrier frequency (7.16 MHz) differential phase measurement system at Macrovision, where he was a Principal Engineer. Also at Macrovision he designed several phase lock loop circuits to provide a regenerated clock signal from the raw EFM (eight-fourteen modulation) data stream from a CD player.

At Hewlett Packard, working in the field of opto-electronics, he developed a family of low-powered bar code readers, which used a fraction of the power consumed by conventional light pen readers.

Currently, he is the holder of more than 450 worldwide patents (which includes over 90 United States patents) in the areas of analog video processing, video signal noise reduction, low-noise amplifier design, low-distortion voltage-controlled amplifiers, wide-band crystal voltage-controlled oscillators, video monitors, audio and video IQ modulation, in-band carrier audio single-sideband modulation and demodulation, audio and video scrambling, bar code reader products, and audio test equipment.

Also, he has served as an Adjunct Lecturer and Lecturer in Stanford University's Electrical Engineering Department.

Troubleshooting Electronic Circuits

Debugging and Improving Your DIY Projects and Experiments

Ronald Quan

New York   Chicago   San Francisco
Athens   London   Madrid
Mexico City   Milan   New Delhi
Singapore   Sydney   Toronto

Library of Congress Control Number: 2019944720

McGraw-Hill Education books are available at special quantity discounts to use as premiums and sales promotions or for use in corporate training programs. To contact a representative, please visit the Contact Us page at www.mhprofessional.com.

Troubleshooting Electronic Circuits:
Debugging and Improving Your DIY Projects and Experiments

1 2 3 4 5 6 7 8 9   LCR   25 24 23 22 21 20

ISBN   978-1-260-14356-0
MHID   1-260-14356-2

This book is printed on acid-free paper.

Sponsoring Editor
Lara Zoble

Editing Supervisor
Donna M. Martone

Production Supervisor
Lynn M. Messina

Acquisitions Coordinator
Elizabeth Houde

Project Manager
Patricia Wallenburg, TypeWriting

Copy Editor
Claire Splan

Proofreader
Alison Shurtz

Indexer
WordCo Indexing Services, Inc.

Art Director, Cover
Jeff Weeks

Composition
TypeWriting

Contents

Preface

This book on troubleshooting analog circuits is intended for the hobbyist, student, educator, or engineer. Because the subject is more into debugging or improving circuits, the book does not require the reader to have an engineering background.

Students who take electronics classes at the technician or engineering level may find this book useful. Some chapters will include basic electronics theory (with some high school algebra). However, we will not be looking for exact calculations when troubleshooting. Instead, approximations will be presented.

Hobbyists who read this book will not only learn troubleshooting techniques. They will also gain insight into why some circuits perform badly. Suggested circuit modifications will be offered to improve the performance of these types of circuits.

Acknowledgments

I would like to thank very much Lara Zoble of McGraw-Hill.

Also, thanks to Lauren Poplawski at McGraw-Hill, and to Michael McCabe, formerly of McGraw-Hill.

In transforming my manuscript and figures into a wonderful book, I owe much gratitude to project manager Patricia Wallenburg, copyeditor Claire Span, proofreader Alison Shurtz, and indexer Stephen Ingle.

Writing this book required many suggestions, and Andrew Mellows not only proofread the entire manuscript but also gave me very helpful comments on the technical matter. Thank you very much, Andrew, for reviewing my work over the past two years.

Some material in this book pertaining to building circuits and entry-level electronics came from working with many students and Professor Robert W. Dutton at Stanford University. I give my greatest thanks to Bob Dutton for inviting me into the Electrical Engineering Department as a mentor and as a lecturer for his classes.

Also, I owe immense gratitude to Professor Thomas H. Lee at Stanford University, who was responsible for my meeting with Bob Dutton in the first place and who throughout the years has encouraged me in my electronics research. And, of course, thank you to Stanford University.

Another person who inspired me is Professor Robert G. Meyer at the University of California, Berkeley. His class notes, books, and lectures played an important role for me in writing about distortion analysis. So thank you, Bob Meyer!

With much appreciation, I also need to mention my two mentors from industry, John O. Ryan and Barrett E. Guisinger. Both gave me a world-class education in analog and video circuits design. One other mentor I need to thank is John Curl.

During this whole process I have been indebted to the following friends for encouraging and supporting me: Alexis DiFirenzi Swale, Germano Belli, Jeri Ellsworth, Phil Sittner, Edison Fong, and Amy Herndon.

I really got into my second career as an author/writer because of Paul Rako. He connected me with Roger Stewart and that started all this. Thanks to you, Paul and Roger.

William K. Schwarze, who was my best teacher at Galileo High School, also deserves thanks for his ability to convey mathematical concepts in a very clear manner to me. Also, my friend James D. Lee continues to be an inspiration to me.

Finally, I acknowledge a big debt of gratitude to members of my family: William, George, Thomas, and Frances, for their support.

And, of course, I dedicate this third book to my parents, Nee and Lai.

Troubleshooting Electronic Circuits

CHAPTER 1

Introduction

This book will cover mostly analog circuits and will include a few timer and logic circuits.

Troubleshooting electronic circuits takes a great deal of detective work. We must have a good idea of how the circuit should work. For example, in an amplifier, we would expect that the amplifier's output provides a larger voltage or current compared to the input signal.

When a circuit does not work as expected, we should check the wiring and solder connections, then double-check the components' value (e.g., correct value resistors and capacitors) along with confirming that the power supply is connected properly. With other test equipment such as a voltmeter or oscilloscope, we begin to trace parts of the circuit for direct current (DC) voltages and alternating current (AC) signals. Eventually we may find where the AC signal disappears, which then allows us to replace a part or confirm whether the original part was connected correctly.

I am writing this book based on observations from my students and colleagues. In most cases, you do not have to be an electrical engineer to troubleshoot circuits. As a matter of fact, many fresh-out-of-school graduates will have limited skill in troubleshooting if they did not have electronics as a hobby most of their lives.

For the hobbyist, just a basic knowledge of using electronic components and test equipment goes a long way. It is more likely that sometimes a hobbyist or a technician will be a better troubleshooter than a young engineering graduate. The reason is that troubleshooting is more of an experience-based skill. If you have been working with many different circuits by building them and probing them (e.g., with a voltmeter and oscilloscope) you will have learned some essential practical knowledge. And if you start learning some electronics theory combined with practical electronics knowledge that includes understanding how to use volt-ohm-meters, oscilloscopes, and signal generators, then you will be able to troubleshoot circuits even better.

To troubleshoot analog circuits, we need to understand components very well via their data sheets (e.g., maximum voltage or power ratings), know how to use test equipment, and understand signals.

Goals of this Book

We will start with components because understanding their specifications and limitations is key to troubleshooting. For example, if you are using a 12-volt circuit but your component is rated at 10 volts, then most likely that component will fail at some point. Also, in some cases, which are rare, an integrated circuit's (IC's) pinouts may be different between its through-hole version (e.g., 8-pin DIP, dual inline package) and its surface mount SO-8 (small outline, 8-pin) counterpart, such as the AD633 analog multiplier chip.

Therefore, the first six chapters are devoted to passive devices, breadboards, and volt ohm milliamp meters. We show how to test these types of components, which is really essential these days because a 10 pf capacitor and a 1 μf capacitor may look nearly identical (e.g., in size) with really small print that is hard to read and easy to mistake one for the other.

When building circuits, we should verify the components' value by inspection and sometimes by measuring them (e.g., for resistance, capacitance, and polarity or confirming NPN or PNP transistors) before soldering or placing them into a circuit board. A variation of the old saying of "measure twice and cut once" can be readapted to "measure twice and solder once."

From Chapters 7 to 15, amplifying devices, integrated circuits, and some electronic systems are presented as examples for troubleshooting audio, RF (radio frequency), timer, and power supply circuits. Some of these circuits come from older hobbyist magazines, books, or postings on the web. We will explore circuits that "kind of work," which can be improved, debugged, or modified for better performance.

Quick Notes: Replacing Electrolytic Capacitors and Soldering

For now, here's a short summary on troubleshooting techniques for the experienced hobbyist concerning (aluminum) electrolytic capacitors.

If you are restoring older electronic devices such as radios, tape recorders, stereos, power supplies, signal generators, etc., you should most likely replace all the electrolytic capacitors with the same (or sometimes larger) capacitance and voltage rating. For example, if there is a 100 μf, 10-volt electrolytic capacitor that has lost its capacitance or became leaky, you can replace it with a new 100 μf, 16-volt version. Of course, make sure that the replacement capacitor is installed correctly in terms of polarity. Also, if the device runs off the power outlet (e.g., 110 volts AC or 220 volts AC), then make sure you completely disconnect its 110v/220v power cord from the wall (or mains) outlet before repairing the device.

Even in many newer devices, you can spot a bad electrolytic capacitor by noticing if the case is bulging. For example, normally the top of an aluminum electrolytic capacitor is flat, but a bad one might be curved up like having a mound added on top of the capacitor. Also, if there are chemical residues near the leads (e.g., white or blue powder, gel, or liquid), these are telltale signs of a bad electrolytic capacitor. See Figure 1-1 that shows a residue leak.

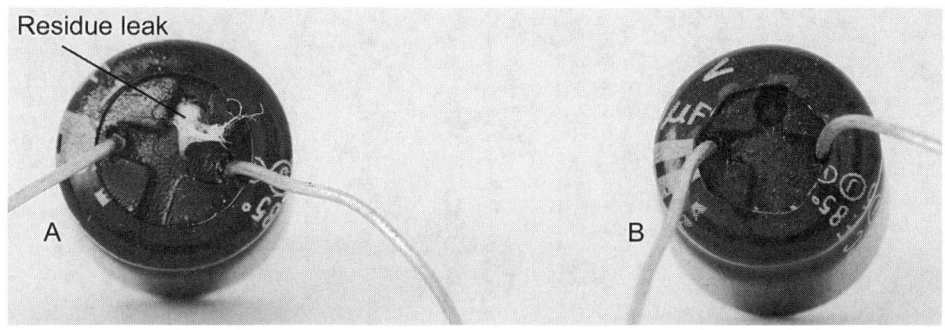

FIGURE 1-1   A bad capacitor on the left side (A), and a good capacitor on the right side (B).

Should you decide to repair or restore older electronic devices from the 1950s to 1980s, keep in mind that aluminum capacitors have about a 15-year life. However, if they are in a warmer environment such as in a TV camera, electrolytic capacitors can fail within 5 years.

Soldering Problems (Cold Solder Connections)

If you are starting out in electronics, then you can build circuits with solderless breadboards, solder your own circuits, or purchase preassembled soldered boards. For troubleshooting, sometimes you will be required to solder when working with new circuits, replacing components, or repairing/modifying printed circuit assemblies. When soldering connections, we must avoid making cold solder joints where the solder may have been "tacked" or "dabbed" onto the circuit, or the solder had not melted sufficiently. See Figure 1-2.

In Figure 1-2, we see that all connections can be improved by just reheating and adding a little more solder until the connections create a smooth mound of solder. If your connections have a "spiky" look similar to number 3 in Figure 1-2, then you may only have a temporary connection that can easily be disconnected with a slight pull on the wire. Generally, spend an extra few seconds on the solder joint to allow the solder to flow all around the connection. See Figure 1-3.

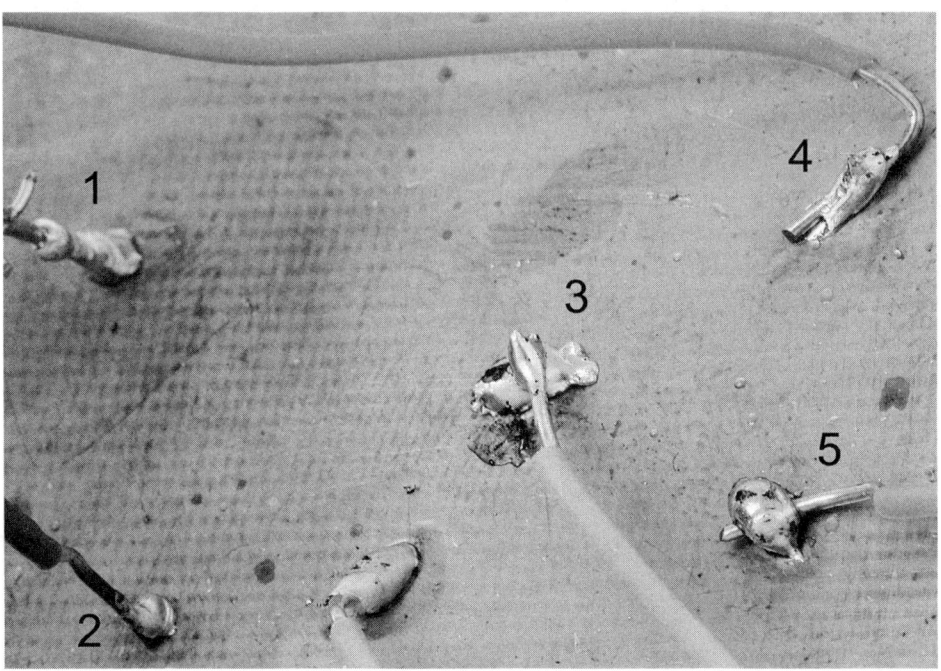

FIGURE 1-2   Bad solder connection examples.

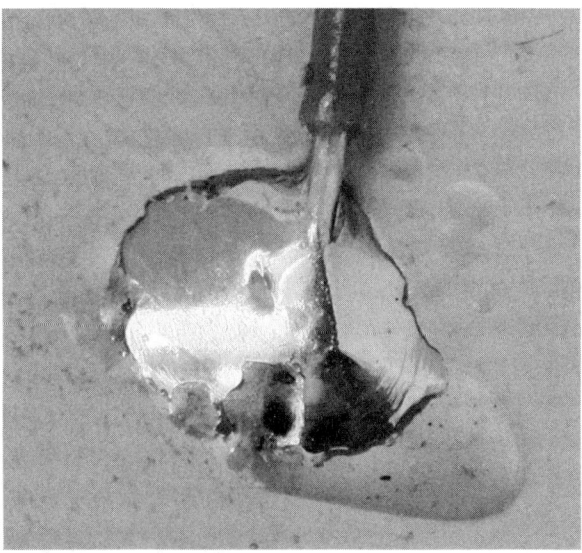

FIGURE 1-3   A good connection where the solder has been heated sufficiently to flow properly.

Summary

Throughout this book we will include basic electronics circuit theory such as Ohm's Law applied to the associated circuits that are discussed.

Finally, do not forget to read Appendix A, which covers choosing test equipment (e.g., power supplies, oscilloscopes, and generators) and shows some of their limitations.

And now let's proceed to Chapter 2.

CHAPTER 2

Basic Breadboards

This chapter will examine various breadboards for constructing circuits. We will cover solderless breadboards, copper clad bare printed circuit boards, and perforated or vector boards.

Let's first look at solderless breadboards, which can vary in size and quality.

Solderless Breadboards

In Figures 2-1 and 2-2, we see two different types of solderless breadboards. Note that each of them has tabs (e.g., Tab1, Tab2, and Tab3) that allow multiple boards of the same type/size to be expanded.

Let's start with Figure 2-1's solderless breadboard. In each column, there is a column number (e.g., 1 to 30) and five rows (a to f and g to l).

Each column (e.g., column 1 to 30) is connected from rows a to f in the upper portion of the breadboard. And on the lower portion each column (e.g., column 1 to 30) from rows g to l is connected.

The columns 1 to 30 from the upper portions with rows a to f are independent and not connected to any of the columns in the lower portion, such as columns 1 to 30 and row g to l.

Stated in other words, each column is independent and not connected to any other column, even if they are on the upper or lower portion of the breadboard. For example, column 1 is insulated and not connected to any other column such as column 2.

For the example in Figure 2-2, each column from 0 to 60 on the top rows, and on the lower rows the five locations in rows A to E and F to J are connected together. For example, the hole in column 1 row A is only connected to column 1 rows B, C, D, and E. The same connections apply to the other columns, for example, column 3 row F is only connected to column 3 rows G, H, I, and J.

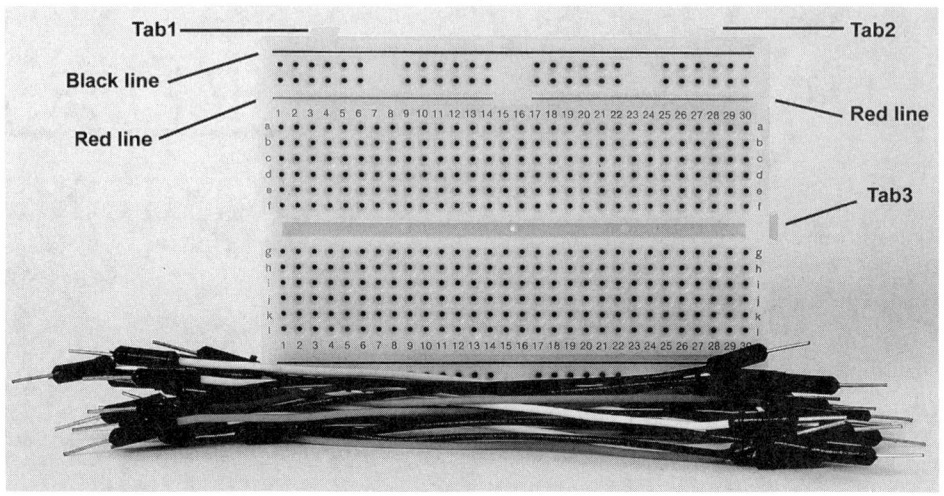

FIGURE 2-1   A small 30-column solderless breadboard with wire jumper connectors.

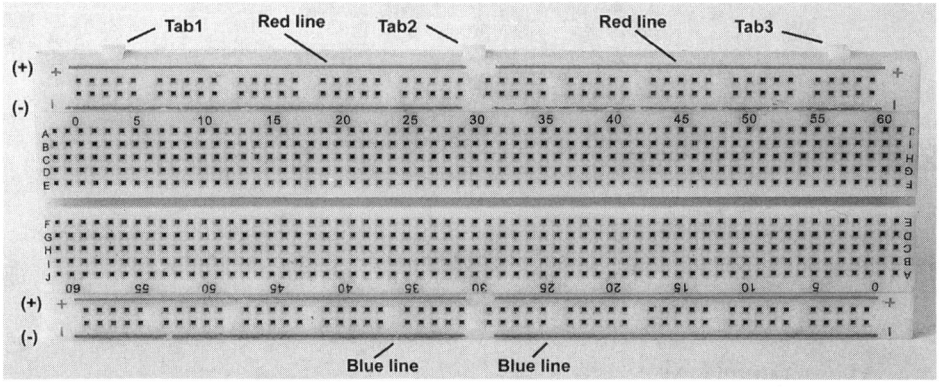

FIGURE 2-2   A larger-sized solderless jumper board with 60 columns. The power buses are denoted by the red and blue lines running across the solderless breadboard.

There are two power buses denoted by red and blue lines toward the edge of the board. We will discuss power buses in more detail later; some are broken into "sectors" within the board, and others are not.

Quality

Figure 2-3 shows a high-quality solderless breadboard where the connectors inside the wells or holes do not cause wire or leads from electronic parts to be jammed.

As we can see in Figure 2-4, the holes or wells labeled "Bad" are closed off, leaving less area to insert wire leads from components. The metal connector clips inside the holes should be expanded outward to allow a larger hole or well size for easier lead or wire insertion.

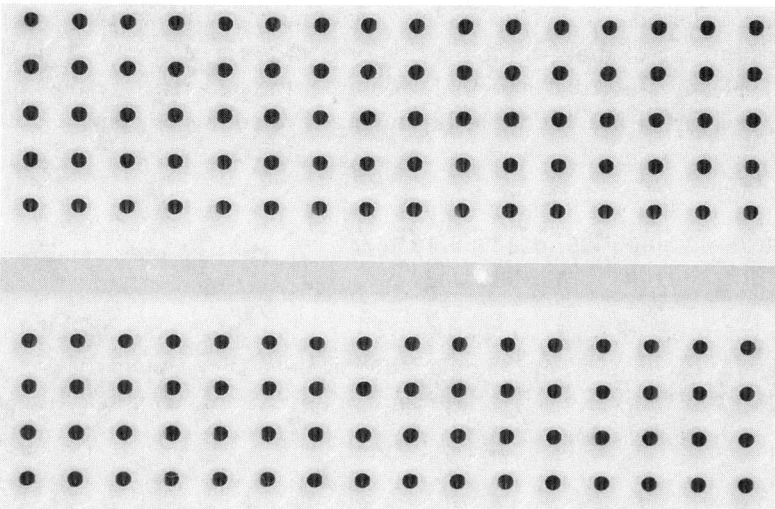

FIGURE 2-3 A high-quality solderless breadboard where the wells or holes are clear and allow for easy insertion of wire leads from electronics components such as resistors, capacitors, etc.

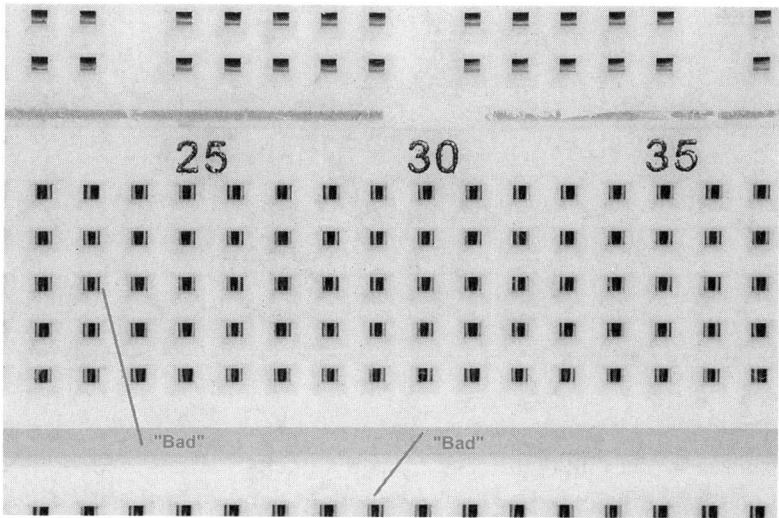

FIGURE 2-4 A lower-quality solderless breadboard with internal connectors or "blades" narrowing the pathways for inserting wires or electronic components.

So be on the lookout for these and avoid them because plugging in resistors, capacitors, and wires will be difficult. One possible workaround on this is to order parts (e.g., resistors and capacitors) with thinner leads. However, you may find that the standard solderless jumper wires may have difficulty plugging in with these partially blocked or narrowed wells.

You can order boards that are made by well-known manufacturers such as Bud Industries, BusBoard, and Twin Industries, or electronics vendors such as Jameco, Mouser, Digi-Key, and Adafruit. Should you order the solderless breadboards from Amazon, look for the user ratings before buying. If you order via eBay, you may find some that are not so good.

Power Buses on Solderless Breadboards . . . Look for Breaks in the Power Bus Lines

Some solderless breadboards will partition one or more power buses to allow different voltages to be applied. For example, a digital and analog system may require +5 volts and +12 volts, respectively. The ground or minus connection may be common to both power supplies. See Figure 2-5.

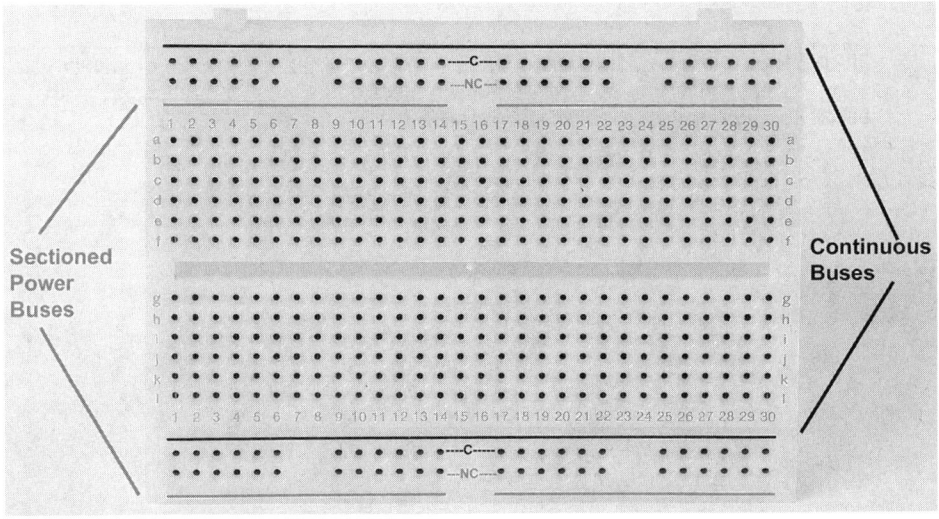

FIGURE 2-5   A solderless breadboard with sectioned or separated power buses as denoted by a break in the red/gray lines.

The solderless board in Figure 2-5 shows that the power bus used for ground or common power supply minus lead is continuous as denoted by the black lines from columns 1 through 30. Columns 14 and 17 are internally connected and are marked in Figure 2-5 by "-----C----" in black font overlays.

However, the power bus with the red/gray lines from columns 1 to 14 is not connected to columns 17 to 30 and there are "no connects" denoted by "---NC---" in red/gray font overlays.

In this example, it is possible to have up to four different power supply voltages with a common connection or ground connection.

To be sure, you should use a continuity tester or ohm meter to confirm no connections between columns that are marked to be separated, which will show "infinite ohms" for the two columns 14 and 17 in red/gray indicated by "---NC---". The two columns 14 and 17 in black with "----C---" should measure continuity or close to 0 ohms.

Figure 2-5 is not indicative of all solderless breadboards in terms of separate buses. For example, Figure 2-6 shows a double-size board with continuous buses throughout.

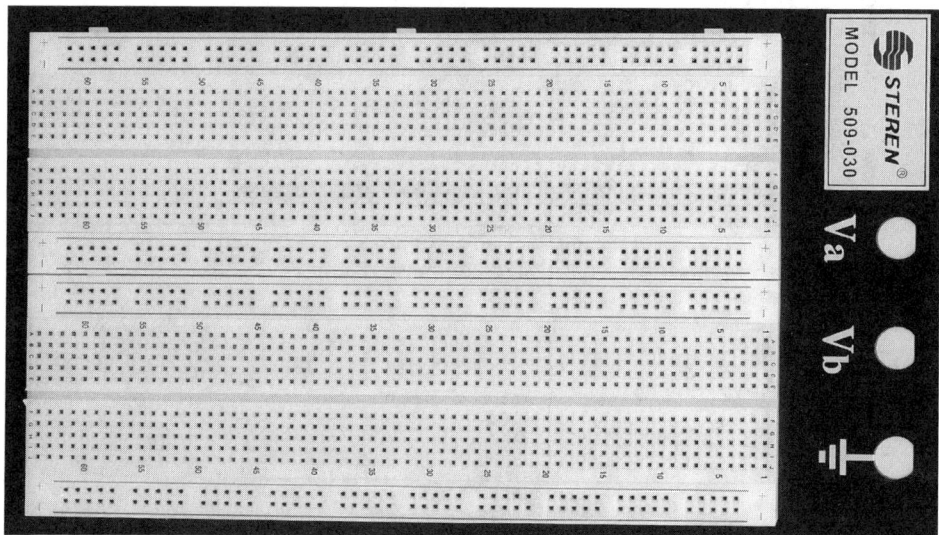

FIGURE 2-6 A solderless breadboard without separations in the "+" and "−" bus lines

One reason for having separate buses is so that a larger board can have multiple sectors with different types of circuits requiring different power supply voltages. In general, you should always test for continuity on all buses around the middle of the board.

Figure 2-7 shows a board with four "independent" sectors.

If you want to connect the power buses together, you can use insulated or uninsulated wires as shown in Figure 2-8.

See Figure 2-9 for a close-up of the bare wires connecting a couple of the power buses.

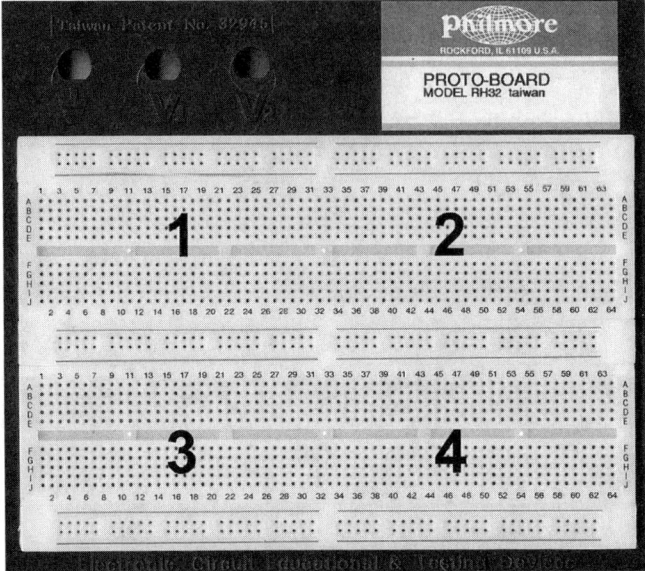

FIGURE 2-7   A board with four sectors and independent power bus lines as denoted by the breaks in the six pairs of lines at columns 32 and 33.

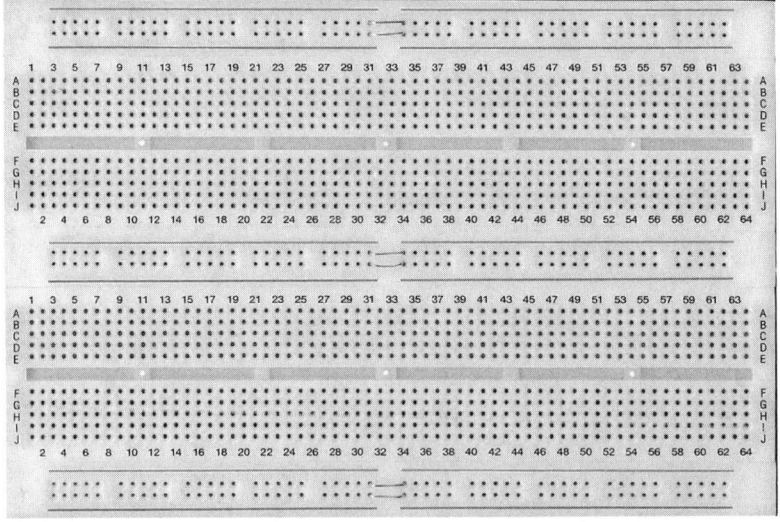

FIGURE 2-8   Bare (un-insulated) wires connecting the power buses.

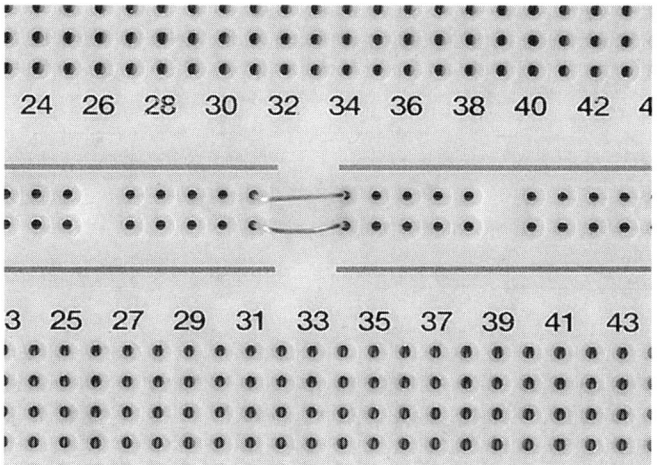

FIGURE 2-9   Bare wires connecting the two power buses.

And Now Some Words of Caution

If the power bus lines are mismarked or misinterpreted in any way, damage to your circuits or power supply may occur. For example, if you think that the power buses are connected in the middle and apply power, you may be only powering half the board.

On the other hand, if the board is marked with a break in the power bus lines, but in fact they are really connected, then when you apply different voltages on the allegedly different power buses, you may short-circuit one supply to another, or send too high of a voltage into circuits that cannot tolerate the high voltage. For example, if you have +5-volt and +12-volt supplies, then the +12-volt supply could send 12 volts throughout the board and into digital circuits that are normally running at 5 volts. With this example, digital chips will be "fried" or damaged, and worse yet it can cause smoke from the chips, or even injury. See Figure 2-10.

The user should confirm continuity on the top board at columns 29 and 31 for the "–" and "+" buses, which were confirmed to be connected for all buses. The top board is incorrectly marked and should have had solid bus lines drawn like in the middle and bottom boards.

This is why before you apply power to your solderless breadboard, you should check the power buses for continuity in the middle of the board.

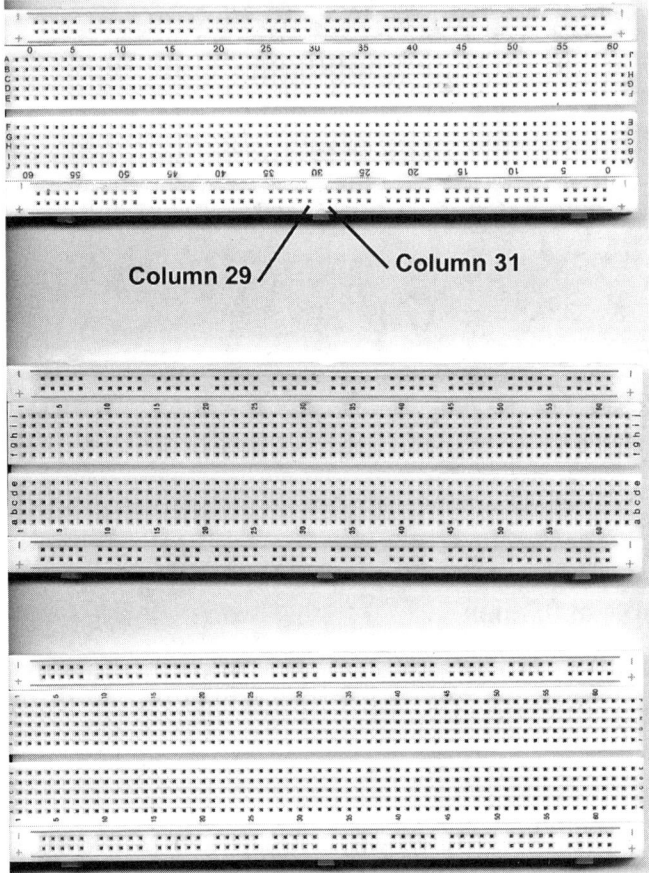

FIGURE 2-10 All three boards have continuous bus connections, including the top board that has breaks in the "−" and "+"bus lines that would have indicated separated "−" and "+" bus lines (red and blue).

Other Breadboards

For analog circuits up to about 10 MHz, including shortwave radios, I have been able to produce satisfactory results with solderless breadboards. For higher frequency circuits, use a copper clad board. See Figure 2-11.

For high-frequency circuits up to 500 MHz, copper clad breadboards work fine. Figure 2-11 shows on the left side a circuit made with through-hole parts. For very high frequencies, sometimes you can use smaller parts, including surface mount resistors, capacitors, and semiconductors. Some of these components will be covered in the subsequent chapters.

FIGURE 2-11   Left side is a "dead bug" style copper clad construction and right side shows a blank copper clad printed circuit board.

If you are building logic circuits, then solderless breadboards can work for low-speed logic gates such as CD4000 series or 74Cxx CMOS gates. If you use faster gates such as 74HCxx gates, you will have to keep track of pulse glitches from the chips propagating through the ground and/or power supply bus. One way to filter out the glitches is to slow down the rise/fall times via an RC filter at the output of each gate. But these RC networks can add up to too many.

The best alternative then is to build on a blank printed circuit board, a vector board with ground plane, or to lay out a PC board. For the hobbyist, usually the vector board (perforated board) approach is a good balance. See Figure 2-12.

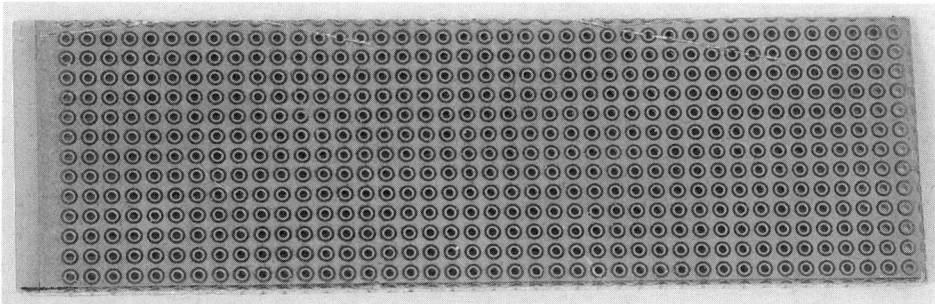

FIGURE 2-12   A perforated board with a ground plane.

Perforated boards can be made with fiberglass such as the one shown in Figure 2-12, or you can buy more inexpensive phenolic types.

The hole spacing in vector or perforated boards is usually 0.100 inch or 100 mils. This is a standard spacing for many through-hole integrated circuits, connectors, LEDs, transistors, and jumper terminals.

CHAPTER 3

Power Sources: Batteries and Battery Holders, Safety Issues, and Voltmeters

We first cover different types of batteries and their holders. In particular, some holders have similar connectors and can cause the wrong voltage to be supplied to your circuits. Also, we will take a look at various digital voltmeters that will be one of your main troubleshooting tools.

Batteries

Many circuits today can be powered by about 3 volts to 9 volts. These voltages can be supplied by batteries connected in series. Although there are different types of 1.5-volt battery sizes, such as AAA, AA, C, and D, the most common ones to use are AA. There are the shorter-life carbon zinc or manganese batteries, and longer-life alkaline cells. All of these are "primary" cells, which must be discarded after they run out. That is, they are *not* rechargeable!

There are "standard" rechargeable batteries, such as nickel cadmium (NiCd) and nickel metal hydride (NiMh). However, these batteries will lose their charges within a few months.

A better investment is to buy "low self-discharge" nickel metal hydride (NiMh) batteries. Low self-discharge rechargeable batteries will retain about 80 percent capacity one year after a full charge. The typical capacity of these cells is about 2000 milliamp-hour for the AA, 4000 milliamp-hour for the C, and 8000 milliamp-hour for the D size.

Battery capacity in milliamp-hour gives us a good estimate of battery life based on current drain. For example, suppose you are powering an LED flashlight with an

AA NiMh rated at 2000 milliamp-hour, and the LED flashlight drains 200 milli-amps. We can then calculate the flashlight's duration to give off light as:

$$\text{time or duration} = (\text{battery capacity})/(\text{current drain})$$

In this example, battery capacity for an AA cell is 2000 milliamp-hour, and current drain is 200 milliamp.

$$\text{time or duration} = (2000 \text{ milliamp-hour})/(200 \text{ milliamp}) = 10 \text{ hours}$$
$$\text{time or duration} = 10 \text{ hours}$$

Rechargeable NiMh 9-volt batteries also come in the low self-discharge variety. Usually, the other key words to look out for in low self-discharge rechargeable batteries are

- "Ready to use"
- "Precharged" or "Pre-Charged" or "Factory Charged"
- "Charge Retention, up to 12 months" or "keeps charged up to 12 months"
- "Long shelf life"

Here are a few manufacturers of rechargeable long–shelf life batteries:

- Eneloop by Panasonic or Sanyo
- energyOn
- Eveready
- Rayovac

Note that NiMh batteries give about 1.25 volts per cell, but fortunately, most electronic circuits will operate in the 2.5-volt (2 cells in series) to 7.2-volt (6 cells in series) range.

See Figure 3-1, which shows various double layered AA cell battery holders.

FIGURE 3-1   AA battery holders for 4, 6, 8, and 10 cells.

Note the battery holders' positive (+) and negative (−) polarity markings. The smaller connector on the left side of each battery holder is the positive terminal, while the larger diameter connector toward the right side is the negative terminal.

See Table 3-1 for a voltage summary on these holders.

TABLE 3-1   Nominal Output Voltages Based on Number of AA Cells

Number of Cells	Voltage Rechargeable NiMh	Voltage Alkaline or Carbon
4	5.0 volts	6.0 volts
6	7.5 volts	9.0 volts
8	10.0 volts	12.0 volts
10	12.5 volts	15.0 volts

Some battery holders come in single layer construction. See Figure 3-2.

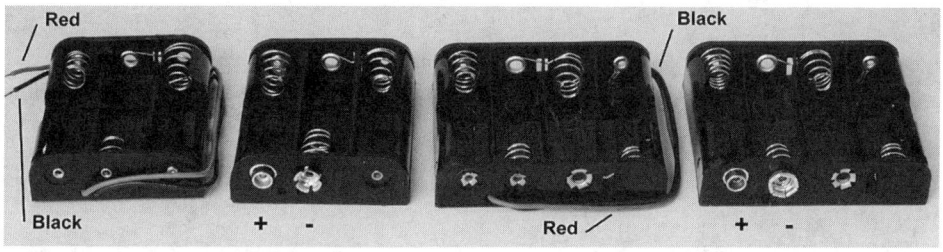

FIGURE 3-2   Flat battery holders with leads (red wire is positive and black wire is negative) and with connectors.

Battery holders with leads are preferred sometimes so that there is no chance of accidentally supplying the wrong voltage to a specific circuit. The reason is because if you set aside a circuit for a while, you may forget what the circuit voltage is. If it has a "9-volt" connector, then the first assumption is to connect 9 volts to the circuit, which may burn it out because the circuit may have 5-volt logic circuits. See Figures 3-3 and 3-4.

Again, a Word of Caution

When working with battery connectors, you must keep track of your operating voltage. If we are not careful, we can mistakenly apply the wrong voltage to the circuit. See Figure 3-4, which shows two different battery voltage sources that have the same connector.

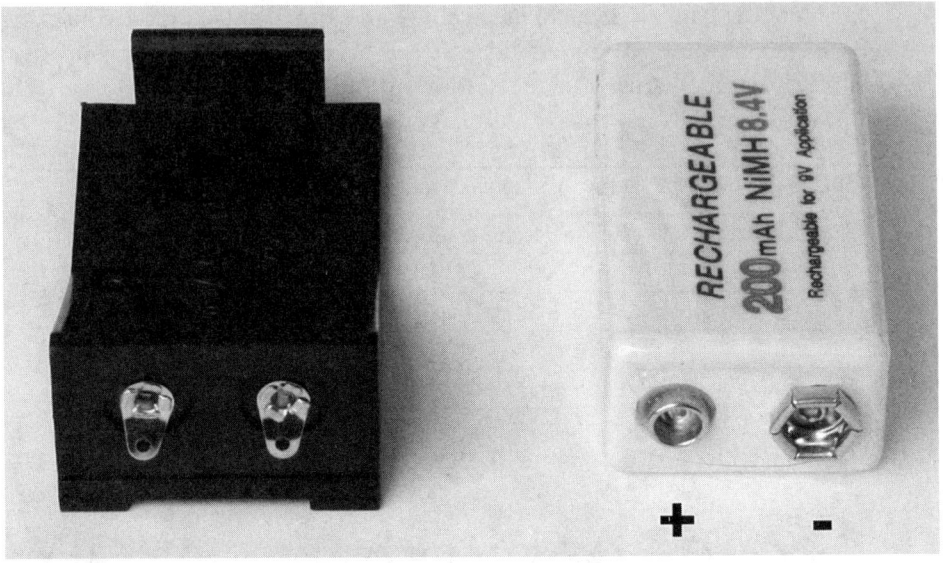

FIGURE 3-3   A 9-volt battery holder with tabs to solder wires to, and a rechargeable 9-volt battery.

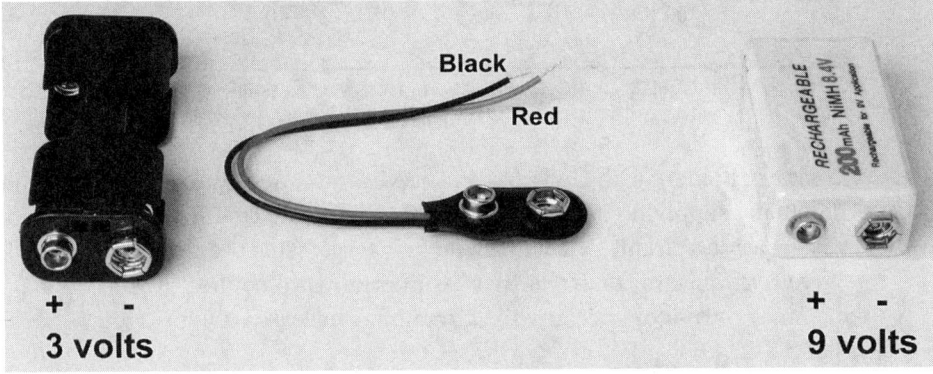

FIGURE 3-4   A 3-volt battery holder, a battery connector where the red lead is positive and the black lead is negative, and a 9-volt battery.

Because AA cell battery holders use the same battery connector as a 9-volt battery, we can inadvertently apply the wrong supply voltage to a circuit, which can destroy some of the electronic components. For example, if you are building a 3-volt (digital) logic circuit, then you can use the 3-volt battery holder with 2 AA cells and the battery connector in Figure 3-4. But it is possible, by accident, that the battery connector is snapped onto a 9-volt battery, which will most likely destroy the 3-volt logic circuit. So, we just need to be careful by labeling the battery connector with the operating voltage.

Expected Battery Capacity

You can use a low self-discharge 9-volt NiMh battery in many electronic circuits. Just be mindful that they have about 200 milliamp-hour capacity, which is about 10 percent of the AA cells that have generally about a 2000 millamp-hour rating. For projects draining less than 50 mA, the rechargeable 9-volt NiMh battery will work fine. Note that most 9-volt NiMh actually deliver about 9 volts (e.g., 7 cells at 1.25 volts each yield 8.75 volts).

Safety Considerations

If you are using rechargeable batteries, be sure to use only the chargers made for these types of batteries. For example, you can usually purchase a charger along with a set of rechargeable batteries.

Battery chargers are *not* to be used with primary batteries such as alkaline, carbon zinc, manganese, and especially lithium cells. For safe practice, do not ever charge any of these batteries!

We will now examine voltmeters, which can confirm the power or battery supply is providing the expected voltage. In particular we will be looking into digital voltmeters (DVMs), which can test voltages from batteries or DC power supplies.

Survey of Digital Voltmeters

The most basic and useful troubleshooting equipment that you can purchase is a volt-ohm-milliamp meter to test for voltage (volts), resistance (ohms), and current (milliamps). Today they are more widely known as digital voltmeters, or DVMs.

One of the basic rules in troubleshooting circuits is to check the power supply with a voltmeter. For example, when a circuit does not turn on, the first thing to check is whether the circuit is receiving the correct voltage. If the voltage is not present, sometimes it's important to make sure that the power switch is turned on, or the wires from the power source are connected to the circuit.

The DVM can measure not only the voltage at the circuit, but also confirm with its ohm meter function that a switch or wire has continuity or low resistance.

For now, let's take a look at some digital voltmeters, which are generally priced from about $6 for the basic ones to $150 for more precision measurements. We will see that because the DVM's performance-to-price ratio has increased steadily over the years, you can invest in a very versatile DVM that measures not only the voltage, resistance, and current, but also measures characteristics in *transistors*, *diodes*, and *capacitors*.

With the DVM, we can test the battery voltage, and in some cases, you can purchase an inexpensive analog volt-ohm-milliamp meter that tests 1.5-volt and 9-volt batteries.

When troubleshooting a circuit, sometimes it helps to monitor voltage and current at the same time, or monitor two different voltage points. Because the prices for certain DVMs are relatively inexpensive, you can obtain two or three of these without breaking your budget.

Let's take a look at two similar-looking DVMs in Figure 3-5.

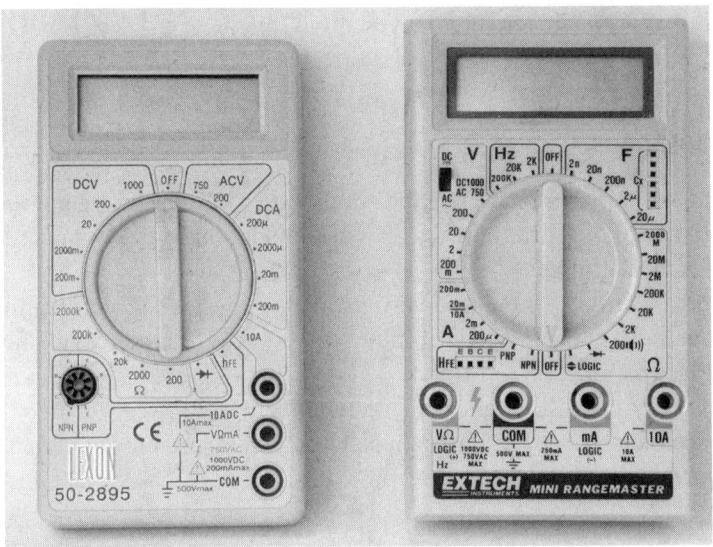

FIGURE 3-5 Two similar DVMs but with vastly different performance levels.

The Lexon (on the left) or the equivalent DT832 DVM (not shown) is rather inexpensive (approximately $5 to $17 as of 2019) and is good for many general measurements. If you compare the Lexon with the Extech, you will notice that the resistance range is limited to 2 million ohms (2000k), whereas the Extech goes up to 2000 million ohms (2000M). Another major difference is in their AC voltage range, which is much more extended down to 200 millivolts AC full scale for the Extech compared to 200 volts AC for the Lexon.

Also, if you notice in Figure 3-5, the Extech will measure capacitors, Cx, (up to 20 uf) and an AC signal's frequency (Hz) up to 200 kHz, whereas the Lexon cannot do any of these measurements.

Note that both DVMs in Figure 3-5 require you to set the measurement range manually. For example, if you are measuring 10 volts DC, you should set the selector knob to "20." Or if you are measuring a 1 million ohm resistor (1 mega ohm), the selector should be turned to "2000" as in 2000 kilo-ohm since 1 million ohms = 1000 kilo-ohms.

For $20 to $40, you can buy auto-ranging DVMs. See Figure 3-6. With these types of DVMs, the range is automatically set for highest accuracy by the DVM. However, you normally need to wait an extra second or two for the DVM to find the correct range.

FIGURE 3-6 Two auto ranging DVMs by the Extech MN26 and General Tools and Instruments model TS04. Note: Extech replaced the MN26 with the MN36 DVM.

In general, most auto ranging DVMs have excellent measuring capabilities for voltage, current, and resistance. The Extech MN26 can also measure frequency (e.g., 1 Hz to 9.99 MHz) of an AC signal and capacitance (approximately 2 pf to 200 µf), along with temperature when a special probe is used.

The General Tools and Instrument DVM will measure voltage, current, resistance, and it has two or three other features. These are

- Battery tester for 1.5-volt and 9-volt batteries
- NCV detector that senses live AC power lines or wires by putting the DVM close by the outlets or walls
- Auto shutoff, which is very useful since some of us forget to turn off our DVMs
- Bluetooth capability to send measurement data to an iPhone or Android smartphone

Still in the $20 to $40 range, you can get larger-sized, manual-range DVMs. See Figure 3-7.

These DVMs are larger and are easier to read and have good overall measurement ranges. However, they do not have as much sensitivity on measuring current (e.g., 2 milliamps versus 200 micro amps for the inexpensive Lexon in Figure 3-5). But if you recall, it's fine to have a second DVM, which can be something like the Lexon if you need to make more sensitive DC current measurements.

FIGURE 3-7   Two larger-sized DVMs, the DT-9205B and UA890C.

Again, note that the UA890C has the auto-shutoff feature, while this version of the DT-9205B does not. When I searched the web, I did find a "Best" DT-9205M that has "Auto Power Off" to save on batteries.

Figure 3-8 shows a comparison with another equivalent Lexon digital voltmeter with the DT-9205B.

The less than $10 Cen-Tech DVM has a slide switch for turning it on or off. It does have a battery tester mode for 1.5-volt and 9-volt batteries that the Lexon in Figure 3-5 does not. Note that it too can measure DC current down to 200 micro amps full scale.

There are more high-end DVMs that generally cost $80 to $150 such as the ones made by Keysight (formerly Agilent or Hewlett Packard) and by Fluke. See Figure 3-9.

Both DVMs have the "True RMS" measurement system to accurately character-ize AC signals.

On the Keysight DVM, there is a U1231A version that does *not* include measur-ing current. I would advise to buy a DVM that includes the three basic measure-ments: voltage, resistance, and current.

FIGURE 3-8   A comparison of a Harbor Freight Cen-Tech DVM with the DT-9205B.

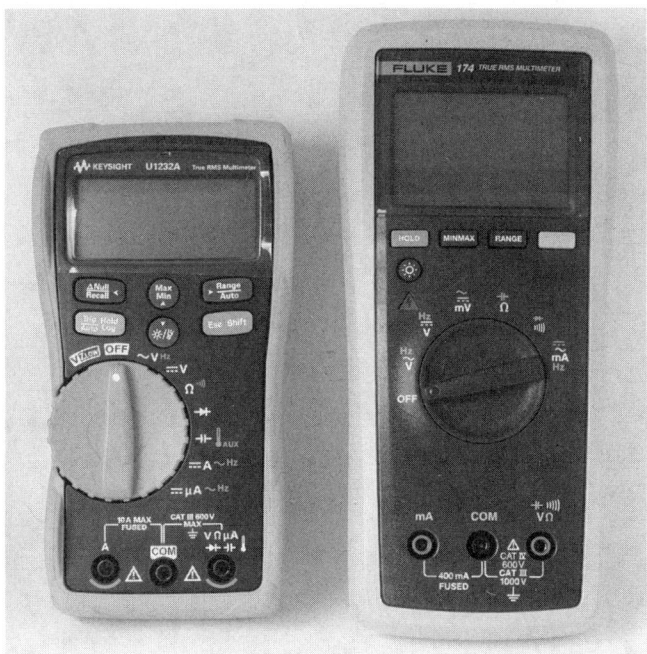

FIGURE 3-9   Keysight U1232A and Fluke 174 precision auto-ranging DVMs.

This finishes the survey on different DVMs that you will be using for trouble-shooting. What I have found is that usually a basic DVM like the Cen-Tech in Figure 3-8 will work for about 90 percent of the time. Once in a while, you may need a DVM that measures capacitance such as the MN26 Extech in Figure 3-6 because it can measure lower value capacitances. Because the prices have come down quite a bit, the basic DVM (e.g., Cen-Tech) can be bought for about $6, and a versatile auto-ranging meter (e.g., Extech MN36, MN26, or MN26T) is in the $44 range. These meters are a good investment for testing parts of your circuits.

In the next chapter, we will look at some basic electronic parts such as resistors, capacitors, and semiconductors (diodes and transistors), which all can be measured and checked with some of these DVMs.

CHAPTER 4

Some Basic Electronic Components

We will now look at two electronic components that are often used in electronics—resistors and capacitors.

Resistors pass electricity with varying amounts of conductivity. For example, a perfect wire has no resistance and has ideal conductivity of electricity. On the other hand, a thin wire has resistance and restricts the flow of the electrons somewhat. The higher resistance results in higher restriction of electron current flow. For example, a 10 million–ohm resistor reduces electron current flow to a "trickle," but it can be useful in many circuits.

However, we will start first with capacitors since they tend to be the ones that can be installed incorrectly and can cause a circuit to fail or work partially.

Capacitors

A capacitor is made of two plates separated by an insulator. So its (DC) resistance is infinite or very high when measured on an ohm meter. See Figure 4-1, which shows a schematic symbol for a non-polarized capacitor.

However, a capacitor does store an electron charge, which makes it like a battery that can quickly charge and discharge DC voltages. If the voltage is an AC signal, it can restrict or pass AC currents depending on its capacitance value, measured in micro farads, nano farads, or pico farads, which are all a fraction of a farad.

$$\text{One pico farad} = 1 \text{ pf} = 1 \text{ farad} \times 10^{-12}$$

$$\text{One nano farad} = 1 \text{ nf} = 1 \text{ farad} \times 10^{-9}$$

$$\text{One micro farad} = 1 \text{ }\mu\text{f} = 1 \text{ farad} \times 10^{-6}$$

FIGURE 4-1   Schematic symbol for a non-polarized capacitor.

Here are some examples of equivalent capacitances:

- 100,000 pf = 100 nf = 0.1 µf
- 10,000 pf = 10 nf = 0.01 µf
- 1000 pf = 1 nf = 0.001 µf

For many capacitors, including ceramic and film type capacitors, the capacitance value is printed in the form of a first and second significant numerical digit and a third multiplier numerical digit, plus a fourth "digit" that is a letter (e.g., K = 10% or M = 20%) to denote the tolerance or accuracy of the capacitance value. The third digit can also be interpreted as the number of zeros after the first two digits.

In general, the three numerical digits tell us the capacitance value in pico farads, which you can equivalently express in nano or micro farads. See the examples below:

- $103 = 10 \times 10^3$ pf = 10,000 pf= 10 nf = 0.01 µf
- $185 = 18 \times 10^5$ pf = 1,800,000 pf (note the five zeros after the "18"), which is really more commonly expressed as 1.8 µf
- $101 = 10 \times 10^1 = 10 \times 10 = 100$ pf and NOT 101 pf
- $270 = 27 \times 10^0 = 27 \times 1 = 27$ pf and NOT 270 pf. However, sometimes a capacitor will be marked by its literal value, and 270 can equal 270 pf. This is why a DVM with a capacitance measurement feature comes in handy.

Many ceramic and film (e.g., polyester, mylar, polycarbonate, or polypropylene) capacitors include a letter to denote the accuracy or tolerance of the capacitance value. The most common letters used are the following list, and the most common tolerances for capacitors are highlighted in the bold fonts:

- **M = ± 20%**
- **K = ± 10%**
- **J = ± 5%**
- G = ± 2%
- F = ± 1%
- Z = –20% to +80

As an example, here is the capacitance range of a 1000 pf 10 percent capacitor, which is marked 102K:

$$[1000 \text{ pf} - 10\% \text{ of } 1000 \text{ pf}] \text{ to } [1000 \text{ pf} + 10\% \text{ of } 1000 \text{ pf}] =$$

$$[1000 \text{ pf} - 0.1 \times 1000 \text{ pf}] \text{ to } [1000 \text{ pf} + 0.1 \times 1000 \text{ pf}] =$$

$$[1000 \text{ pf} - 100 \text{ pf}] \text{ to } [1000 \text{ pf} + 100 \text{ pf}] =$$

$$900 \text{ pf to } 1100 \text{ pf}$$

Therefore, it is rare that the capacitor's capacitance will be exactly as marked. For example, if you measure an off-the-shelf capacitor marked 2.2 µf, it will most likely read below or above that value on the capacitance meter.

Sometimes the capacitor is marked with a "K" after the third numerical digit, which may be confusing as to what the capacitance value really is. This is because the K may also mean kilo instead of 10 percent tolerance. See Figure 4-2.

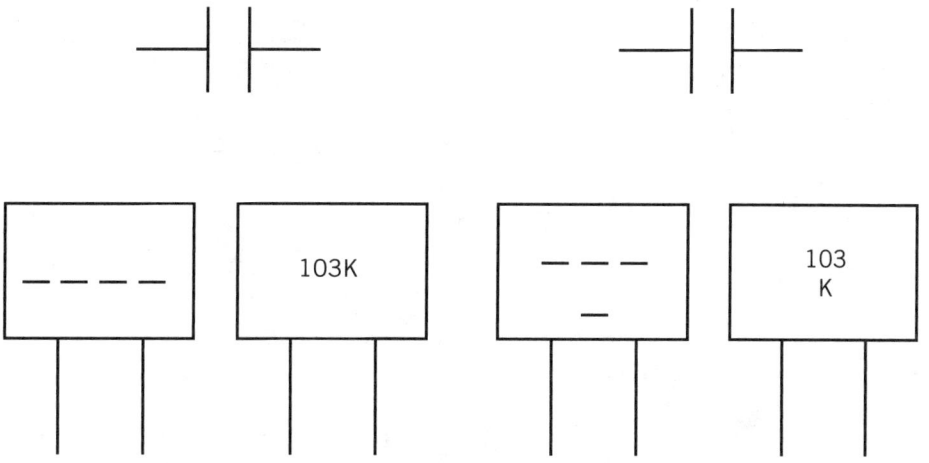

FIGURE 4-2 Capacitor symbols with typical markings either 4 consecutive "digits" on the left side, or 3 digits plus the tolerance letter below as shown on the right side.

On the left side the 0.01 µf (10,000 pf), 10 percent tolerance cap marked "103K" may be misinterpreted because K may be (incorrectly, in this case) thought of as equal to 1000. See examples A and B below, which are incorrect or misinterpreted:

$$103 \times 1000 \text{ pf} = 103{,}000 \text{ pf} = 0.103 \text{ µf}$$

or as:

$$10{,}000 \text{ pf} \times 1000 = 10 \text{ million pf} = 10 \text{ µf}$$

Again the correct value of a capacitor marked as 103K is a 0.01 µf (or 10 nf) capacitor with 10 percent tolerance.

Generally, for electronics work, a farad is too large a value, and capacitors that have capacitances in the 1-farad range are considered to be "super capacitors." These super capacitors are used in some power supply filter circuits or as a battery to keep memory circuits working when power is cut off.

In many circuits, the capacitors are non-polarized, which means that you can connect them in any manner without worrying about the polarity of the DC voltage across them. For example, a non-polarized capacitor will work properly whether a negative or positive voltage is applied across it.

One of the major "problems" constructing circuits with capacitors is that these days they look "all the same." In the past, it was easier to spot a larger value capacitor by just looking at the physical size. See Figure 4-3.

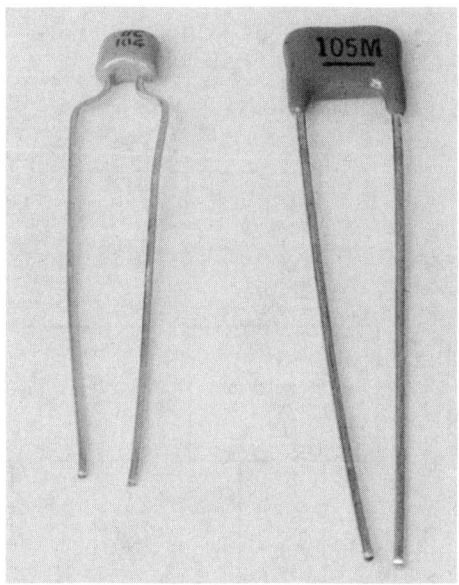

FIGURE 4-3   A smaller dimension 0.1 µf (104) capacitor on the left with large physical size 1 µf (105M = 1 µf at 20 percent tolerance) capacitor on the right side.

As we can see in Figure 4-3, we can easily spot the larger capacitance value capacitor by observing both capacitors' dimensions.

However, with the advent of newer technology, capacitors of small and large values have "equal" sizes. Let's look at ceramic capacitors that have essentially the same size for a very wide range of capacitances. See Figure 4-4.

In Figure 4-4, the capacitors from left to right are: 10 pf, 100 pf, 1000 pf, 100,000 pf (0.1 µf), 1,000,000 pf (1 µf), and lastly 10,000,000 pf (10 µf).

Because the capacitors of various values look the same size, I would recommend measuring the capacitance function on an appropriate multimeter. For example, see Figure 4-5.

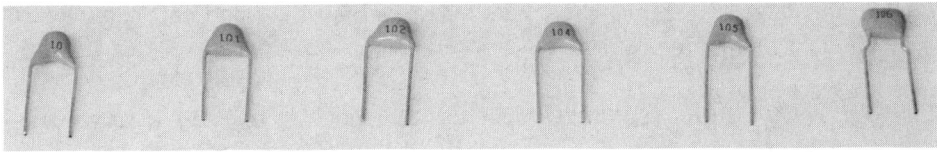

FIGURE 4-4   Ceramic capacitors ranging from 10 pf to 10,000,000 pf (10 μf) that are essentially the same size.

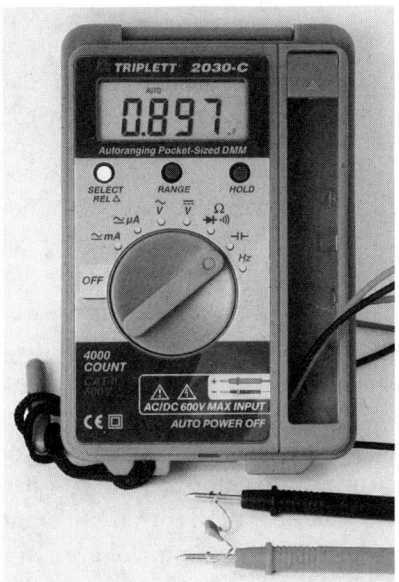

FIGURE 4-5   Measuring a 1 μf (105M) capacitor with an auto-ranging DVM that reads 0.897 μf.

In Figure 4-5, the 105M (1 μf 20 percent tolerance) capacitor shown on the right side of Figure 4-3 measures 0.897 μf, which is well within the 20 percent tolerance since a 1 μf 20 percent capacitor ranges from 0.8 μf to 1.2 μf. Also note that the capacitor is connected only to the test leads, and not to a circuit. When testing a capacitor, it should be out of the circuit. If you have to measure a capacitor in a circuit, *first shut off the circuit by turning off the power supply or source*, then remove one lead from the circuit and apply the test leads to the capacitor. This method works reasonably well for capacitor values > 1000 pf. For measuring smaller value capacitors < 1000 pf, it's better to remove the capacitor from the circuit and measure the capacitor with a DVM that has a 2 nf (2000 pf) full scale range such as those shown in Chapter 3, Figure 3-7.

NOTE: Beware of large-value, small-sized capacitors with low working voltages; their capacitance can "depreciate" as more DC voltage is applied across them. For example, some ceramic capacitors may have half the rated capacitance at close to the maximum rated voltage.

We now turn to polarized capacitors that have positive and negative leads. When a polarized capacitor is in a circuit, the voltage across the positive to negative lead must read a positive value at all times. See Figure 4-6 on the schematic symbol.

Polarized

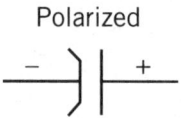

FIGURE 4-6   A polarized capacitor's schematic symbol with a curved or bent negative plate to distinguish it from a non-polarized capacitor symbol (Figure 4-1).

A polarized capacitor is usually reserved for larger values, and polarity must be observed. If the voltage across it is not the correct polarity, the capacitor can start drawing DC current and the capacitance value may drop. Worse yet, if the polarized capacitor is connected incorrectly polarity-wise, it may cause injury. As a safety issue, always check the wiring of a polarized capacitor before turning on the power. If there is a problem, rewire the polarized capacitor.

All polarized capacitors have a maximum voltage rating. In general, use a maximum voltage rating of twice the supply voltage if possible. For example, if you have a circuit that runs on 12 volts DC, the electrolytic capacitors should be rated at 25 volts. This 100 percent safety margin allows safe operation for any tolerance and surge voltages from the power supply. Also, the extra maximum voltage rating normally provides longer service life of the capacitor.

For example, if you have a 9-volt powered circuit and you put a 10-volt electrolytic capacitor across the 9-volt supply, chances are that this 10-volt capacitor will lose capacitance faster over time than if you used a 16-volt or 25-volt electrolytic capacitor.

Larger value capacitors from 0.1 μf to > 10,000 μf are generally polarized electrolytic capacitors. Generally polarized electrolytic capacitors have tolerances of ± 20 percent, or in some cases like the "Z" tolerance of –20 percent to +80 percent.

However, it is possible to purchase ***non-polarized electrolytic capacitors*** that range from 0.47 μf to 6800 μf. Non-polarized electrolytic capacitors are also known as bi-polarized or bi-polar electrolytic capacitors.

NOTE: You can make a non-polarized (NP) capacitor by connecting two equal value capacitance polarized capacitors back to back in series with the two negative terminals connected, or with the two positive terminals connected. The final capacitance is half of the value of one capacitor. See Figure 4-7.

For example, if you connect two polarized 100 μf electrolytic capacitors in series back to back, the final capacitance will be half of 100 μf, or 50 μf.

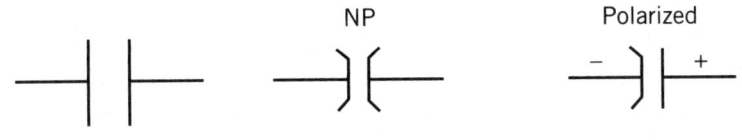

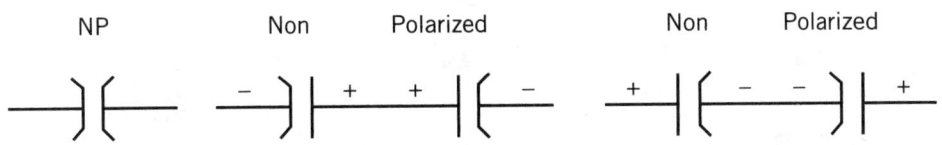

FIGURE 4-7   Schematic symbols on the top row for regular, non-polarized (NP), and polarized capacitors, with examples on the bottom row on how to make a non-polarized capacitor with two polarized capacitors in series.

Again often for electronics work, a farad or 1,000,000 µf is too large a value, and capacitors that have capacitances in the 1-farad range are considered to be "super capacitors." These super capacitors are used in some power supply filter circuits in car audio systems, or as a battery backup DC voltage source to keep low current memory circuits working when the main power is turned off.

Radial and Axial Electrolytic Capacitors

When the two leads from the capacitor protrude from one side (e.g., bottom side), they are radial leaded types. See Figure 4-8 and note the polarity markings and their working voltage.

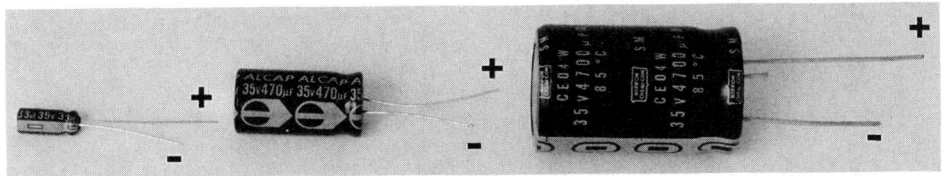

FIGURE 4-8   Radial lead capacitors at 33 µf, 470 µf, and 4700 µf, all at 35 volts maximum working voltage. Note the (–) stripes and that the negative leads are shorter than the positive leads.

Again, in an aluminum electrolytic capacitor, the negative lead is denoted by a stripe, usually with a minus sign (–).

When the capacitor has leads from both sides, it is an axial lead capacitor. See Figure 4-9.

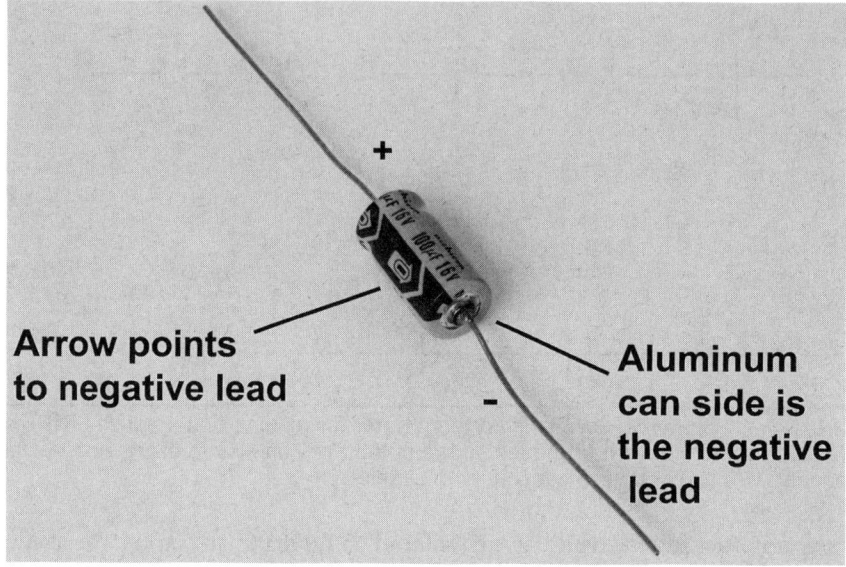

FIGURE 4-9   A 100-µf, 16-volt axial lead electrolytic capacitor.

Note in the axial lead capacitor, there is an arrow that points to the negative lead. Often the negative lead is connected to the aluminum case.

Another way to identify the polarity is that the lead or wire from the insulated side is the positive lead. See Figure 4-10.

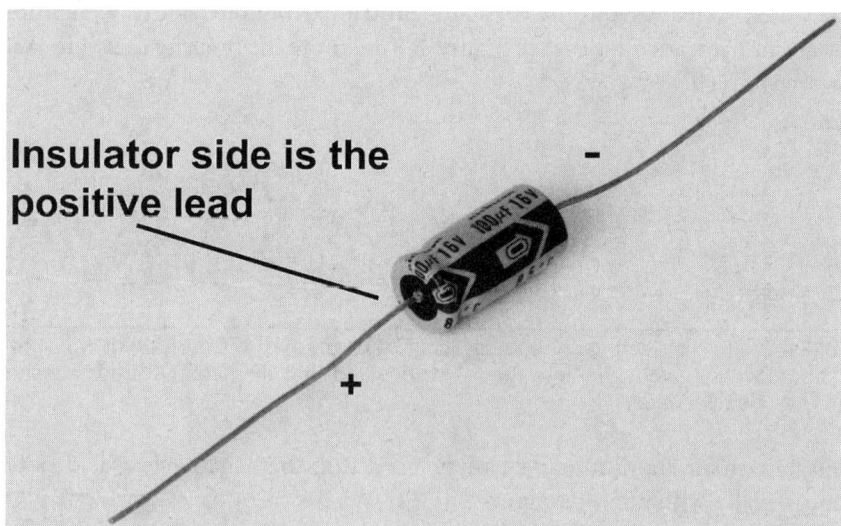

FIGURE 4-10   A view of the 100-µf, 16-volt electrolytic capacitor showing its positive lead.

Finally for this section, we will show how to make non-polarized electrolytic capacitors from polarized ones. See Figure 4-11.

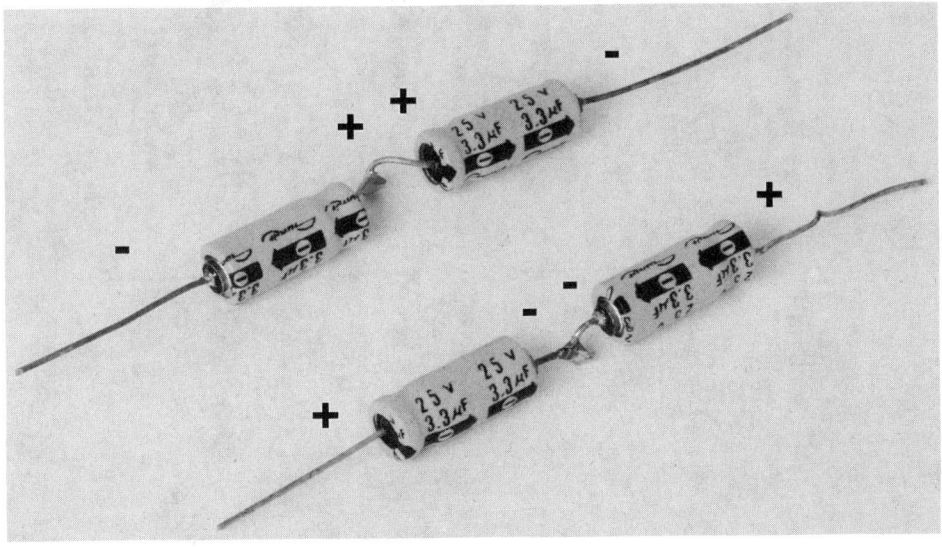

FIGURE 4-11   Examples of making non-polarized capacitors via a series connection.

As illustrated in Figure 4-7 schematically, we see in Figure 4-11 that two 3.3-µf electrolytic 25-volt capacitors are soldered in series back to back. In the top pair the two positive (+) leads are soldered. And in the bottom pair, the two negative (–) leads are connected. In both cases the result is a non-polarized capacitor whose capacitance is one-half of 3.3 µf or about 1.65 µf.

To be on the safe side, I would rate the maximum voltage at 25 volts and not 50 volts because there is no guarantee that the voltage is equal across each capacitor.

Measure Twice, Install Once: Erroneously Marked Capacitors

Once in a while, it is possible to run into an incorrectly marked capacitor. Figure 4-12 shows such an example.

As we can see from Figure 4-12, the capacitor is marked 0.68 µf, but measures instead 0.157 µf. Thus, this 0.68-µf capacitor is mismarked or defective. In fact, it may be a 0.15-µf capacitor printed incorrectly.

A Proper Method for Measuring Capacitors

1.   Measure the capacitor with a DVM or capacitance meter when the circuit is shut off and at least one of the capacitor's leads is completely disconnected from the circuit. The DVM or capacitance meter sends out a signal to the capacitor during its measurement. Any "extraneous signal" added to the

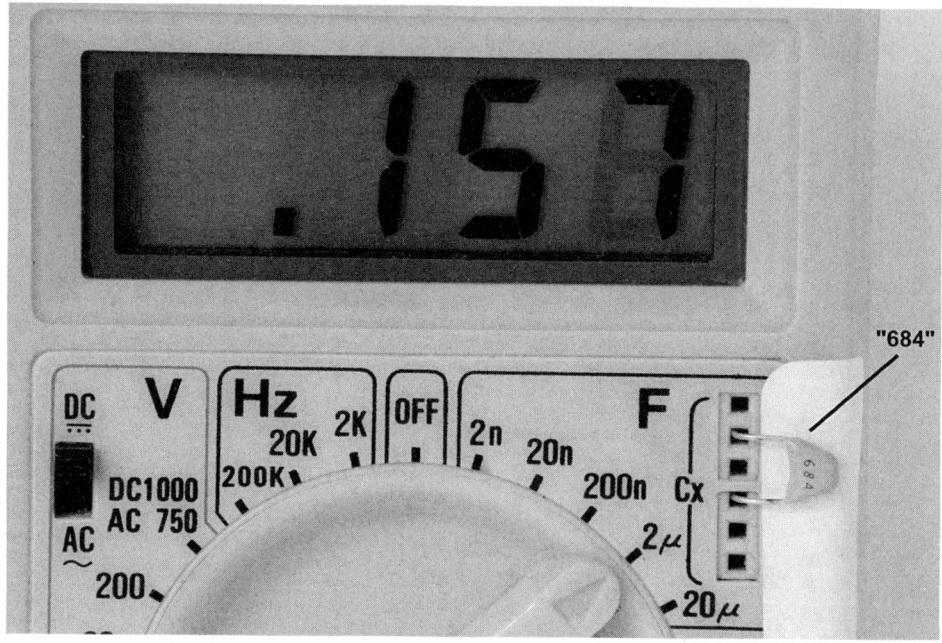

FIGURE 4-12   A ceramic capacitor marked 684 = 0.68 μf reads instead 0.157 μf.

capacitor via being connected to the circuit will give erroneous capacitance readings, or worse, can cause damage to the DVM or capacitance meter.

2.   Make sure to not touch either of the capacitor's lead with your hands. For example, if you hold the capacitor with both hands to make the measurement, your hands can cause an erroneous measurement. This is true when you are measuring low capacitances such as 10 pf. Use alligator or EZ hook test leads to connect the capacitor to the DVM or capacitance meter.

3.   If you are measuring low-value capacitances < 100 pf, then you must subtract off any residual capacitance reading from the DVM or capacitance meter. For example, take note of the capacitance read when the test capacitor is not connected. Usually, this residual capacitance will be in the < 2-pf range. Connect the capacitor and note the reading, and then subtract the residual capacitance. It's very much like stepping on a bathroom scale that is not zeroed, such as it showing 5 pounds with no weight on it. You then step on the scale, look at the reading, and to get the true weight, you subtract 5 pounds in this example.

Resistors

Resistors restrict the flow of electric current. But they are used in many circuits to set up a DC voltage to an electronic component such as a transistor or integrated circuit. In other uses, resistors can be used to set the gain of an amplifier or set the voltage output of a DC power supply.

In terms of troubleshooting, the most common "error" in using resistors is misreading the color code on the resistor. In some cases, resistors come prepackaged with the resistance value, in ohms, which can be mismarked.

Back in the days before the Internet and apps, most people had to memorize the resistor color code to read the resistance value of the resistor. Today, one can learn the resistor color code, but it is easier to just measure it with the ohm meter that is in virtually all digital voltmeters. A resistor is measured in ohms and has a "shorthand" symbol using the Greek letter omega (Ω). So whenever you see Ω, it means ohm or ohms.

For example, instead of writing 120 ohms, we can equivalently restate it as 120Ω.

Figure 4-13 shows a drawing of a standard four-band (e.g., 5 percent tolerance) resistor.

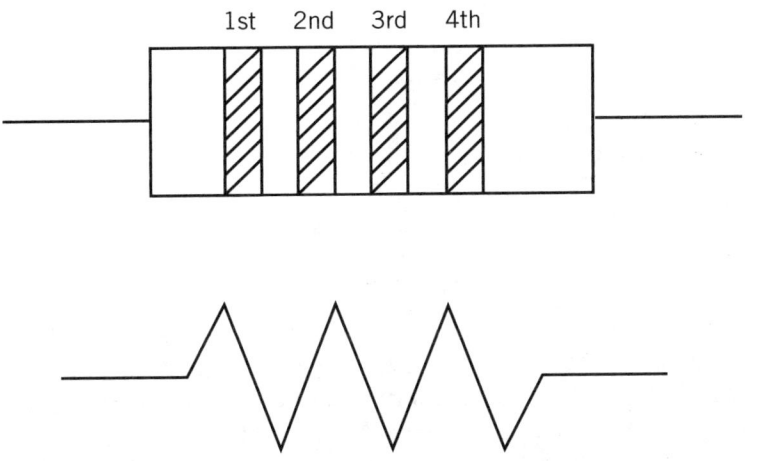

FIGURE 4-13 Standard tolerance four-band resistor and its associated schematic diagram symbol.

NOTE: A resistor is a non-polarized electronic component. Put another way, there are no polarities in resistors. Thus, you can connect their two leads to a circuit in any order.

To read the four-band resistor, we look at the first band, where the first band is located closer to the edge of the resistor than the fourth band. We start reading the resistor from first and second bands to get the first two digits, then read the third band to multiply by a number such as 0.01, 0.10, 1, 10, 100, 1000, etc. Sometimes people read the third band to add the number of zeroes after the first two digits.

Since each band is a color including black and white, let's take a look at the code that is based on colors of the rainbow. The colors below are for the first, second and third band, but not the fourth band. Note that the third band denotes the number of zeroes after the first two digits. So if the third band is red, for example, it is not the number 2, but two zeroes added after the first two digits.

The fourth band represents the tolerance, which is either gold = 5 percent or silver = 10 percent. There are no other colors than these two for the fourth band.

For example, a 5600-ohm 5 percent tolerance resistor will show the following colors from first to fourth bands:

Green, blue, red, and gold
0 = black
1 = brown
2 = red
3 = orange
4 = yellow
5 = green
6 = blue
7 = violet
8 = gray
9 = white
Divide by 10 or multiply by 0.10 = gold
Divide by 100 or multiply by 0.01 = silver

When the third band has gold or silver "colors," these are lower-value resistors (< 10Ω). For example, a 2.7Ω 5 percent resistor is denoted as red, violet, gold, and gold. And a 0.39Ω 5 percent resistor has its first to fourth bands as: orange, white, silver, and gold. See Figure 4-14 for examples of four-band color-coded resistors.

Five percent resistor values are incremented in about every 7 to 10 percent. For example, starting from a 220Ω resistor, the next value is 240Ω. If we want to have a 5 percent resistor that is in between, like 226Ω, there is no such value available because there are only two significant digits and "226" requires three significant digits.

Thus we need a resistor with five bands. The first three bands provide the three significant digits, the fourth band gives the number of zeroes, and the fifth band denotes the tolerance that is either brown for 1 percent tolerance, or red for 2 percent tolerance. More commonly, almost all five-band resistors have accuracy to within 1 percent with a brown fifth band. See Figure 4-15.

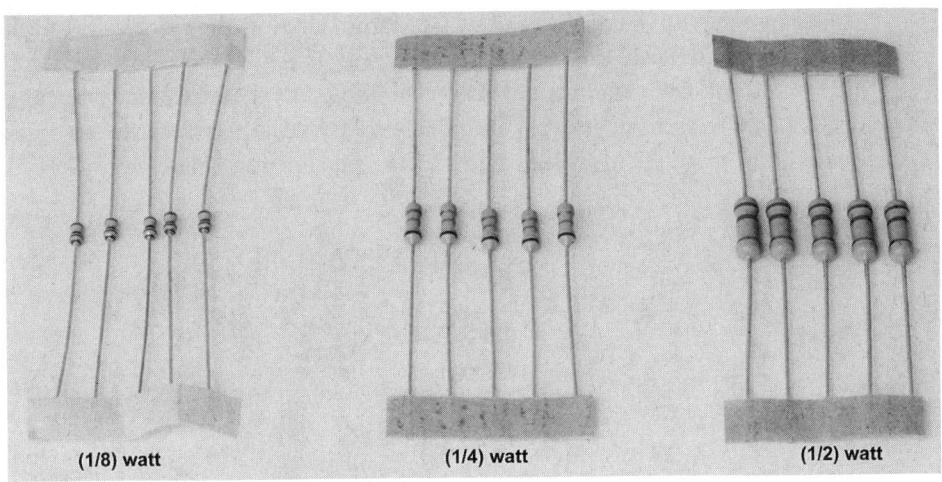

(1/8) watt   (1/4) watt   (1/2) watt

FIGURE 4-14 Four-band color-coded eighth-watt, quarter-watt, and half-watt resistors.

NOTE: There are some 2 percent five-band resistors where the fifth band is red.

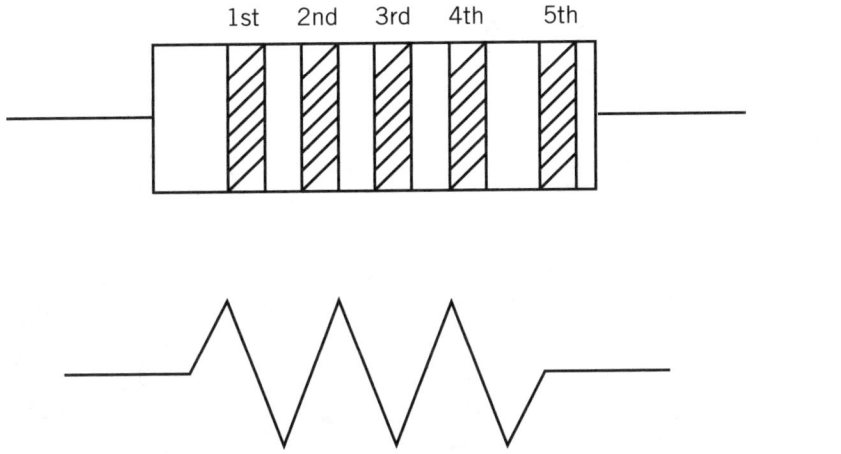

FIGURE 4-15 A five-band precision resistor with its schematic symbol below.

Reading five-band resistors can be confusing because you have to look carefully to find where the first digit is. The first band is always a little bit farther from the edge of the resistor than the fifth band. The fifth band denotes the tolerance rating, which is usually brown for 1 percent. Note in Figure 4-14 the fifth band is almost right on the edge of the resistor, whereas the first band is not. Also notice that the spacing between the first four bands are the same, but the spacing is wider from the fourth to fifth bands.

Because the fifth band and the first band from far away may look the same distance from the end or edge of the resistor, you may have to read the resistor both left to right and right to left. Each reading will often give a different resistance value. So my recommendation is to take out your DVM and just measure the precision resistor in the ohm meter mode of the DVM. Figure 4-16 shows 5 percent and 1 percent resistors.

FIGURE 4-16   A 22KΩ, 5 percent resistor on the top and 22.1KΩ, 1 percent resistor on the bottom. Note that the fifth band that denotes tolerance rating is closest to the edge in the 1 percent resistor, and it has wider spacing between the adjacent fourth band.

Precision 1 percent resistors and even more accurate resistors with 0.5, 0.25, and 0.10 percent tolerances will have four numbers printed to determine the resistance value, followed by a letter to provide tolerance information. For example, the tolerance code is

F = 1 percent
D = 0.5 percent
C = 0.25 percent
B = 0.1 percent

A 1 percent 4750Ω resistor will read 4751F. Remember that the fourth number denotes the number of zeroes after the first three digits.

Another example having a 0.25 percent 10,000Ω resistor will read 1002B. That's 100 + 2 zeroes afterward from the first 3 digits. This gives 10,000Ω or 10KΩ, where K is a thousand or 1000. See Figure 4-17.

Using a DVM to Measure Resistance Values

Suppose we have a DVM whose maximum resistance ranges are:

- 200Ω
- 2000Ω
- 20kΩ
- 200kΩ
- 2000kΩ (or 2MΩ)

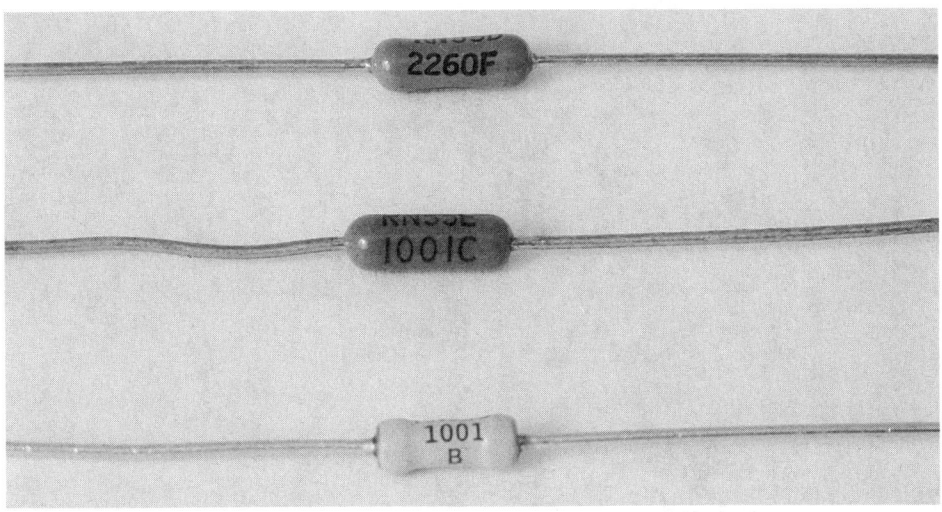

FIGURE 4-17   From top to bottom: 226Ω 1 percent (F), 1000Ω 0.25 percent (C), and a 1000Ω 0.10 percent (B) precision resistors.

Which setting should we use to measure for the most accuracy for a 390Ω 5 percent (orange/white/brown/gold) resistor? The answer is the 2000Ω maximum resistance setting.

Before we measure resistance using a DVM, remember the following:

- Turn off the circuit's power.
- Disconnect at least one resistor lead from the circuit.

Do not hold the resistor with the DVMs test probes/wires with both your hands. The resistance from your hands may give an erroneous lower resistance measurement. This is common when the resistance measurements are > 100kΩ. As a matter of fact, if you have sweaty hands or if the humidity is high in your area, then measuring a high resistance value (e.g., > 2000kΩ) will result in the DVM measuring your hand-to-hand resistance in combination with the resistor. This will result in a lower and inaccurate reading. You can hold one probe and one resistor lead together with one hand. And the insulated portion of the second probe is handled by the other hand while the metallic tip of the second probe touches the second resistor lead. You will now get an accurate reading without the DVM measuring your hand-to-hand resistance.

Now let's take a look at measuring the 390Ω with all the resistance settings in the DVM. See Figures 4-18 to 4-22.

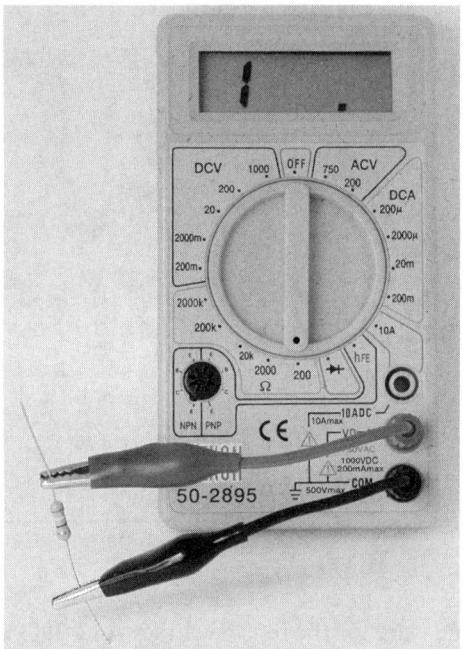

FIGURE 4-18   The 390Ω resistor shows a "1" and blank digits indicating "out of range" at the DVM's 200Ω setting.

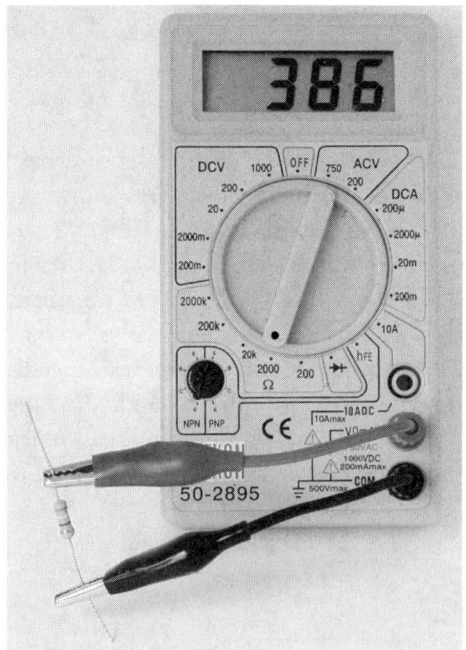

FIGURE 4-19   At the DVM's 2000Ω setting we see a reading of 386Ω for the 390Ω resistor.

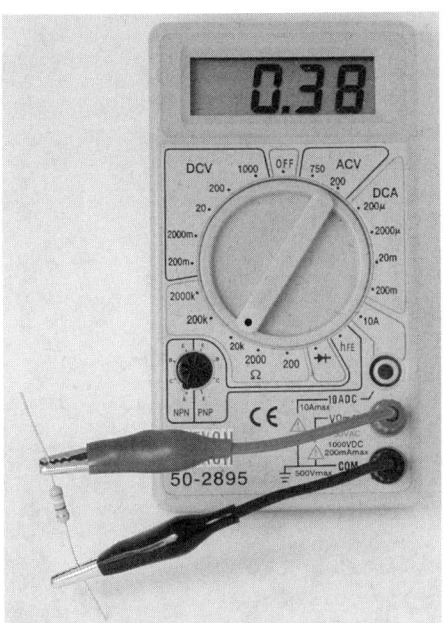

FIGURE 4-20 With the DVM's 20kΩ setting, the DVM is displaying in kilo-ohms. The 390Ω resistor is measured as 0.38kΩ = 380Ω. Note that we just lost some accuracy compared to Figure 4-19, which measured 386Ω.

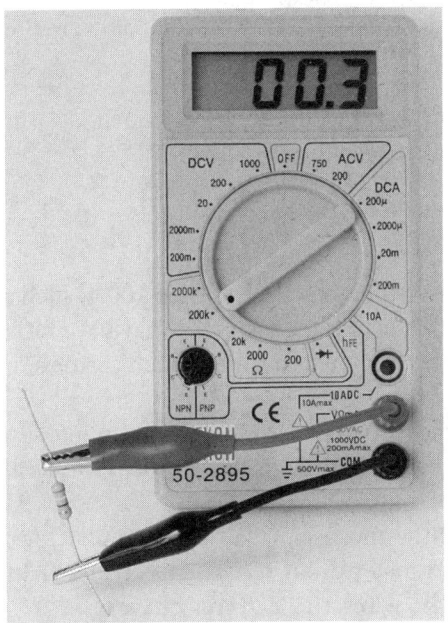

FIGURE 4-21 At the 200kΩ setting, the DVM is "looking" for resistors in the 10kΩ range. So the 390Ω resistor is measured as 0.3kΩ = 300Ω. Note that we are further losing measurement accuracy.

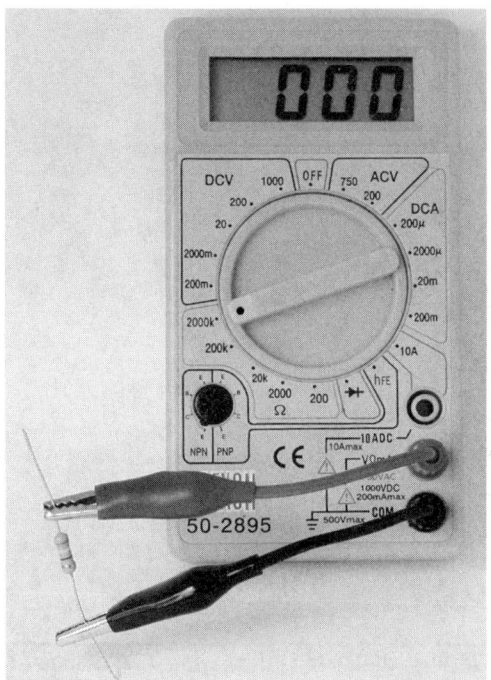

FIGURE 4-22   With the DVM's resistance set to 2000kΩ or 2MΩ, it is expecting resistance values in the 100kΩ range. And with this setting it cannot resolve measuring resistances below 1000Ω. Any resistor value below 1000Ω, such as 390Ω, will register as zero as shown above.

One quick way to measure resistors without thinking too much is to turn the selector until you see the most significant digits displayed.

Measuring Low Resistance Values

In some cases, you will be measuring resistors that are less than 100Ω, such as 10Ω to 1Ω. What you will find in many DVMs is that the test leads may have resistances in the order of 1Ω. It depends on the length of the test leads, which can be short (< 2 inches) or longer (in the 2-foot range).

To make an accurate reading, touch the two test leads together to measure their resistance. Then measure the resistor, and to get the correct value, subtract the test leads' resistance. See Figures 4-23 and 4-24.

In Figure 4-23 the test leads' length is short. Normally, the standard test leads that come with a DVM are longer, and may exhibit resistances greater than 0.2Ω. However, if the test leads use heavy gauge wires, the leads may have < 0.2Ω.

Now we will measure a 2.2Ω resistor. See Figure 4-24.

FIGURE 4-23   Shorting the test leads to read their resistance.

FIGURE 4-24   The combined test leads' resistance (0.2Ω) with the 2.2Ω resistor is 2.4Ω.

To obtain the correct measurement of the resistor, subtract the test leads' 0.2Ω from the 2.4Ω measurement in Figure 4-24. The answer is then:

$$2.4\Omega - 0.2\Omega = 2.2\Omega$$

Note that in Figures 4-23 and 4-24, the test leads are very short. Standard test leads are generally 18 inches or more, which can exhibit more than 0.2Ω.

Some DVMs have a zero calibration mode, which subtracts out the test leads' resistance. For example, in the Keysight U1232A DVM short or connect the two test leads together until the resistance value settles to a constant value such as 2.2Ω (or some other resistance value depending on the resistance of the test leads), keep the test leads connected or shorted together, then hit once the ΔNull/Recall button to zero out the resistance value of the test leads. You are now ready to measure resistances with the DVM calibrated.

NOTE: If you shut off the DVM and later turn it on, you should repeat the calibration procedure.

See Figure 4-25 for a picture of the Keysight U1232A DVM with ΔNull/Recall button.

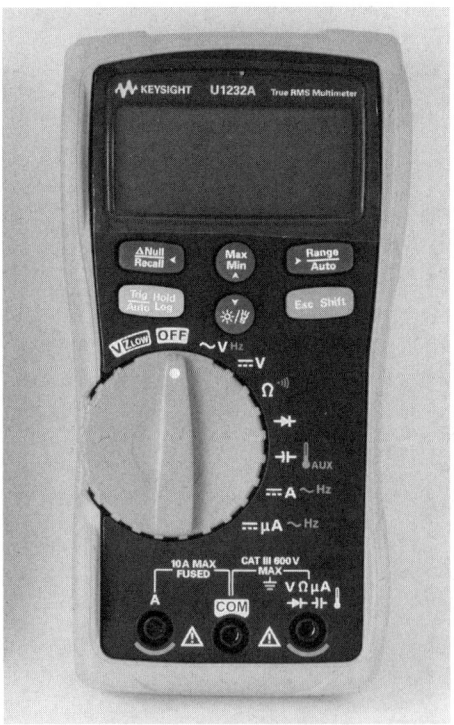

FIGURE 4-25   Keysight U1232A DVM with the ΔNull/Recall button and to the right of it, the Max Min key.

To calibrate and zero out the test leads' resistance, first connect or short the two test leads until a resistance reading is stable. Then keep the two leads connected or shorted, and then hit/tap once the ΔNull/Recall button that is to the left of the Max Min key. You should now see zero ohms on the display with the test leads connected or shorted together.

We have now covered resistors and capacitors. In the next chapter we will examine semiconductors such as diodes, rectifiers, and zener diodes.

CHAPTER 5

Diodes, Rectifiers, and Zener Diodes

Resistors can work properly without regard to "polarity." That is, in a circuit a resistor can be wired either way. For example, you can desolder the resistor from the circuit and flip the resistor leads and solder them back into the circuit and the circuit should work the same.

In this chapter, we will look at some basic two-terminal *semiconductors*. These are devices that require proper attention to how their leads are wired because their terminal leads have polarities. If you reverse a semiconductor's lead, chances are that the circuit will not work at all, not work well, or not work the same (e.g., symmetrically).

Diodes and Rectifiers

Devices that conduct electricity from one terminal to another, while not conducting electrical current when the terminals are reversed, are diodes and rectifiers.

Before we see this in action, let's take a look at them in terms of types, connection terminals (e.g., cathode and anode), and schematic diagram symbols. See Figure 5-1.

The schematic symbol and terminal markings are shown in Figure 5-2.

There are two important characteristics in diodes and rectifiers: peak reverse voltage (PRV) and maximum conduction or forward current (I_F).

Semiconductors such as diodes, rectifiers, transistors, and integrated circuits usually have a specific part number printed on them. This is different than the markings on a capacitor where there is the capacitance in micro farads or pico farads and the voltage rating.

Thus, you will not find a diode or rectifier that is marked "50 volts PRV, 1 Amp." Instead, you will need to look up (e.g., via Google search or data books) the part number based on your diode's PRV and maximum current rating. In this example, you will find a 50-volt PRV, 1-amp diode has a part number as 1N4001. If you need

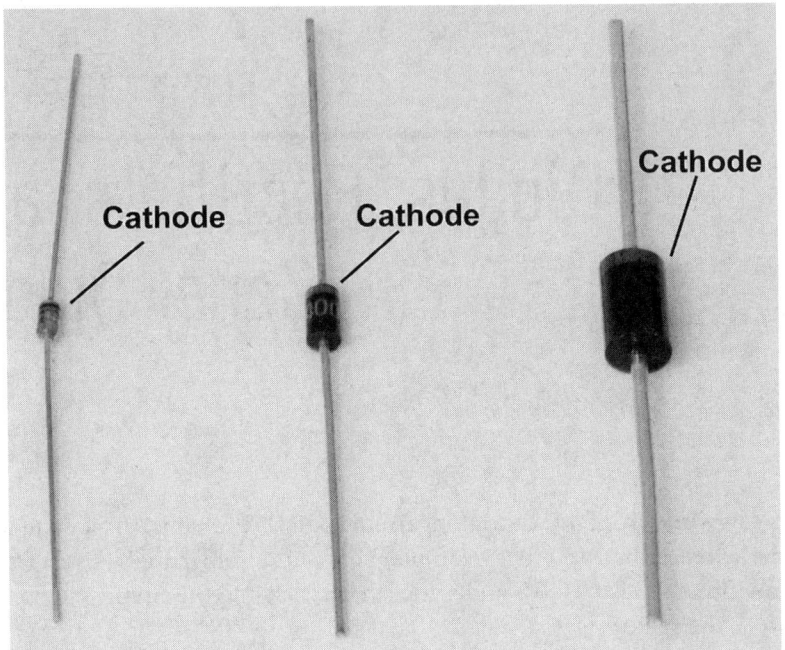

FIGURE 5-1   From left to right, a small signal diode (1N914), a 1 amp rectifier (1N4002), and a 3 amp rectifier (1N5401), all with cathode terminals as marked with stripes or bands.

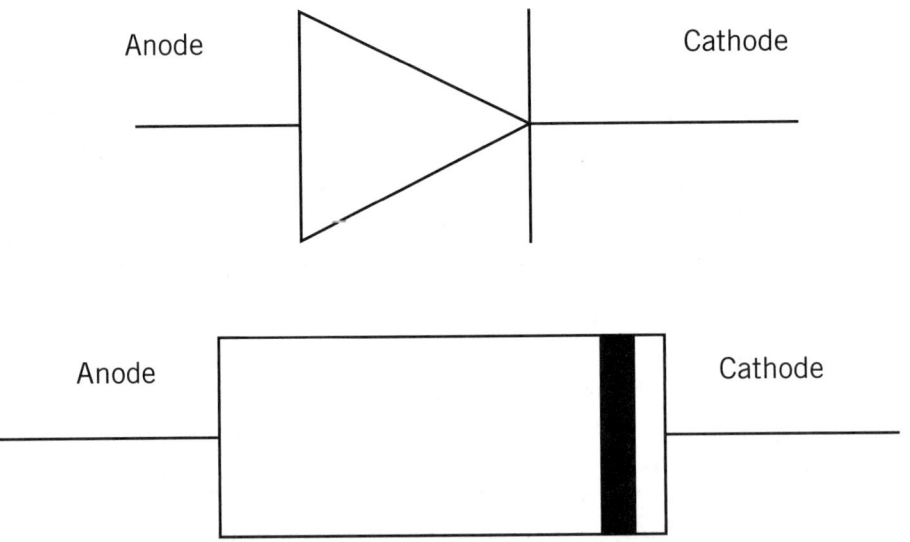

FIGURE 5-2   The schematic symbol of a diode or rectifier is shown on the top and its associated drawing with its cathode marking band is shown on the bottom.

a 3-amp version at 50 volts, the part number will be 1N5400 (e.g., from On Semiconductor).

A typical small signal diode, such as a 1N914 or 1N4148, has a peak reverse voltage of about 75 volts to 100 volts and a maximum forward current of about 100 mA. Fortunately, in most low voltage circuits today, we do not worry much about the PRV since the power supply voltages generally work at less than 25 volts. However, even low-voltage circuits can damage small signal diodes if the forward current, I_F, is exceeded. For example, if a 1N914 small signal diode is used as a power supply rectifier providing 500 milliamps, then the diode may be damaged by excessive diode current flowing into a circuit. When the diode is damaged, often it will conduct electricity both ways or directions like a low-resistance resistor, usually in the order of less than 1Ω.

To "correct" the situation, you can replace the 1N914 or 1N4148 small signal diode with a 1N4002 that is a 1-ampere, 100 PRV rectifier. Table 5-1 shows a short list of common small signal diodes and power supply rectifier/diodes with their PRV and maximum current rating.

TABLE 5-1   Silicon Small Signal Diode Part Numbers

Small Signal Diodes	PRV (Peak Reverse Voltage)	Maximum Forward Current (IF)
1N914	100 volts	300 milliamps
1N4148	100 volts	300 milliamps
1N4153	75 volts	200 milliamps

For large signal diodes or rectifiers, see Table 5-2, where these devices are used primarily for power supplies. Large signal diodes/rectifiers may also be used to protect other devices from high voltages, or to prevent circuits from accidental reverse voltages.

TABLE 5-2   Large Signal Diode/Rectifier Part Numbers

Large Signal Diode/Rectifier	PRV (Peak Reverse Voltage)	Maximum Forward Current
1N4001	50 volts	1 Amp
1N4002	100 volts	1 Amp
1N4003	200 volts	1 Amp
1N4004	400 volts	1 Amp
1N4005	600 volts	1 Amp
1N4006	800 volts	1 Amp
1N4007	1000 volts	1 Amp
3 Amp Devices Below		
1N5400	50 volts	3 Amps

(continues on next page)

Large Signal Diode/Rectifier	PRV (Peak Reverse Voltage)	Maximum Forward Current
1N5401	100 volts	3 Amps
1N5402	200 volts	3 Amps
1N5403	300 volts	3 Amps
1N5404	400 volts	3 Amps
1N5405	500 volts	3 Amps
1N5406	600 volts	3 Amps
1N5407	800 volts	3 Amps
1N5408	1000 volts	3 Amps

Large signal diodes such as the ones listed in Table 5-2 can be read relatively easily because the printing is gray or white on a black or dark epoxy diode body. For glass diodes such as the small signal diodes listed in Table 5-1, or other glass body diodes, the part numbers are imprinted on a shorter body, and the numbers are spanned out over three lines or rows. See Figure 5-3.

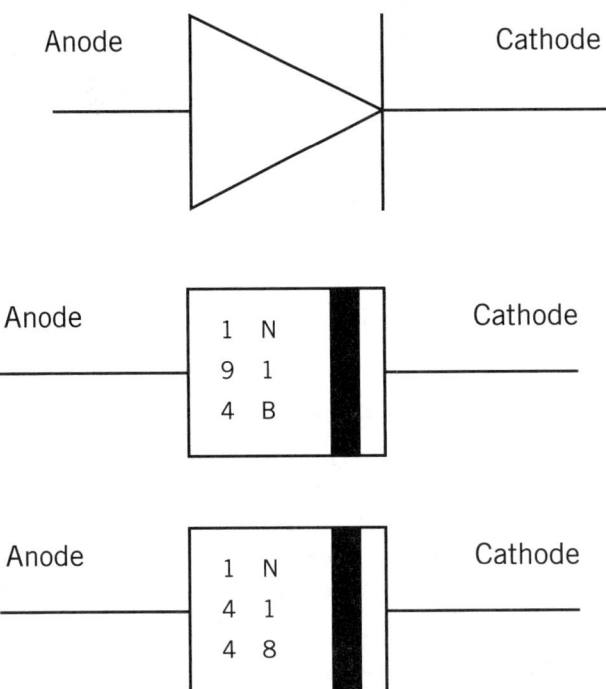

FIGURE 5-3 "Short body" small signal diodes, such as 1N914B and 1N4148, with their part numbers divided into a succession of three rows with two characters per row.

To identify these smaller glass diodes, we often need a magnifying glass or loupe to read their part numbers. See Figure 5-4.

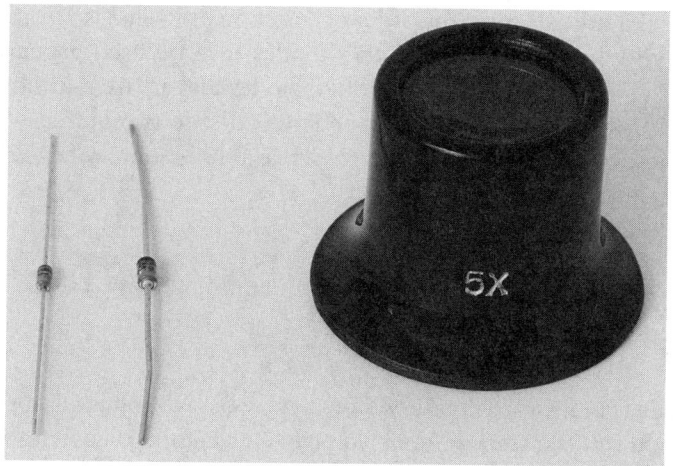

FIGURE 5-4   Glass diodes and a 5X magnifying loupe to read their markings.

One of the most common mistakes is putting in the wrong diode due to not reading its part number. Smaller glass diodes can include small signal diodes, but also include other types such as Schottky and Zener types. See Figure 5-5 for their schematic symbols compared to a standard diode.

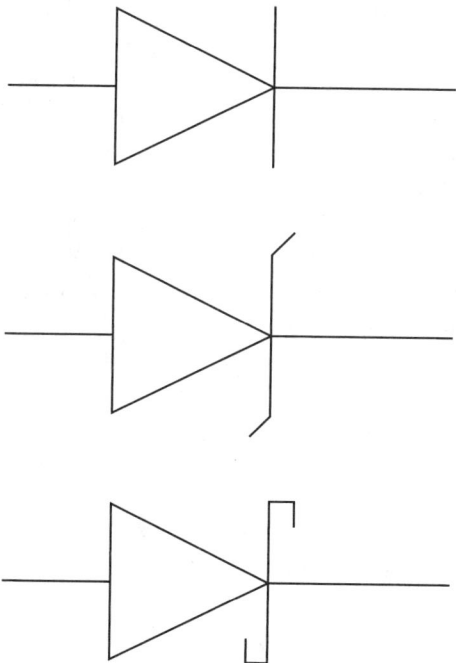

FIGURE 5-5   Top: A standard diode with a straight bar for its cathode; Middle: A Zener diode with a "bent" cathode; Bottom: A Schottky diode—notice that the cathode resembles an "S" (for Schottky).

The drawings in Figure 5-5 are in the order of commonality. That is, we are more familiar with standard diodes first, then Zener diodes that are used primarily for establishing a specific voltage, and finally Schottky diodes that are used in situations where low forward voltage drop across the anode to cathode is required. We will look further into these two devices later. For now, let's examine some voltage characteristics of standard diodes.

Forward Voltage Across Anode to Cathode and Reverse Voltage Effects

When an ideal diode is in forward conduction where a positive voltage is applied to the anode, this diode will transfer the same voltage to its cathode. Therefore, in an ideal diode there is no voltage drop across the anode to cathode when the ideal diode is conducting current. Put in another way, when an ideal diode is conducting current, it behaves like a wire. If the voltage is reverse biased where the anode voltage is negative, for example, the ideal diode will not pass any electricity and become an insulator or open circuit. The ideal diode has a forward voltage (V_F) equal to zero volts.

However, standard silicon diodes have a forward voltage (V_F) in the 0.5-volt to 0.7-volt range when measured across the anode to cathode leads. What this means is that when these common silicon diodes and rectifiers conduct in one direction, they develop a "loss" of about + 0.5 volts to + 0.7 volts across the **anode to cathode** terminals. The diode's forward conduction happens when a positive voltage is applied to the anode with respect to the cathode voltage.

When a standard diode is reverse biased where the **cathode to anode voltage** is a positive voltage, there is essentially no current flow through the diode until the reverse bias voltage is equal to or exceeds the peak reverse voltage (PRV). Although essentially no current flows during reverse bias, the diode may leak through a very small amount of current, usually a micro-amp or less.

For example, a 50-volt PRV diode (e.g., 1N4001) will exhibit a very high resistance when the voltage is less than or equal to +50 volts across the cathode to anode leads. Beyond +50 volts, the 50-volt PRV diode will start conducting current.

When they are not conducting due to reverse voltage applied to the anode and cathode, the resistance is in the order of a million ohms. This reverse bias voltage occurs when the positive voltage is applied to the cathode with respect to the anode voltage, or put in another way, when the negative voltage is applied to the anode with respect to the cathode.

To illustrate forward conduction and reverse voltage non-conduction of a diode, see Figures 5-6 and 5-7.

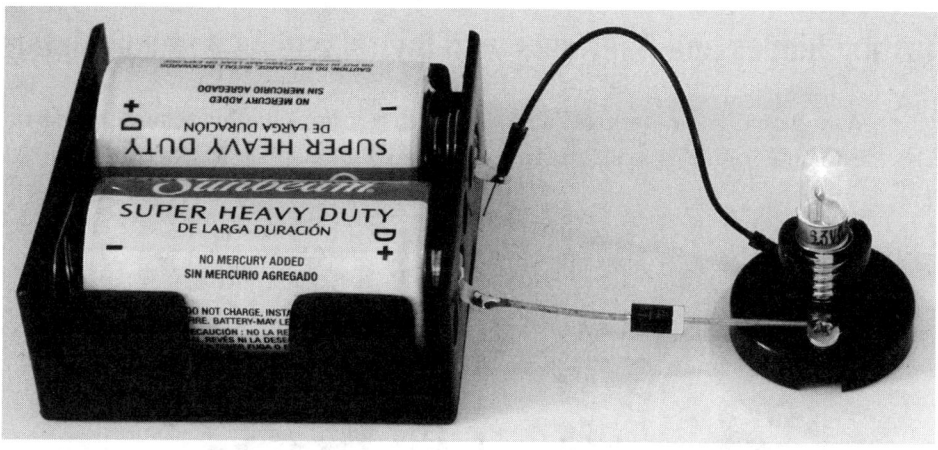

FIGURE 5-6 Positive terminal of battery is connected to the diode's anode to enable the diode into forward bias conduction that supplies current to the lit lamp.

In Figure 5-6, we see that a 1N5401 (3 amp) diode supplies current into the lamp.

Although a small signal 1N914 diode could have been used, the lamp's current would have been close to the maximum current rating of the small signal diode. It's always good to play it safe and use a higher current rating diode. Of course, the 1-amp 1N4002 device would have worked as well.

Now let's look at what happens when the diode is reversed such that the battery's positive terminal connects to the diode's cathode lead. See Figure 5-7.

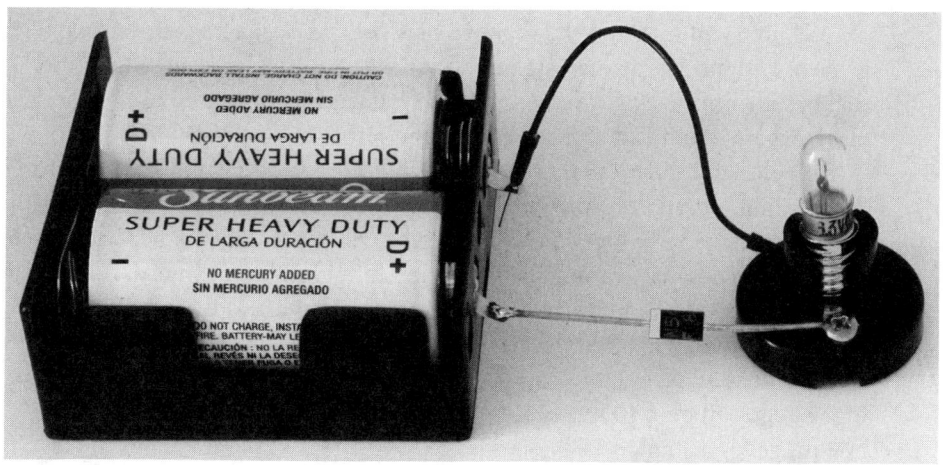

FIGURE 5-7 Diode connected in reversed bias mode with the battery's positive terminal connected to the diode's cathode. No current is flowing and the lamp is off.

Testing Diodes and Rectifiers with Digital and Analog Volt Meters

With digital voltmeters (DVMs), we can determine whether a diode works or not by using the "diode" test feature that is available in most DVMs. See Figure 5-8.

FIGURE 5-8   A typical DVM with its switch set to the "diode" test function that is represented by the schematic symbol of a diode.

We now will measure the forward bias voltage (V_F) across the anode to cathode leads of two diodes, a small signal diode 1N914, and a large signal rectifier 1N4002 that has a 1-amp rating. The red positive test lead is connected to the anode and the black negative test lead is connected to the cathode. See Figure 5-9.

In general, higher current diodes or rectifiers will have lower forward voltages than a small signal version. One reason is that the higher current diodes and rectifiers must have very low equivalent series resistance. This is because when larger currents are flowing through a diode, any lossy resistance internal to the diode or rectifier will add as a forward voltage drop across the anode to cathode leads.

A bad diode will often have shorted out and exhibit the characteristics of a wire, which will conduct both ways. Also, when tested as shown in Figure 5-9, the resulting voltage will be < 10 mV or 0.010 volt instead of the voltage in the 500 mV to 700 mV range for a standard silicon diode.

When the test leads are reversed where the black test leads are connected to the anodes, and the red positive leads are connected to the cathodes, the DVMs will read a "1" and blank digits to show an "out of range" reading. This out-of-range reading means that the diodes connected this way look like a very high resistance resistor or an open circuit. See Figure 5-10.

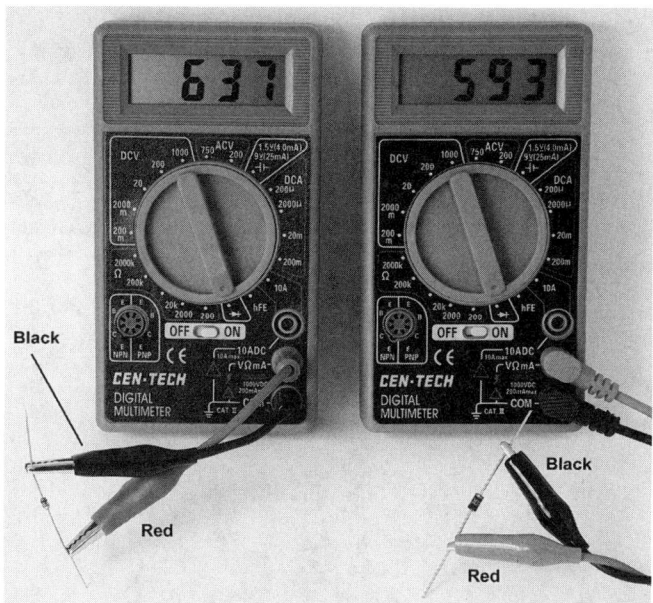

FIGURE 5-9   With the DVM's test leads connected to the anodes via the red positive leads, and the cathodes connected to the black negative leads, we see that the forward voltage (V_F) for the 1N914 is +0.637 volt, and the 1N4002 has voltage of +0.593 volt.

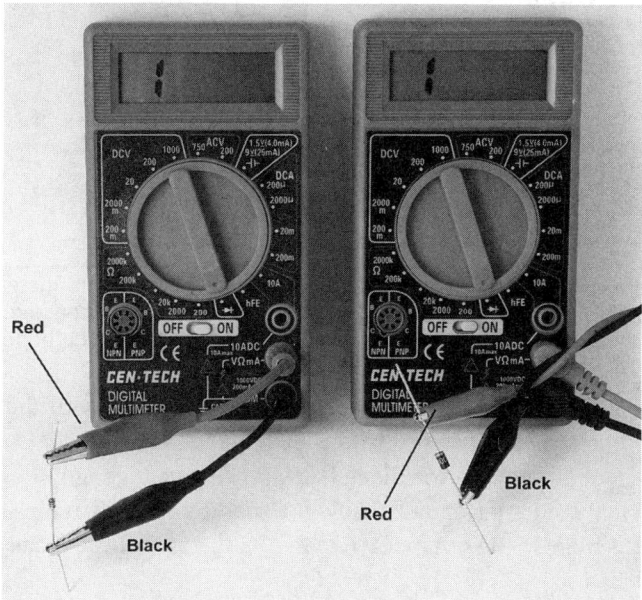

FIGURE 5-10   With the red test leads connected to the cathodes (e.g., striped side), the DVMs show an out of range display with only a "1" with blank digits elsewhere. This means that the diodes connected this way are like open circuits or insulators.

NOTE: With most DVMs, the red test lead is like supplying electric current from the positive terminal of a battery. However, this is not always the case with some analog meters.

Suppose we now try to test diodes with the DVM configured as an ohm meter? What we would expect is that in the forward conduction mode, the DVM's ohm meter will display some resistance value. However, this may not be the case at all. See Figure 5-11 where we compare measuring a rectifier with an analog VOM (volt-ohm-milliamp meter) and a DVM.

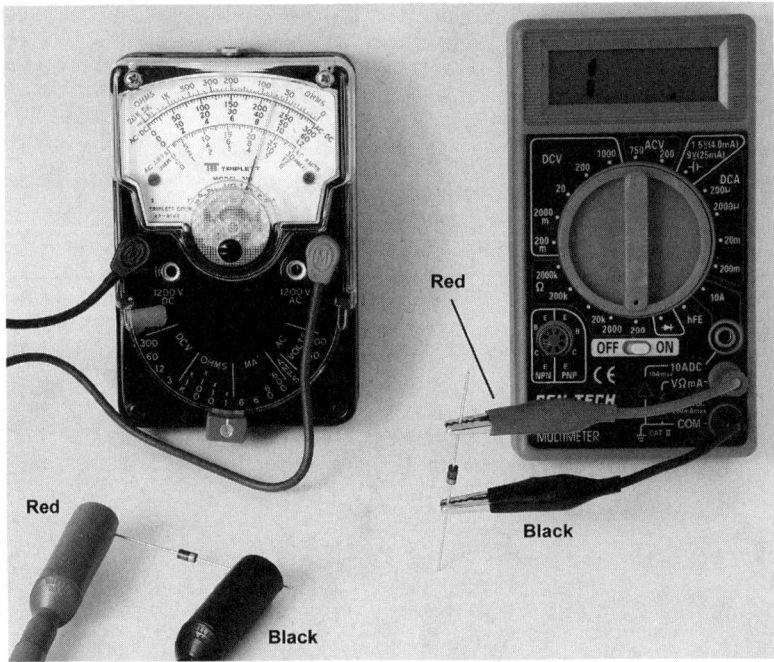

FIGURE 5-11 1N4002 rectifiers being tested on the left with an analog ohm meter that shows about 70Ω. The DVM on the right side shows an out-of-range condition signifying an open circuit, which is erroneous. Black leads of both meters are on the cathodes.

NOTE: The Triplett 310 analog ohm meter shows a 70Ω resistance with the diode at forward conduction when the positive test lead is connected to the anode. However, the actual resistance is lower because the diode has a forward voltage drop (e.g., 0.6 volt) that results in an inaccurate resistance reading. Usually, a reading in the forward conduction mode of less than 150Ω but more than 2Ω indicates that the diode is OK.

Many DVMs may not be able to test diodes in their ohm meter mode. Virtually all analog VOM meters will be able to test conduction of diodes in one direction while showing very high resistance in the reverse direction.

In Figure 5-11, the red positive test lead of the **analog meter** is connected to the anode of the diode-rectifier for testing forward conduction resistance. If the leads were reversed where the black lead is connected to the anode and the red lead is connected to the cathode, the meter's needle will not move and stay on the side of the top scale, which would show infinite resistance.

With analog meters in the ohm meter mode, the red test lead does not always relate to the positive terminal of a battery. For those older meters made in United States, such as the Simpson 260 or the Triplett 310, 620, and 630 series, these analog VOMs have their positive test leads relating to the positive terminal of a battery.

However, some but not all analog VOMs elsewhere have their polarities reversed. That is, the black negative test lead is connected to the positive terminal of a battery. See Figure 5-12.

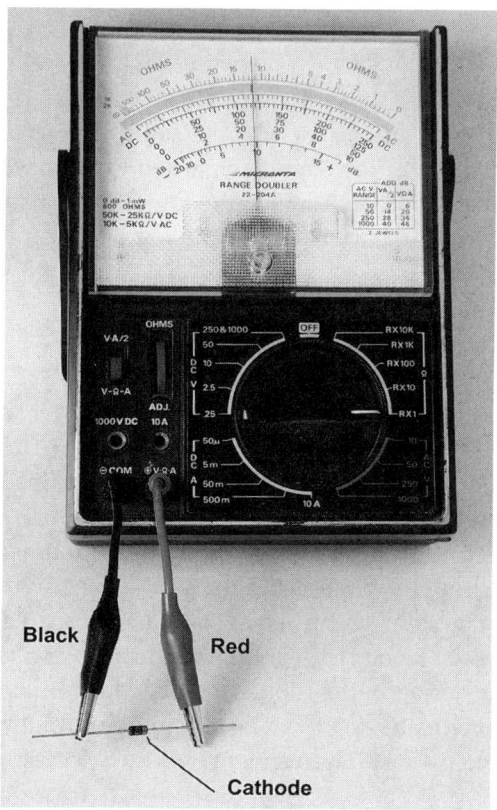

FIGURE 5-12   A "Micronta" analog VOM tested shows forward conduction (11.5Ω on top scale) with the negative black test lead connected to the rectifier's anode.

Notice that the diode's forward conduction occurs when Micronta meter's *negative test lead* is connected to the anode. This is different from Figure 5-11 where the

Triplett ohm meter shows forward conduction when its **positive red test lead** is connected to the anode of the rectifier.

Again in Figure 5-12, if the leads were reversed such that the red positive test lead is connected to the anode and the black test lead of the Micronta meter is connected to the cathode, then the diode will show essentially infinite resistance.

Testing a diode is normally done with analog VOM set to Rx1 (Ohms x 1) or Rx10 (Ohms x 10). Also, the actual resistance value for the diode in the forward conduction mode may vary from one analog voltmeter to another. The reason for this variation is because each meter may supply different test current or voltage to the diode. Usually, if the diode's forward conduction resistance is less than 150Ω with Rx1 selected, the diode is OK.

Usually, a bad diode will short out and the resistance will read close to 0Ω. If you see this, then replace the diode. But also inspect or test other electronic components connected to the diode.

NOTE: Always test diodes with the power shut off and disconnect one lead of the diode from the circuit. If a diode is tested "in the circuit," then an erroneous reading may occur. The reason is that other parts of the circuit connected to the diode may add a parallel resistance across the anode and cathode leads of the diode. This parallel resistance can cause the diode's measurement in the reverse bias mode to show an unusually low resistance.

Schottky Diodes

Recall the schematic symbol for a Schottky diode has the cathode resembling an "S."

Schottky diodes have lower voltage drop across their anode to cathode terminals compared to standard silicon diodes. With this lower voltage drop, Schottky diodes work well for lower voltage power supplies as to not "take away" some of the raw supply voltage. For example, if the raw supply is furnishing only 1.5 volts, a standard diode's forward voltage drop (V_F) is 0.55 volt that will subtract from 1.5 volts. As a result, about 0.95 volt will be at the output of the standard diode. A Schottky diode has a lower forward voltage drop, such as $V_F = 0.15$ volt, and thus the output voltage will be 1.5 volts – 0.15 volt = 1.35 volt, a vast improvement over the 0.95 volt output from the standard silicon diode.

In general, you can replace standard silicon diodes with Schottky diodes, but be aware that the PRV rating may be lower. Schottky diodes are commonly used for switching power supplies that typically provide voltages below 24 volts DC but at very high currents.

For example, Table 5-3 shows the voltage ratings for the following 3-amp Schottky diodes. Note the lower PRV ratings compared to the 1N5400 series devices.

TABLE 5-3 Schottky Diodes Characteristics for the 1N5820 Series 3-Amp Devices

Schottky Diodes	PRV (Peak Reverse Voltage)	Maximum Forward Current, IF
1N5820	20 volts	3 amps
1N5821	30 volts	3 amps
1N5822	40 volts	3 amps

In contrast, the standard silicon 3-amp diode 1N5408 has a 1000 volt PRV.

Now let's take a look at the forward conduction voltage drops of Schottky diodes. See Figures 5-13 and 5-14.

FIGURE 5-13 Testing for V_F a Schottky diode (1N5822) on the left side, and a standard silicon diode (1N5402) on the right. Black leads of both meters are on the cathodes.

Note that the Schottky and silicon diodes look physically identical but the Schottky has a 0.153 forward voltage (V_F) compared to a larger 0.533 voltage drop for the standard version.

Now let's turn to small signal diodes. See Figure 5-14 for measurements of V_F.

FIGURE 5-14   A Schottky small signal diode (1N5711) on the left and a standard small signal 1N914 diode on the right. Black leads of both meters are on the cathodes.

Note that the Schottky diode here has a 0.346-volt drop versus the 0.636-volt drop for the standard silicon 1N914 small signal diode.

In small signal circuits such as those with radio frequencies in a crystal radio, a standard silicon diode like the 1N914 or 1N4148 is unsuitable as the crystal detector because the radio frequency signals are weak, typically < 300 mV. A small signal Schottky is better but will still not pass through weaker radio signals in the crystal radio.

A better detector for a crystal radio is the germanium diode. It is based on very old technology (before 1950) and the germanium diode has a lower forward conduction voltage (V_F), typically less than 300 millivolts. See Figure 5-15.

As we can see here, the germanium diode has a 0.233-volt drop versus the 0.638 volt in the silicon diode. Note that the Schottky diode measures 0.346 volt in Figure 5-14.

Also, beware of vendors selling "germanium" diodes that are of the size of a 1N914. These are unlikely to be (e.g., 1N34) germanium diodes. It should be noted that germanium diodes are generally packaged in larger (physical) sizes than small signal silicon diodes. However, a few vendors on the web may try to sell their germanium (e.g., 1N34) type diodes that are silicon and have the same larger size as a germanium diode. Thus, beware and (to be certain) always measure the forward diode voltage as shown in Figure 5-15.

FIGURE 5-15 A germanium 1N34 diode on the left and a silicon 1N914 diode on the right. Note that the physical size of the germanium diode is much larger than the silicon device. Black leads on both meters are on the cathodes.

For high frequency or radio frequency circuits (e.g., whose frequencies > 400 kHz), small signal diodes also require very low internal diode capacitance between the anode and cathode leads. The reason is that if there is a large capacitance across the anode and cathode, this large capacitance will cause signals to leak through the diode and bypass the diode as a signal detector.

For example, if we look back at Figure 5-14, then the 3-amp Schottky has a 0.152-volt voltage drop, even lower than the 1N34 germanium diode at 0.233 volt. So why not use the Schottky diode instead? The reason is that the 1N5822 Schottky rectifier has hundreds of pico farads of internal capacitance across the anode and cathode leads. This will cause the radio signal to bypass the 1N5822 rectifier, and thus the radio signal will not be demodulated correctly. However, if a radio signal's frequency is in the 50 kHz or lower range, then the 1N5822 Schottky rectifier may just work.

A Brief Look at Zener Diodes

Zener diodes are very similar to standard silicon diodes in that they can pass current in one direction such as having a positive voltage applied to the anode. But this is not how Zener diodes are used. Instead, Zener diodes have a set or predetermined and accurate PRV (peak reverse voltage). This type of PRV in Zener diodes is known as the Zener voltage.

Some Zener voltages are:

- 3.3 volts
- 3.9 volts
- 4.3 volts
- 4.7 volts
- 5.1 volts
- 5.6 volts
- 6.2 volts
- 6.8 volts
- 8.2 volts
- 12 volts
- 15 volts

A Zener diode is also rated in wattage, which allows you to determine the maximum current flowing into it. Some of the common wattage ratings are 400 milliwatts, 500 milliwatts, 1 watt, and 5 watt.

The maximum Zener diode current, I_{Zener} = Wattage rating/Zener voltage

For example, if we have a 5.6-volt, 5-watt Zener diode (1N5339B), the maximum current is:

I_{Zener} = Wattage rating/Zener voltage

I_{Zener} = (5/5.6) amp = 0.89 amp = 890 mA

In another example, suppose we have a lower wattage Zener diode such as the 1N5237B, rated at 500 mW (0.50 watt) and 8.2 volts.

I_{Zener} = (0.50/8.2) amp = 0.061 amp = 61 mA

One thing to keep in mind as the Zener voltage is increased with the same wattage, the Zener diode's maximum current goes down in proportion. But if the Zener voltage is decreased, then the Zener diode's maximum current will increase given the same wattage rating.

TABLE 5-4   Example Maximum Zener Diode Current Based on Wattage and Zener Voltage

Zener Diodes	Wattage	Zener Voltage	Maximum Zener Current I_{Zener}
1N4737	1 watt	7.5 volts	133 mA
1N4744	1 watt	15 volts	66 mA
1N4751	1 watt	30 volts	33 mA

From Table 5-4 we first look at the 1N4744 Zener diode at 15 volts with a Zener current of 66 mA. If we decrease the Zener voltage by 2, which is a 7.5-volt Zener diode (1N4737), we have twice the Zener current at 133 mA. And increasing the 15-volt Zener voltage to 30 volts (1N4751) leads to half the Zener current at 33 mA.

In general, you should try to keep bias on the Zener diode to not more than 75 percent of the maximum Zener current. A good starting point is running the Zener diode at somewhere in the 15 percent to 50 percent maximum of I_{Zener}. For example, the 1N4744 device has a 66 mA maximum Zener current, and you can bias it at 50% or 33 mA.

Now let's take a look at how various Zener diodes are labeled. See Figure 5-16.

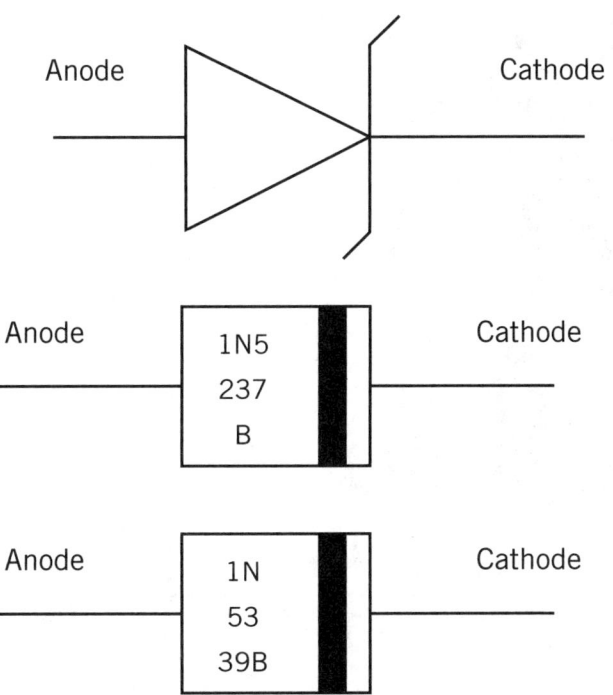

FIGURE 5-16   An example of a 0.50-watt 1N5237B Zener diode, and of a 5-watt 1N5339B Zener diode with their associated labeling.

Note in Figure 5-16 that the markings can span two or three lines for identifying the Zener part number. In the example of the 1N5237B, actually the first two lines (rows) will give your correct part number. You can usually ignore the suffix "B."

For the 1N5339B, we see in Figure 5-16 that the part number is spanned over three lines (rows) to identify the Zener diode.

Just a reminder, a positive voltage greater than the Zener voltage will be connected to the cathode of the Zener diode. But the Zener current will be limited via resistor or a current limiting circuit.

So a pure voltage source such as a power supply or battery that has a higher voltage than the Zener voltage is **NEVER** connected across the Zener diode.

Now let's look at two examples of Zener diodes in simple circuits. See Figures 5-17 and 5-18. Figure 5-17 shows that the red lead is on the anode.

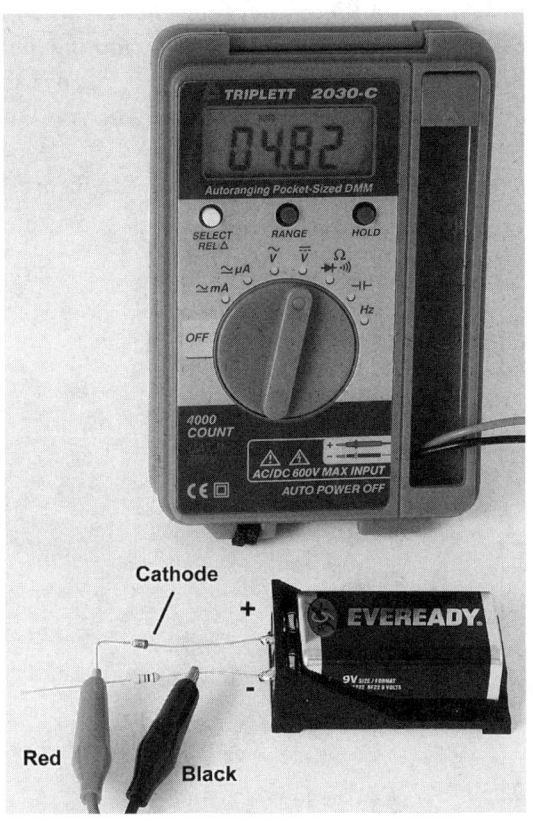

FIGURE 5-17 A 4.3-volt 1N749A, 400 mW Zener diode is connected in series with a resistor to drop or subtract 4.3 volts from the 9-volt battery.

In Figure 5-17 we see the 9-volt battery's positive terminal connect to the cathode of the 4.3-volt Zener diode. Had this been a regular diode such as a 1N914, we would have a reverse bias condition and the voltmeter would have shown 0 volts. But because this is a Zener diode, it "breaks down" at about 4.3 volts and forms a 4.3 positive voltage from the cathode to the anode. This cathode to anode voltage, which is the Zener voltage, subtracts from the battery voltage when the voltage is measured from the anode of the Zener diode and the negative terminal of the battery. The resistor value shown is 1000Ω, and 4.82 volts is measured on the voltmeter. Note that 9 volts – 4.3 volts = 4.7 volts, which is approximately 4.82 volts. If the resistor's value was changed from 1000Ω to 2000Ω or to 510Ω, the measured voltage would still be about 4.82 volts.

Let's take a quick look at what happens if a Zener diode is accidentally reversed in wiring.

See Figure 5-18 for a Zener diode's forward voltage, which is similar to a regular silicon diode.

FIGURE 5-18   A 4.3-volt 1N749A, 400 mW Zener diode on the left and a standard silicon diode, 1N914 on the right. Both black leads are on the cathodes.

As shown in Figure 5-18, both diodes are forward biased by having the red positive test leads connected to the anode of the Zener diode and standard diode. The Zener diode shows 0.789 volt while the standard small signal diode (1N914) shows 0.633 volt.

Thus, if you reverse a Zener diode's connections you will get a voltage lower than the Zener diode voltage. For example, the 4.3-volt Zener diode provides about + 4.3 volts (across cathode and anode) when connected correctly, but will give only about + 0.789 volt when the leads are reversed.

Now let's see if there is a better way to use the Zener diode. Figure 5-17 shows a series connection method of using a Zener diode. But it is not necessarily the best method to provide a stable voltage, or in an accident, a safe voltage. For example, suppose the 4.82-volt output voltage at the resistor is connected to a 5-volt logic circuit. If the Zener diode is inadvertently reversed in connection where the positive terminal is connected to the anode, the resistor will instead have 9 volts – 0.7 volt = 8.3 volts.

If 8.3 volts is applied to a 5-volt logic circuit, most likely there will be damage. Also, if the 9-volt battery drops in voltage, the voltage at the anode of the Zener will drop as well below 5 volts. For example, suppose the 9-volt battery drains down to 8 volts. Then the output voltage in Figure 5-17 will be 8 volts − 4.3 volts = 3.7 volts. In the next example we will show a different circuit known as a *shunt regulator* that will maintain essentially constant voltage as long as the battery or power supply voltage is greater than the Zener diode voltage.

A better circuit is shown in Figure 5-19. Should the Zener diode be inadvertently connected in reverse or if the Zener diode shorts out, the voltage will be 0.7 volts or close to zero volts. This will be below the nominal 5 volts and will provide an "under voltage" that does not harm a circuit.

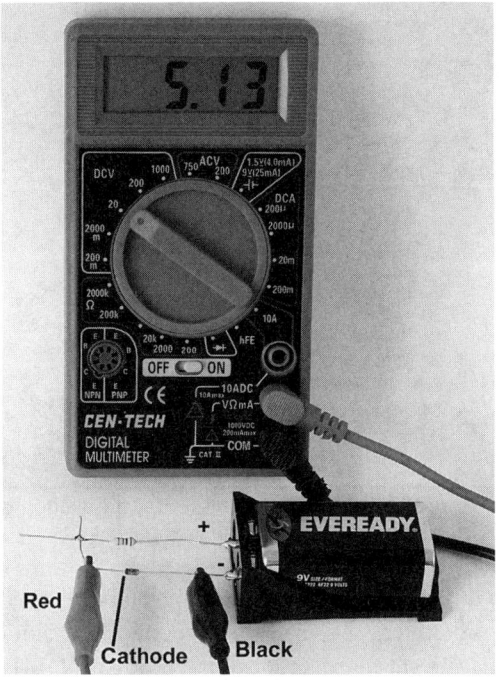

FIGURE 5-19   A 5.1-volt 1N751 400 mW Zener diode configured as a "shunt" regulator circuit that provides 5.13 volts as shown in the DVM. The red lead is on the cathode.

In Figure 5-19, the output voltage is taken across cathode to anode of the Zener diode. This output voltage is the Zener voltage. The series resistor (e.g., 1000Ω) is connected to the battery's positive terminal and furnishes current to the Zener diode's cathode. A positive voltage is provided at the cathode with the anode connected to the battery's negative terminal. Figure 5-19 shows a shunt regulator circuit because the Zener diode is "shunted" across the output terminals.

Suppose the Zener diode is accidentally reversed. This accident results in the anode connected to the resistor and cathode connected to the battery's negative terminal. The voltage will be the forward voltage of the diode, which is about + 0.7 volt. So, if a 5-volt logic circuit is connected to this shunt regulator circuit, there will be no damage to the circuit due to over voltage. And if the diode should short out, the voltage will be close to zero volts, which again will not damage the 5-volt logic circuit.

Some General Rules About Diodes

- They should not be wired in parallel for greater current capability because diodes bought off the shelf are not matched. If you measure the forward diode voltage from anode to cathode and the diodes are "matched" within 5 mV (e.g., Diode #1 = 0.608 volt, Diode # 2 = 0.603 volt), then they can be paralleled in close proximity to each other.
- If you buy the diodes in "tape" form, you have a much better chance of having matched diodes. The matched diodes can be used in circuits requiring nearly identical forward voltage characteristics and is not limited to paralleling diodes. See Figure 5-20 and be sure to double-check the diode's forward voltages (V_F) with a DVM. Both black leads are on the cathodes.

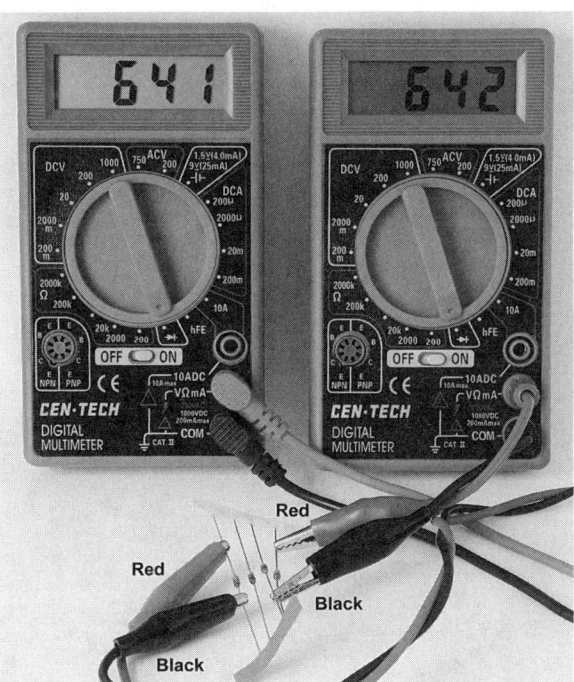

FIGURE 5-20   Two 1N914 diodes measured from a "tape" of 4 successive diodes with very close matching of V_F within 5 mV. In this example the two diodes' forward voltage readings of 0.641 volt and 0.642 volt are only 1 mV apart from each other.

So this ends a lengthier than normal chapter with introductions on standard, Schottky, and Zener diodes.

We will take a "breather" with a slightly shorter Chapter 6 that deals with light-emitting diodes (LEDs).

CHAPTER 6
Light-Emitting Diodes

In this chapter we will explore light-emitting diodes (LEDs) of different emitted wavelengths (e.g., red, green, and blue), and how their "turn-on" voltages (V_F) are related. We will also discuss common problems that can cause an LED to fail, such as over current or "excessive" reverse voltage.

The LED's Light Output

Light-emitting diodes are measured for luminous intensity (light output) in milli-candelas (mcd) at a particular LED current. There are various descriptions of LEDs such as "high efficiency," "high intensity," or "high performance." However, the most important specification to look for is the light output for a given LED current, which is usually done at 10 mA or 20 mA.

With the exception of using older LEDs that give out about < 10 milli-candelas at 10 mA current, which are suitable for indicator lamps such as power-on lights, you will find that most LEDs are much brighter.

For example, typical LEDs will provide over 1000 mcd at 20 mA. In keychain LED flashlights or multiple LED flashlights, the rating is in the order of > 10,000 mcd at 20 mA. The extra brightness works well in products such as light cubes, photographers' studio or portable lighting, emergency lights, home light bulbs and light fixtures, or automobile headlights.

Here are some basic rules of thumb on LEDs:

- For indicator lamps, generally with today's LEDs of any color—red, green, yellow, blue, or white—the LED current can be in the 1 mA to 5 mA range. With very high-brightness LEDs such as a 20,000 mcd unit, the drive current can be as low as 0.10 mA or 0.25 mA.
- If you are using LEDs to generate intense lighting such as making a flashlight or a light cube, you should not exceed 20 mA per LED. Otherwise, the LED may take in excessive current, which causes it to burn out.
- For those situations where intense LED light is generated, *do not look directly into the LED light source*. You can damage your eyesight.
- The light output is generally proportional to drive current. For example, if you are using an LED that is providing 20,000 mcd at 20 mA. Suppose you want to turn it down so that it can be used as an indicator lamp. If the LED drive current is reduced by 200 times from 20 mA to 0.10 mA, the LED will output 100 mcd, which is still pretty bright for an indicator lamp.

NOTE: 20,000 mcd divided by 200 = 100 mcd.

Standard LEDs come in 3mm, 5mm, and 10mm sizes. In this chapter, we will be looking at the common 5mm types that have 100 mil (0.1 inch) lead spacing.

Now, let's take a look at some LEDs.

Figure 6-1 shows a red, green, and yellow 5mm size LED from left to right. Notice that each of the LEDs comes with a longer lead, which is identified as the anode. Each of these LEDs will turn on at less than 2 volts DC, typically about 1.7 volts. Also, they have a "frosted" lens that allows for a wider viewing angle and is easier to look at because the light is more diffused.

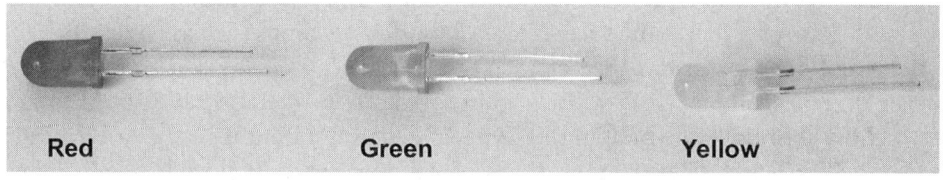

Red   **Green**   **Yellow**

FIGURE 6-1   Red, green, and yellow LEDs give out "diffused" light due to their encapsulations with their associated colors.

In Figure 6-2, the blue and white LEDs turn on at a higher voltage, around 2.5 volts to 3.0 volts DC. The blue and red LEDs have a frosted lens that allows for a wider viewing angle. The middle clear lens, white LED is made for maximum brightness but over a narrower viewing angle. Clear lens LEDs are used for flashlights or for efficient projection/illumination of a smaller area.

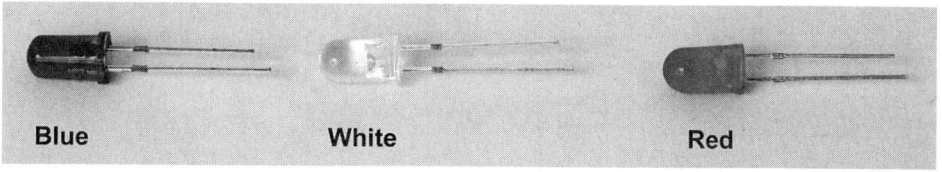

| Blue | White | Red |

FIGURE 6-2 A blue, white, and again a red LED (shown from left to right) that are 5mm in size.

Again, the spacing between the two leads in 5mm LEDs is 0.1 inch. To identify the LED's anode and cathode leads see Figure 6-3.

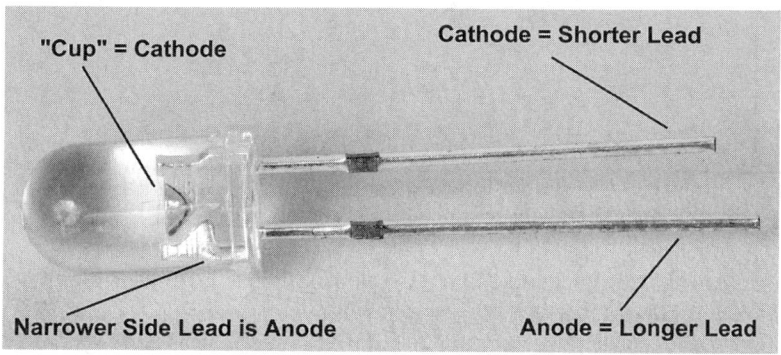

"Cup" = Cathode

Cathode = Shorter Lead

Narrower Side Lead is Anode

Anode = Longer Lead

FIGURE 6-3 Identifying the anode and cathode leads by length or by locating the "cup" that is connected to the cathode lead. From the side of the LED, the narrower lead as shown is the anode.

We now turn to the LED's schematic symbol. See Figure 6-4.

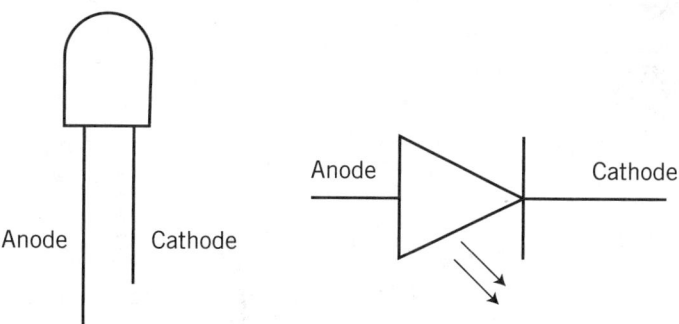

Anode   Cathode

Anode   Cathode

FIGURE 6-4 An illustration identifying the anode and cathode via the lead lengths (e.g., longer lead is the anode) and its schematic symbol with two outward arrows signifying light emission.

It should be noted that the schematic symbol is often drawn vertically. See Figure 6-5.

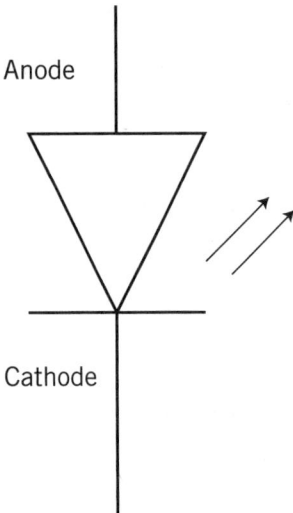

Anode

Cathode

FIGURE 6-5   Often the LED symbol is drawn vertically with the cathode grounded and the anode connected via a resistor to a positive power supply.

Before we look at lighting up LEDs, there is one "trick" to always identify a white light LED. See Figures 6-6 and 6-7.

FIGURE 6-6   A red (R), yellow (Y), green (G), or infrared (IR) LED on the left, and a white LED on the right.

In Figure 6-6 you can always identify a white light LED by its yellow phosphor when looking directly into it without power applied to it. If you see "gray," it is an LED other than white.

Figure 6-7 shows a closer look at the white LED's yellow phosphor.

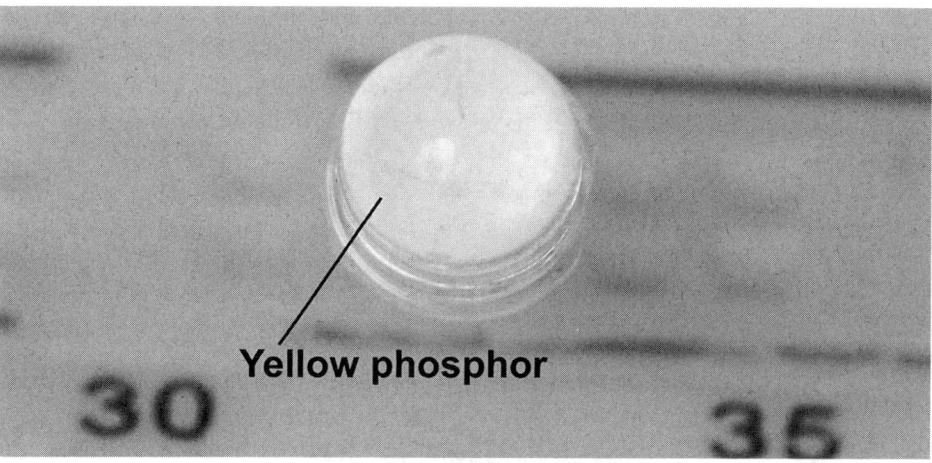

Yellow phosphor

FIGURE 6-7 Viewing straight from the top, you will see yellow throughout, which signifies that this is a white LED.

NOTE: A burnt-out white LED may be missing the yellow phosphor.

The yellow phosphor is excited by a blue LED below it, which, when combined, forms a white light. Note that on the color wheel, blue and yellow are at opposite ends. In fact, if your photo has a yellow color cast, you add blue to remove the yellow cast. Conversely, if the photo has a blue cast to it, you add yellow to return the picture to normal colors.

LED "Minimum Turn On" Voltages

In this next section we will measure the turn-on voltages for LEDs. The turn-on voltages of particular LEDs are relatively constant, almost regardless of LED current. This means that a slight voltage increase (e.g., +100 mV DC more than the previous voltage) across the anode to cathode leads can result in an increase in LED current by tenfold or more. This is why generally LEDs are driven by a resistor or some type of current limiting device such as a transistor.

When the LED voltage and current characteristics are well known, then sometimes a voltage source can be applied to the anode and cathode terminals. For example, we see this in some three AAA cell multiple LED flashlights. See Figure 6-8. In this specific case, there is somewhere between 3.6 volts to about 4.0 volts directly across the anode and cathode of each of the white LEDs. The minimum turn-on voltage of white LEDs is between 2.5 volts and 3.0 volts. Although there are three 1.5-volt AAA cells in series to provide nominally 4.5 volts, by the time all the LEDs are lit, there is enough current draw to lower the three AAA batteries' 4.5 volts to about 4.0 volts or lower (measured 3.93 volts with three alkaline AAA cells when the LEDs are turned on), which is *just* safe enough to prevent the LEDs from burning out. In this example, the three series connected AAA cells have sufficient internal current limiting resistance. This allows the LEDs to be powered on safely.

NOTE: If you use higher capacity batteries that have lower internal resistance than the AAA batteries such as C or D cells, the LEDs' life may be shortened significantly due to excessive current. The reason is that the LEDs will essentially receive the full 4.5 volts from the three series connected C or D alkaline cells.

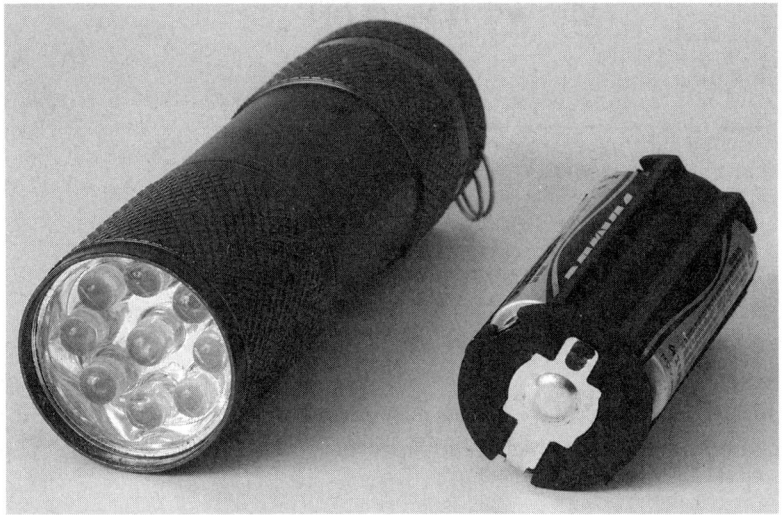

FIGURE 6-8   A three AAA cell 9 LED flashlight, where the LEDs are matched in terms of turn-on voltage.

In Figure 6-8, the three AAA cells are connected in series and provide sufficient internal series resistance for LED current limiting. The nine matched LEDs are connected in parallel (e.g., all nine anode leads and all nine cathode leads are connected) that provide equal brightness among each of the LEDs.

If you randomly take nine white LEDs and connect them in parallel, the brightness will vary between some of these nine LEDs due to different turn-on voltages.

For long LED life, you will need to drive them with batteries that have some internal resistance. Using three AAA batteries will be fine for driving the LEDs within their current limits. But to reiterate, if you drive the LEDs with three C or three D cells, you can burn out the LEDs. The reason is that the C and D cells have lower internal resistance (than the AAA cells) and can deliver much more current into the LEDs.

Now let's take a look at LEDs whose turn-on voltages are typically less than 2 volts DC from anode to cathode. See Figure 6-9.

FIGURE 6-9   LED turn-on voltages (V_F) of 1.786 v = Red, 1.872 v = Green, and 1.842 v = Yellow. The drive voltage is 9 volts with an 8060Ω 1 percent series resistor for each LED.

See Table 6-1 for color and wavelength related to Figure 6-9.

TABLE 6-1   LED Turn-On Voltages and Their Wavelengths

LED	Turn-on Voltage (V_F)	Wavelength
Red	1.786 v	~ 640 nanometers
Yellow	1.842 v	~ 583 nanometers
Green	1.872 v	~ 565 nanometers

Notice that the turn-on voltages increase as the wavelengths get shorter. For example, the longer wavelength red LED has 1.786 volts turn-on voltage, whereas the shorter wavelength green LED has a turn-on voltage of 1.872 volts.

See Figure 6-10 for a schematic diagram of the LEDs being lit in Figure 6-9. In this schematic the LEDs have one arrow emanating from them to denote light emission. Generally, whether there is one or two arrows emanating from the diode symbol, the meaning is the same—the diode is an LED.

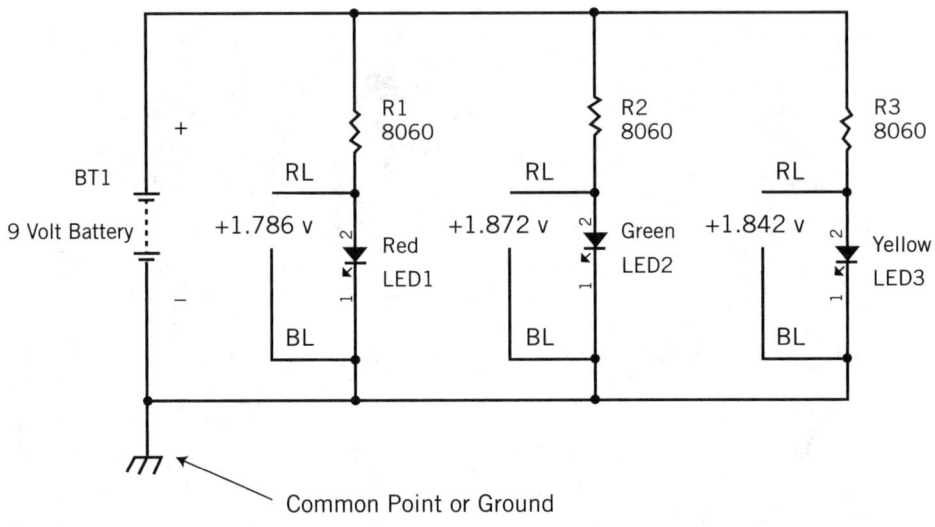

RL = Red Lead of DVM
BL = Black Lead of DVM

FIGURE 6-10   Schematic diagram of the red, green and yellow LEDs with their respective forward voltages (V_F) at 1.786 v, 1.872 v, and 1.842 v.

The forward LED current (I_F) is the battery voltage minus V_F divided by the drive resistor, such as R1, R2, or R3.

For example, the current flowing through the red LED is

I_F = (9 volts – 1.786 volts) / R1 = (9 volts – 1.786 volts) / 8060Ω

I_F = 7.214 volts / 8060Ω

I_F = 0.895 milliamp

Notice in Figure 6-10, a "ground" or common point terminal has been added that is connected to the negative terminal of the 9-volt battery. Often, the DVM's negative or black test lead is connected to the ground or common point. Then the DVM's red test lead is "free" to probe or measure for voltages at different parts of a circuit.

Now let's take a look at turn-on voltages for blue, white, and red LEDs.

FIGURE 6-11   White (W), blue (B), and red (R) LEDs lit via a 9-volt battery and 8060Ω resistors.

Both white and blue LEDs are very bright, and the blue LED is actually overexposing the camera's sensor. See Figure 6-12 for another look at these three LEDs lit.

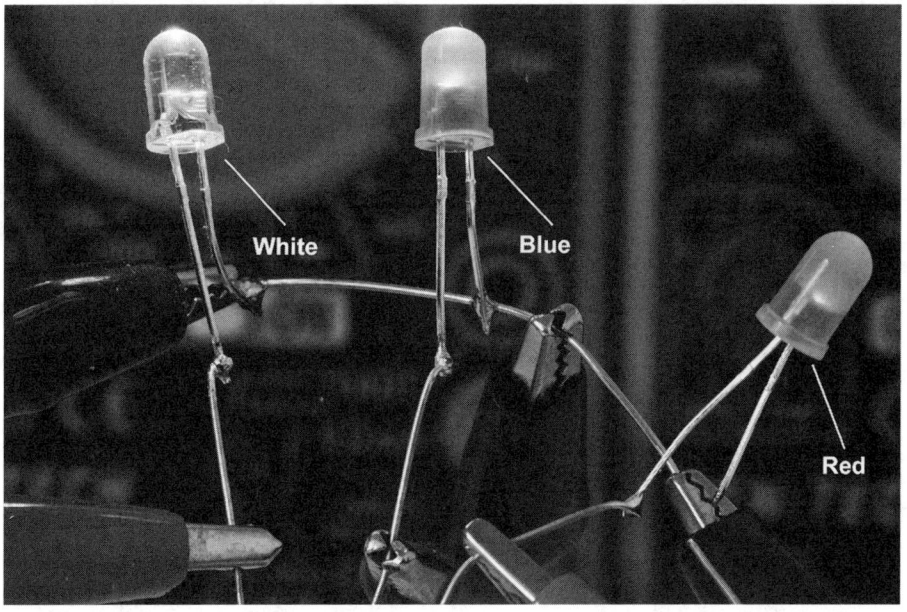

FIGURE 6-12   The white (W), blue (B), and red (R) LEDs lit.

Now let's take a look at Table 6-2, a summary of the turn on-voltages for white, blue, and red LEDs.

TABLE 6-2   Turn-On Voltages for White, Blue, and Red LEDs

LED	Turn-on Voltage (V_F)	Wavelength
White	2.67 v	Not applicable to white light
Blue	2.60 v	~ 470 nanometers
Red	1.78 v	~ 640 nanometers

Note that since a white LED really uses a blue LED inside with a yellow phosphor, both white and blue LEDs have essentially identical turn-on voltages, 2.67 volts and 2.60 volts.

A schematic diagram is shown in Figure 6-13.

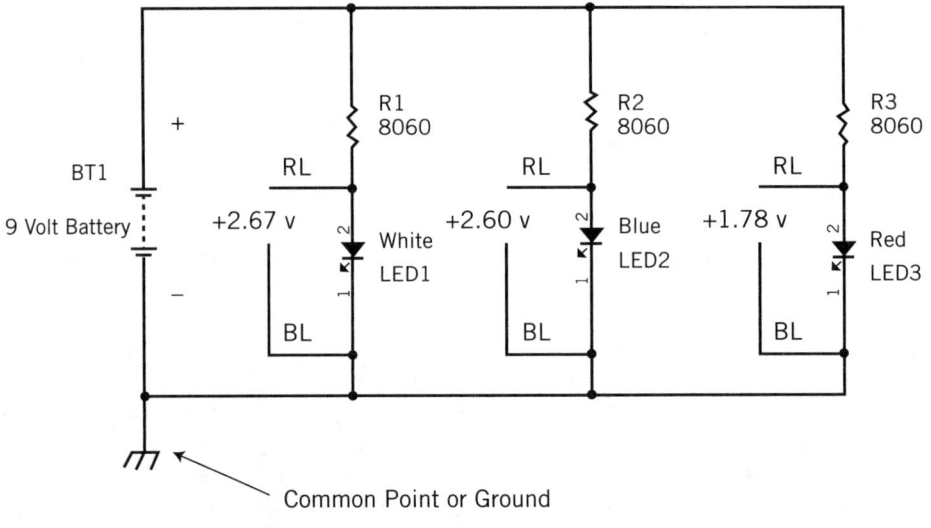

RL = Red Lead of DVM
BL = Black Lead of DVM

FIGURE 6-13   A schematic diagram showing the turn-on voltages (V_F) measured for white, blue, and red LEDs.

Again, we can determine the forward LED current (I_F) by the following:

I_F = (supply voltage – V_F) / drive resistance

For the blue LED, we have:

$I_F = (9 \text{ volts} - 2.60 \text{ volts}) / R2$

$I_F = (9 \text{ volts} - 2.60 \text{ volts}) / 8060\Omega = 6.40 \text{ volts} / 8060\Omega$

$I_F = 0.794 \text{ milliamp}$

Other Types of Green LEDs

In Table 6-1, we found that the green and yellow LEDs had very close turn-on voltages at 1.872 and 1.842 volts. And if you look carefully at these types of green and yellow LEDs, we notice that the green LED gives closer to a green-yellow color instead of a darker or deeper green hue without much yellow. (See Figure 6-9 for an example). It turns out that there are green LEDs that are closer to green-blue in tint, which is a "truer" rendition of what we think of as green color.

Also, there are LEDs that give out more light such as the green 1-watt LED in Figure 6-14.

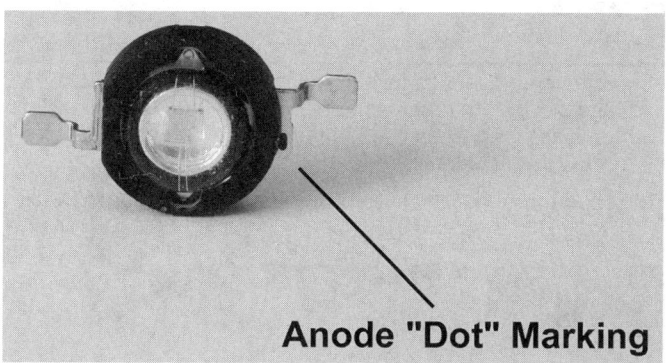

Anode "Dot" Marking

FIGURE 6-14 A 1-watt LED with its anode marking as shown with the dot on one lead. At 1 watt, the maximum current for this LED is about 300 mA. The LED has to be heat-sinked properly by soldering the two leads to large areas of copper that dissipates the heat away.

Let's take a look at the turn-on voltages between a standard green-yellow LED and a green-blue LED. See Figure 6-15.

FIGURE 6-15 The 2.58-volt turn-on voltage for the (phosphor) green-blue (G-B) LED is closer to the turn-on voltage of LEDs that are white (2.67 v) or blue (2.60 v). Note that a standard green LED, which gives green-yellow (G-Y) light has a turn on voltage of 1.88 v.

We can see a difference in colors of the two "green" LEDs in Figure 6-16.

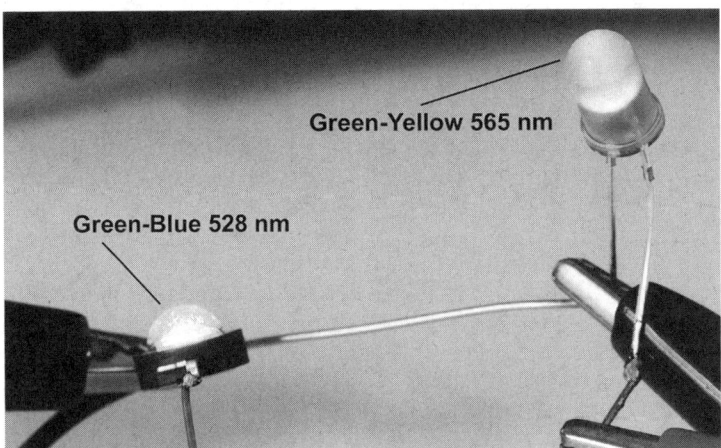

FIGURE 6-16 A shorter wavelength green-blue (528 nm) LED on the left shows a deeper green compared to the longer wavelength green-yellow (565 nm) LED on the right. Note: nm = nanometers.

Although the example in Figure 6-16 shows a 1-watt LED, the green-blue LED is available in the same (5 mm) size as the green-yellow LED.

Problems with Paralleling Two LEDs with Different Turn-On Voltages

In Figures 6-9 to 6-13, 6-15, and 6-16 the LEDs are driven via a series resistor so that the LED current can be controlled more precisely. For example, see Figure 6-17.

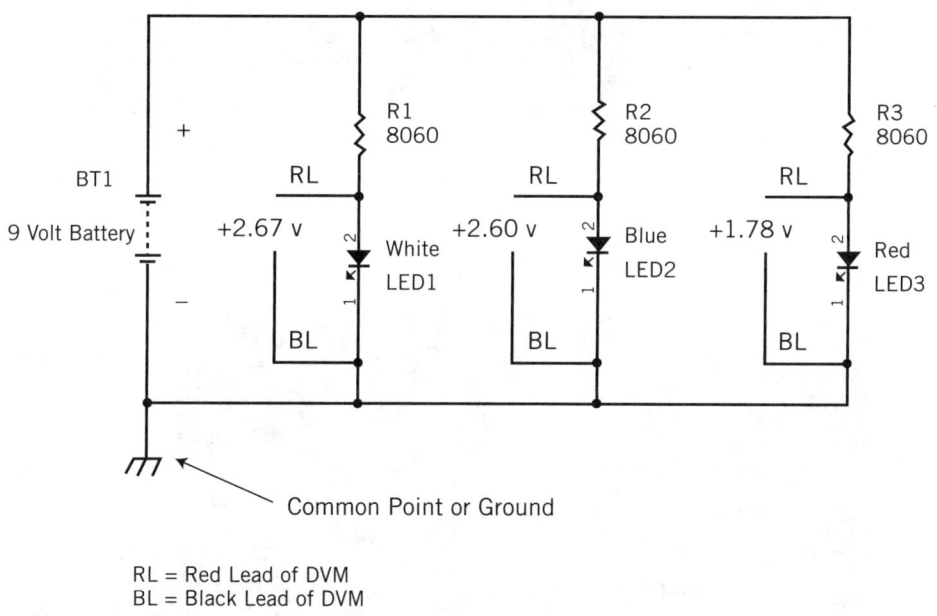

FIGURE 6-17   These LEDs include series current-limiting resistors R1, R2, and R3.

However, what happens if we only drive 2 paralleled LEDs with just one current-limiting resistor? See Figure 6-18.

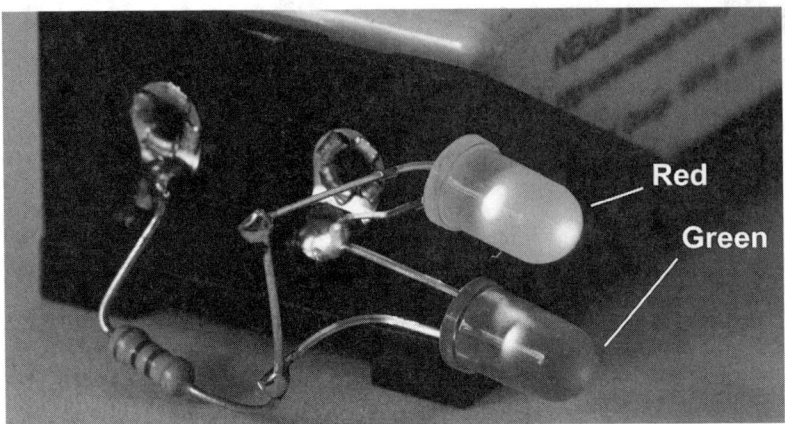

FIGURE 6-18 A red LED and a green LED connected in parallel are driven via one series current-limiting resistor. Note that the red LED is much brighter than the green LED.

For a close-up of the two LEDs, see Figure 6-19.

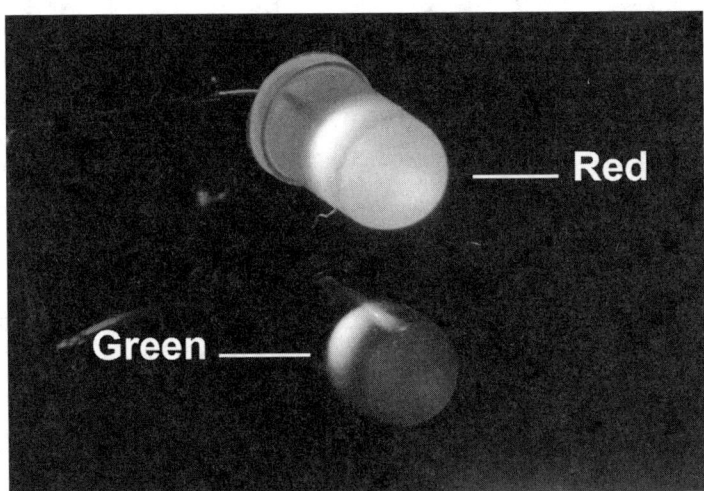

FIGURE 6-19 A close-up view of the two LEDs in parallel that shows the green LED being much dimmer than the red one.

The circuit is shown schematically in Figure 6-20.

The measured voltage across the two LEDs is 1.81 volts. However, from Figure 6-9, the different turn-on voltages with 8060Ω driving resistors resulted in 1.786 volts for red and 1.872 volts for green. Hence, there is about an 86 mV difference with the red LED that turns on with a lower voltage. As shown in Figure 6-20, the measured voltage at the anode is 1.81 volts with a 4300Ω resistor. Note that the red LED's 1.81-volt VF is still lower than the 1.872-volt turn-on voltage of the green LED in Figure 6-9. Hence, the red LED will divert or hog most of the current and it will light up brighter than the green LED (see Figure 6-19).

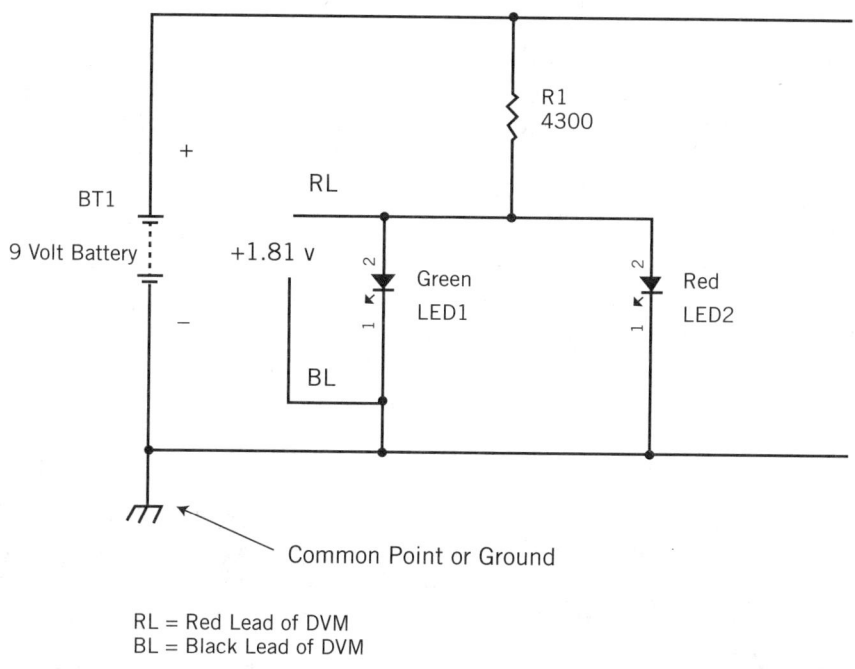

RL = Red Lead of DVM
BL = Black Lead of DVM

FIGURE 6-20 Two LEDs are wired in parallel with a single series resistor R1.

In general, if you are going to parallel LEDs together, first they should be of the same color emission. Second, they should be matched, which, for example, can be done by purchasing the LEDs in "tape" form. See Figure 6-21.

FIGURE 6-21 LEDs bought in tape form are generally better matched in terms of turn-on voltages than those bought in a bag or off the shelf.

Protecting LEDs from Damage Due to Reverse Voltage Across the Anode and Cathode

If you are going to connect LEDs to an AC (alternating current) voltage source such as an AC adapter that gives AC voltage (e.g., 6VAC, 9VAC, 12VAC, etc.) you will need to protect the LEDs from damaging reverse voltage. Recall that AC voltages provide positive voltages at one time followed by a negative voltage at another time.

For example, a 60 Hz AC voltage source has a repetition rate of (1/60) second or 16.666 millisecond. In the first half of that period, the voltage is positive for 8.333 milliseconds (e.g., half cycle). And in the next 8.333 milliseconds (e.g., the remaining half cycle) the voltage is negative.

LEDs only have a peak reverse "breakdown" voltage of about 5 volts. This means for all practical purposes, they must be protected should AC voltage sources be connected to them. Figure 6-22 shows an example with a rectifier (e.g., 1N4004) connected in series with the LED.

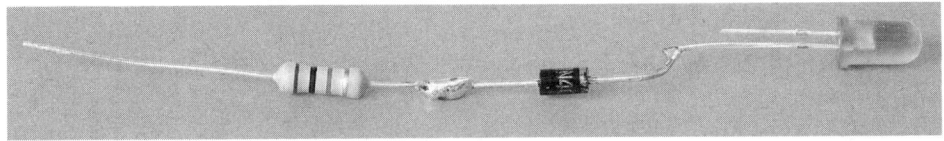

FIGURE 6-22   A limiting resistor (e.g., 1000Ω) is connected to the anode of the rectifier, and the rectifier's cathode is connected to the anode of the LED.

A schematic diagram is shown in Figure 6-23 on the left side.

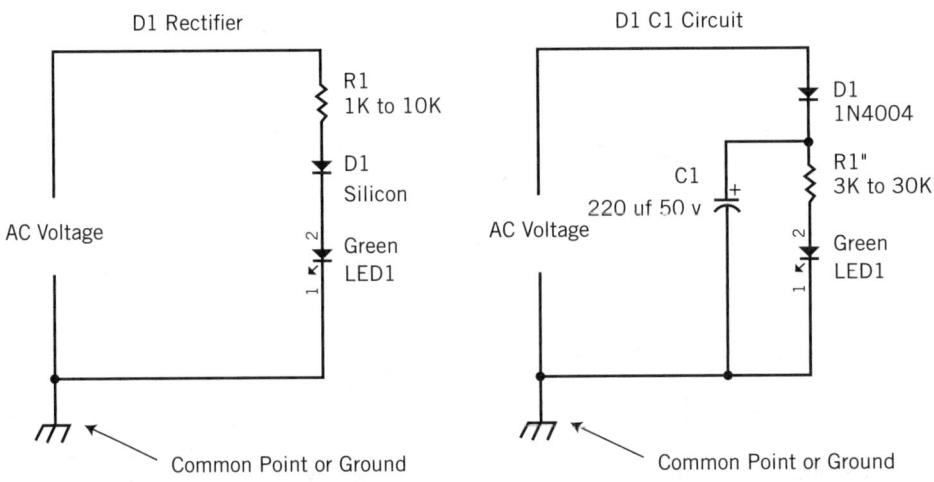

FIGURE 6-23   Two examples of using a "half wave" rectifier D1 to ensure that the LED does not experience any reverse bias voltage from the AC voltage source.

When using a peak hold capacitor, as shown in the "D1C1" circuit in Figure 6-23, be sure to increase the resistance value for R1 by about 3 times. The reason is that the average DC voltage at the positive terminal of C1 is higher than the half wave rectifier D1 circuit by about threefold. Thus, we see that the D1 C1 circuit's LED current-limiting resistor R1" = 3KΩ to 30KΩ, versus R1 = 1KΩ to 10KΩ in the circuit on the left side "D1 Rectifier."

Alternatively, one can connect an LED or diode in "counter" parallel to the LED as shown in Figure 6-24 to safely limit the reverse LED voltage. Note that the protection LED2 is connected with its cathode to the anode of LED1. In the D1 protection diode example, D1's cathode is connected to the anode of LED1, and D1's anode is connected to LED1's cathode. In either example, the reverse bias voltage is limited to about –2 to –3 volts via LED2, and –0.7 volts via D1, which are both safe reverse bias voltages for LED1.

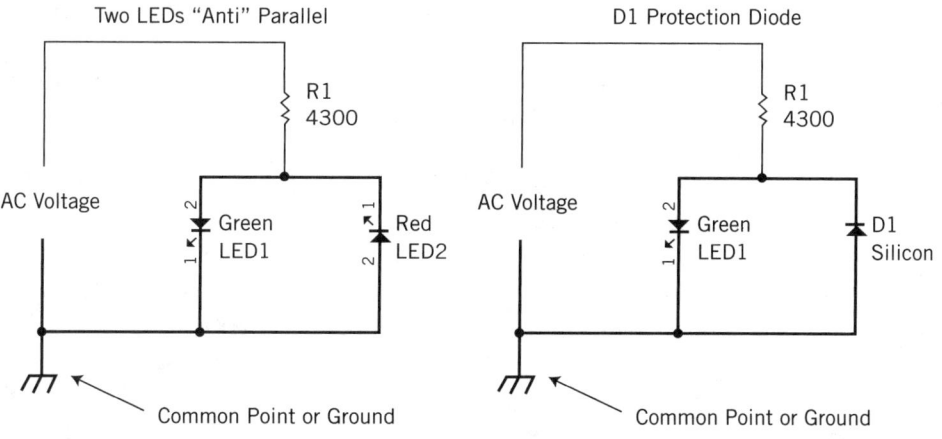

FIGURE 6-24 Two examples of protecting LED1 via LED2 and D1 wired in "anti-parallel."

It should be noted that in Figure 6-24 LED2 lights up only during the negative cycle of the input AC voltage, and LED1 turns on only during the positive cycle of the AC voltage. This circuit can be used to identify the polarity of any incoming voltage. For example, if the AC voltage source is replaced with a DC voltage, the two LEDs "anti-parallel" circuit will readily identify the polarity.

On the left-side circuit of Figure 6-24, LED2 can be any type of LED, such as IR (infrared), red, green, yellow, blue, or white.

And on the right-side circuit of Figure 6-24, D1 can generally be almost any diode including small signal, Schottky, or rectifier diode. If in doubt, just use a 1N4000 series rectifier (e.g., 1N4002).

Some Keys Points About Light Emitting Diodes

- Light emitting diodes (LEDs) can be damaged by excessive forward bias current. Normally, the LEDs are driven at about 20 mA or less using a current-limiting resistor or some other circuit. See Chapter 7.
- LEDs can also be damaged by reverse voltage by having a negative voltage from anode to cathode. If the negative voltage is limited to between 0 volts to −4 volts, this will be OK. An "anti-parallel" connected diode or LED can be used to protect the LED in question.
- Different color LEDs generally cannot be connected in parallel because their turn-on voltages are different, which will cause "current hogging" for the LED that turns on with the lowest voltage. For example, if blue and yellow LEDs are connected in parallel, almost all the current will go to the yellow LED. This is because a yellow LED has a turn-on voltage around 1.8 volts, whereas a blue LED requires a higher 2.4 volts.
- If you need to have matched turn-on voltage LEDs, purchase them on a tape, as shown in Figure 6-21.

This concludes Chapter 6. In Chapter 7, we will explore transistors, where they can be used as current source and amplifying devices.

Bipolar Junction Transistors

In previous chapters, we have dealt with two terminal electronic components such as resistors, capacitors, diodes, and LEDs. We will now explore the commonly used bipolar junction transistors (BJTs, also known as *bipolar transistors*), which are three-terminal devices. They are in many do-it-yourself (DIY) projects and manufactured products.

Bipolar Junction Transistors

The first transistor was invented at Bell Laboratories in 1948. It had a ***point contact*** type of construction that was difficult to manufacture. The point contact transistor technology was replaced by the (alloy) junction transistor that was more easily manufactured. Alloy junction germanium transistors such as the CK722, 2N107, 2N109, 2N408, 2SB32, 2SB54, 2SB56, 2SB75, 2SB77, 2SB113, 2SB115, AC125, AC126, OC71, and OC72 were quite prevalent from the mid-1950s to the late 1960s.

So, what is a transistor? The bipolar transistor is a three-terminal device that allows controlling large amounts of output current by a small change in input current or voltage. Thus, a bipolar transistor can be used as an amplifier where a small input signal can provide a larger signal at its output, with the "help" of a power supply. That is, transistors by themselves cannot amplify or control large electrical current unless there is a supply voltage included in the transistor circuit.

The three terminals to a bipolar transistor are:

- The emitter as identified by the letter "E" or "e"
- The base as identified by the letter "B" or "b"
- The collector as identified by the letter "C" or "c"

See Figure 7-1 for bipolar transistors with different plastic packages. These range from small signal transistors (TO-92) to power transistors (TO-220 and TO-247).

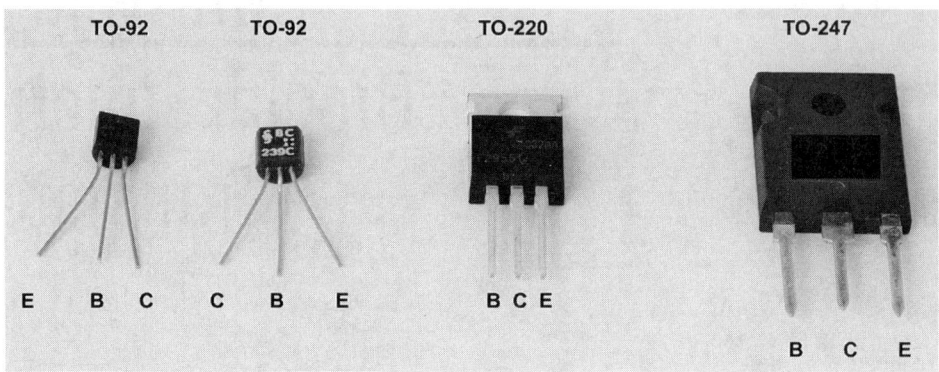

FIGURE 7-1   Various transistors. Note the different pin outs for the two TO-92 transistors.

Most plastic small-signal transistors (e.g., < 500 mA collector current) with 2N or PN prefixes, such as 2N3904, 2N4124, 2N4401, PN2222 or 2N5089, have the pin out sequence E, B, C as shown in the first transistor on the left in Figure 7-1. This left-to-right emitter, base, collector sequence is taken as you look at the transistor with the flat side up.

In Europe, transistors that generally have a prefix of BC, such as BC548, BC556, or BC239C (as shown in Figure 7-1, second transistor from the left side), and other part numbers with a P2N prefix, such as P2N2222A and P2N2907A, have a "reversed" pin out sequence of C, B, E. The left-to-right collector, base, emitter sequence is taken when you look at the transistor flat side up.

For the two power transistors shown in Figure 7-1, the sequence, taken left to right with the part number side up or facing you, is base, collector, emitter, or B, C, E. Note that the metal tab in the TO-220 transistor actually extends further down in the back of the transistor. This metal tab is connected to the middle lead or collector of the TO-220 power transistor. In general, with very few exceptions, it is very important to insulate or to ensure that the metal tab is not connected to ground or to some other point of a circuit. TO-220 transistors have maximum collector currents of about 10 amps. However, for many power applications such as driving loudspeakers or motors, the TO-220 transistor should be mounted on a heat sink with a thermally conductive insulated sheet or washer.

The TO-247 transistor in Figure 7-1 can handle collector currents greater than or equal to 15 amps. Not shown but in its back side is an exposed metal piece that is also connected to the middle lead or collector of the power transistor. Again, if the transistor is delivering high power (> 5 watts) to a load such as a motor, loudspeaker, or lamp assembly, a heat sink is required. The TO-247 transistor would be mounted on a heat sink via a thermally conductive insulated sheet or washer. See Figure 7-2 for various TO-220 heat sinks.

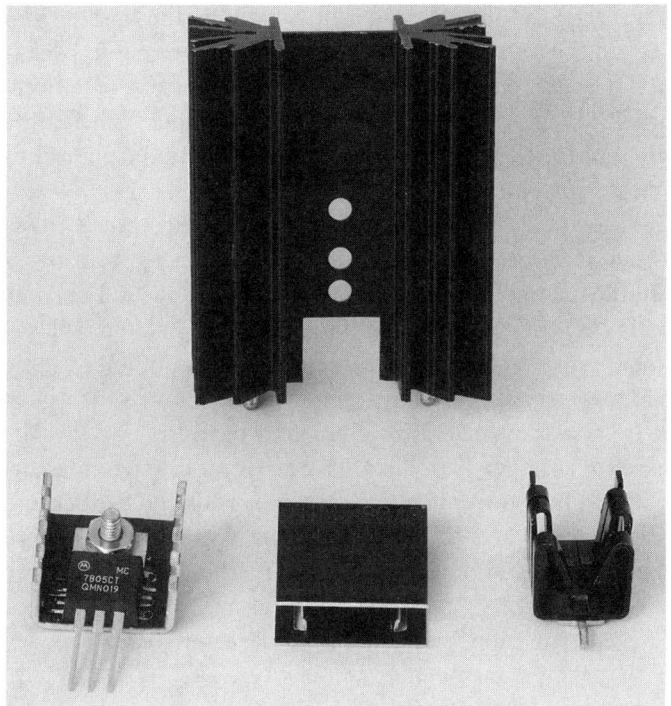

FIGURE 7-2   Various TO-220 heat sinks with an example TO-220 voltage regulator mounted on the bottom-left side.

Transistors also come in metal packages, and the two most common ones are TO-18 and TO-5 (see Figure 7-3). These transistors have protruding tabs, which identify the emitter terminals.

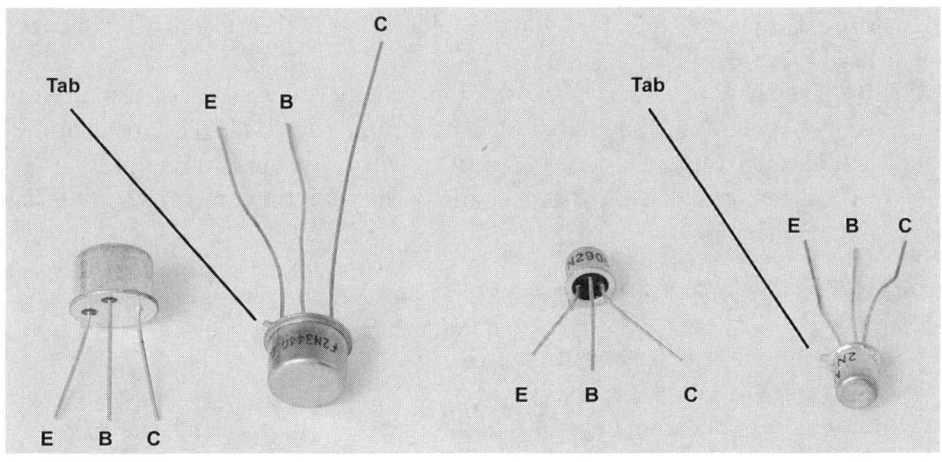

FIGURE 7-3   TO-5 transistors on the left and the smaller TO-18 transistors on the right.

In metal case transistors the middle lead will usually be the base terminal. To find the emitter lead, find the tab that is located next to it. The remaining lead opposite the emitter terminal is the collector. Another way is to look at the bottom of the metal case transistor with the middle lead pointing to "12 o'clock." The emitter lead will be to the left of "9 o'clock," and the collector lead will be on the right that is analogous to a "3 o'clock" position.

Generally, metal case small and medium collector current transistors are not used as much, and they have been replaced with plastic package transistors. One reason is cost, and another is that they only offer a little more collector current capability than some plastic case transistors of the similar sizes. What you will see, however, is that metal case transistors were used often in older equipment (e.g., older Hewlett Packard solid state signal generators, vintage stereo amplifiers, tuners, and receivers, or ham radio transceivers, transmitters, and receivers). So, it is good to know about metal case transistors when you are fixing or modifying older circuits.

It is very important to know the pin out for each transistor you use because none of the three terminals is interchangeable. For example, if you reverse the collector and emitter leads, your circuit will not work properly. In other cases, mis-wired transistor leads can result in damage.

In general, a damaged transistor usually has two characteristics:

1.   Shorted collector to emitter junction, which when measured with an ohm meter, the normally "open" or infinite ohms reading goes to a low resistance reading of typically $< 10\Omega$.
2.   Open base to emitter junction, which means if you use your DVM's diode voltage testing function, you do not get typically 0.6 volts to 0.7 volts, but instead get an "error" reading.

However, because most DVMs have a transistor tester, you can just test the transistor. For power transistors, you can solder wires to their terminals for easier testing in a DVM.

If you have never seen a transistor in a circuit and want to just test it, what would you look for? The DVM tests for current gain known as H_{fe}, h_{FE}, beta, or the Greek letter beta, β. Generally, a good transistor will have a current gain of 10 or more.

Current gain is defined as the ratio of the collector current, IC, to the base current, IB. That is:

$$H_{fe} = h_{FE} = beta = \beta = \frac{IC}{IB}$$

Or put in another way, $IC = \beta\ IB$.

Typically, $15 < \beta < 500$.

For example, if we have 1 µA of base current, IB, and the $\beta = 250$, then the collector current:

$IC = 250 \times 1\ \mu A$

or

$IC = 250\ \mu A$

Typically, small signal (TO-92) and metal case transistors (e.g., TO-5 and TO-18) have ≥ 20 current gain. If the transistor is mis-wired and connected backwards in terms of the collector and emitter terminals, then you will see a substantial current gain decrease to typical numbers of $0 < \beta < 4$.

See Figures 7-4 and 7-5, which show testing a 2N2906 (TO-18) transistor.

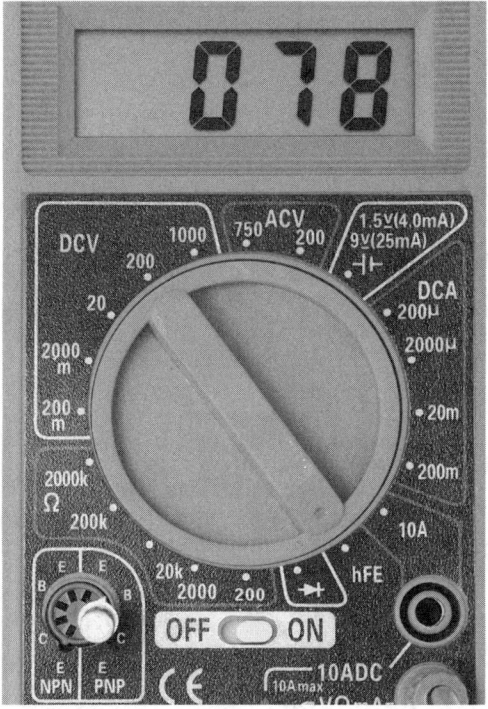

FIGURE 7-4 A TO-18 transistor with its emitter tab pointed up toward "E" with the DVM selector knob set to h_{FE}. Here, we see the transistor has a current gain, $h_{FE} = 78$.

There are two types of bipolar transistors, NPN and PNP. To identify the transistor, you can plug the test transistor into the NPN socket and test, and then do the same in the PNP socket. Only one socket will give you a correct reading, which is usually a number between 15 and 1000. The other way to identify the transistor is to read the part number printed on the case and look up the data sheet. For example, the transistor being tested is a 2N2906, which its data sheet identifies as a PNP transistor.

Now, should you swap or reverse the collector and emitter leads, while having the base lead correct, you will find that the current gain is $h_{FE} = 1$. See Figure 7-5.

FIGURE 7-5   The same transistor tested "backwards" with the emitter inserted in the collector, C, portion of the socket and the transistor's collector lead inserted into the E portion of the socket. Note that the tab that identifies the emitter is pointing down. The resulting current gain is very low at $h_{FE} = 1$ when the transistor is wired backwards this way.

What Happens When a Transistor Is Damaged

Transistors, just like diodes, share a couple of common specifications. One is the maximum voltage, and the other is the maximum current. When either of these specifications is exceeded, the transistor can be damaged. The typical blown or damaged transistor suffers from any combination of a short circuit or low resistance between the collector and emitter that can be verified with an ohm meter. Make sure to test the collector and emitter leads with the test leads one way and then the other—for example, the red test lead connected to the collector and black test connected to the emitter. Then switch the ohm meter's leads where the black test lead is connected to the collector and the red test lead is connected to the emitter. Note the two resistance readings, which should be <20Ω if the transistor is damaged via a short from collector to emitter.

A damaged transistor can have an open circuit between the base and emitter terminals. This can be verified by using the DVM's diode voltage test as shown in Figure 7-6.

FIGURE 7-6   Testing the base-to-emitter voltage with the DVM where the one on the left is good with a forward base-to-emitter voltage reading of ~0.768 volt in three digits. But the one on the right is damaged with a single digit "1" that indicates an open circuit. If the reading is not in the 0.7-volt range, reverse the test leads just in case you are testing a transistor of the opposite polarity (e.g., NPN versus PNP, or PNP versus NPN). It is sometimes easy to incorrectly think of which type of transistor you are testing because of a misread transistor part number.

Schematic Symbol of NPN and PNP Transistors

In most circuit schematic diagrams, transistors are designated with a "Q" prefix such as Q1, Q2, Q3, etc. (See Figure 7-7.) Alternatively, sometimes the transistors are labeled with a prefix of "TR," which results in TR1, TR2, TR3, etc., or sometimes with a dash in between like TR-1, TR-2, TR-3, and so on.

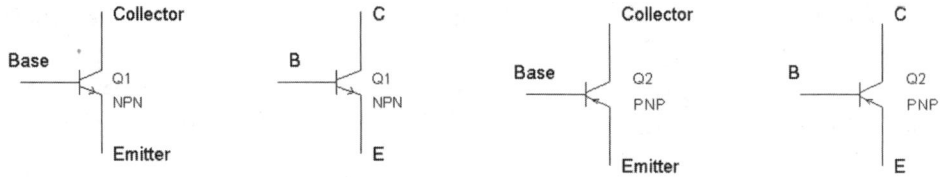

FIGURE 7-7   Schematic symbols for NPN (Q1) and PNP (Q2) with their terminal's emitter, base, and collector labeled equivalently as E, B, and C.

NOTE: The arrows in the emitter terminals show the direction of positive current flow.

Applying a DC Voltage to the Base of the Transistor to Provide a Constant Current Source

To reiterate, the transistor is an electrical current controlling device. In most cases, it provides current gain and/or voltage gain from an input to an output. Let's take a look at several circuits where the transistor works as a constant current source. In the first examples, a small base current, IB, at the "input" results in a large collector current, IC, at the "output." See Figure 7-8.

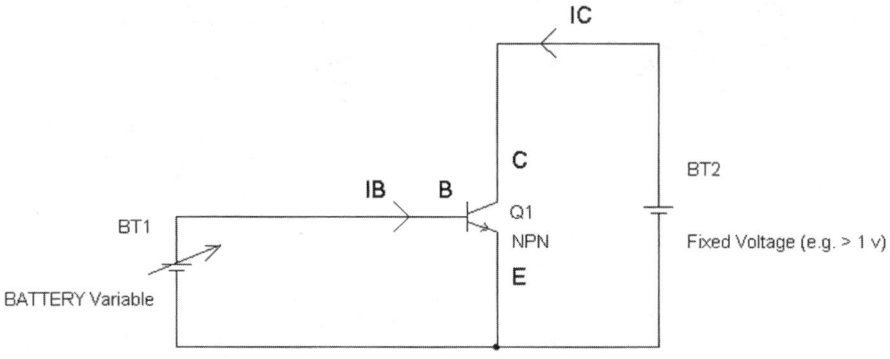

FIGURE 7-8   A constant current source via collector output current IC. A variable power supply voltage source, BT1, is applied directly to the base-emitter junction, which can damage the transistor Q1.

In Figure 7-8 a positive voltage, BT1, is applied to the base with respect to the emitter. And the output current, IC, requires a positive voltage to the collector with respect to the emitter.

A basic principle of biasing a transistor is shown in Figure 7-8, which is not normally done in practice because the transistor can be easily damaged this way. Biasing is a way of turning on the transistor Q1 via applying a bias voltage, BT1, across the base and emitter terminals. In this case variable voltage source BT1 is turned up to a voltage generally in the range of 0.6 volts to 0.7 volts to generate a collector current IC at the output. However, in practice this method is rarely used because if BT1 exceeds 0.8 volts, the transistor may burn out and either the collector and emitter terminals get shorted out due to excessive collector currents and/or the base-emitter junction becomes like an open circuit. This happens because the bipolar transistor has an ***exponential relationship*** between the base-emitter voltage and collector current. So, what does this really mean? It means that a small increase in base-emitter voltage (e.g., from BT1) will result in a large increase in collector current.

For example, if a transistor has 1 mA collector current for a BT1 set to 0.600 volt, the collector current will increase to 10 mA for BT1 = 0.660 volt, and 100 mA when BT1 = 0.720 volt. As you can see, every increase in 60 mV causes a change in collector current by 10-fold.

Another reason not to directly drive the base-emitter junction directly with a voltage source is that every transistor has slightly different turn-on voltage. Generally, a power transistor will have slightly lower turn-on voltage. But even transistors of the same type (e.g., small-signal) or same part number can have base-emitter turn-on voltages that are different by 60 mV or more due to the transistor manufacturing process. This means that without sorting or measuring each transistor, you can get a 10:1 variation in collector current for the same base-emitter bias voltage.

Just like LEDs, transistors have maximum base current and collector current ratings. Generally, we look for the maximum collector current specification that is found in a data sheet.

For example, a very popular general purpose NPN transistor, 2N3904 (TO-92 plastic case, see Figure 7-1), has a maximum collector current rating of about 200 mA. For higher currents, a popular NPN power transistor is the MJE3055T transistor (TO-220 case, see Figure 7-1) that has a 10-amp collector current rating.

In practical transistor circuits, generally the base and/or emitter will include a resistor to help "stabilize" the collector current, which allows for plugging in different transistors while providing essentially a reliable or known collector current. Figure 7-9 shows simple series base resistor biasing. This type of circuit was used often in the early days of transistor circuits back in the 1950s. It can still be used today but there are problems with this circuit because β can vary from one transistor to another, which means the collector current will vary for a given fixed base current.

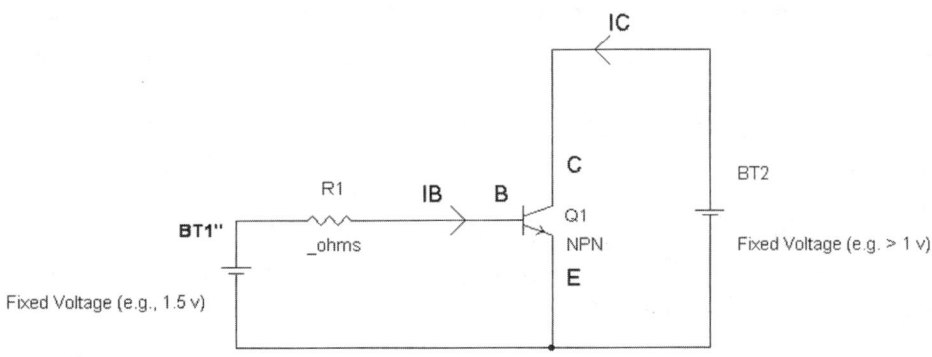

FIGURE 7-9   Biasing a transistor via a series base resistor, R1, to set a collector current IC.

To prevent accidental damage to the transistor due to excessive collector current, the series base resistor R1 isolates BT1 from connecting its voltage directly to the base-emitter terminals. The collector current IC then can be calculated in the following manner.

Again, we need to reiterate that current gain is the ratio of IC (collector current) to IB (base current), or:

$$\text{current gain} = \frac{IC}{IB}$$

There are two commonly used symbols for transistor current gain, β and h_{FE}, which are one and the same. Thus:

$$\beta = h_{FE} = \frac{IC}{IB}$$

With a little algebra, we can find the collector current if we know the current gain and base current. In this book β and h_{FE} are interchangeable. Thus:

IC = β IB

For example, if $\beta = h_{FE} = 100$ and IB = 10 µA, then IC = 100 × 10 µA or IC = 1000 µA = 1 mA.

In Figure 7-9, the base current, IB, is the current flowing through R1. We can determine the current flowing through R1 by finding the potential difference across R1.

IB = (BT1 – VBE)/R1

If we know the battery voltage BT1, and approximate that VBE is in a narrow range of 0.6 volt to 0.8 volt, we can use a value in the middle such as 0.7 volt for VBE. Thus, IB = (BT1 – 0.7 volt)/R1.

IC = β IB so:

IC = β (BT1 – 0.7 volt)/R1

For example, suppose R1 = 100KΩ and β = 100, with BT1 = 1.5 volts.

IC = 100 (1.5 volts – 0.7 volt)/100KΩ = 100 (0.8 volt)/ 100KΩ = 80 volts/100KΩ

IC = 0.0008 amp or 0.8 mA or 800 µA

If the β = 300, then IC = 300 (1.5 volts – 0.7 volt)/100KΩ = 0.0024 amp or 2.4 mA.

The current gain β or h_{FE} is usually specified by the transistor's data or specification sheet. Also, β or h_{FE} normally has a wide range such as 3:1. For example, the β can vary from 50 to 150 for some transistors, and 100 to 300 for others.

NOTE: In Chapter 6 we stated that an LED is generally not to be driven directly with a voltage source such as a power supply. This applies the same to any base-emitter junction. A base-emitter junction directly driven with a low-resistance power supply can result in damaging the transistor. Often, when excessive current has been applied to the base, the base-emitter junction becomes an open circuit and thus no longer conducts or allows base current to flow. This open circuit between the base and emitter, due to damage, will usually result in zero collector current.

We will now show examples of biasing a transistor via a series base resistor. See Figures 7-10 and 7-11, which illustrate driving LEDs.

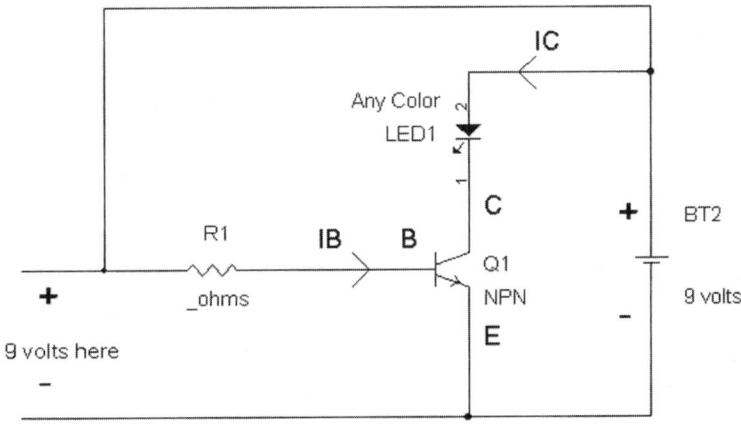

FIGURE 7-10   A simple LED driver circuit using an NPN transistor current source.

The simple LED drive circuit in Figure 7-10 uses the existing collector supply voltage source, BT2, to drive the base resistor, R1. The base current with BT2 = 9 volts is then IB = (BT2 – 0.7 volt)/R1 and the LED current, which is the collector current, is then $IC = I_{LED} = \beta IB = \beta(BT2 – 0.7 \text{ volt})/R1$.
For example, if $\beta = 50$ and R1 = 50KΩ, then:

$I_{LED} = \beta(BT2 – 0.7 \text{ volt})/R1 = [50 (9 – 0.7)/50,000] \text{ amp} = [8.3/1000] \text{ amp} = 0.0083 \text{ amp}$

or

$I_{LED} = 8.3 \text{ mA}$

The current gain in the simple circuit in Figure 7-10 can vary if you have the same part number transistors, such as 2N3904. Figure 7-11 shows an illustration where the β for Q1 = 129, and for Q2 β = 239.

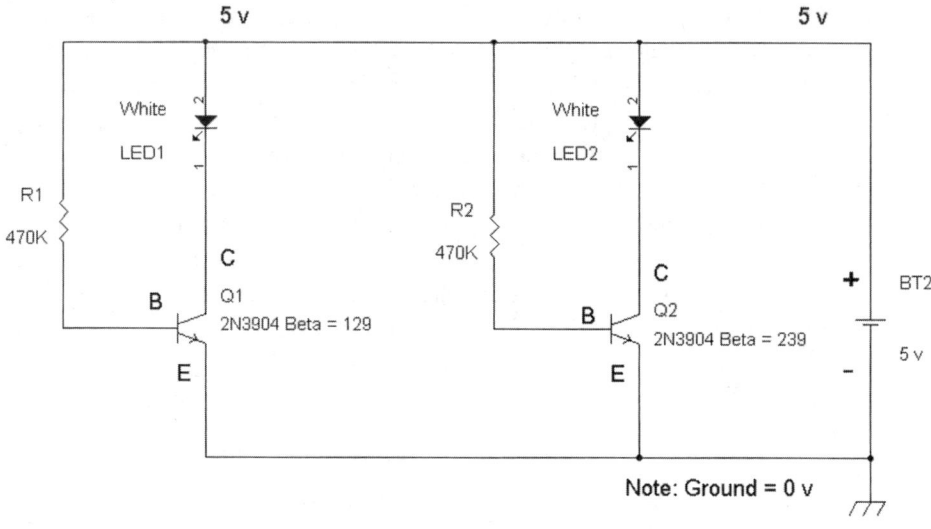

FIGURE 7-11   A simple LED drive circuit with a different current gain, β, for transistors Q1 and Q2.

Because of the two different current gains, β = 129 and β = 239 for transistors Q1 and Q2, the LED currents for LED1 and LED2, I_{LED1} and I_{LED2}, are proportionally different as well.

I_{LED1} = β(BT2 – 0.7 volt)/R1 = [**139**(5.0 – 0.7 volt)/470,000] amp or
I_{LED1} = 0.00127 amp = 1.27 mA

I_{LED2} = β(BT2 – 0.7 volt)/R1 = [**239**(5.0 – 0.7 volt)/470,000] amp or
I_{LED2} = 0.00218 amp = 2.18 mA

The result of the current gain variation causes a change in LED brightness. See Figure 7-12.

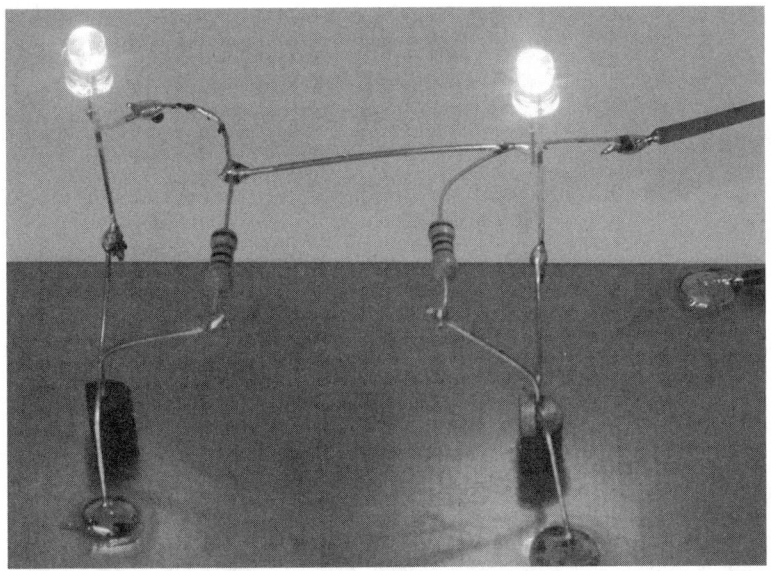

FIGURE 7-12   The circuit built from Figure 7-11 shows that the LED on the left side via Q1where β = 139 is dimmer than the LED on the right side via Q2 where β = 239.

Improved Current Source Circuits

We can improve current source circuits to be less dependent on the current gain β by adding an emitter series resistor, RE, as shown Figure 7-13.

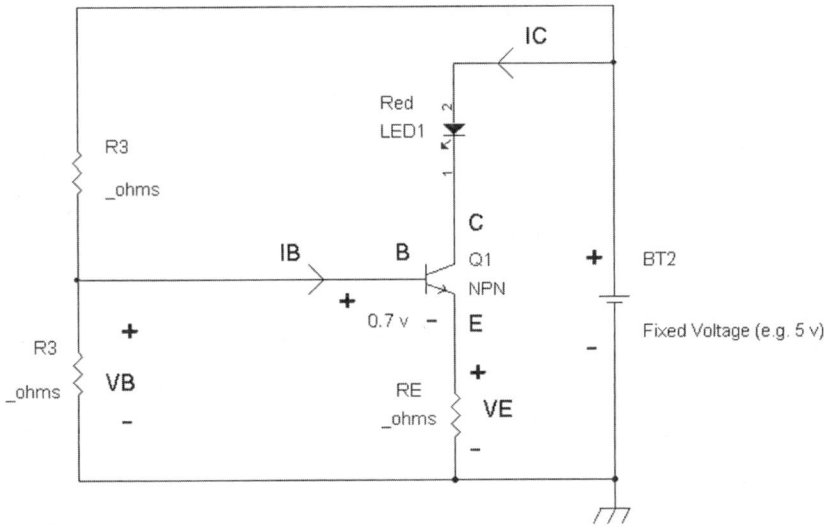

FIGURE 7-13   An improved circuit that is not β dependent to set the collector or LED current.

The collector current or LED current is just the emitter current IE = VE/RE because in a transistor the collector and emitter currents are essentially equal when used as a current source or amplifier circuit.

I_{LED1} = VE/RE

Resistors R2 and R3 form a voltage divider circuit that provide a base biasing voltage.

VB = BT2 [R2/(R2 + R3)]

To find VE, note that:

VB = VBE + VE

VBE is the voltage across the base and emitter of Q1. As shown in Figure 7-13, **VBE = 0.7 volt**.

VB = 0.7 volt + VE

Or put in another way in terms of VE:

VE = VB − VBE

VE = VB − 0.7 volt

We need to have VB greater than the turn-on voltage of Q1, which is 0.7 volt. But how much greater should we set it to? It depends on how much voltage we set for the emitter to ground voltage, VE. Typically, we want at least 0.5 volts so that any different lot of transistors with a 0.5-volt to 0.8-volt base emitter turn-on voltage will still work. But a rule of thumb would be something like VE ≥ 0.7 volt.

Because VB = 0.7 volts + VE, and VE ≥ 0.7 volt:

VB = 0.7 volt + ≥ 0.7 volt

VB ≥ 1.4 volts

For example, suppose we want VB = 1.7 volts with BT2 = 5 volts, which means VE = VB − 0.7 volt or VE = 1.0 volts. Or in general:

I_{LED1} = VE/RE

Or put in another way:

RE = VE/ I_{LED1}

For example, if we want 1 mA = 0.001 amp = I_{LED1}, and VE = 1 volt, then:

RE = 1 volt/0.001amp = 1000Ω

Now let's get back to solving for R2 and R3 when VE = 1.0 volt.

VB = 0.7 volt + VE

That is, if we measure the voltage across the base and emitter (VBE) and then measure the voltage (VE) at the emitter and ground, we will find VB = VBE + VE. VBE ~ 0.7 volt, so for 1-volt VE:

VB = 0.7 volt + 1.0 volt

VB = 1.7 volts

Resistors R2 and R3 form a voltage divider circuit that provides a base biasing voltage with **BT2 = 5.0 volts and RE = 1000Ω**. Again, in general:

VB = BT2 [R2/(R2 + R3)]

However, via some algebra a more convenient equation can be derived as: [(BT2/VB) – 1] = R3/R2, and if BT2 = 5.0 volts and VB = 1.7 volts, then we have:

[(5.0/1.7) – 1] = R3/R2

[2.94 – 1] = 1.94 = R3/R2

Or

R3 = 1.94 × R2

Now the question is, which resistors do we choose?
Can we make R2 = 1Ω and R3 = 1.94Ω? Yes, but you will be either burning up the resistors because the current through the resistors will be BT2/(R2 + R3) = 5 volts/2.94Ω = 1.7 amps, or you will be wasting power like crazy.

Power = P = 5 volts × 1.7 amps = 8.5 watts

A general rule of thumb would be to have R2 have the following range:

RE ≤ R2 ≤ 10 RE

So, let's take an example setting R2 = 10RE:

R2 = 10 × 1000Ω = 10KΩ

and

R3 = 1.94 × R2 = 1.94 × 10KΩ = 19.4KΩ

or

R3 ~ 20KΩ

With the resistor values R2 = 10KΩ, R2 = 20KΩ, and RE = 1000Ω, Figure 7-13's current source will deliver a reasonably stable 1 mA given the battery voltage is 5.0 volts, even though the β of the transistor may vary 3:1 with a minimum β of 50. This is because the base current of Q1 no longer predominately determines the collector current or LED current. Instead, it is the voltage VE (via VB) that sets the LED current for a given RE resistance value.

We can improve the current source further by "regulating" VB so that it is a constant voltage. See Figure 7-14 that uses a Zener diode as a reference voltage source, Vref.

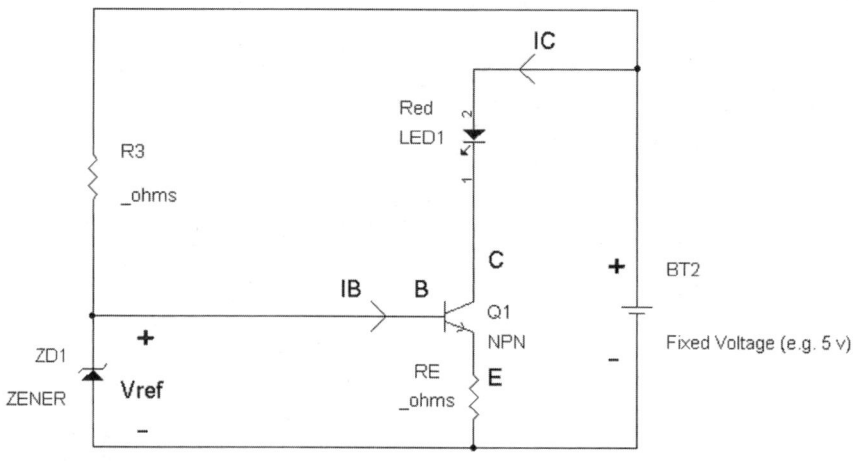

FIGURE 7-14   A Zener diode current source circuit.

In this circuit, first determine the LED current, which is:

$$VE = Vref - VBE$$

With

$$VBE = 0.7 \text{ volt}$$

$$VE = Vref - 0.7 \text{ volt}$$

And

$$I_{LED1} = (VE)/RE$$

$$I_{LED1} = (Vref - 0.7 \text{ volt})/RE$$

$$RE = (Vref - 0.7 \text{ volt})/I_{LED1}$$

As a general rule of thumb, we want the current through R3 to be $\geq 10\%$ of I_{LED1}. For example, we can set the current through resistor R3, $I_{R3} = (BT2 - Vref)/R3 = 10\% \, I_{LED1}$, then:

$$R3 = (BT2 - Vref)/(10\% \, I_{LED1})$$

For example, suppose if we want $I_{LED1} = 20$ mA, and ZD1 = 1N4371 with a 2.7-volt Zener voltage Vref, then:

$RE = (Vref - 0.7 \text{ volt})/ I_{LED1} = (2.7 \text{ volts} - 0.7 \text{ volt})/0.02 \text{ amp or}$

$RE = 2.0 \text{ volts}/0.02 \text{ amp} = RE$, which leads to:

$RE = 100\Omega$

$BT2 = 5.0 \text{ volts}, Vref = 2.7 \text{ volts}, I_{LED1} = 0.02 \text{ amp}$

$R3 = (BT2 - Vref)/(10\% I_{LED1}) = (5.0 \text{ volts} - 2.7 \text{ volts})/(10\% \, 0.02 \text{ amp})$

$R3 = 3.3 \text{ volts}/0.002 \text{ amp} = 1650\Omega \sim 1.6K\Omega$

$R3 \sim 1.6K\Omega$

In Figure 7-14, there is one precaution to take in terms of making sure the transistor is not in saturation. To ensure that it is really working in the **forward active region**, we need to have the collector voltage ≥ base voltage, or where the collector voltage is ≥ Vref in this case.

The collector voltage in Figure 7-14 is $VC = (BT2 - V_{LED})$.

From Chapter 6, we can summarize the following:

- For red, yellow, and most green LEDs, $V_{LED} \sim 1.8$ volts to 2.0 volts (2.0 volts worse case)
- For blue, white, and green phosphor LEDs, $V_{LED} \sim 2.7$ volts to 3.0 volts (3.0 volts worse case)
- If Vref = 2.7 volts, then VC ≥ 2.7 volts for a red LED and BT2 = 5.0 volts. (We will use worse case numbers to be on the safe side.)
- Thus, VC = (5.0 volts – 2.0 volts) = 3 volts ≥ VB = Vref = 2.7 volts.

So, we are "fine" using the circuit for red, yellow, and most standard green LEDs.

However, if we use a white or blue LED, VC = (5.0 volts – 3.0 volts) = 2.0 volts, which is less than Vref at 2.7 volts, then the transistor will most likely be in the saturation region. This means that excessive base currents forming a larger than expected voltage drop across R3 will cause Vref to drop and the desired LED current will be lower than expected. Essentially, when the transistor is in the saturation region by having VC < VB, the current gain is no longer β, but some number much less. So how do we fix the problem if we want to put in a white LED for LED1 while not saturating the transistor Q1? One way is by using a 1N4678 1.8-volt Zener diode, which results in Vref = 1.8 volts and RE → 56Ω and R3 can still be 1.6KΩ. The other way with Vref = 2.7 volts is just to raise the battery or power supply voltage to 6 volts or more.

However, in many cases we can replace the 1.8-volt Zener diode with three 1N914 small-signal diodes in series, or use a red LED as shown in Figure 7-15.

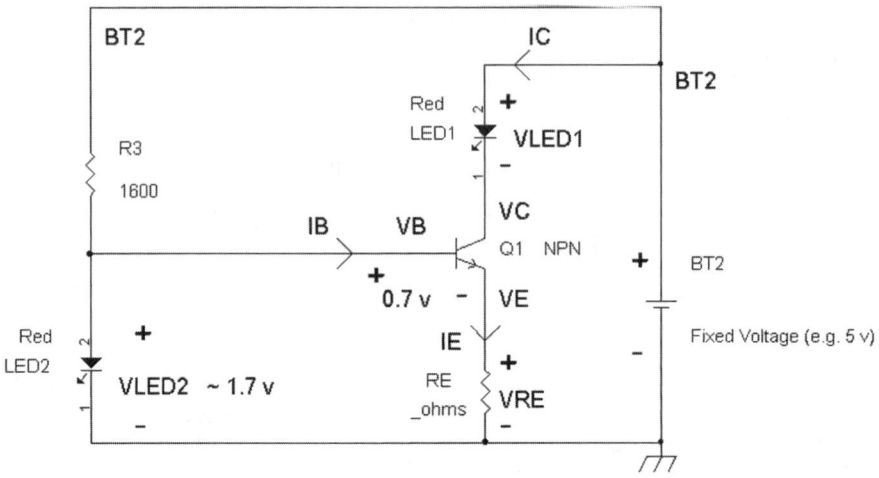

FIGURE 7-15   A practical circuit using a red light-emitting diode, LED2 in place of a Zener diode for Vref, a voltage reference source.

Almost any LED can be used as a voltage reference for biasing purposes. Just remember that standard infrared, red, yellow, and green (non-phosphor) LEDs have turn-on voltages of about 1.6 volts to about 2.0 volts. And the blue and white LEDs have about a 2.7-volt to 3.0-volt turn-on voltage. The LED turn-on voltage is then used to provide a reference voltage for biasing.

Figure 7-15 shows a red LED as a voltage reference source to supply about 1.7 volts = Vref. The emitter voltage, VRE = 1.7 volts – 0.7 volt (VBE) ~ 1.0 volt across resistor RE. The voltage, VRE, across resistor RE then sets up the transistor's emitter current. And essentially the emitter current is equal to the collector current, which in turn is the current flowing through LED1.

If RE = 100Ω, then the LED current is 1.0 volt/100Ω = 10 mA = IC = I_{LED1}. Now if we work backwards to determine the base current, IB, that is equal to IC/β. And if β = 50, IB = 0.01amp/50 = 0.2 mA = IB. With BT2 = 5 volts, the current flowing through R3 is [5.0 volts – V_{LED2}] divided by 1600Ω. With V_{LED2} = 1.7 volts, we have 3.3 volts/1600Ω = 2.06 mA = current flowing through R3. We can think of the current flowing through R3 branches out to two DC currents:

- A current flowing into LED2, I_{LED2}
- And a current flowing into the base of transistor Q1, IB

We know that the base current, IB = 0.2 mA = IC/β = 0.01amp/50, when β = 50.

Of that 2.06 mA flowing through R3, 0.2 mA goes to the base of Q1 via IB, while the remaining current, I_{LED2}, flows into LED2. To find I_{LED2}, we know that:

3.3 volts/1600Ω = 2.06 mA = current flowing through R3 = IB + I_{LED2}

$2.06 \text{ mA} = IB + I_{LED2}$

But IB = 0.2 mA, so:

$2.06 \text{ mA} = 0.2 \text{ mA} + I_{LED2}$

By algebra, subtracting 0.2 mA from both sides of the equation:

$2.06 \text{ mA} - 0.2 \text{ mA} = I_{LED2}$

$ILED2 = 1.86 \text{ mA}$

The main thing to take away from this is that there is sufficient current to turn on LED2. Generally, any LED current between 0.5 mA and about 20 mA will work fine.

What Happens When Things Go Wrong

Probably the most common mistakes in LED, diode, or transistor circuits, are reversed component connections. We will look at some examples and see what this will do to other parts of the circuits, starting with Figure 7-16.

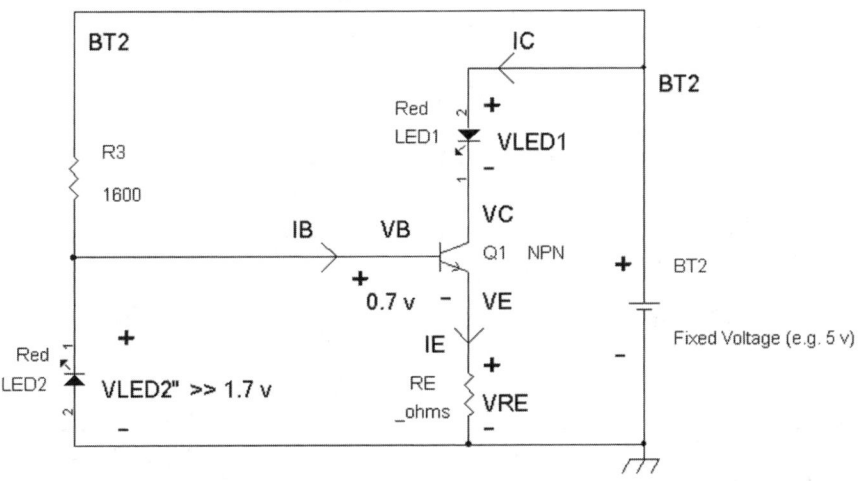

FIGURE 7-16   Voltage reference LED2 is reversed and can cause LED1 to burn out.

With LED2 connected in "reverse bias" mode, it will not light up and it is essentially an open circuit that causes VLED2" >> 1.7 volts. This means that BT2's voltage via resistor R3 will supply Q1's base voltage, VB, which will be much higher than 1.7 volts. With VB >> 1.7 volts, the LED1 current $I_{LED1} = (VB - 0.7 \text{ v})/RE$ and LED1 will have excessive current flowing through it. For example, if RE = 75Ω and VB is supposed to be 1.7 volts but someone inadvertently reversed the leads and VB → 3.3 volts via R3, then:

I_{LED1} = (VB – 0.7 v)/RE = (3.3 v – 0.7 v)/75Ω = 34 mA, for the reversed LED2 connection

The actual voltage VB when LED1 is reverse biased is difficult to calculate off-hand because we often do not know the actual β of Q1. Also, the voltage of BT2 enters into this calculation. As a troubleshooter, we do not have to "bother" with trying to figure out the circuit analysis for an incorrectly wired circuit. We just need to solve the problem and restore the circuit back to proper operation. That is, don't waste too much time analyzing circuits with mistakes. We have to get it working and move on.

To fix this, you connect LED2 properly (LED2 anode to base Q1, and LED2 cathode to ground; the LED's long lead = anode) as shown in Figure 7-15, and then VB ~ 1.7 volts so that:

I_{LED1} = (VB – 0.7 v)/RE = (1.7 v – 0.7 v)/75Ω = 13.3 mA

Now let's look at Figure 7-17 with Q1's collector and emitter leads reversed.

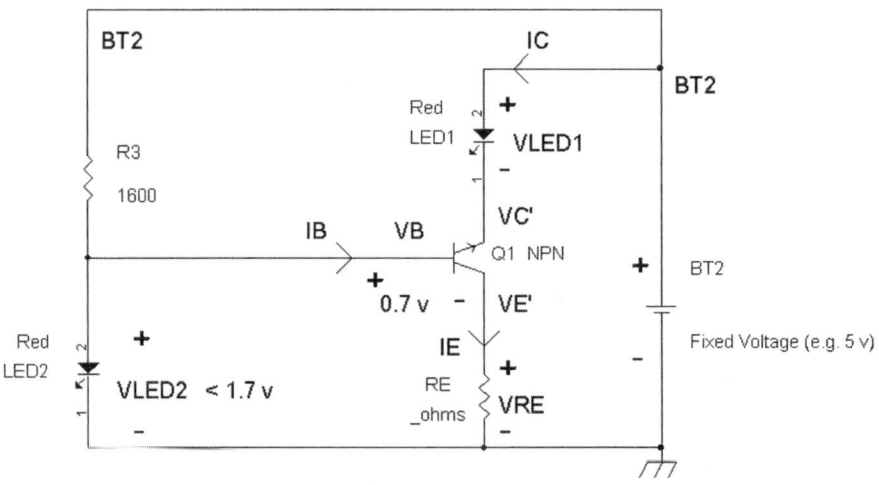

FIGURE 7-17 With Q1's emitter and collector leads reversed, the current gain is very low with β ≤ 4.

With the collector and emitter leads reversed, transistor Q1 has low β and starts pulling extra current via resistor R3. This causes a large voltage drop across R3, which is IB × R3. For example, if **RE = 100Ω** and β =2 (due to having the emitter collector terminals reversed), then what is VB? It has to be lower than 1.7 volts, which will mean that LED2 does not turn on due to the low $\beta = 2$. We can approximate VB as follows: If the desired voltage, VB = 1.7 volts, then VRE = 1.7 volts – 0.7 volt = 1 volt, then the desired emitter current is 1 volt/100Ω = 10 mA.

NOTE: For small values of β, generally $\beta < 10$, we will need to use the more exact equations relating IB and IC and IE. They are: IE = $(\beta + 1)$IB, IB = IE/$(\beta + 1)$, and we still have IB = IC/β , which leads to: IC = IE $[\beta/(\beta + 1)]$. For example, if $\beta = 9$, IC = IE $[9/(9 + 1)]$ = IE(0.90).

If IE = 10 mA and $\beta = 3$, then the base current:

IB $= 10$ mA/$(\beta + 1)$ = 10 mA/$(3 + 1)$ = 10 mA/(4)

IB = 2.5 mA

Thus, the voltage drop across R3 = $1600\Omega \times 2.5$ mA = 4.0 volts. VB = BT2 – (voltage drop across R3). VB = 5 volts – (4.0 volts) = 1 volt. At 1 volt = VB, this means that LED2 does not turn on due to LED2's turn-on voltage being 1.7 volts. This leaves VRE = VB – 0.7 volt = 0.3 volt.

The emitter current IE = 0.3 volt/100Ω = 3 mA.

The "exact" collector current IC = IE $[\beta/(\beta + 1)]$ = 3 mA $[3/(3 + 1)]$ IC = 3 mA $[3/4]$ = 3 mA (0.75).

IC = I_{LED1} = 3 mA (0.75) = 2.25 mA, which is smaller than the 10 mA of LED1 current if the transistor Q1 was correctly connected.

There are two things to take away from connecting the transistor incorrectly in Figure 7-17.

- The reference voltage light-emitting diode LED2 may go dim or turn off.
- The LED1, which is expected to be driven with 10 mA, will be driven with a lower current and will give out much less light.

There are times when we assume that the transistor's pin out has an order of emitter, base, and collector (EBC). However, there are some transistors such as PN2222 (EBC) that has a similarly named part number P2N2222 (CBE) that has reversed emitter and collector leads. Generally, in the United States common TO-92 plastic case transistors such as 2N3904, 2N3906, 2N4124, and 2N4126 have emitter, base, collector pin outs (EBC). Whereas often used TO-92 transistors in Europe such as BC213, BC238, BC308, BC547, BC557 have pin outs of collector, base, emitter (CBE).

Now let's look at another example where a Zener reference diode is reversed (see Figure 7-18).

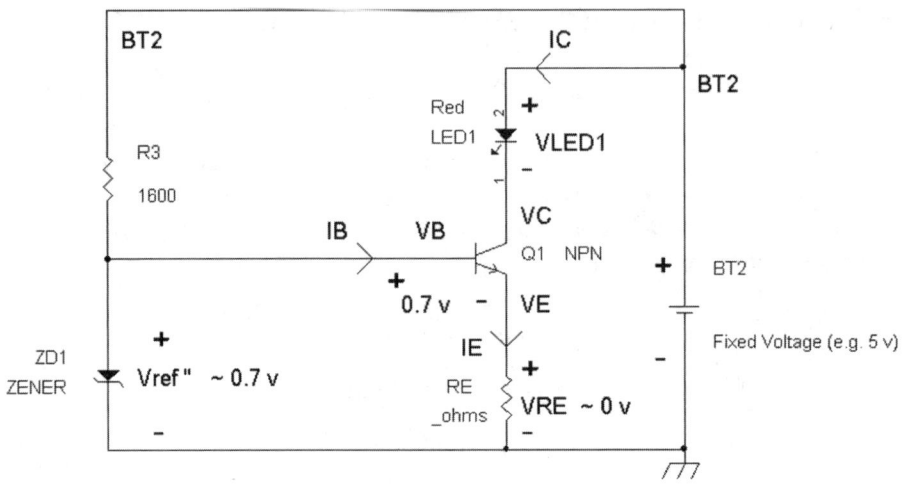

FIGURE 7-18   Zener diode ZD1 is reverse connected, which results in a dimmed LED1.

With a Zener diode connected in reverse, it gives a voltage, 0.7 volt, similar to a forward biased diode. Because of this, VB ~ 0.7 volt, and VRE = VB – VBE, with VBE = 0.7 volt. So VRE = VB – 0.7 volt = 0.7 volt – 0.7 volt ~ 0 volt = VRE. As a result, LED1 is either turned off or being driven with a very small current.

What happens if we reverse LED1's terminals? We know that it will not light up. But what happens to the rest of the circuit? See Figure 7-19.

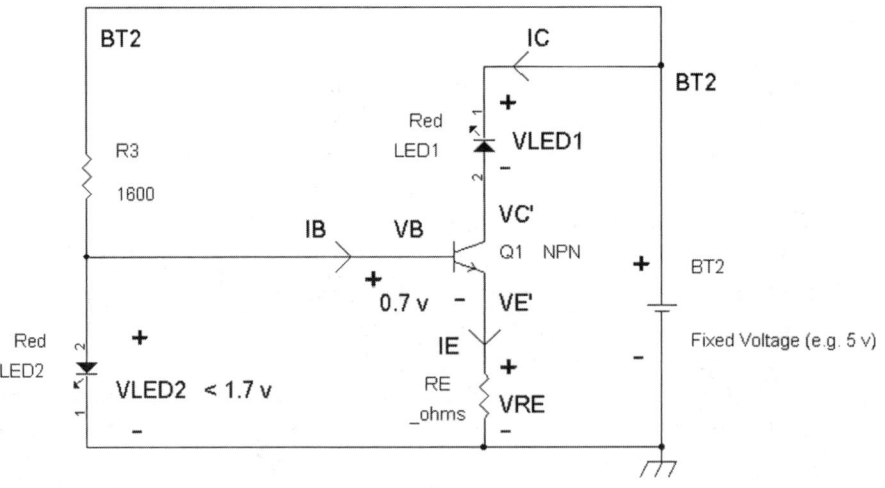

FIGURE 7-19   LED1's anode and cathode leads are reversed, which leads to some very interesting analysis for VB and VE.

We see that with LED1 wired incorrectly, it will not light up. This means it's essentially an open circuit, which disconnects Q1's collector lead to any DC path to BT2. See Figure 7-20.

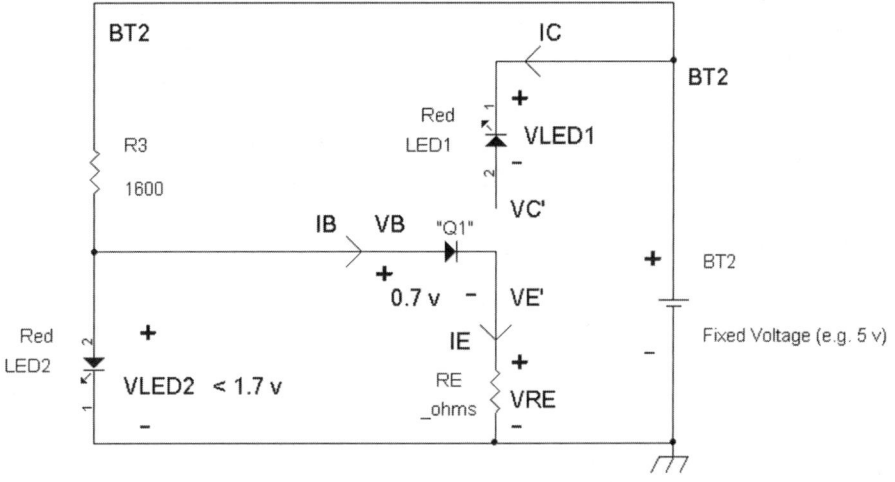

FIGURE 7-20   An equivalent circuit to Figure 7-19 using a diode "model" in place of Q1.

To determine VB, the base-emitter junction of Q1 now acts like a diode. There is no such thing as β or current gain because the collector lead is disconnected. If we make an estimate that VLED2 is less than its turn-on voltage (1.7 volts), then we can safely say that there is no current flowing into LED2. This means the current is flowing from BT2 via resistor R3, diode "Q1", and RE to ground; and no current flows into LED2.

One way to calculate VB is by finding the current IB flowing though RE. The current, IB, flows into diode "Q1" and it also flows through RE.

Let's work out a variation of a loop equation. That is the voltage:

BT2 = Voltage across R3 + diode voltage of "Q1" + voltage across RE. The diode voltage of Q1 can be approximated as 0.7 volt.

If you use a volt meter and try measuring BT2 first and then measure the voltages across R3, "Q1" diode, and RE, you will find that summing up all the voltages across R3, "Q1" diode, and RE will equal the voltage of BT2 = 5 volts.

With the following:

- Voltage across R3 = IB × R3
- Voltage across RE = IB × RE
- Diode voltage of "Q1" = 0.7 volt (see Figure 7-20)

we can now solve for IB, and once that is done we can find IB × RE = VRE, which will lead to VB = 0.7 volt + VRE.

Let BT2 = 5.0 volts, R3 = 1600Ω, and RE = 100Ω.

BT2 = IB × R3 + 0.7 volt + IB × RE

5.0 volts = IB × 1600Ω + 0.7 volt + IB × 100Ω

5.0 volts – 0.7 volt = IB 1600Ω + IB × 100Ω = IB (1600Ω + 100Ω) = IB × (1700Ω) = 5.0 volts – 0.7 volt = 4.3 volts

4.3 volts = IB (1700Ω) or by dividing by 1700Ω on both sides of the equation:

(4.3 volts/1700Ω) = 2.53 mA = IB

IB × RE = VRE = 2.53 mA × 100Ω = 0.253 volt = VRE

VB = 0.7 volt + VRE = 0.7 volt + 0.253 volt

VB = 0.953 volt

And as confirmed, VB is less that the 1.7 volts that is needed to turn on LED2.

Insufficient "Headroom Voltage" for the Transistor

We now need to discuss the limitations of transistor current sources. This means we need to pay attention to the collector-base voltage. See Figure 7-21.

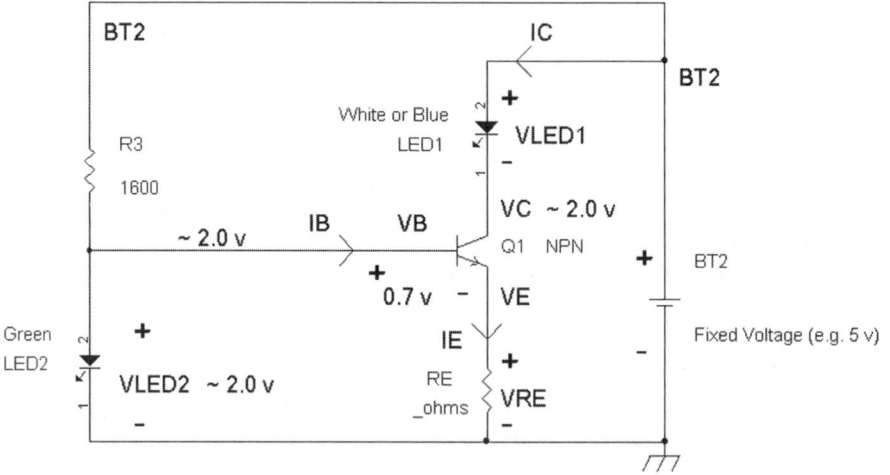

FIGURE 7-21   An LED drive circuit on the "edge of disaster" due to VC ~ VB. If BT2's voltage drops slightly (e.g., by ~ 0.5 volt), the circuit will start to fail. For proper operation, it is preferable that the collector voltage VC is greater than the base voltage VB, or in other words VC > VB.

In Figure 7-21, as shown, the LED2 voltage is about 2.0 volts = VB, and we replaced LED1 with a blue or white LED, which has a turn-on voltage of 3.0 volts. This means the collector voltage:

VC = BT2 – VLED1 = 5.0 volts – 3.0 volts

VC = 2.0 volts

Generally, to play it safe, VC ≥ VB. In this case, the LED circuit is still workable. However, if the battery voltage, BT2, drops from 5.0 volts to 4.5 volts, then VC = 4.5 volts – 3.0 volts = 1.5 volts. This means VC < VB, or in other words, VB > VC. When VB > VC, the transistor Q1 is either approaching the saturation region, or is already in the saturation region. For current source and amplifying circuits, we want to avoid transistors with these conditions because the β will start dropping.

So, one fix to Figure 7-21 is to lower VB while maintaining a "reasonable voltage" at VE (e.g., VE ≥ 0.5 volt). See Figure 7-22.

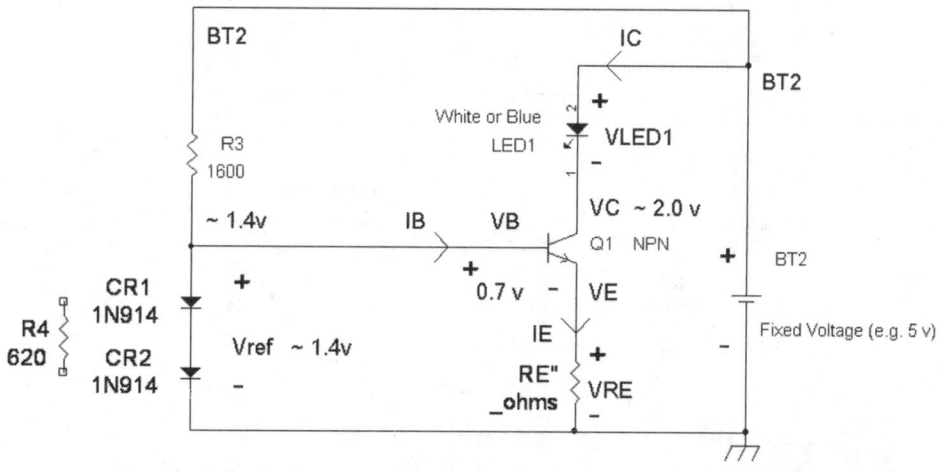

FIGURE 7-22   Base voltage VB is lowered by replacing the green LED 2-volt reference voltage with a lower one at 1.4 volts via two series connected diodes, CR1 and CR2.

With this modification, VB = 1.4 volts, while VC = 2.0 volts. Note that the 1.4 volts developed by these two diodes stay relatively constant should BT2 drop in voltage to 4.5 volts.

Hence, even if BT2 drops to 4.5 volts, VC = 1.5 volts > VB = 1.4 volts, which provides a little safety margin.

Alternatively, the two diodes may be replaced with a 620Ω resistor, R4, to provide 1.4 volts for BT2 = 5.0 volts. However, VB will drop proportionally with BT2. For example:

BT2 = 5.0 volts → VB = 1.4 volts

BT2 = 90% 5.0 volts = 4.5 volts $\rightarrow$ VB = 90% 1.4 volts = 1.26 volts

This is because VB ~ BT2[R4/(R3 + R4)], neglecting base current IB, which is reasonable for a $\beta \geq 50$.

Another option to avoid transistor Q1 from going into the saturation region is to simply increase the supply voltage BT2, if possible. For example, in Figure 7-21, where BT2 = 5.0 volt and VB = VC = 2.0 volts, having BT2 $\rightarrow$ 6.0 volts will result in VC = 6.0 volts – 3.0 volts = 3.0 volts, which is greater than the 2.0 volts of VB.

Sometimes Even a Correct Circuit Goes Bad

So far, we looked at only DC conditions, which is very important in troubleshooting circuits. However, even when the DC bias points should work, there can be other problems. This transistor LED circuit discussed so far acts as both current source circuit, but also as amplifier. See Figure 7-23, which can self-oscillate at a very high frequency (e.g., > 100 MHz).

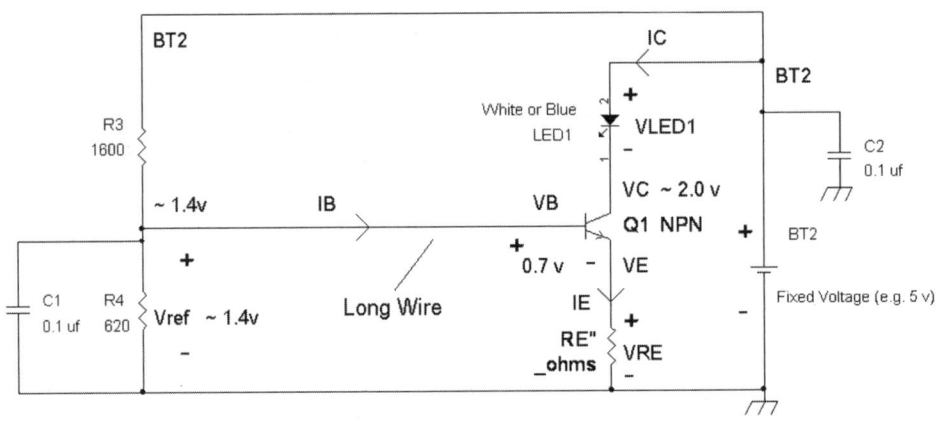

FIGURE 7-23   A current source circuit that can oscillate at very high frequencies due to a long base lead.

Usually, this circuit will work fine. But if you replace the transistor Q1 with a much higher frequency device, such as a VHF (very high frequency) transistor (e.g., MPSH10 whose f_t > 600 MHz), the circuit may oscillate at a very high frequency that causes a variation in LED1's drive current when you touch the base lead.

A long wire connected to Q1's base behaves like an inductor, which forms a resonant circuit with stray and parasitic capacitances included in the transistor. This can inadvertently form an oscillator circuit. To stop the "parasitic" oscillation, we have to make a lossy inductor by adding a series resistor. See Figure 7-24.

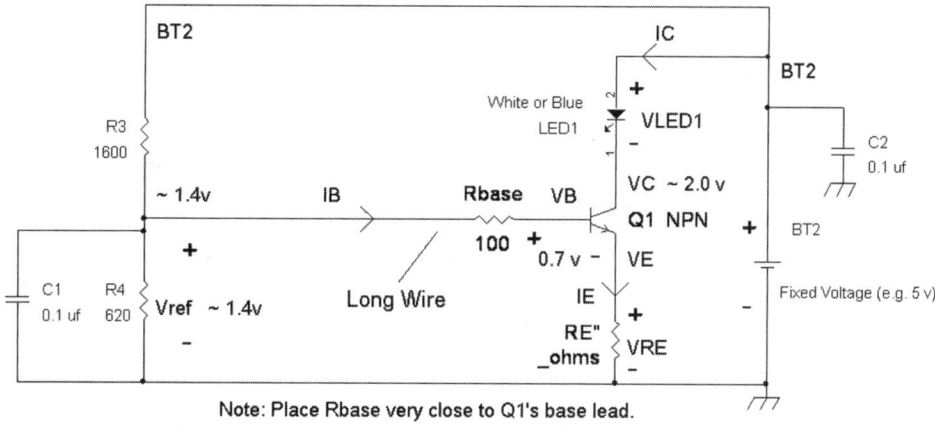

FIGURE 7-24 A series base resistor, Rbase is added in series to the terminal of Q1 to prevent parasitic oscillations.

Rbase usually has a value of 22Ω to 220Ω and should have very short leads itself. And Rbase should be soldered or placed very close to Q1's base lead. The objective is to prevent the base lead of Q1 from being connected to a long wire.

Figure 7-24 also shows decoupling capacitors to ground, C1 and C2. Generally, you put a decoupling capacitor at your circuit that is receiving the power supply voltage. C2 is added at the circuit near LED1, and not at BT2, which can be connected via long wires that behave like inductors. Decoupling capacitor C1 is placed across the voltage divider circuit R4 and R3 to reduce any high-frequency noise. Generally, most decoupling capacitors have values between 0.01 μf to 2.2 μf.

Summary

We have presented many circuits showing how the transistor works in a current source circuit. Also, several troubleshooting techniques have been illustrated. The gist of troubleshooting is to learn by experience more so than try to be an engineer and analyze everything. Basically, troubleshooting is detective work. You will need to look up data sheets to find the pins outs, but also other specifications like maximum current or voltage that the device can take before breakdown.

Chapter 8 will continue with simple transistor amplifiers based on the LED drive circuits.

Troubleshooting
Discrete Circuits
(Simple Transistor Amplifiers)

For this chapter we will be introducing transistor amplifiers that are "derived" from the LED circuits from Chapter 7. We shall be showing that these amplifiers have limitations such as amplitude gain (ratio of output signal to input signal amplitudes), waveform distortion (e.g., a sine wave input signal with a distorted waveform at the output), output swing, and output current. In general, an amplifier is a circuit that provides a larger voltage signal or larger current signal at the output when compared to the input signal. All amplifier circuits require some type of supply voltage such as a 9-volt battery or a 12-volt power supply.

Important Practical Transistor Specifications

It should be noted that amplifiers are imperfect in that they cannot supply infinite current nor infinite voltage at their outputs. The transistors themselves have limitations on maximum collector-to-emitter voltage and maximum collector current, along with maximum power dissipation. For example, the ubiquitous 2N3904 NPN transistor has a typical maximum collector-to-emitter voltage (V_{CE}) of 40 volts, a 200 mA maximum collector current (I_C), and a 200 mW power dissipation rating. Power dissipation is usually given by $P_d = V_{CE} \times I_C$. For example in Figure 8-1, suppose the collector current driving the light-emitting diode LED1 is **20 mA** $= I_C$, and we have 9 volts as the supply BT2 with a red LED1 with a 2-volt forward turn-on voltage $= V_F$. BT2 = 9 volts and V_F = 2 volts so BT2 $= V_F + V_{CE}$, and $V_{CE} = $ BT2 $- V_F$ or $V_{CE} = $ 9 volts $-$ 2 volts, so:

$V_{CE} =$ **7 volts**

$P_d = V_{CE} \times I_C$, which is the power dissipation of transistor Q1

$P_d = 7$ volts $\times$ 20 mA

$P_d = 140$ mW, which is fortunately below the 200 mW power dissipation rating of the 2N3904

Although this example will work, as a rule of thumb, you want the power dissipation to be less than 50 percent of the maximum rating. In this example, the transistor is dissipating at 70 percent of maximum power dissipation (via 140 mW/200 mW = 70%), so we may want to use a higher wattage transistor, such as a 2N4401where $P_d = 500$ mW maximum.

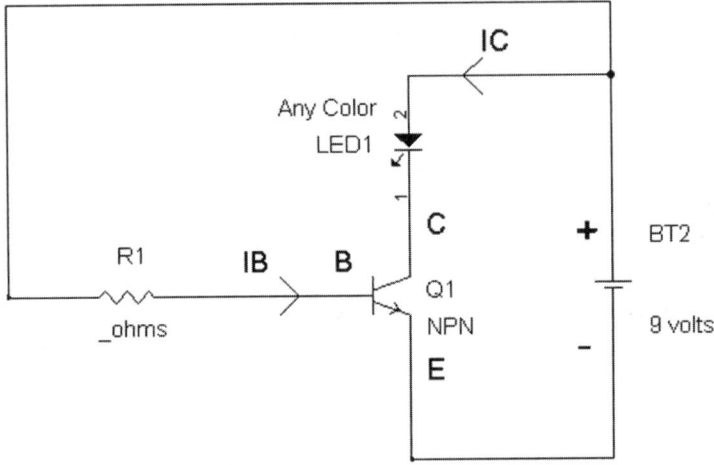

FIGURE 8-1   An LED drive circuit used for determining transistor Q1's power dissipation.

Note that both $V_{CE} = 7$ volts and $I_C = 20$ mA are well within the maximum specifications of 40 volts and 200 mA, respectively, for the 2N3904 transistor. However, this example shows that we have to keep track of power dissipation even if the voltages and currents are well within specifications.

Simple Transistor Amplifier Circuits

The basic LED constant current source drive circuit shown in Figure 8-2 can be reconfigured to operate as an amplifier as shown in Figure 8-3.

NOTE: The base driving resistor is renamed as RB.

In Figure 8-2 the transistor's base-to-emitter voltage is held constant via the base current driving resistor RB that is connected to the power supply. With this constant base-to-emitter voltage, a constant collector current is provided to the LED. However, if we somehow are able to vary the base-to-emitter voltage by just a little, such as a variation of a few millivolts (mV), then the collector current will vary as well. See Figure 8-3, which is essentially the same circuit with a couple of modifications.

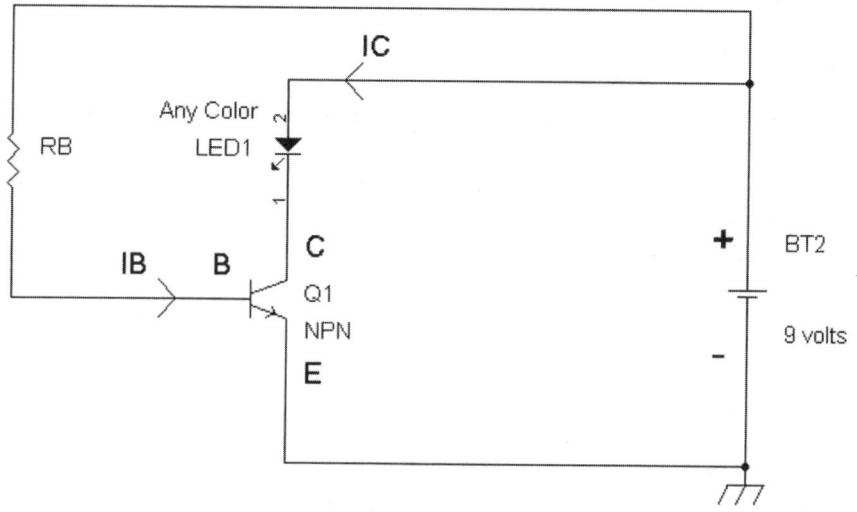

FIGURE 8-2   Constant current source to the LED via Q1's collector current.

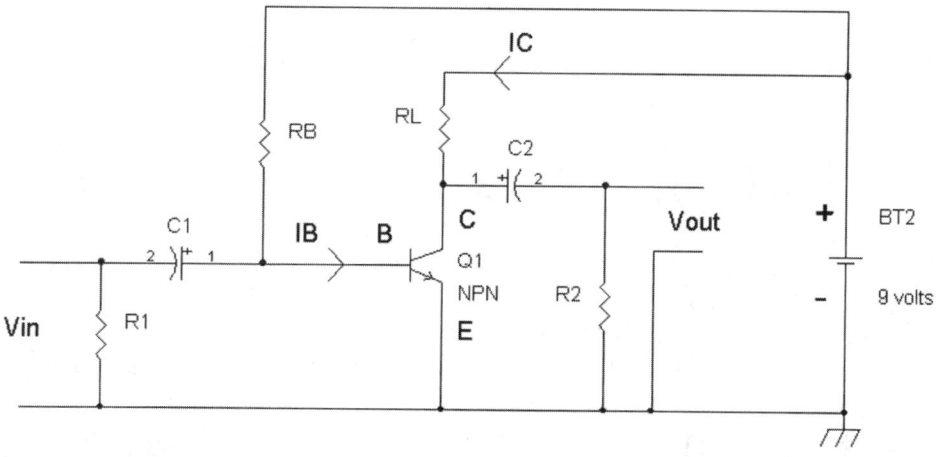

FIGURE 8-3   The constant current source LED drive circuit modified into an amplifier.

For this type of amplifier, the signals must be low level like the types of signals from microphones that produce < 10 mV to avoid distorted output signals. If larger signals are used such as an audio signal from your digital or CD player, then the input should be attenuated with a voltage divider circuit. Typical attenuation would lower the amplitude by 10 to 100 fold.

Although this circuit is "easy" to analyze, it has limited purposes on its own as an amplifier. However, some circuits do not require low distortion. This circuit can be used for a fuzz pedal distortion amplifier used in an electric guitar, or be used in an oscillator circuit, or be used as a mixer that deliberately generates distortion products, such as in an RF mixer.

First DC Analysis: Capacitors = Batteries with Self Adjusting Voltages

Capacitors C1 and C2 can be viewed as "fast charge" batteries that charge to a voltage such that no DC current flows through them. Let's take a closer look at the one-transistor amplifier with some voltages labeled. See Figure 8-4 first.

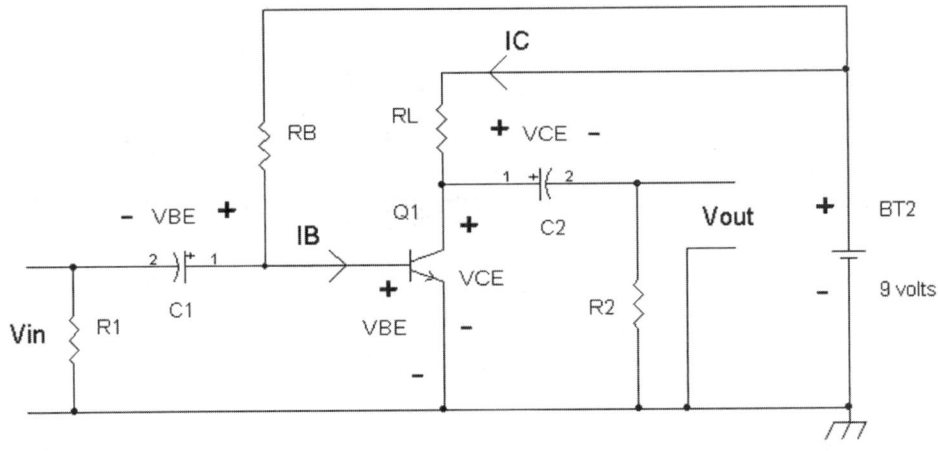

FIGURE 8-4 Capacitors C1 and C2 are charged to voltages "VBE" and "VCE" as marked.

When we look at Figure 8-5, the DC voltage at Vin is equal to the series voltages of BT_C1 and the voltage at the base-emitter junction of Q1. Because the polarities of BT_C1 and VBE are the same with equal voltages via the + side of BT_C1 connected to the + side of VBE, the voltage at Vin is 0 volts DC. This amounts to two back-to-back equal voltage sources connected in series such that the net voltage is zero. Here is another way to look at this. Suppose you have a two-cell flashlight. Instead of installing the 1.5-volt batteries correctly (in series) to provide 3 volts, you install them back to back, which gives zero volts instead.

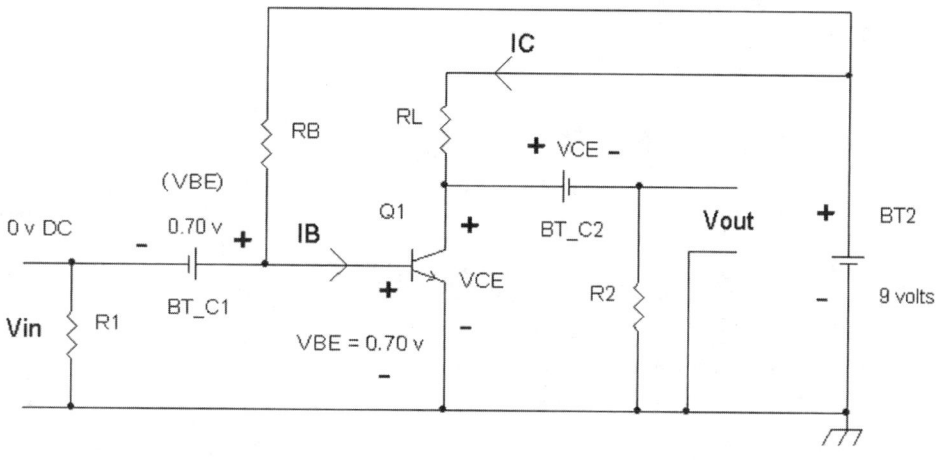

FIGURE 8-5 Batteries BT_C1 and BT_C2 can be thought of as replacing C1 and C2 as voltage sources.

In the amplifier, if the VBE of Q1 = 0.70 volts, capacitor C1 charges up to exactly 0.70 volts in Figure 8-4. In Figure 8-5, we see if BT_C1 is 0.70 volt as labeled, then the DC voltage at Vin has to be zero volts due to series connection of back-to-back VBE voltages from the Q1 and C1.

Likewise, if C2 in Figure 8-4 is charged up to a voltage of VCE, we see in Figure 8-5 that the DC voltage at Vout has to be zero volts. The reason is that BT_C2 has an equal and opposing voltage to VCE, the voltage at the collector of Q1. Thus, the DC voltage at Vout has to be zero volts due to the back-to-back series connection of VCE from the transistor and VCE in C2 or BT_C2.

The capacitors C1 and C2 play two important roles. First, they block out DC currents from Vin and Vout. Second, they pass through an AC signal such as an audio signal.

Second DC Analysis: Take Out the Capacitors to Find the DC Currents and DC Voltages

With the capacitors removed we can now more clearly see for DC analysis. Not only the capacitors C1 and C2 have been removed but the adjoining components R1 and R2 are now also no longer part of the circuit. What we are left with is just three components, RL, R2, and Q1, plus the power supply, BT2.

There are two reasons for determining the DC base and collector currents, IB and IC. Finding the expected base and collector currents will allow us to determine the collector-to-emitter voltage, which has to be > 0.7 volts in most cases. For example, if the collector current is too high or the voltage drop across RL is too high, then the transistor is in the saturation region, and the circuit will not amplify.

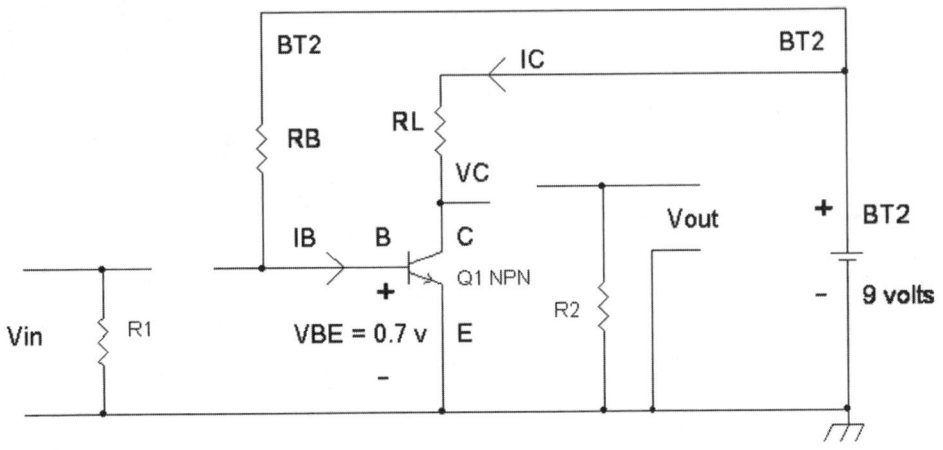

FIGURE 8-6   To calculate the DC currents and voltages, remove all capacitors that are C1 and C2.

To determine the collector current IC, first, we calculate for the base current IB.

IB = (BT2 – VBE)/RB

For most good approximations, VBE = 0.7 volt.

IB = (BT2 – 0.7 v)/RB

$$\mathbf{IB} = \frac{\mathbf{(BT2 - 0.7\ v)}}{\mathbf{RB}}$$

Let's take some examples pertaining to Figures 8-4 or 8-6:

If BT2 = 9 volts and RB = 56kΩ, then IB = (9 v – 0.7 v)/56kΩ = (8.3 v)/56kΩ

IB = 0.148 mA or IB = 148 μA

IC = β IB

For this example, we use a lower current gain transistor such as a 2N3903 where β = 50. Since we already calculated the base current as IB = 0.148 mA then:

IC = β IB = 50 × 0.148 mA

IC = 7.4 mA

If we have RL = 1000Ω, then we can now find VC, Q1's collector voltage reference to ground.

VC = BT2 – IC (RL)

VC = 9 v – 7.4 mA (1000Ω) = 9 v – 7.4 v

VC = 1.6 volts

What happens if β → 100?

Then **IC = β IB** = 100 × 0.148 mA = 14.8 mA.

We find that VC = 9 v – 14.8 mA (1000Ω) = 9 v – 14.8 v = –5.8 volts = VC???.

This is not possible since we only have a positive power supply and thus no negative volts can be generated. Instead, VC ~ 0 volts, and usually VC ~ 0.2 volt. That is, if the calculated value for VC is < 0.5 volt, usually the amplifier will not work because the transistor is now a switch that "shorts" the collector terminal to ground.

What we have shown in this example is that if the current gain is too high, the transistor amplifier in Figure 8-4 can cause the transistor to turn into a switch. That is why a circuit like Figure 8-4 requires selecting the β of transistors in a narrow range, such as 30 < β < 50, or "tweaking" the resistance value of the base driving resistor, RB.

For example, if we want to use a transistor whose β = 100, then we should approximately double the value of RB from 56kΩ to 110kΩ with BT2 = 9 v and RL = 1000Ω. This leads to:

IB = (BT2 – 0.7 v)/RB = (9 v – 0.7 v)/110kΩ = 8.3 v/110kΩ = 75.455 μA

IC = 100 × 75.455 μA = 7.545 mA

VC = BT2 – IC RL = 9 v – 7.545 mA × 1000Ω = 9 v – 7.545 v = VC

VC ~ 1.45 volts, which is pretty close to VC = 1.6 volts for RB = 56kΩ and β = 50.

It would be safer to aim for a slightly higher VC. We can have RB = 120kΩ with β = 100 and RL = 1000Ω, which results in IB = 69.167 μA and IC = 100 × 69.167 μA or IC = 6.9167 mA.

VC = 9 v – 6.9167 mA × 1000Ω = 9 v – 6.9167 v = VC

VC = 2.08 v

In the real world, the specific transistor part number (e.g., 2N3904, 2N2222, etc.) will not only have a range of β, but the current gain β will change with temperature. So, the DC collector current and DC collector voltage will vary. In general, the transistor amplifier in Figure 8-4 is more of a hobbyist or DIY (do-it-yourself) amplifier where you have to individually tweak it to a desired collector current. We will find in the next section that the DC collector current IC also determines the AC signal's gain at Vout, and also the input loading resistance at Vin.

For now, we have:

VC = BT2 – IC (RL)

$$IC = \beta \frac{(BT2 - 0.7\ v)}{RB}$$

A more general formula would be:

$$VC = BT2 - \beta\frac{(BT2 - 0.7\text{ v})}{RB}(RL)$$

For a current gain $\beta = 50$, BT2 = 9 volts, RB = 56kΩ or 56,000Ω, and RL = 1000Ω.

$$VC = 9\text{ v} - 50\frac{(9\text{ v} - 0.7\text{ v})}{RB}(1000) = 9\text{ v} - 7.4\text{ v} = VC$$

VC = 1.6 volts

And note the collector current:

$$IC = \beta\frac{(BT2 - 0.7\text{ v})}{RB} = 50 \times 0.148\text{ μA} = 7.4\text{ mA} = IC$$

Alternatively, instead of using the general formula for VC, sometimes it's more logical to find the base current, IB, first and then calculate the collector current, IC to find VC, the collector voltage.

If BT2 = 5 volts and RB = 100kΩ, then IB = (5 v – 0.7 v)/100kΩ = (4.3 v)/100kΩ.

IB = 0.043 mA or IB = 43 μA

If β = 78, then IC = β IB = 78 (43 μA).

IC = 3.225 mA

Let RL = 820Ω

VC = BT2 – IC (RL) = 5 volts – 3.225 mA (820Ω) = 5 volts – 2.6445 volts = **VC**

VC = 2.3555 volts

Finding the AC Signal Gain

The DC collector current is the most important DC characteristic to find for calculating the AC signal gain. Via the DC collector current we can also calculate the input resistance.

Input resistance to an amplifier is important since the input signal source itself usually has an output source resistance or an optimum input load resistance. For example, an antenna may have a 50Ω source resistance. To achieve maximum power transfer, the input resistance of the amplifier should be 50Ω as well.

In another example, if your signal source is a dynamic microphone, then the input resistance should be typically 1000Ω or more. If you should connect the dynamic microphone to an amplifier with a 50Ω input resistance, then you will lose signal amplitude from the microphone.

Thus, we have to keep in mind the amplifier's input resistance based on the application (e.g., RF amplifier, microphone preamplifier, sensor amplifier, etc.). This is particularly important when we have one amplifier's output connected to the input of a second amplifier.

We can find the AC signal gain of this amplifier by noting the following:

- Capacitors with "sufficiently large" capacitances to be like batteries are AC short circuits or zero ohm resistors for AC signals.
- Power supplies are also treated as AC short circuits or zero ohm (0Ω) resistors with respect to AC signals.

At first glance, the second item may seem absurd. But it you think about it, if you probe for AC signals in a clean power supply or battery, you will find no AC signal. By definition, power supplies and batteries are DC voltage sources and thus cannot include any AC signal.

Likewise, if a capacitor has a large enough capacitance that it acts like a "battery" (e.g., see Figure 8-5 again), then again by definition a battery cannot include an AC signal; it only produces a DC voltage. Let's take a look again at the "original" schematic in Figure 8-7.

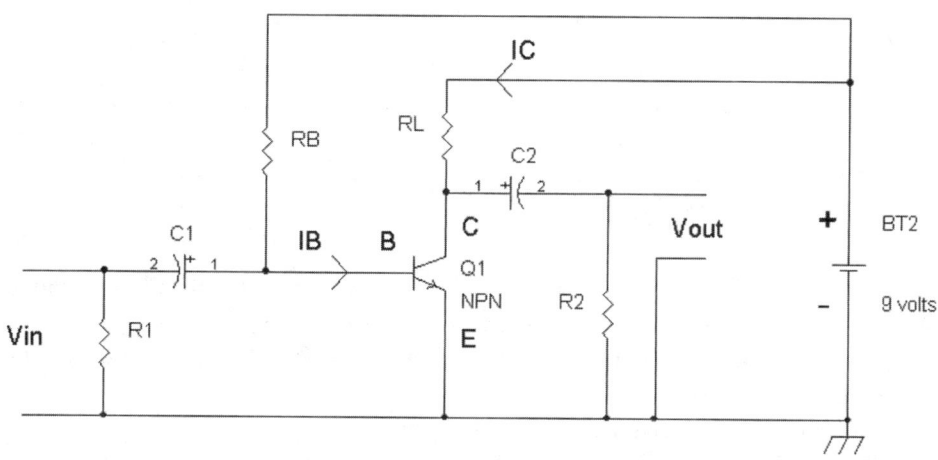

FIGURE 8-7   The one-transistor amplifier from Figure 8-5.

Now let's take a look at the AC analysis model shown in Figure 8-8.

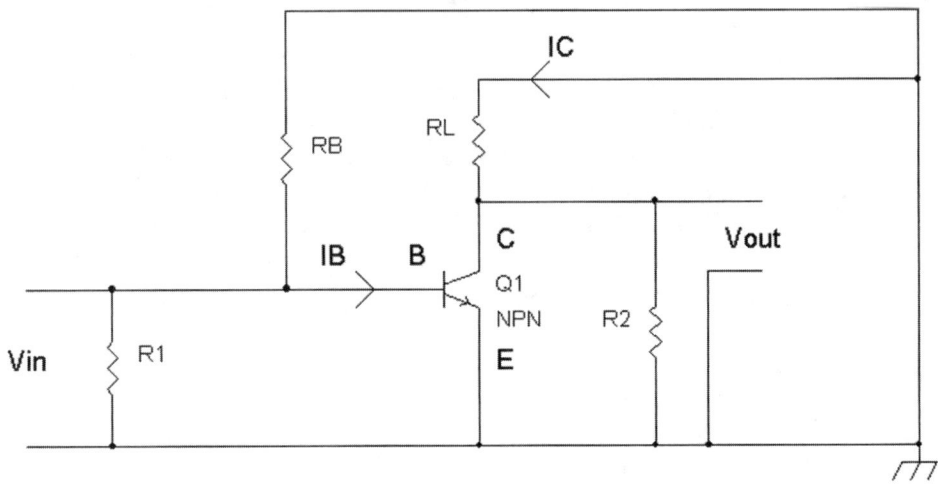

FIGURE 8-8   The capacitors C1, C2, and battery (DC power supply) BT2 are "modeled" as AC short circuits or AC zero ohm wires.

Figure 8-8 can be further redrawn to look simpler or more intuitively familiar. See Figure 8-9.

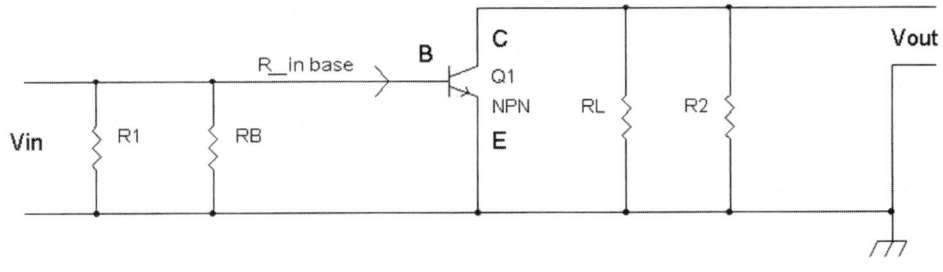

FIGURE 8-9   A redrawn version of Figure 8-8 to show a more intuitive idea of the one-transistor amplifier's AC analysis circuit.

As you can see in Figure 8-9, the circuit looks a little "funny" in that there is no power supply. However, the schematic gives us an idea of how the AC signal at Vin will be amplified. Note that there is a resistance into Q1's base, which is referenced to ground.

Finally, we can model Figure 8-9 in terms of a block diagram with the resistors R1, RB, RL, and R2. See Figure 8-10. Note that R_in base is now "modeled" as a resistor to ground along with R1 and RB.

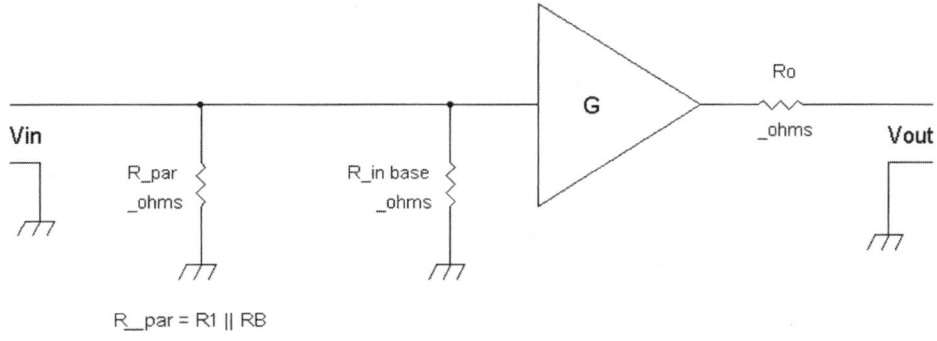

FIGURE 8-10 A block diagram of Figure 8-9's AC analysis circuit.

In order to complete the AC analysis circuit, we have to find two "unknowns," resistor, R_in base, and the gain, "G." See below for the formulas given that IC = DC collector current.

R_in base = β (0.026 v/IC)

G = -(IC/0.026 v) × RL || R2

Where $RL \parallel R2 = \dfrac{RL \times R2}{RL + R2} = Ro$ and

$R\_par = R1 \parallel RB = \dfrac{R1 \times RB}{R1 + RB}$

Sometimes it is convenient to express the term (0.026 v/IC) = r_e, where a handy starting point of IC = 1 mA results in:

(0.026 v/0.001A) = 26Ω = r_e

From this 1 mA starting point we can find r_e for any other collector by a scaling factor such that r_e is inversely proportional to collector current. That is, for a given current:

$$r_e = \frac{0.001\ A}{IC} \times 26\Omega$$

Since we often set collector currents in milliamps, mA, you can express the formula this way where the collector current IC is in milliamps.

$$r_e = \frac{0.001\ A}{IC\ mA} \times 26\Omega$$

For example, if IC = 2 mA, then $r_e = \dfrac{1\ mA}{2\ mA} \times 26\Omega = 13\Omega$.

And if IC = 0.1 mA = 100 μA, then $r_e = \dfrac{1\ mA}{0.1\ mA} \times 26\Omega = 260\Omega$.

For the input resistance R_in base:

R_in base = β (0.026 v/IC) = $\beta\, r_e$

NOTE: R_in base is inversely proportional to IC. For example, if you decrease IC by tenfold, then R_in base increases by tenfold.

For example, if β =100 and IC = 1 mA, then $r_e = \dfrac{1\ mA}{1\ mA} \times 26\Omega$.

r_e = 26 Ω

R_in base = $\beta\, r_e$ = 100 × 26Ω = 2.6KΩ

If β =100 and we have IC = 0.1 mA, then R_in base = $\beta\, r_e$ = 26KΩ.
For the gain factor G:

G = –(IC/0.026 v) × RL || R2

NOTE: G is proportional to the DC collector current IC. So, if you increase IC by tenfold (e.g., 0.1 mA to 1 mA), the gain, G, goes up by tenfold. And if you decrease IC by tenfold, G, decreases by tenfold.

In this amplifier given current gain β, if you increase the DC collector current the input resistance to the base, R_in base, decreases while the gain, G, increases.

A decreasing R_in base can cause the input signal to be loaded down and thus attenuated when we increase the DC collector current IC. One way to offset this is to pick a transistor that has a higher β. For example, a 2N3904 may have a typical β of 150, but a 2N5088 has higher typical β of 300. So, if you want to increase the collector current twofold but keep the same R_in base resistance, you can select a transistor with twice the β. To keep the same DC collector current IC, this also means that RB would be increased about twofold in resistance because of the higher current gain (e.g., twice the β).

Now let's take one of the previous examples concerning DC analysis:

BT2 = 5 volts, RB = 100kΩ, then IB = (5 v – 0.7 v)/100kΩ = (4.3 v)/100kΩ

IB = 0.043 mA or IB = 43 μA

If β = 78, then IC = β IB = 78 (43 μA) = 3.225 mA.

IC = 3.225 mA = 0.003225A

RL = 820Ω

R1 = R2 = 220 KΩ

Let C1 = C2 = 33 μf (with C1and C2 rated at ≥ 16 volts), RL = 820Ω, R2 = 100KΩ, with IC = 0.003225 A.

Gain Calculation

G = –(IC/0.026 v) × RL ∥ R2

$$G = -(0.003225A/0.026\ v) \times 820\Omega\ \|100K\Omega = -(0.124A/v) \times \frac{820 \times 100K}{820 + 100K}\Omega$$

$$820\Omega\ \|100K\Omega = \frac{820 \times 100K}{820 + 100K}\Omega = 813\Omega$$

NOTE: The units A/v or amps per volt has a unit of (1/resistance) or 1/Ω.

$$G = -(0.124\ A/v) \times 813\Omega = -100.85 = \frac{Vout}{Vin}$$

$$\frac{Vout}{Vin} = G = -100.8$$

The minus sign indicates an inverted AC signal, or a signal that is 180 degrees out of phase.

For example, if Vin is a 1 mV peak sinewave, then Vout provides an inverted sinewave at 100.8 mV peak.

Base Resistance, R_in Base Calculation

R_in base = β (0.026 v/IC) = 78 (0.026 v/0.003225A) = 78 (8.06Ω)

NOTE that the units v/A or volts per amp has a unit of resistance or Ω.

R_in base = 628.8Ω

Limitations of This One-Transistor Amplifier

We will now look at the one transistor amplifier in terms of input amplitude range. See Figure 8-11.

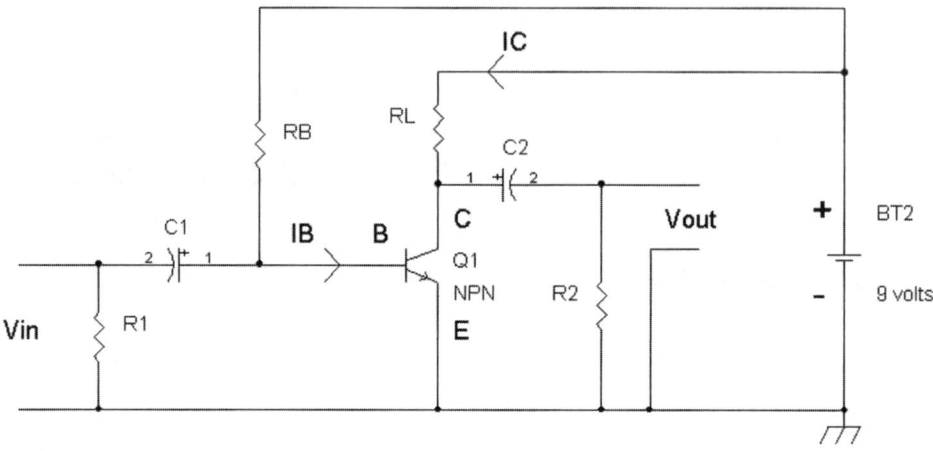

FIGURE 8-11   The one-transistor amplifier that requires a low-level input signal to avoid distortion.

Limited Input Amplitude Range

When Vin is driven with a low-impedance generator such as one that has a 50Ω source resistance, which is common in many function generators, this amplifier will start to distort with signals starting at about 1-mV peak sine wave. For example, a 10-mV peak signal that is 20 mV peak to peak at the input will result in Vout having a generally larger amplitude but also a waveform that is distorted with about 10 percent second order harmonic distortion. As an example, if the frequency is 1000Hz, the output signal will provide a signal at 1000Hz, and another one at 2000Hz at 10 percent of the amplitude at 1000Hz.

Because this amplifier can produce so much distortion with signal sources from portable music players or smartphones (> 100 mV peak), you would need to place a voltage divider circuit such as a volume control to adjust the level into Vin, the input terminal.

See Figure 8-12 for an amplifier with the following:

R1 = 10MegΩ, R2 = 10MegΩ, RB = 4.7MegΩ, RL = 12KΩ, IC = 0.3375 mA
IC = 0.0003375 A, or IC = 337.5 μA

Q1 = 2N3904, BT = 9 volts

G ~ −145 as measured with 20 mV peak to peak sinewave at Vin with a 50Ω source resistance generator.

G = −144 as calculated. That is, G = −(0.0003375A/0.026 v) 12KΩ ~ −144.

NOTE: 12KΩ ‖ 10MegΩ = 12KΩ within about 0.12%.

See below for the input and output signals.

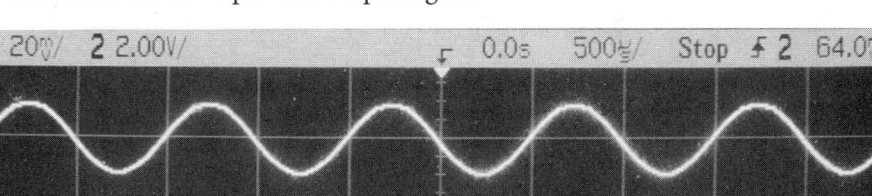

FIGURE 8-12   Amplifier 20 mV peak to peak signal input on top trace. Bottom trace shows a 2.7-volt peak to peak (inverted) output signal with 10 percent second harmonic distortion; and notice the rounder positive cycle and narrower negative cycle.

We can mitigate the distortion problem by using a series resistor, R_ser, as shown in Figure 8-13.

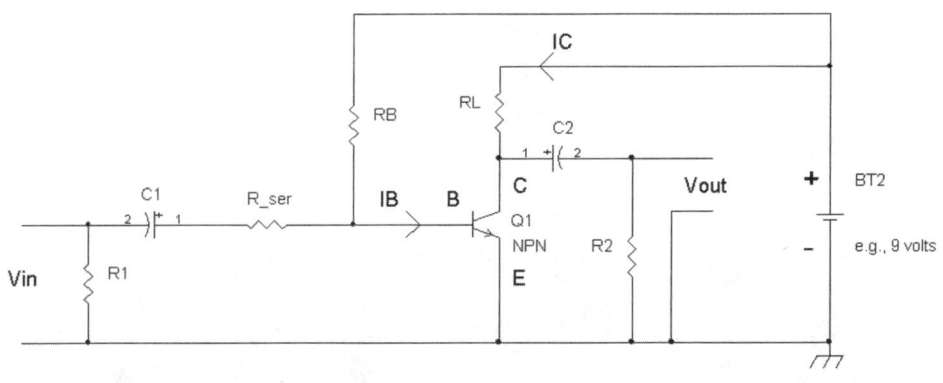

FIGURE 8-13   Lower distortion for the same output swing.

Typically, if we want lower distortion from an amplifier, we have to give up some of the voltage gain.

For example, if the gain of the amplifier is about 145, we can have lower distortion for the same output voltage if we set the series base resistor at about $10 \times$ R_in base. Again, if we use the previous example with:

R1 = 10MegΩ, R2 = 10MegΩ, RB = 4.7MegΩ, RL = 12KΩ, IC = 0.3375 mA or IC = 337.5 μA

Q1 = 2N3904, BT = 9 volts

IB is then calculated as about $(9 \text{ v} - 0.7 \text{ v})\ 4.7\text{Meg}\Omega = 1.766\ \mu\text{A}$. We can now find $\beta = \text{IC}/\text{IB} = (337.5\ \mu\text{A}/1.766\ \mu\text{A})$:

$\beta = 191$

This leads to R_in base = $\beta(0.026 \text{ v}/\text{IC})$ or R_in base = $191(77\Omega)$.

R_in base = $14.7\text{K}\Omega$. If we make R_ser about 10x of R_in base, then R_ser ~ $150\text{K}\Omega$. The resulting distortion can now be compared with R_ser = 0 Ω and R_ser = $150\text{K}\Omega$ for gain and distortion at Vout for the same output voltage. See Table 8-1 that shows the effect on distortion with different values for input series resistor, R_ser. There is a trade-off in that you can have lower distortion but at the expense of lower voltage gain.

TABLE 8-1   Distortion Measurements for the Same Amplitude Output at Vout

R_ser	Vout pk to pk	Vin pk to pk	Gain	Vout Distortion
~ 0 Ω	5.18 volts p-p	38.9 mV p-p	−133	~ 20%
150KΩ	5.18 volts p-p	429 mV p-p	−12	~ 2%

NOTE that with the gain reduced by about tenfold, the distortion is reduced by ~10 as well. Also it is preferred that RB >> R_ser for reducing distortion.

See Figures 8-14 and 8-15, where the top trace waveforms are the input signals, and the bottom traces show the output signals. Note the phase inversions between output and input signals.

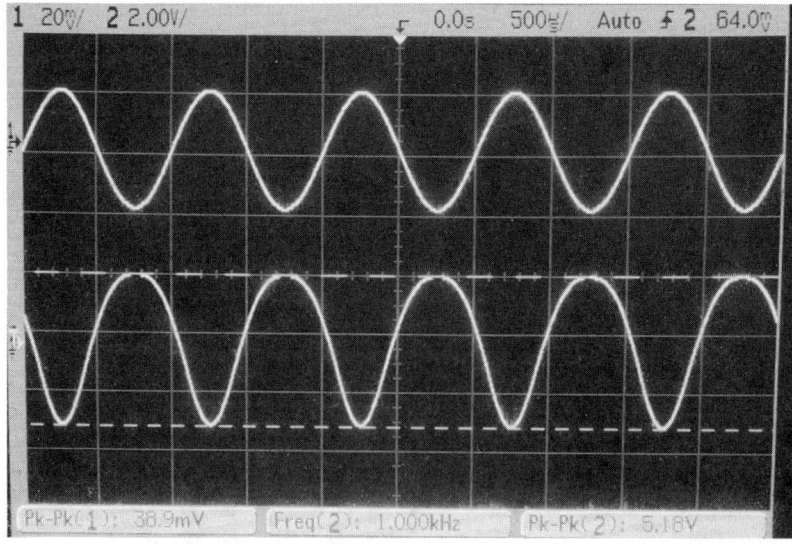

FIGURE 8-14   Output waveform on the bottom trace with R_ser = ~ 0 Ω, and note the compression on the positive cycle and narrowing on the negative cycle that denotes 20 percent harmonic distortion.

Figure 8-15 shows when a "linearizing" series 150K resistor is added.

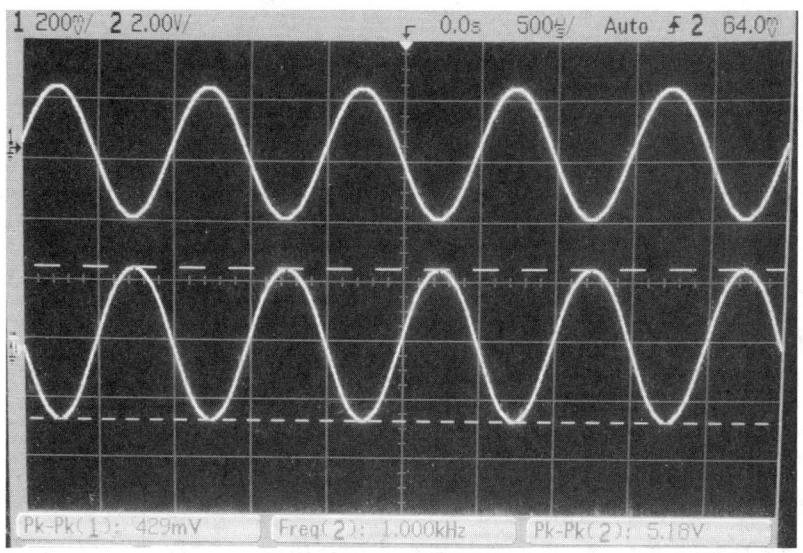

FIGURE 8-15   Output waveform on bottom trace with R_ser = 150KΩ, which shows almost no compression or narrowing distortion compared to Figure 8-14. Harmonic distortion was measured at approximately 2 percent.

Output Swing Determined by IC and RL ∥ R2

To maximize output voltage swing, we should bias the collector current, IC, such that the DC collector voltage is in a range of 40 to 60 percent of the supply voltage. For example, if the supply voltage such as BT2 in Figure 8-13 is 10 volts (e.g., 8 AA rechargeable 1.25-volt batteries in series), then a good starting point will be at the 50 percent of 10 volts or 5 volts DC at the collector of Q1. RB would be selected for this. The reason we want the DC collector voltage close to 50 percent of the supply voltage is to provide maximum output AC voltage swing. However, depending on the application, sometimes we just need a few volts of AC swing, peak to peak.

The maximum peak voltage swing without distortion is calculated as:

Vout max peak to peak = 2 × IC × RL∥R5, where VC ≥ 50% of power supply voltage

For example, suppose you have BT2 = 6 volts and RL = 3KΩ, and you bias RB such that:

IC = 1 mA DC

This means that VC = 6 volts – IC(RL) = 6 volts – 1 mA (3KΩ) = 6 volts – 3 volts or VC = 3 volts.

VC = 3 volts, which is 50 percent of BT2 = 6 volts. This is a good starting point. If R5 is >> RL, such as RL = 1MegΩ, then the output AC voltage swing will be close to 6 volts peak to peak.

This is because with R5 >> RL, RL || R5 ~ RL:

Vout max peak to peak = 2 × IC × RL||R5 ~ 2 × IC × RL = 2 × 1 mA × 3KΩ = 2 × 3 volts

Vout max peak to peak ~ 6 volts peak to peak

However, if R5 = RL, then RL || R5 = 0.5 RL and the voltage swing will be reduced by 50 percent. For example, when RL = 3KΩ and R5 = 3KΩ, the output voltage swing is reduced by 50 percent.

Vout max peak to peak = 2 × IC × RL||R5 = 2 × IC × RL||RL = 2 × IC × 0.5 RL = IC × RL = Vout max peak to peak

Vout max peak to peak = 1 mA × 3KΩ = 3 volts peak to peak

If the 3 volts peak to peak Vout max is sufficient, then you do not need to change the circuit. However, if R5 is lower in value such as 2KΩ or 1KΩ and you still want 3 volts peak to peak output, then you have to make RL a smaller resistance such as lowering it to 2KΩ or 1KΩ while reselecting RB to have VC = 3 volts DC such that the DC collector current, IC is increased.

So far, the one-transistor amplifier circuits presented have many variables in terms of gain, input resistance, and DC operating points such as collector current IC and collector voltage VC due to the variation in β, the current gain. If you need to build many of these amplifiers and make them repeatable, you can buy the transistors in tape form as shown in Figure 8-16. They should be reasonably matched in terms of turn-on voltage and β, but check them with a DVM for confirmation.

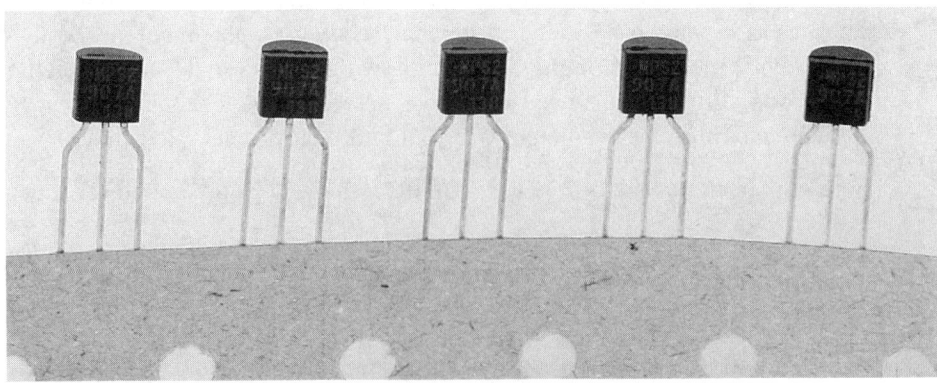

FIGURE 8-16 Transistors on tape are reasonably matched for β and VBE turn-on voltage.

Troubleshooting the One-Transistor Amplifier

The one-transistor amplifier may not work for the following reasons:

- If the transistor is put in backwards with the collector and emitter swapped, then the collector current will be very low because $\beta \rightarrow 1$ or 2. This will result in very low gain since the collector current will be too small. So, expect VC to be almost the same voltage as the power supply, which says that the voltage across RL is close to zero volts.
- If the collector and emitter terminals are reversed, and the power supply voltage is > 6 volts DC, there's a good chance that the transistor will act like a Zener diode, and the voltage at the collector will be in the range of 5 or 7 volts DC, but there will be very little signal output at Vout.
- If you replace the transistor, but find that the output signal is low, check the DC collector voltage, VC. If the collector's voltage VC is close to 0 volts (e.g., < 0.5 volt), the transistor is in saturation because there is too much base current. Reselect the base driving resistor, RB, to have VC at about half the supply voltage, or select a transistor with lower current gain, β. You may also have to confirm the proper resistance values for RL and RB. If RL is too high in resistance and or if RB is too low in resistance, it can cause the transistor to be in saturation.
- If you think you connected everything correctly but you find strange DC readings and output signals, make sure the transistor is the correct polarity (e.g., NPN or PNP). Sometimes it's easy to inadvertently put in the wrong polarity transistor and/or put it in with swapped leads.
- Check power supply voltage, and always add a bypass capacitor such as 0.1 µf to 1 µf across the power supply voltage source close to the transistor amplifier. Also note the supply voltage and choose the bypass capacitor to have at least twice the voltage rating. For example, if the supply voltage is 12 volts, choose a 25-volt to 100-volt capacitor.

Using Negative Feedback to Build "Mass Production" Amplifiers

If we look at the one-transistor amplifier schematic in Figure 8-13 with its output waveform in Figure 8-14, we will notice that the AC signal going into the base produces a signal at the collector that is in negative phase or opposite phase of the input.

This makes sense because the DC collector voltage, **VC = BT2 – IC (RL)**. Since IC is proportional to base current, IB, and IB is some function of the base voltage, then an increase in base current causes an increase in collector current. But an increase in collector current, IC, causes a decrease in VC due to VC = BT2 – IC (RL). For example, if BT2 = 10 volts, RL = 1KΩ, then VC = 5 volts – IC 1KΩ. If we start out with IC = 1 mA, then VC = 5 volts – 1 mA 1KΩ = 5 volts – 1 volt.

VC = 4 volts

Suppose we increase the collector current such that IC → 3 mA, then:

VC = 5 volts – 3 mA 1KΩ = 5 volts – 3 volts = 2 volts

VC = 2 volts

As we can see, an increase in collector current results in a decrease in collector voltage. Since an increase in base current causes an increase in base voltage, then it follows that an increase in base voltage causes an increase in collector current, which causes a decrease in collector voltage. Thus, we have a "negative" phase relationship between the base and collector terminals. If we have this relationship, we can apply a self-biasing resistor between the base and collector as shown in Figure 8-17. If this resistor is chosen properly, we can set a DC collector voltage and collector current that is not as sensitive to β variations. As long as β >> 1 such as β ≥ 20, the DC collector current that is set by this circuit will not vary much whether the transistor has a β of 50 or a β of 500. Also, as long as the AC signal's gain is kept to ≤ 10, the gain will be set by two resistors and it again will be insensitive to β variations.

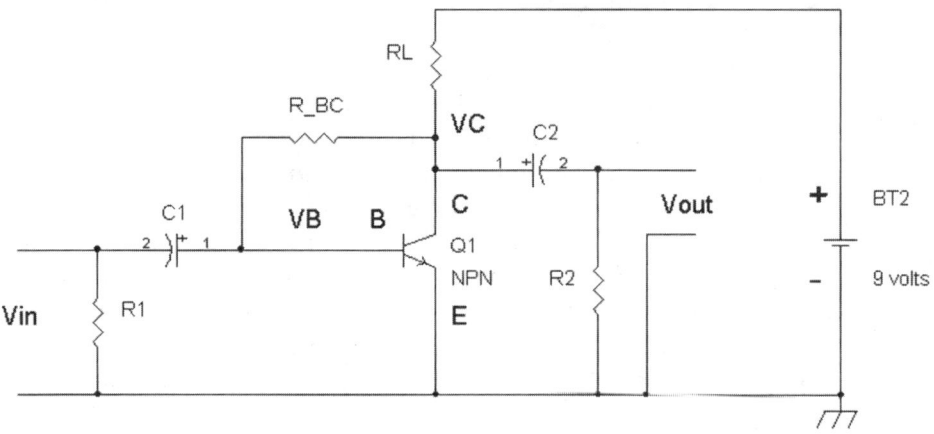

FIGURE 8-17   A simple self-biasing one-transistor amplifier via R_BC for "insensitivity" to β.

DC Analysis of Self-Biasing Amplifier

With a negative feedback resistor R_BC that is connected from the collector to base, a DC voltage is established at VC. For the DC analysis, see Figure 8-18.

Typically, the base-emitter voltage at VB is ~ 0.7 volt. If Q1's β is greater than 50, then there's a good chance that Q1's base current, IB, is very small such that there is a very small voltage across R_BC, V_R_BC = IB × R_BC. (Refer to Figure 8-18.)

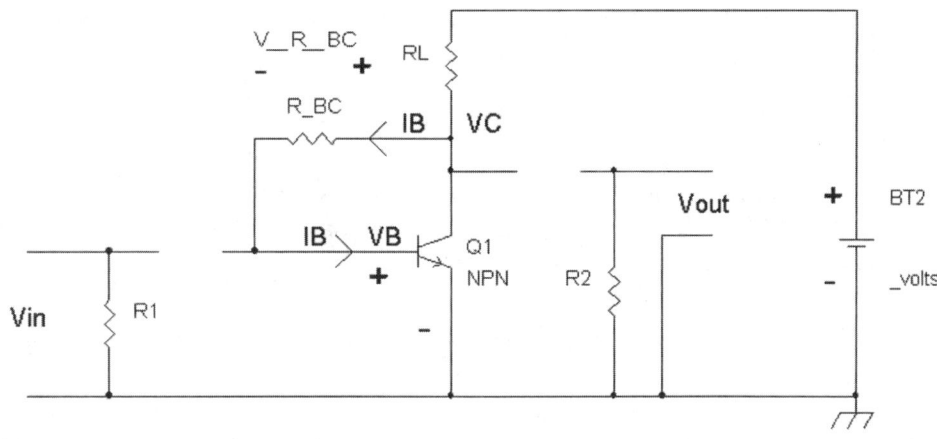

FIGURE 8-18 DC analysis circuit for the self-biasing one-transistor amplifier via removing the DC blocking capacitors.

NOTE that VC = V_R_BC + VB or VC = (IB x R_BC) + VB.

If this voltage across resistor R_BC, V_R_BC << 0.7 volt, then VC ~ VB. A commonly used transistor, such as2N4124 or equivalent, generally has $\beta \geq 50$ at currents from 100 µA to about 20 mA.

For power supply voltages that are >> 0.7 volts (e.g., $\geq$ 5 volts) even with base current IB flowing through R_BC in the order of V_R_BC = 0.7 volts such that VC = 1.4 volts versus VC = 0.7 volts, the collector current will vary < 25 percent for BT2 $\geq$ 5 volts.

For example, with BT2 = 5 volts, **IC ~ (BT2 – VC)/RL.**

With VC = 0.7 volt, IC ~ (5 volts – 0.7 volt)/RL or $IC_{0.7v}$ ~ 4.3 volts/RL.

With VC = 1.4 volts, IC ~ (5 volts – 1.4 volts)/RL or $IC_{1.4v}$ ~ 3.6 volts/RL.

We can take the ratio of the two collector currents for VC = 0.7 volt and 1.4 volts to determine the change in collector current.

$(IC_{0.7v}/IC_{1.4v})$ = [(4.3 volts/RL)/(3.6 volts/RL)], the RL's cancel out, leaving $(IC_{0.7v}/IC_{1.4v})$ = (4.3 volts/ 3.6 volts) = 1.1944, which is a 19.44 percent change in collector current.

Let's take a look at some typical component values. Q1 = 2N4124 or 2N3904, BT2 = 5 volts, RL = 3300Ω, R_BC = 100KΩ, with β = 170, the measured VC = 1.3 volts. The collector current with β = 170 is:

IC ~ (5 volts – 1.3 volts)/3300Ω = 1.12 mA = IC

If we replace Q1 with a 2N5089 where β = 461, VC = 0.9 volt, and IC = (5 volts – 0.9 volt)/3300Ω = 1.24 mA for β = 461, which is only about a 10.8 percent change with 1.12 mA for β = 171. So even when β increased from 170 to 461, a 2.77-fold increase, the collector current, IC, only increased by about 10.8 percent.

Having VC biased at a DC voltage of about 1 volt will limit the voltage swing from about 0 volt to about 2 volts or 2 volts peak to peak before clipping of the output waveform occurs.

To maximize output voltage swing, it is generally better to have VC set to about one half the supply voltage. For example, if BT2 = 5 volts, and VC is set to 2.5 volts DC, then the output swing can be from about 0 volts to nearly 5 volts, or close to 5 volts peak to peak. By adding a resistor (R_BE) across the base-emitter terminals, we can raise VC's DC voltage. See Figure 8-19.

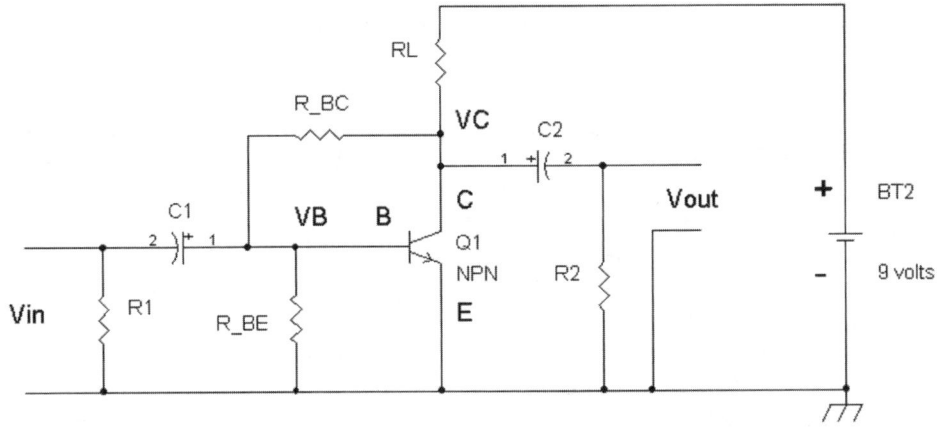

FIGURE 8-19   Adding an extra resistor R_BE to form a voltage divider with R_BC.

By adding R_BE, we can raise the DC collector voltage. See Figure 8-20 for the DC analysis schematic.

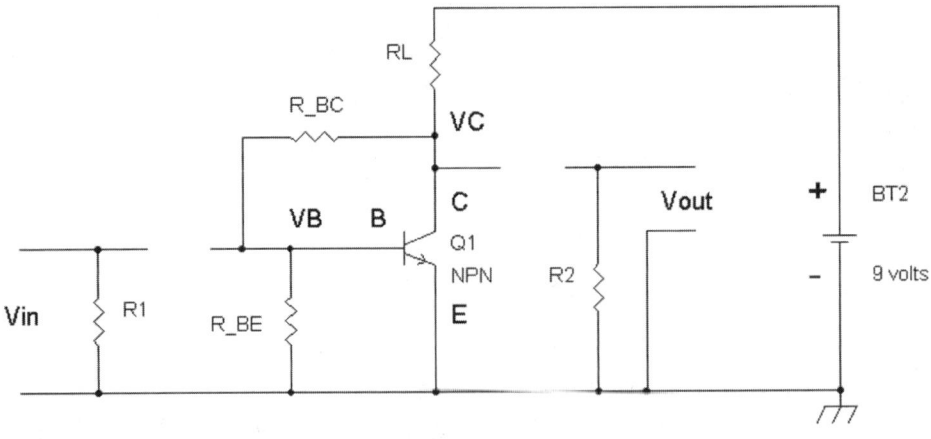

FIGURE 8-20   DC analysis circuit for Figure 8-19.

Since VB ~ 0.7 volt, we can work backwards to find VC since the voltage across R_BE is ~ 0.7 volt. By using a voltage divider formula and neglecting the base current from Q1, we have:

0.7 volt = VB = VC [R_BE/(R_BE + R_BC)]

0.7 volt/[R_BE/(R_BE + R_BC)] = VC

VC = 0.7 volt [(R_BE + R_BC)/R_BE]

In essence the DC voltage at the collector, VC, is a scaled "up" voltage of VBE. For example, if we want **VC = 1.4 volts**, then R_BC = R_BE.

VC = 0.7 volt [(R_BE + R_BE)/R_BE] = 0.7 volt [2R_BE/R_BE] = 0.7 volt × 2

VC = 1.4 volts

A general formula where it is easier to just find the ratio of R_BC to R_BE based on having a specified VC is shown here:

R_BC/R_BE = (VC/0.7 v) – 1

For example, if BT2 = 5 volts, RL = 3300Ω, and we want VC = 2.5 volts, then:

R_BC/R_BE = (2.5/0.7 v) – 1 = 3.57 – 1 = 2.57

R_BC = 2.57 (R_BE)

We can make **R_BE = 39KΩ** so that R_BC = 2.57 (39KΩ)

R_BC = 100KΩ

In general VC will be a bit higher than calculated using a general-purpose transistor such as with a 2N4124 transistor. With a higher β transistor such as a 2N5089, the calculated VC voltage will be closer in practice.

To preserve good voltage swing, R_BC >> RL. For example, if RL = 1KΩ, R_BC ≥ 10KΩ. However, we will see that the simple 1 transistor amplifier has a "drawback" in terms of input resistance being too low. But not to worry—we use the lower input resistance to our advantage by adding a series input resistor to set the gain of the amplifier (see R1" in Figure 8-25 if you are curious)..

AC Analysis of a Self-Biased Amplifier

Adding the feedback resistor R_BC causes a lower total input resistance as "seen" by Vin (see Figure 8-21).

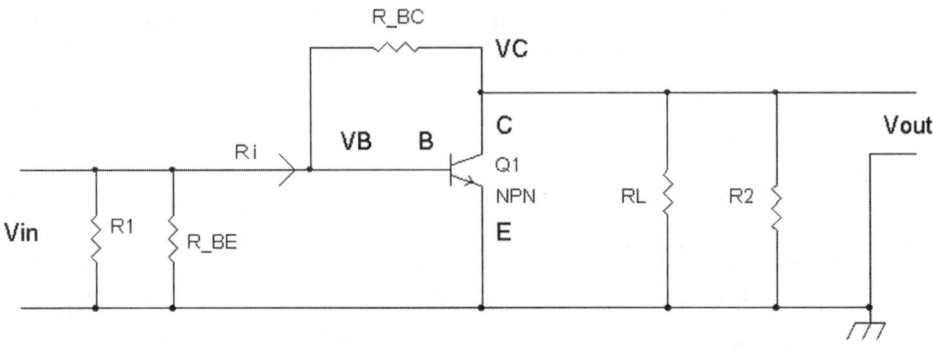

FIGURE 8-21   AC analysis circuit for the one-transistor feedback amplifier.

With negative feedback via the collector to base resistor, R_BC, this resistor and with the amplifier's gain, G, will result in a low-value resistor, Ri, referenced to ground, where Ri is in parallel with R1 and R_BE.

Ri ~ [R_BC/(1 – G)] || R_in base, where G = Vout/Vin, and since this is an inverting amplifier, G will be a negative number.

G ~ –(IC/0.026 v) × (RL || R2 || R_BC)

The gain, G, is calculated to be the same as the simple transistor amplifier shown in Figure 8-10 when R_BC >> RL || R2. For a good enough approximation where R_BC ≥ 10 × (RL || R2), use:

G ~ –(IC/0.026 v) × (RL || R2)

For example, if the supply voltage BT2 = 12 volts, Q1 = 2N5089 where β = 400, RL = 4700Ω, R_BC = 100KΩ, R_BE = 20KΩ and VB = 0.7 volt, then VC = 0.7 v [(R_BC/R_BE) + 1].

VC = 0.7 volt [(100K/20K) + 1] = 0.7 volt [6]

VC = 4.2 volts

IC ~ (BT2 – VC)/RL = (12 volts – 4.2 volts)/4700Ω = 7.8 volts/4700Ω

IC ~ 1.66 mA = 0.00166 A

Let R2 = 1MegΩ so that RL || R2 = 4700Ω || 1MegΩ ~ 4700Ω = RL || R2.

G ~ –(IC/0.026 v) × (RL || R2)

G ~ –(0.00166 A/0.026 v) × (4700Ω)

G ~ –300

R_in base = β(0.026 v/IC) = 400(0.026 v/0.00166 A)

R_in base $= 6.26\text{K}\Omega =$ internal resistance across the base and emitter of Q1

Ri $\sim$ [R_BC/(1 – G)] || R_in base

Ri $\sim$ [100KΩ/(1 – –300)] || 6.26KΩ = [100KΩ/(1 + 300)] || 6.26KΩ
Ri $\sim$ 332Ω || 6.26KΩ

Since 332Ω << 6.26K, 332Ω || 6.26KΩ $\sim$ 332Ω, which leads to an approximation of:

Ri $\sim$ 332Ω

R_in base is the resistance into the base-emitter junction of the transistor that does not include any of the other external resistors such as R_BC, R_BE, and R1.

The resistance [R_BC/(1 – G)] is due entirely to Vin's signal current flowing through R_BC, which results in a lower equivalent resistor, R_BC/(1 – G) referenced to ground. So, the question arises as to how we get such a lower equivalent resistor that is related to the amplifier's gain, G? See Figure 8-22.

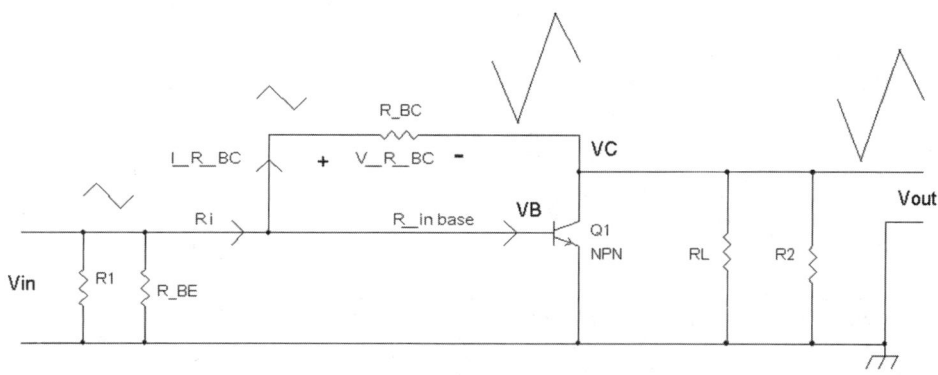

FIGURE 8-22   Input and output waveforms showing increased AC voltage across R_BC.

The amplifier produces an amplified (e.g., a larger amplitude) signal at the output, Vout, which is in opposite phase of the smaller amplitude input signal, Vin. This opposite phase signal at Vout then pulls extra current through the resistor R_BC. By pulling extra current, it makes it appear that Vin is driving an equivalently lower-value resistor that is referenced to ground like R1.

An example of this lowered resistance effect is if we imagine that the gain, G = –1 and R_BC = 1KΩ, then what happens when Vin = +1 volt? The output voltage is then – 1volt due to G = –1. However, the current flowing through the 1KΩ has a potential difference of +1 v – –1v or 2 volts. This means the resistor current is now 2 v/1KΩ = 2 mA. A +1-volt input across an equivalent resistor referenced to ground that will drain 2 mA would result in a 500Ω resistor since 1 v/500Ω = 2 mA.

In general, the voltage across resistor R_BC is V_R_BC, which has Vin on one side and Vout on the other side. Thus, V_R_BC = Vin – Vout, but Vout = –G Vin, so:

V_R_BC = Vin – –GVin = Vin + G Vin

V_R_BC = Vin (1 + G)

The current through R_BC is then:

I_R_BC = V_R_BC/R_BC

and by substituting Vin (1 + G) for V_R_BC:

I_R_BC = Vin (1 + G)/R_BC

We want to now model an equivalent resistor referenced to ground (as shown in Figure 8-23) that will drain the same amount of current as **I_R_BC**, which is **I_R_BC** = I_equiv = Vin(1 + G)/R_BC.

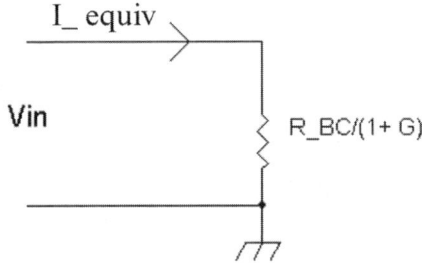

FIGURE 8-23   Equivalent resistor referenced to ground that drains the same current as I_R_BC.

Let's see if the current, I_equiv, is the same as I_R_BC.

$$I\_equiv = \frac{Vin}{R\_BC/(1 + G)} = \frac{Vin\,(1 + G)}{R\_BC} = Vin\,(1 + G)/R\_BC = \mathbf{I\_R\_BC}$$

I_equiv = **I_R_BC**

This confirms that the equivalent resistor referenced to ground is R_BC/(1 + G).

This makes sense because it would take a smaller-value resistor referenced to ground with just Vin applied to it to drain the same amount of (higher) current that would have to be a resistor divided in value by (1 + G). See Figure 8-24.

Note if the amplifier has no gain where G = 0 that causes Vout = 0 volt, then essentially R_BC is shorted to ground on the collector of Q1, and we do indeed have R_BC as the equivalent resistor reference to ground.

R_BC/(1 + G)

with G = 0

R_BC/(1 + 0) = R_BC

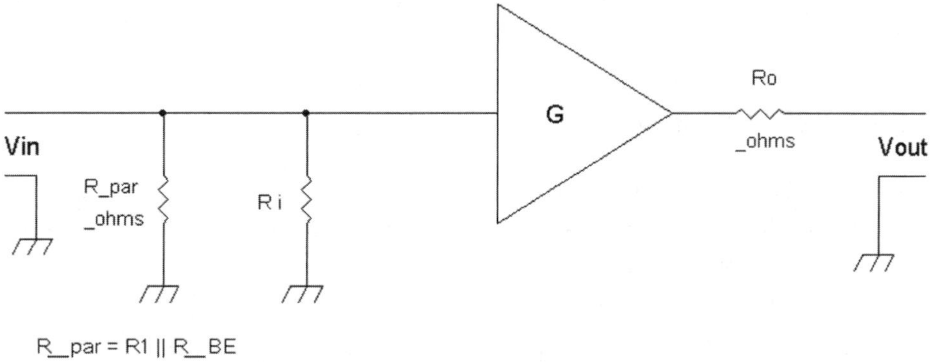

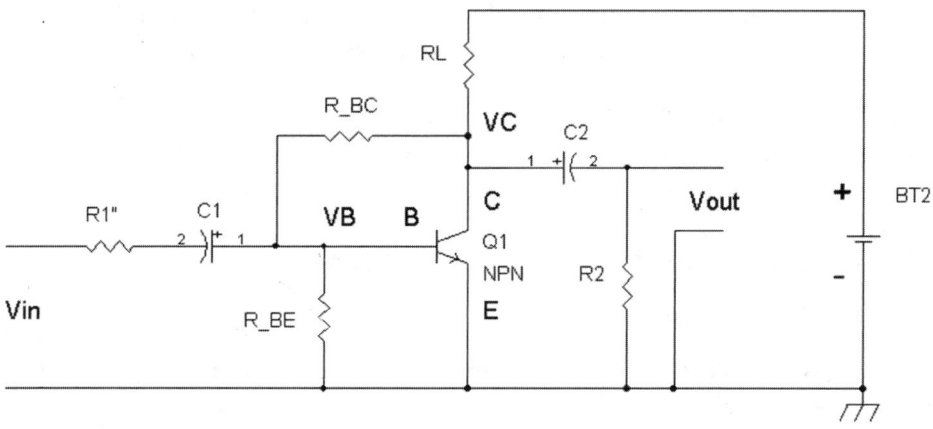

R_par = R1 || R_BE

FIGURE 8-24   AC model where Ri = [R_BC/(1 − G)] || [R_in base].

The output resistance, Ro, is actually dependent on the input signal's series source resistance. If Vin is driven by a pure voltage source or very low output imped-ance amplifier, Ro = RL||R2||R_BC, which is similar to the simple amplifier in Figures 8-3 and 8-7. Normally, R_BC >> RL||R2, so Ro ~ RL||R2.

Because R_BC lowers the overall input resistance seen by Vin to almost "short" circuit compared to the other resistors, R1, R_BE, and R_in base, generally, we do not drive the input directly with a signal source with low source resistance.

However, we can improve upon the amplifier in Figure 8-19 by adding a series resistor that will lower the gain, Vout/Vin, but then improves on raising the input resistance, lowering the output resistance, and also lowering the distortion for the same amplitude output. See Figure 8-25.

FIGURE 8-25   A modified one-transistor feedback amplifier with a series input resistor, R1".

We can define the gain $G' = \dfrac{Vout}{VB} = -(IC/0.026 \text{ v}) \times (RL \parallel R\_BC)$ that is relative to the signal voltage at the base of Q1, VB and Vout, and not taking R2 into account for now.

And then we can define the actual gain $\dfrac{Vout}{Vin} \sim -\dfrac{R\_BC}{R1''}$ where $-\dfrac{R\_BC}{R1''} \leq 5\%$ of G'.

Output Resistance Ro'

The reader can skip over this section pertaining to the output resistance, Ro', if desired. These calculations are long. The gist is having R_BC as a feedback resistor results in a lower output resistance than RL||R2.

The output resistance, Ro', not including R2 is approximately:

Ro' = {[(R_BC + R_in base || R_BE || R1'')/(R_in base || R_BE || R1'')] × (0.026 v/IC)} || RL

Here are a couple of calculations with their measured results.

With BT2 = 5 volts, RL = 2700Ω, R_BC = 100KΩ, R_BE = 40.2KΩ, and R1" = 20KΩ, the calculated gain (Vout/Vin) = –(R_BC/R1") = –100K/20K.

Vout/Vin = –5.0

Measured gain is –4.5, within 10 percent of the expected gain of –5.

Now let's calculate the output resistance, with R2 removed from the circuit in Figure 8-25.

The calculated VC ~ VBE (1 + R_BC/R_BE) where VBE ~ 0.7 v:

VC ~ 0.7 v(1 + 100K/40K) = 0.7 volt (3.5)

VC ~ 2.45 volts. The measured VC was 2.3 volts.

With BT2 = 5 volts and VC = 2.45 volts with RL = 2.7KΩ, the calculated IC ~ (BT2 – VC)/RL since R_BC = 100KΩ >> RL = 2.7KΩ:

IC ~ (5 v – 2.45 v)/2.7KΩ

IC ~ 0.907 mA

The current gain β ~ 450 for the 2N5089.

R_in base = β (0.026 v/0.000907A) = 450(286Ω)

R_in base = 12.9KΩ

With R_BC = 100KΩ, R_in base = 12.9KΩ, R_BE = 40.2KΩ, R1" = 20KΩ, and RL = 2700Ω

Ro' ~ {[(R_BC + R_in base || R_BE || R1") /(R_in base || R_BE || R1")] × (0.026 v/IC)} || RL

where R_BC in parallel with RL is neglected since R_BC = 100KΩ >> RL = 2.7KΩ.
Therefore:Ro'~{[(100KΩ + (12.9KΩ||40.2KΩ||20KΩ)/(12.9KΩ||40.2KΩ||20KΩ)] × (0.026 v/0.000907 A)}|| 2.7KΩ

NOTE: (12.9KΩ||40.2KΩ||20KΩ) = 6.56KΩ.

Ro' ~ {[(100KΩ + 6.56KΩ)/(6.56KΩ)] × (0.026 v/0.000907 A)}|| 2.7KΩ

Ro' ~ {16.24 × (0.026 v/0.000907 A)}|| 2.7KΩ

Ro' ~ {16.24 × (28.66Ω)}|| 2.7KΩ

Ro' ~ 465Ω|| 2.7KΩ

Ro' ~ 465Ω|| 2.7KΩ

Ro' ~ 397Ω calculated

Ro' ~ 390Ω measured via setting R2 = 390Ω and noticing that the signal dropped by 50 percent.

The output resistance, Ro', calculation is rather long, and sometimes we do not have time to do this. Here are some rules of thumb concerning the transistor amplifier.

1. Use low-value gains that are typically less than 5 or 10, such that R_BC/R1" ≤ 10.
2. Make sure that R_BC >> RL by at least tenfold. Thus, R_BC does not lower the gain G by much since as "seen" by the collector of Q1, RL and R_BC are essentially in parallel. The reason is that the resistors R_BE and R_in base that are coupled to ground and connected to R_BC at the base are << R_BC, that is (R_BE || R_in base) << R_BC.
3. Bias the DC voltage at the collector, VC, to be at one-half the power supply voltage.
4. Use as high β transistor as you can, such as a 2N5089. If you use a lower β transistor such as a general-purpose type, 2N3904, then you may have to set for lower gains such as R_BC/R1" ≤ 5.
5. Generally, bias the collector current, IC, in the range of 0.2 mA to 5 mA with the appropriate value RL resistor that is scaled from the example at ~ 1 mA where RL = 2700Ω. For instance, if you are running IC at 200 µA, then RL ~ 5 × 2700Ω or about 12KΩ, and R_BC ~ 5 × 100KΩ = 510KΩ, R_BE ~ 5 × 40.2KΩ = 200KΩ, and R1" = 100KΩ. Going the other direction with IC = 5 mA, all the resistors that were used for IC ~ 1 mA are

scaled by (1/5) or 20 percent or 0.2. Thus, RL ~ 0.2 × 2700Ω ~ 560Ω, R_BC ~ 0.2 × 100KΩ ~ 20KΩ, R_BE ~ 0.2 × 40.2KΩ = 8.2KΩ, R1" ~ 0.2 × 20KΩ ~ 3.9KΩ.

Note the input capacitor, C1, can be scaled according to the collector currents. For example, if a 1 µf is used for 1 mA collector, then 4.7 µf is used for 5 mA IC, and 0.22 µf is used for 0.2 mA collector current. However for C1, you can always just use the highest value capacitor for all cases such as 4.7 µf or greater capacitance. The output capacitor C2 should be greater than or equal to 4.7 µf to ensure good low frequency response.

6.   One feature of this amplifier is that the performance increases as you set the gain lower via lower ratios of R_BC_ to R1". This is normally done by increasing the value of R1". The increase in performance parameters result in lower output resistance, Ro', and lower harmonic distortion at Vout for the same output amplitude.

7.   Another feature of this amplifier is that you can set the inverting gain to attenuate the signal. That is, 0 < |R_BC/R1"| < 1. For example, if R_BC = 100KΩ, we can set R1" = 300KΩ for a gain of – (1/3) or – 0.33. This can be useful if the amplifier is interfacing with a signal source whose amplitude exceeds the amplifier's power supply voltage, BT2. For example if your generator or signal source provides 10 volts peak to peak, and your amplifier can only deliver about 4 volts peak to peak, having the gain set to –0.33 will keep the output from clipping since the output voltage will be 10 volts × (–0.33) peak to peak or 3.3 volts peak to peak.

Referring to #6, see Figure 8-26, which shows the output waveform for the circuit in Figure 8-25, where the gain is ~ –94 via input series resistor R1" = 25Ω. In comparison see Figure 8-27 where the distortion is lower for the same output voltage because R1" is increased to 20KΩ for a (lower) gain of – 4.2.

To lower output signal distortion R1" is increased to 20kΩ. See Figure 8-27 bottom trace.

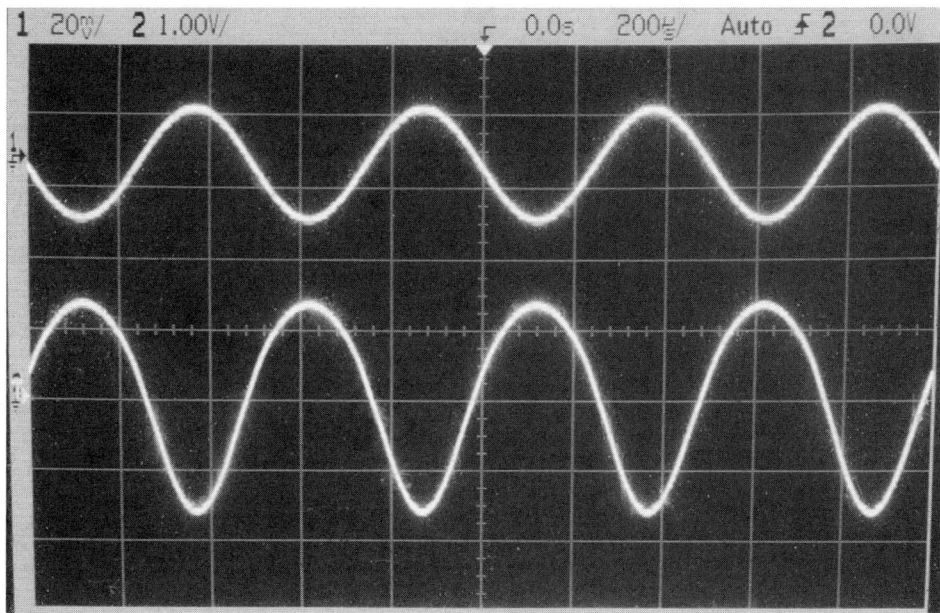

FIGURE 8-26 R1" = 25Ω, Vin = 32 mV peak to peak (top waveform), and bottom waveform Vout = 3 v peak to peak with 12 percent harmonic distortion. Gain ~ −94 = Vout/Vin. Also note the phase inversion between the output signal Vout and the input signal Vin.

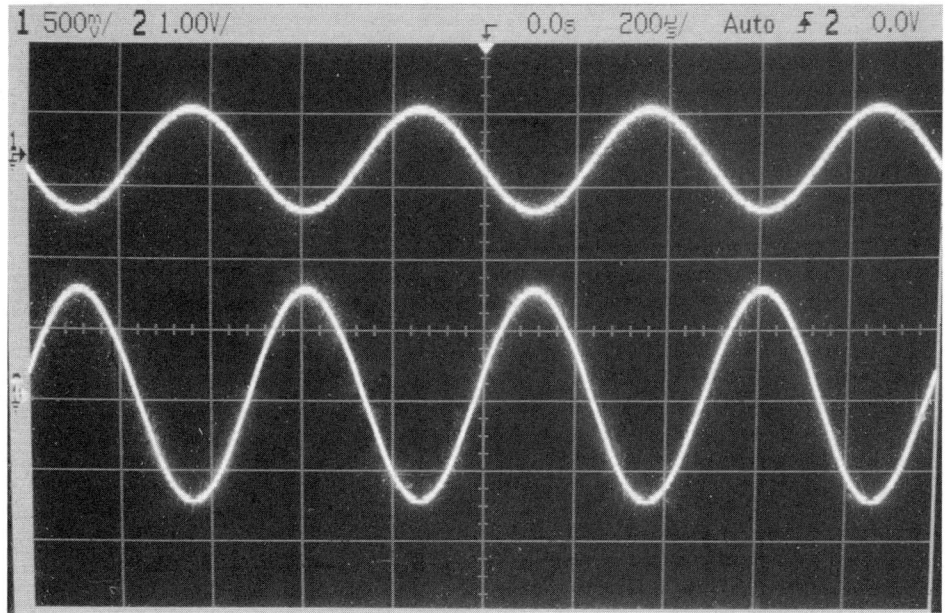

FIGURE 8-27 R1" = 20KΩ, Vin = 720 mV peak to peak (top trace), and Vout = 3 v peak to peak for a gain of − 4.2 (bottom trace). Harmonic distortion measured at 1 percent. Again, note the phase inversion of Vout with respect to Vin.

Another Common Emitter Amplifier

We will now look at another type of common emitter amplifier with a series emitter resistor derived from an LED drive circuit. See Figure 8-28, which shows a constant current source LED drive circuit, and then Figure 8-29, which converts the LED drive circuit into an amplifier.

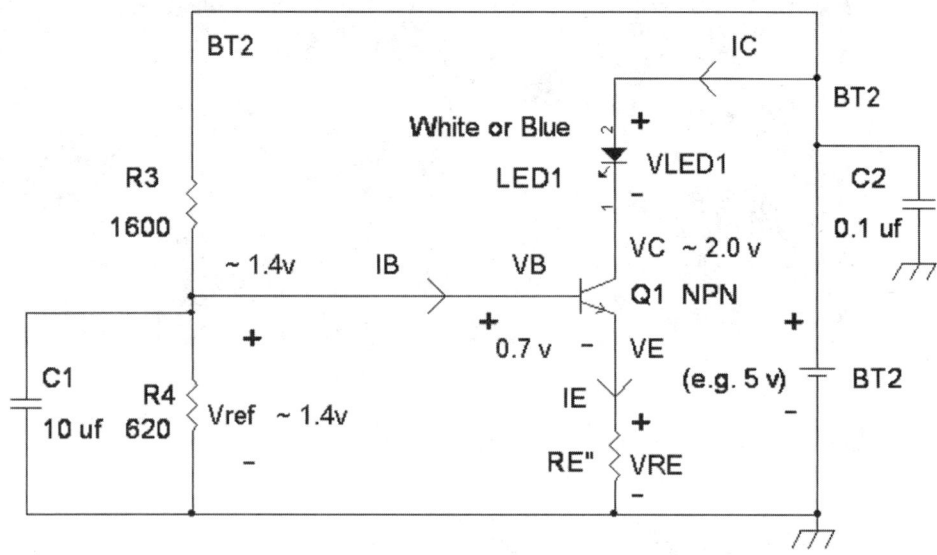

FIGURE 8-28   A constant current source amplifier where the LED drive current = VE/RE".

We can reconfigure the constant current source circuit to an amplifier by replacing the LED with a load resistor, RL, and coupling an input signal voltage to the base with a capacitor. See Figure 8-29.

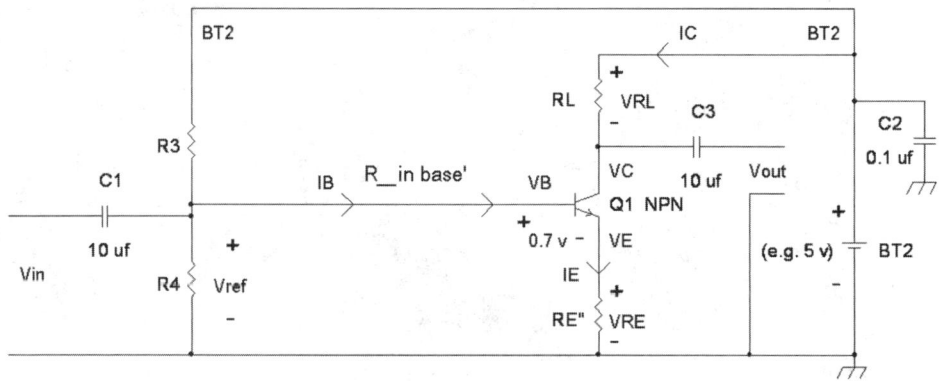

FIGURE 8-29   A simple common emitter amplifier with a series emitter resistor RE".

To bias this correctly, generally "Vref" = BT2[R4/(R3 + R4)] is less than half the supply but ≥ 0.5 volts. Typical collector currents can range from 50 μA to about 20 mA for a small signal transistor such as the 2N3904. Generally, the voltage divider resistors ≥ 470Ω, and the emitter resistor RE" that sets the DC collector current can be anywhere from about 100Ω to 10KΩ. Collector current IC = VE/RE", where β >> 1, which is usually the case. In terms of gain, a first approximation has Vout/Vin ~ – RL/(r$_e$ + RE"), where r$_e$ = (0.026 v)/IC. Again, this is an inverting amplifier. Generally, it's harder to provide very high gains from this configuration unless very high supply voltages are used. For example, if we want a gain of –100, and IC = 1 mA, with VE = 1 volt and RE" = 1000Ω, then RL has to be ~100KΩ. This would require BT2 >100 volts. For example, BT2 = 200 volts so that VC is operating at 100 volts DC. Also, a general-purpose transistor will have insufficient breakdown voltage and a high-voltage transistor such as a 2N3439 or MPSA42 would be used instead. If we want instead a gain ~ –10, then we can have RL = 10KΩ, BT2 = 20 volts, Q1 = 2N3904, with VC ~10 volts when IC = 1 mA.

An emitter bypass capacitor CE2 provides higher gain (Vout/Vin) at the expense of higher distortion. But this circuit allows for lower supply voltages. See Figure 8-30.

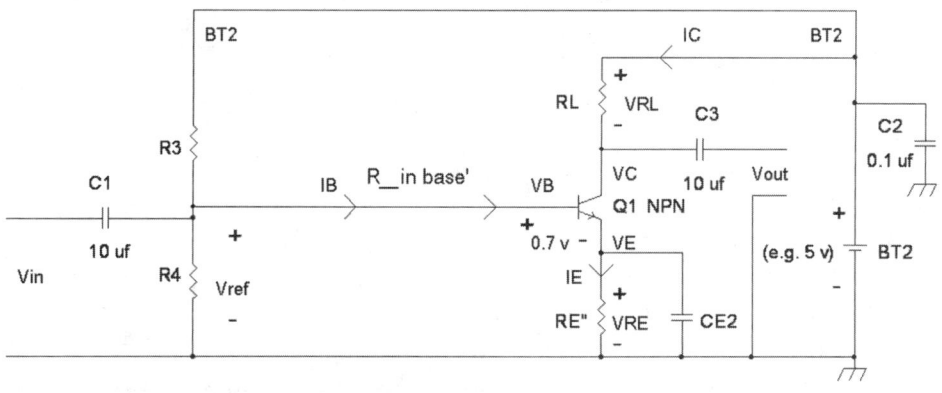

FIGURE 8-30   Capacitor CE2 effectively AC grounds Q1's emitter.

By adding a resistor (RE2) in series with capacitor CE2, the AC gain can be set without upsetting Q1's DC bias points (e.g., VB and IC). See Figure 8-31, which includes RE2.

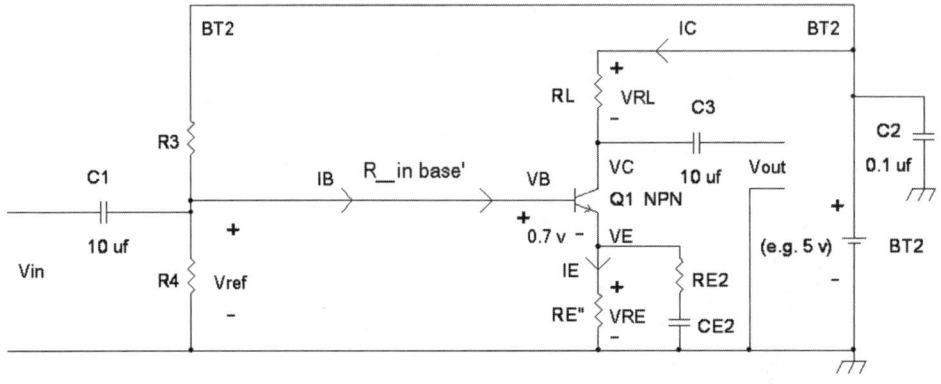

FIGURE 8-31   A second series emitter resistor, RE2, allows for increasing the gain without changing the DC bias conditions. Capacitor CE2 has a large capacitance such that it is an AC short circuit.

In Figure 8-31, with RE2 in series with CE2, the gain G = −[RL/(r_e + RE" || RE2)], where r_e = (0.026 v/IC). To find the DC operating points, VC, IC, and VE, we remove the capacitors as shown in Figure 8-32.

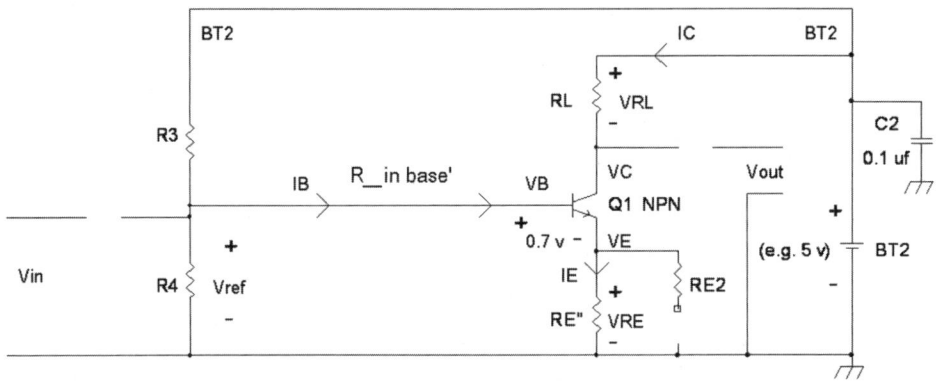

FIGURE 8-32   A DC analysis circuit of Figure 8-31 where capacitors are removed.

With the capacitors removed and neglecting DC base currents due to IB, VB ~ BT2 [R4/(R3 + R4)] and with Q1's turn-on voltage, VBE ~ 0.7 volt DC, VE ~ VB − 0.7 volt. The collector current IC ~ IE, the emitter current when β >>1, which is usually the case. We now have:

IE = VE/RE"

IC ~ VE/RE"

We now have to also find AC resistance, R_in base, which can be found via IC, β, r_e, RE", and RE2.

R_in base = β r_e + (β + 1) (RE" || RE2), where r_e = (0.26 v/IC).

For the AC analysis model, see Figure 8-33.

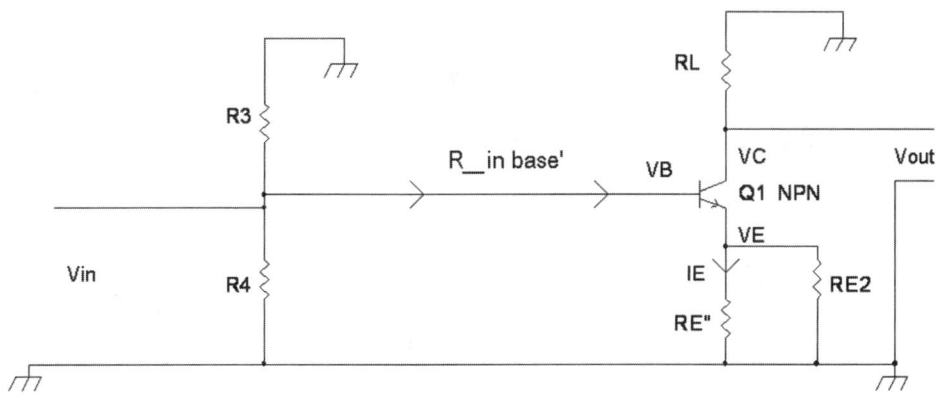

FIGURE 8-33 An AC signal analysis circuit.

For example, if BT2 = 6 volts, R3 = 40.2KΩ, R4 = 20KΩ, RE" = 3900Ω, RL = 10KΩ, RE2 = 1KΩ, β = 150, then VB ~ 6 v (20K/(40.2K + 20K)) ~ 6 v (20/60.2K) VB ~ 2 volts.

VE ~ VB – 0.7 v = (2 – 0.7) v or VB = 1.3 volts. IC ~ VE/RE" = 1.3 v/3900Ω or IC ~ 0.00033A.

To find the gain, G, we calculate for:

r_e = (0.026 v/IC) = (0.026 v)/(0.00033A) or r_e = 78.8Ω

G = –[RL/(r_e + RE" ∥ RE2)

RE" ∥ RE2 = 3900Ω ∥ 1KΩ = 796Ω, thus

G = –[10K/(78.8 + 796)] or G ~ –11.4 , and with β = 150 we have:

R_in base = β r_e + (β + 1) (RE" ∥ RE2) = 150 (78.8Ω) + (150 + 1) (796Ω)

R_in base ~ (11.82KΩ + 120.2KΩ)

R_in base = 132.02KΩ

When RE" ∥ RE2 >> r_e, the emitter resistors RE" and RE2 are the main contributors for R_in base that leads to: R_in base ~ (β + 1) (RE" ∥ RE2).

From Figure 8-34, to find the AC input resistance to the amplifier, it is (R_par ∥ R_in base'), where R_par = R3 ∥ R4. With the example where R3 = 40.2KΩ, R4 = 20KΩ, and the calculated R_in base' = 132.02KΩ, the input resistance is:

R3 ∥ R4 ∥ R_in base' = (40.2KΩ ∥ 20KΩ) ∥ 132.02KΩ = (13.33KΩ) ∥ 132.02KΩ
R3 ∥ R4 ∥ R_in base' = 12.13KΩ

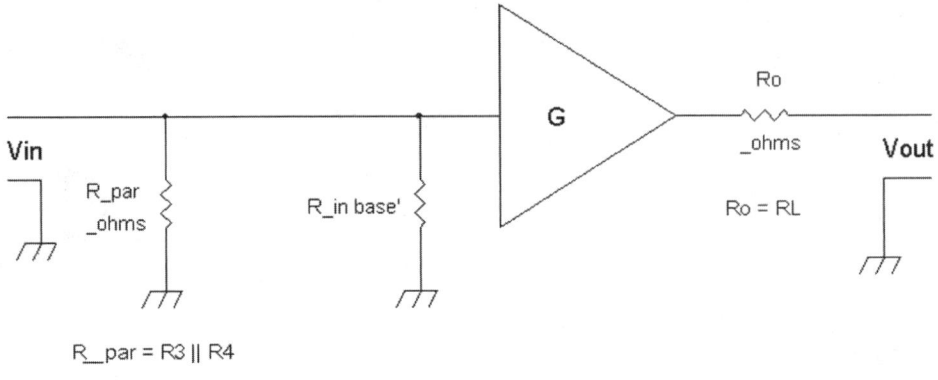

R_par = R3 || R4

FIGURE 8-34   AC signal analysis block diagram with output resistance Ro = RL.

The input resistance = 12.13KΩ.

We can reiterate the gain calculation for G from knowing the collector current and the values of the collector resistor RL and emitter resistors, RE" and RE2 along with r_e.

$$G = -[RL/(r_e + RE" || RE2)$$

r_e = 78.8Ω = (0.026 v/IC), where IC = 0.00033 A as calculated previously.

RE" || RE2 = 3900Ω || 1KΩ = 796Ω, RL = 10KΩ and thus G = –[10K/(78.8 + 796)]

G ~ –11.4

The output resistance, Ro = RL, and with RL = 10KΩ in this example, Ro = 10KΩ.

NOTE: An amplifier having a high output resistance such as 10KΩ results in a gain reduction of G when driving a subsequent device (e.g., input stage of another amplifier). If Vout is loading into another device that has an effective 10KΩ to ground will cause the gain, G, drop from –11.4 to half or G → –5.7.

Typical input resistances of audio power amplifiers are between 1KΩ to 100KΩ.

For example, an amplified computer stereo loudspeaker with a 3.5 mm connector has a typical input resistance of 10KΩ.

Troubleshooting the Amplifier in Figure 8-31

1. For measuring the DC conditions, turn off the AC signal source and measure the collector, base, and emitter DC voltages, VC, VB, and VE. Make sure that the transistor is in the amplifying region (forward active region) and not in the saturation region. This can be confirmed by measuring with a DVM that

the voltage across the collector and emitter (VCE) > +1 volt DC, or that the collector to base (VCB) voltage > 0.7 volt. Generally, we should expect that VCE or VCB is in the range of at least a quarter of power supply voltage. For example, if BT2 = 12 volts, then VCE ≥ 3 volts. In another example, if BT2 = 5 volts, then the DC voltages across the collector to base (VCB) and across the collector to emitter (VCE) should be ≥ 25 percent (5 volts) or ≥ 1.25 volts.

2. Confirm that R3, R4, RE", and RL are correct. Measure the resistors with the power turned off and with one lead of each resistor disconnected from the circuit. In a special case, you can measure RE2 in circuit because it has a series DC blocking capacitor CE2, but you may have to keep the ohm-meter's probes across RE2 for about 10 seconds to let the resistance measurement settle.

3. With power turned back on, confirm that the voltages comply with the expected voltage from the voltage divider circuit with R3 and R4. That is, VB = BT2[R4/(R3 + R4)], and VE = VB – 0.7 volt.

4. Confirm that the emitter current is approximately equal to the collector current by measuring the voltage across RL = VRL and voltage across RE" = VRE. Then confirm via calculation from the resistance values of RL and RE" and the measured voltages, VRL and VRE that VRL/RL ~ VE/RE" is within 15 percent if 5 percent resistors are used.

5. If the DC voltages seem unstable, the amplifier may be oscillating at >100MHz. Try inserting a 100Ω series base resistor close to Q1 and see if the DC voltages stabilize and measure to their expected values.

6. With the signal generator turned on having a small amplitude such as 100 mV peak to peak at Vin, measure the AC signal at VE with an *oscilloscope*. The AC signal at VE should be approximately 100 mV peak to peak within about 25 percent on the low side (e.g., VE's output AC voltage is in the range of 75 mV to 100 mV peak to peak). Note that the AC voltage at the emitter is the same phase as the base since the emitter terminal acts like having a 0.7 volt DC source voltage between the base and the emitter. *A DC voltage source such as the VBE turn-on voltage in series with an AC signal source at the base cannot change the phase of the AC signal at the emitter.* The output signal at the collector, VC, should be G × 100 mV peak to peak, or in this example, where G = –11.4, Vout AC should be about 1.14 volts peak to peak that is out of phase with the input signal at the base.

7. In Figures 8-30 and 8-31 emitter capacitors CE and CE2 are chosen to determine the low-frequency response. The 10 μf input and output capacitors, C1 and C3, are designed with sufficient capacitance for most applications with a 20 Hz or lower cut-off frequency. For example, for audio applications, where a low-frequency response of ≤ 20Hz is required, C1 and C3 generally will work because input and output resistances are generally greater than 1KΩ.

For example, the cut-off frequency with a 10 µf capacitor and 1KΩ resistor is:

$1/[2\pi\ 10\ uf\ (1K\Omega)] \sim 16\ Hz$

To determine the worst-case capacitance values for CE and CE2, choose the low-end frequency you want such as 20Hz. Then use the following formula:

$CE\ or\ CE2 = 1/[2\pi\ f_c\ r_e]$

Let f_c = 20 Hz = low-frequency cut-off frequency for an audio application that responds to bass note frequencies.

For example, if IC = 1 mA, then r_e = (0.026 v/0.001A) or r_e = 26Ω.

$CE\ or\ CE2 = 1/[2\pi\ 20Hz\ 26\Omega\ ] = 3.06 \times 10^{-4}$ Farad or 306 µf

The closest commercially available value is 330 µf, but to ensure a safety margin just in case, use a 470 µf unit if possible. Since the emitter voltage is generally less than 25 volts, you can use at least a 16-volt electrolytic capacitor. Make sure the (–) terminal of CE or CE2 is grounded.

8. Confirm that any electrolytic capacitors in the circuit are correctly connected (polarity-wise) to avoid reverse biasing. Reverse biasing electrolytic capacitors usually cause the DC bias points to be wrong, due to current leakage through the electrolytic capacitors.

9. The common emitter amplifier in Figure 8-30 where capacitor CE is an AC short circuit from emitter to ground has the same input amplitude limitation as the simple bias circuit in Figure 8-11 for tolerable distortion at the output, Vout. By tolerable distortion, an example would be for intelligible voice signals, < 10 percent harmonic distortion is workable. That is, generally the input signal, Vin, is limited to less than 10 mV or 20 mV peak to peak when driven by a low impedance source signal such as a 50Ω generator. If Vin is driven with a larger value source resistance such as a ≥ 10KΩ series resistor, the distortion will be lower but the voltage gain will be lower as well due to the input resistance forming a voltage divider circuit.

For higher-fidelity applications that require lower distortion at the output less than 1 percent, use the circuit in Figure 8-31. See #10 below concerning distortion calculations.

10. In Figure 8-31 where CE2 is an AC short circuit (e.g., CE = 470 µf), we can estimate the input signal's amplitude for second order harmonic distortion at the output. Note that $Vin_{\_mV\_peak}$ is measured in peak amplitude sinewave.

$$\text{Harmonic distortion in percent} = (Vin_{\_mV\_peak}\ [\frac{1}{1 + (RE''||RE2/re)}]^2)\ \%$$

Where $r_e = re$ = (0.026 v/IC).

For example, for Figure 8-31:

BT2 = 6 volts, R3 = 40.2KΩ, R4 = 20KΩ, RE" = 3900Ω, RL = 10KΩ, RE2 = 1KΩ, G = –11.4

r_e = (0.026 v/IC) = (0.026 v)/(0.00033A) or r_e = 78.8Ω

RE" || RE2 = 3900Ω || 1KΩ = 796Ω

(RE" || RE2/r_e) = 796/78.8 ~ 10

1 + (RE" || RE2/r_e) = 1 + 10 = 11

Second order harmonic distortion in percent =

$$(\text{Vin}_{\_mV\_peak} \left[\frac{1}{1 + (RE"||RE2/re)} \right]^2) \%$$

Second order harmonic distortion in percent = $(\text{Vin}_{\_mV\_peak} \left[\frac{1}{11} \right]^2) \%$

Second order harmonic distortion in percent = $(\text{Vin}_{\_mV\_peak} \times \frac{1}{121}) \%$

So, if Vin = 121 mV peakpeak (which is also 242 mV peak to peak), then:

Second order harmonic distortion in percent = $(121 \times \frac{1}{121}) \% = 1\%$

Vin's peak to peak voltage for 1 percent second order distortion is then 2 × 121 mV peak to peak or 242 mV peak to peak. This 242 mV peak to peak input level assumes that the output has not clipped. With a gain G = –11.4, the output will be Vout = 2.76 volts peak to peak (11.4 × 242 mV p-p).

Note that the 2nd order harmonic distortion formula is proportion to input level. That is doubling the signal level results in twice the distortion, or halving the input level gives half the distortion at the output.

A safer estimate to ensure that there is no clipping is to reduce Vin to the 0.5 percent level. This is just one half of 242 mV peak to peak input level for 121 mV peak to peak that results in the output.

Vout = 1.36 volts peak to peak for 0.5 percent second harmonic distortion

Maximum Output Voltage Swing

We will now examine how to achieve maximum output swing in differently configured common emitter amplifiers.

Amplifier's Emitter AC Grounded via CE

The DC collector voltage will determine output voltage swing. For the circuit in Figure 8-30, where capacitor CE is an AC short circuit, the maximum voltage swing will be from VE to BT2. For example, if VB is 2 volts, then VE = 1.3 volts due to the 0.7-volt base – emitter voltage. If BT2 = 6 volts, then the peak to peak output swing will be from +1.3 volts to +6 volts, which will be (6 volts – 1.3 volts) peak to peak or 4.7 volts peak to peak. However, often the DC collector voltage is not centered to give maximum voltage swing without clipping prematurely on one-half of a sine wave cycle before the other half of the sine-wave. To maximize, you can set the DC collector voltage at the average of BT2 and VE, which is VC = (BT2 + VE)/2.

For this example:

VC = (6 volts + 1.3 volts)/2 or VC = 3.65 volts for "symmetrical" clipping

Amplifier's Emitter Partially AC Grounded via Series RE2 and CE2

In Figure 8-31, we have a little bit more going on because both emitter and collector terminals have AC output signals. Clipping occurs when either the collector voltage approaches the power supply voltage (e.g., BT2), or when the collector voltage matches the (same) voltage at the emitter. That is during the positive sine-wave cycle of the emitter and the negative cycle of the collector matches the voltage across the collector and emitter (e.g., VCE) is approximately 0 volt. Put in other words, the collector voltage equals (or matches) the emitter voltage.

For Figure 8-31 where RE2 is generally > r_e, the input signal range for "undistorted" output can be approximated when RE" || RE2 >> r_e as follows: Turn off the input signal and measure DC voltages for VC and VE. Let's call these DC voltages, VC_{dc} and VE_{dc}. The total collector voltage swinging downward toward the emitter's voltage is $VC_{dc+ac} = VC_{dc} + G\ Vin_{peak}$ and the total voltage at the emitter swinging up toward the collector voltage is $VE_{dc+ac} \sim VE_{dc} + Vin_{peak}$.

The input voltage that can be found by the collector voltage equals the emitter voltage:

$VC_{dc+ac} = VE_{dc+ac}$

$VC_{dc} + G\ Vin_{peak} = VE_{dc} + Vin_{peak}$

To find Vin_{peak}, this can be summarized as:

$Vin_{peak} = (VC_{dc} - VE_{dc})/(1 - G)$

A step-by-step calculation is now shown.

For example, BT2 = 6 volts, RL = 10KΩ, R3 = 40.2KΩ, R4 = 20KΩ so
VB = BT2 [R4/(R3 + R4)] = 6 volts [20K/(40.2K + 20K)] = 6 volts [20K/60.2K]
VB ~ 6 volts [2/6] or VB ~ 2 volts. IC ~ VE/RE" = (VB – 0.7 volt)/3900Ω or
IC ~ 1.3 v/3900Ω or IC = 0.00033 A. This makes VC_{dc} = 6 volts – IC(RL) or:

VC_{dc} = 2.7 volts. VE_{dc} = VB – 0.7 volts = 2 volts – 0.7 volt = 1.3 volts

VE_{dc} = 1.3 volts

We need to equate VC_{dc+ac} = VE_{dc+ac} to find the maximum peak voltage swing at clipping.

$VC_{dc} + G\ Vin_{peak} = VE_{dc} + Vin_{peak}$

$VC_{dc} + VE_{dc} = G\ Vin_{peak} + Vin_{peak} = (1 - G)\ Vin_{peak} = (VC_{dc} - VE_{dc})$ or
$Vin_{peak} = (VC_{dc} - VE_{dc})/(1 - G)$, to get a positive value in terms of peak input voltage:

$Vin_{peak} = (VC_{dc} - VE_{dc})/(1 - G)$

For this example:

VC_{dc} = 2.7 volts

VE_{dc} = 1.3 volts

G = –11.4

$Vin_{peak} = (VC_{dc} - VE_{dc})/(1 - G)$

Vin_{peak} = (2.7 v – 1.3 v)/(1 – –11.4) = 1.4 v/(12.4)

Vin_{peak} = 0.1129-volt peak input

The output peak voltage:

$Vout_{peak} = Vin_{peak} \times |G|$

G = –11.4 → |G| = 11.4

$Vout_{peak}$ = 0.1129-volt peak × 11.4

$Vout_{peak}$ = 1.287 volts peak

$Vout_{peak\ to\ peak}$ = 2.574 volts peak to peak

In practice both Vin_{peak} and $Vout_{peak}$ will be slightly smaller to avoid clipping or distortion.

Finding an Optimum Bias Point for Maximum Output Swing with Just an Emitter Resistor

With a common emitter amplifier with just a series emitter resistor (e.g., RE) and where the series emitter resistor $>> r_e = (0.026 \text{ v/IC})$, we can find an optimal bias voltage based on the collector load resistor, RL and series emitter resistor, RE. See Figure 8-35.

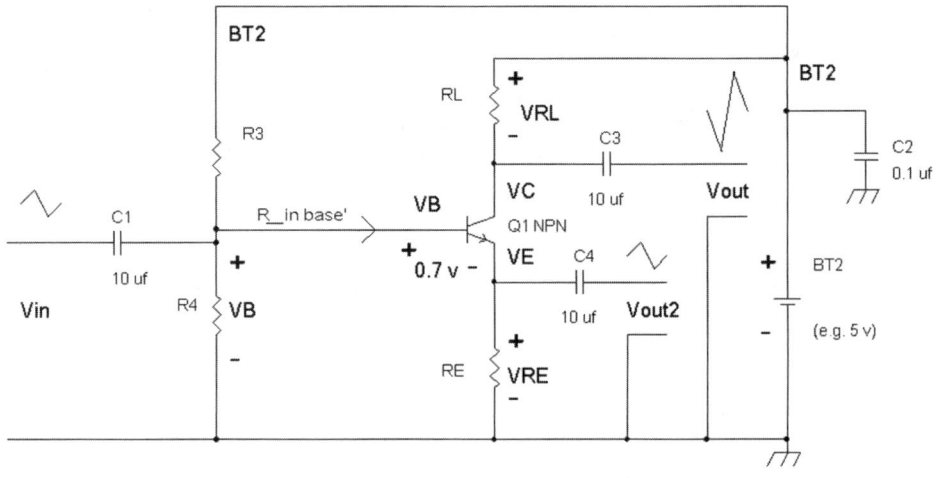

FIGURE 8-35   A common emitter amplifier with an emitter series resistor, RE.

To find an optimal biasing voltage, VB, given the values for RL and RE, we have:

$$VB = \frac{BT2}{2} \frac{1}{1+\frac{RL}{RE}} + 0.7 \text{ volt}$$

For example if we want to make a unity gain phase splitter amplifier where amplitudes of the AC signals are about equal via the collector and emitter terminals (Vout and Vout2), RL = RE.

NOTE: For a phase splitter amplifier circuit, output terminals Vout and Vout2 are generally loaded with equal resistance values of at least 10 x RL.

If BT2 = 12 volts, and RL = RE = 1KΩ, then:

$$VB = \frac{12v}{2} \frac{1}{1+\frac{1K}{1K}} + 0.7 \text{ v} = 6 \text{ v} \times \frac{1}{2} + 0.7 \text{ v}$$

VB = 3 v + 0.7 v or VB = 3.7 volts

This would mean VE = VB – 0.7 v or VE = 3 volts.

If there is a 3-volt drop across RE, then there should be a 3-volt drop across RL since RE = RL, given that the emitter current = collector current for β >>1. That is VRL = VRE.

Thus, VC = BT2 – VRL = BT2 – VRE = 12 v – 3 v or VC = 9 volts.

This makes sense because at VE = 3 volts, the maximum swing at the emitter is 3 volts ± 3 volts or 0 volt to 6 volts at the emitter. If there is 6 volts at the emitter for VE, then the collector voltage is VC = BT2 – VRE = 12 v – 6 v, or VC = 6 volts. This then satisfies the condition that maximum swing is when the collector and emitter voltages are equal.

In practice the collector-to-emitter voltage is rarely 0 volt but close, such as 0.2 volt. And this will reduce the maximum calculated swing by a slight amount. But this equation is fine for a first approximation.

$$VB = \frac{BT2}{2} \frac{1}{1+\frac{RL}{RE}} + 0.7 \text{ volt}$$

With RL = RE = 1KΩ, and BT2 = 12 volts, here are example resistor values for R3 = 39KΩ and R4 = 18KΩ. And we can confirm that VB is close to 3.7 volts.

VB = 12 v [R4/(R3 + R4)] = 12 v [18K/(39K + 18K)] = 12 v [18K/57K]

VB = 12 v [0.3158] or

VB = 3.789 volts ~ 3.7 volts

Because RE = 1KΩ and R3||R4 < 20RE → 39KΩ || 18KΩ < 20KΩ = 20RE, we can use a transistor of β ≥ 100 such as a 2N3904.

Summary

The amplifiers presented have limitations in terms of input amplitudes and output swing. Distortion can be a problem unless there is a series input resistor or a series emitter resistor that reduces gain and distortion. Be aware of the capacitors' capacitance values at the input, at the output, and especially connected to the emitter (e.g., CE and CE2), which will determine the amplifier's low-frequency performance.

Also, make sure electrolytic capacitors are biased correctly; otherwise, they will be reverse biased. Having electrolytic capacitors wired backwards results in leakage currents that will cause the expected DC bias points to shift up or down. You can use a voltmeter to confirm the correct polarity voltage across the electrolytic capacitors.

This concludes Chapter 8 concerning simple one-transistor amplifiers. In Chapter 9 we will explore some linear integrated circuits such as operational amplifiers and voltage regulators.

Analog Integrated Circuits Including Amplifiers and Voltage Regulators

In Chapter 8, we examined some simple discrete amplifier circuits. We noted that these amplifiers include characteristics of input resistance, output resistance, voltage gain, output swing, and distortion. In particular, these amplifiers have limitations of output swing when connected to a low-resistance load at Vout, and distortion with large output voltage swings.

Many of the limitations of discrete amplifiers can be reduced by using integrated circuit (IC) operational amplifiers. Also, IC amplifiers take less space and can provide lower current draw than discrete circuits. This chapter will explore some of the IC operational amplifier basic characteristics such as gain, drive capability, and power supply requirements.

In addition to IC amplifiers, we will look at voltage regulators. A voltage reference (e.g., "super Zener diode") with a power operational amplifier generally makes up a voltage regulator that provides a constant voltage even when the current load is large (e.g., > 100 mA).

Operational Amplifiers

Operational amplifiers, or "op amps," historically were used to solve operational math operations such as adding, subtracting, integrating (e.g., averaging), and other operations. When op amps are used with multiplying or nonlinear circuits, they can be used to find magnitudes of vectors (e.g., $c = \sqrt{a^2 + b^2}$), or they can be used to find an inverse function such as a log function where the original function is an exponential function.

161

In general, an op amp, when used as an amplifier has the following characteristics:

- A non-inverting input terminal.
- An inverting input terminal.
- An output terminal.
- A positive voltage power supply terminal.
- A negative voltage power supply terminal or a ground terminal.
- A negative feedback element (e.g., resistor or capacitor) that is connected to the output terminal and the inverting input terminal.
- When an op amp includes negative feedback, the voltage at the non-inverting input terminal is "equal" to the voltage at the inverting input terminal.
- The non-inverting input terminal should be connected in such a way that it provides a DC path. Generally, the non-inverting input terminal should be connected to a voltage source via a straight connection or via a resistor or inductor. Otherwise, the op amp will drift up or down in voltage until the output voltage is clipped to one of the power supplies or ground voltages.
- The inverting input terminal is connected via a resistor, inductor, and/or capacitor to the output terminal to form a negative feedback arrangement.

Operational amplifier U1 in Figure 9-1 shows an LM741 single package op amp with the non-inverting input (+) at pin 3, the inverting input (−) at pin 2, the output at pin 6 with the +V supply at pin 7 and the −V supply at pin 4. The other pins 1 and 5 are generally not used unless the user wants to adjust the input offset voltage to zero. Ideally, the op amp will have perfectly matched input transistors that result in zero volts offset at its input. Zero volts offset means if you ground both inputs, the output voltage is 0 volts. However, in practice the input offset voltage can be within ± 10 mV DC, which means if you ground the two inputs, the output pin will not be zero volts. Generally, input offset voltages are not a problem except for instrumentation and other measuring circuits where zero input offset voltage is required.

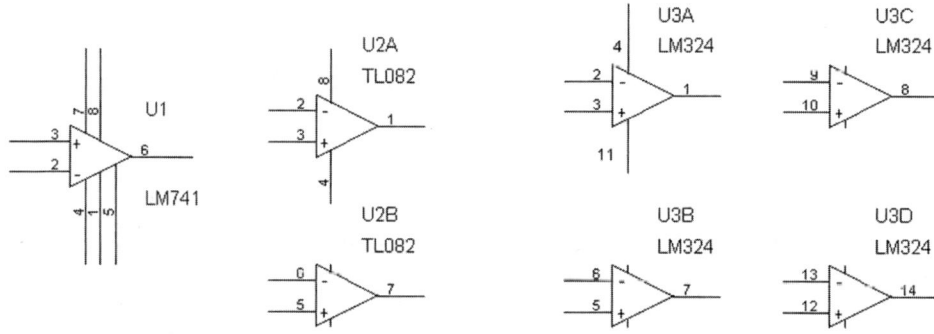

FIGURE 9-1   Op amp schematic symbols showing single, dual, and quad op amps in single packages of 8 pins (single and dual) and 14 pins (quad).

The LM741's pin 8 is not connected. Note that U1's power pins are 7 (+ v) and 4 (– v), with output pin 6, and input pins 2 (– input) and 3 (+ input). These pin assignments are standard for other single-package op amps such as TL081, TLC271, LF356, LF441, MC1456, and AD797.

U2A and U2B have standard pin outs for dual op amps in an 8-pin package. Here, pin 8 is the +V supply pin and pin 4 is the –V supply. *Note that the power pins are at the corners of the 8-pin package, and if the IC is installed backwards (e.g., the IC is turned around), the power pins 8 and 4 will receive a reverse voltage that will damage it.* If this happens, you should replace the IC and ensure that the IC's installation orientation is correct. For the first section, pins 2 and 3 are the inverting and non-inverting input pins, while pin 1 is the output. For the second section, pins 6 and 5 are the inverting and non-inverting input pins with output pin 7. Again, these pin outs are standard for dual op amps in an 8-pin package. Thus, dual 8-pin op amps such as the LM358, LM1458, LM4558, TL082, TLC272, LF353, LF412, ISL28208, ISL28218, NJM4560, and AD712 will have the same pin outs.

For a 14-pin package, op amp sections U3A, U3B, U3C, and U3D show standard pin outs for the inverting and non-inverting input pins and corresponding output pins. Pin 4 is the +V power pin, while pin 11 is the (negative) – V power pin. Again, for the most part these are standard power pins for a quad op amp in a 14-pin package. Op amps with this pin out are LM324, TLC274, TL084, LM348. But there are exceptions such as the MC3301 quad op amp with +V at pin 14, –V at pin 7. Therefore, if you are using a quad op amp, be sure to look up the data sheet.

NOTE: Again, if the IC is turned around, the power pins will be reversed and damage will be done to the quad op amp. You will have to replace the op amp and ensure that it is installed correctly.

Figure 9-2 shows dual inline packages (DIP) for through hole and surface mount 8-pin (SO-8, small outline 8-pin) and surface mount 14-pin (SO-14, small outline 14-pin) ICs.

FIGURE 9-2　8- and 14-pin DIP packages on the top row and 8- and 14-pin surface mount ICs on the bottom row.

We will now show a simple unity gain amplifier known as the voltage follower amplifier. See Figures 9-3 and 9-4. The voltage follower has close to zero ohms output resistance.

A voltage follower works on the negative feedback principle (Vout is connected to the inverting input). The signal voltage at the non-inverting input causes inverting input to have the same voltage. For example, in Figure 9-3 with U1, Vin1 is at pin 3 and at pin 2. But pin 6, the output pin (Vout1) is connected to pin 2 (inverting input). Thus, Vout1 = Vin1. With dual power supplies such as ± 5 volts, the voltage follower has a gain of 1 for DC and AC signals. If the input signal is a DC signal (e.g., for instrumentation), the gain is 1. For example, if Vin1 = +1.56 volt DC, then Vout1 = +1.56 volt DC. Using bipolar (plus and minus) power supplies allows a lower-frequency response down to DC because a coupling capacitor is not required. These circuits will amplify AC signals as well. So why is a voltage follower useful? If the input voltage has little capability to drive larger currents, such as driving a headphone or low-impedance load (e.g., < 1KΩ), using the voltage follower circuit allows for that. For example, if your input signal has an equivalent source resistance of 10KΩ, then only a small current (e.g., < 1 mA) will drive a 64Ω headphone, which will result in low volume. However, if you use an NJM4556 that can output up to 70 mA, driving the headphone will not be a problem.

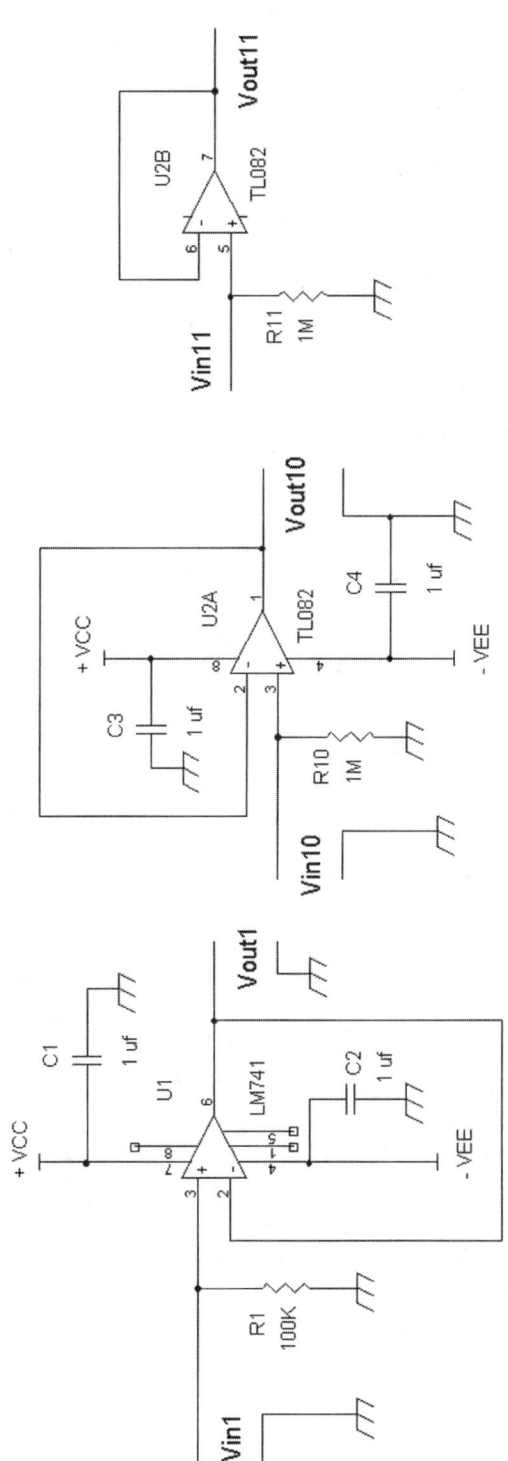

FIGURE 9-3 Voltage follower amplifiers with dual supplies with decoupling capacitors (C1, C2, C3, and C4) connected directly to the power pins to avoid oscillations from the op amps.

The input resistances for each of the op amp voltage follower circuits are R1, R10, and R11, which can be changed to virtually any practical value. Thus, the input resistances of the voltage follower circuits as shown in Figure 9-3 are R1 = 100KΩ, R10 = 1MΩ, and R11 = 1MΩ. Input resistors R1, R10, and R11 can have different value resistors but we sometimes have to keep in mind the "tiny" input bias currents of the inverting and non-inverting input terminals. Input bias currents can cause a noticeable DC shift at the output when the product of the input bias and input resistors exceed 10 mV or more. If you choose an op amp that has an FET (Field Effect Transistor) input stage (e.g., TL082, LF353, TLC272), the input bias current is negligible and input resistors up to 10MΩ can be used without creating offset voltage problems.

Figure 9-4 shows a voltage follower circuit when using a single positive supply.

When using a single power supply, generally, we need to create a half voltage supply to bias the input terminals and output terminal to half the supply voltage. The bias voltage, +Vb = +VCC [R10/(R10 + R11)]. Since R10 = R11, +Vb = +VCC [½]. Thus, pin 3 has the bias voltage +Vb combined with the AC portion of the Vin14. For example, if Vin14 is a 1-volt peak to peak sine wave and +VCC = 9 volts, then +Vb = 4.5 volts, and pins 3, 2, and 1 will have a composite signal of +4.5 volts DC + 1 volt peak to peak AC sinewave. The bias voltage of one half +VCC provides for the largest AC signal swing at the output (e.g., pin 1 or 7). In Figure 9-4, one of the first things to look for in single supply op amp circuits is to measure referenced to ground that the two input terminals and the output terminal are at about one-half the supply voltage.

Figure 9-4 also shows the polarity connections for polarized electrolytic capacitors C40, C50, and C60. Note the positive terminal is connected to the input and output terminals that have positive DC voltages of +VCC/2. Decoupling capacitor C30 is connected closely to pin 8, for example, within an inch to avoid parasitic oscillations from the op amp.

Also shown in Figure 9-4 is the second section of U2B that has its non-inverting input terminal pin 5 connected to U2A's output pin 1. We can do this because pin 1 has the needed bias voltage +Vb = +VCC/2, which "automatically" correctly DC biases pin 5 of U2B. Pin 7 then provides a second or extra output signal Vout15 via C60 essentially identical to Vout14. For example, if +VCC = 12 volts, then +Vb = 6 volts DC, and if U2 = NJM4556, then we can drive two 64Ω earphones separately via Vout14 and Vout 15.

The AC input resistance with C40 being an AC short circuit, is R10 ∥ R11, and with R10 = R11 = 200KΩ, the AC input resistance = 200KΩ ∥ 200KΩ = 100KΩ. Again, as a reminder R11 is connected to +VCC, which serves as an AC ground. If you are only using a negative supply, see Figure 9-5.

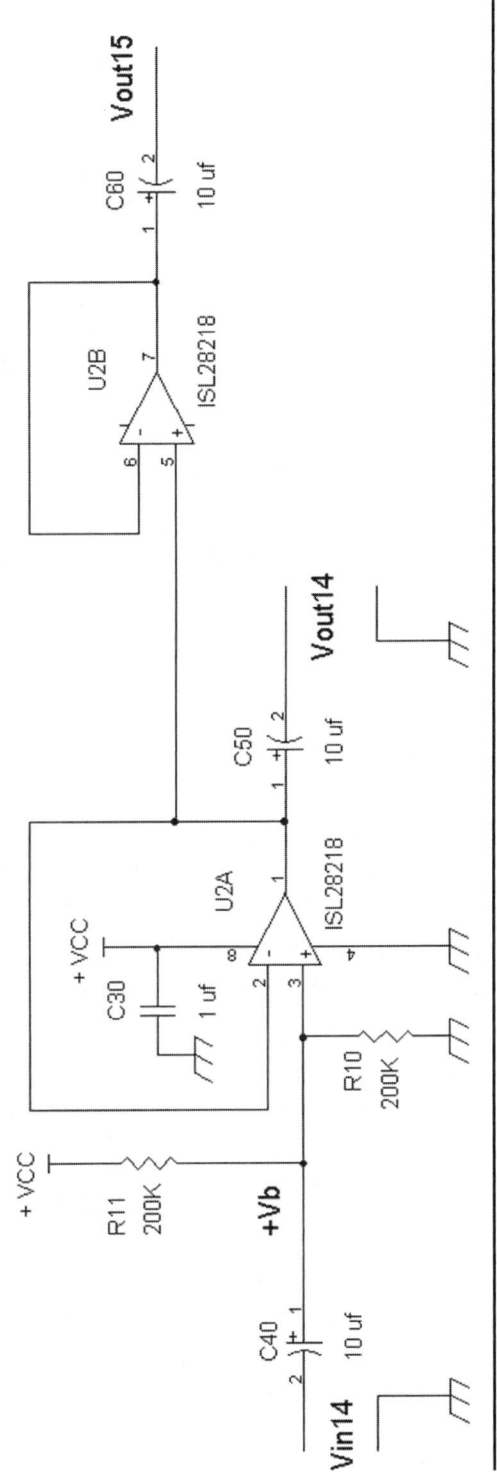

FIGURE 9-4 A voltage follower amplifier with a single positive voltage supply to work with AC signals.

167

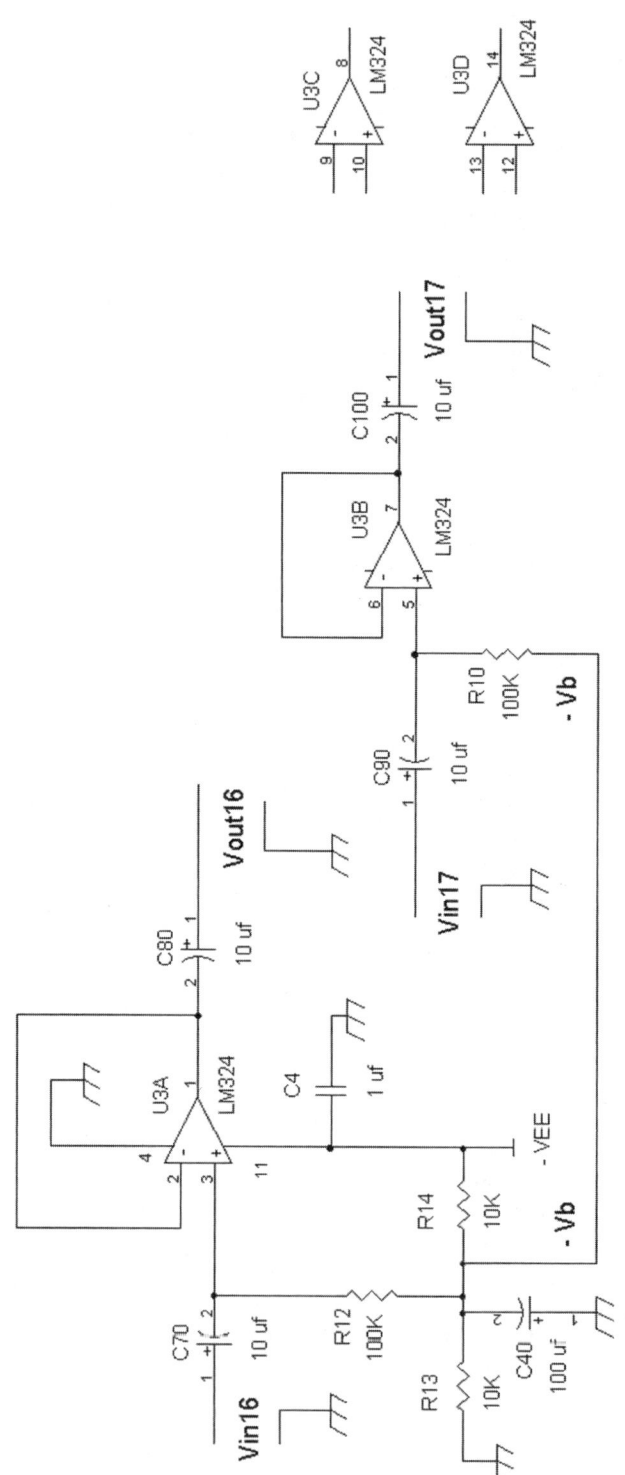

FIGURE 9-5 A voltage follower circuit with only a single negative power supply.

In this circuit, note that the +V power pin 4 is connected to ground. We need to bias the input terminals to a voltage, –Vb such that –Vb = –VEE/2 . For example, if –VEE = –5 volts, then –Vb = –2.5 volts DC.

Also, to avoid oscillations, decoupling capacitor C4 connects directly to pin 11.

A common –VEE/2 voltage source can be made with a simple voltage divider R13 and R14 and with a large value decoupling capacitor C40 as shown in Figure 9-5. C40 provides an AC short circuit to ground even at low frequencies such as 20 Hz, and it also provides an AC short circuit compared to the 100KΩ bias resistors R12 and R10. Any AC signal voltages coupled to R12 and R10 via input capacitors C70 and C90 at the input terminals of the op amps will be isolated from each other. This is because both lower connections of R12 and R10 go to C40, which is an AC short to ground. And because C40 is an AC short to ground, the AC input resistance to Vin16 is R12 (100KΩ), and the AC input resistance to Vin17 is R10 (100KΩ). Again, the (total) voltage at pins 3, 2, and 1 of U3A is –Vb plus the AC signal from Vin16. Likewise, the (total) voltage at pins 5, 6, and 7 of U3B is –Vb plus the AC signal from Vin17. Thus, the AC gain is 1. Note that –Vb can supply bias voltages via input bias resistors (e.g., 100KΩ) to the two other op amps U3C and U3D at pins 10 and 12.

In general, when making a voltage source from a resistive divider, you can scale the resistors from about 10 percent to 50 percent of the input resistors' values, and scale the decoupling capacitor (e.g., C40) to at least ten times the capacitance of the input capacitors (e.g., C70 and C90).

Again, note the polarity of the electrolytic capacitors (e.g., C70, C80, C90, and C100) where the (–) terminals are connected to the input and output terminals of the op amps. Also the decoupling capacitor C40 has its (–) terminal connected to the voltage divider circuit R13 and R14 because R14 is connected to a negative voltage, –VEE.

The voltage follower is a unity gain amplifier that can be characterized as an amplifier with almost infinite input resistance, Rin (not counting its input biasing resistor), and nearly zero output resistance, Ro. See Figure 9-6.

With plus and minus power supplies, the block diagram is valid for both AC and DC signals. The input resistance is just the input resistor RA that also serves to provide a DC bias voltage (e.g., 0 volts DC) to the non-inverting input terminal of U2A. A voltage follower circuit has close to 0Ω output resistance. However, all op amps include output current limiting usually in the tens of milliamps to protect the op amp from destruction should the output terminal (e.g., pin 1) be accidentally shorted to ground.

For a single power supply where the bias voltage is provided by a voltage divider, see Figure 9-7 for the block diagram.

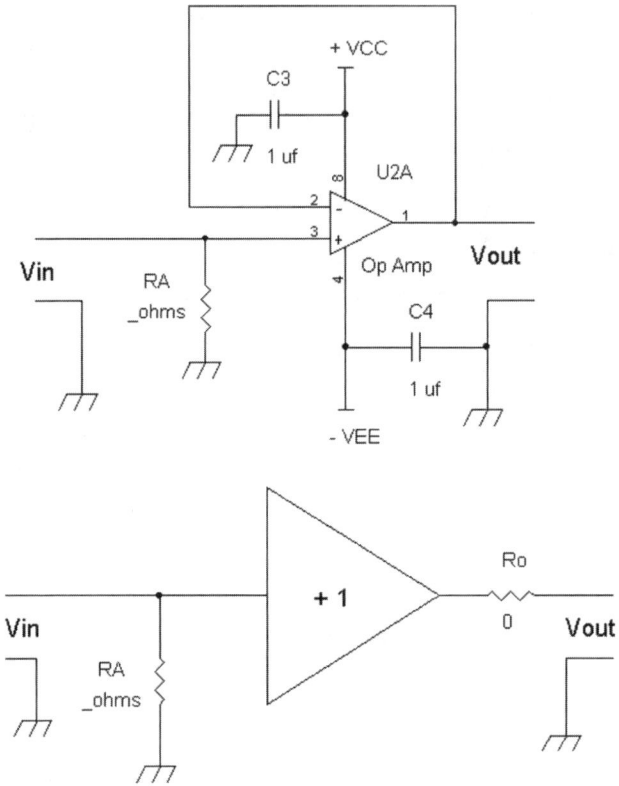

FIGURE 9-6   Model of a voltage follower with plus and minus supplies.

Because the circuit is AC coupled via input and output capacitors, C40 and C50, the block diagram refers only to AC signals for Vin and Vout. See Figure 9-7.

And as can be seen R11 is connected to the DC supply voltage +VCC that is an AC short circuit to ground. Recall that by definition a pure DC supply voltage does not include any AC voltage. Thus, the AC input resistance is R10 ∥ R11. When +Vb = +VCC/2, R10 = R11, so the input resistance = R10 ∥ R10 = (½) R10 or (½) R11.

The AC output impedance provided by C50 is nearly an AC short circuit and is very small when compared to the subsequent load resistance at Vout. For example, if the load resistance, RL, at Vout to ground is 2kΩ and you need the amplifier to have a good low frequency response down to 20 Hz, a good rule of thumb is to set the cut-off frequency to about one-half to one-tenth of the lowest frequency desired. In this example, we can set the cut-off frequency between 2 Hz and 10 Hz. The low-frequency cut-off frequency = $1/[2\pi(C50)RL]$ = $1/[2\pi(10\ \mu f)(2k\Omega)]$ = 7.96 Hz, which is within the range. However, electrolytic capacitors are relatively inexpensive and the cost of a 10 µf capacitor and a 33 µf capacitor is about the same. So, having C50 → 33 µf will result in a low-frequency cut-off frequency of:

(7.96 Hz)/3.3 = 2.4 Hz. The 3.3 factor comes in because 33 µf is 3.3 times 10 µf.

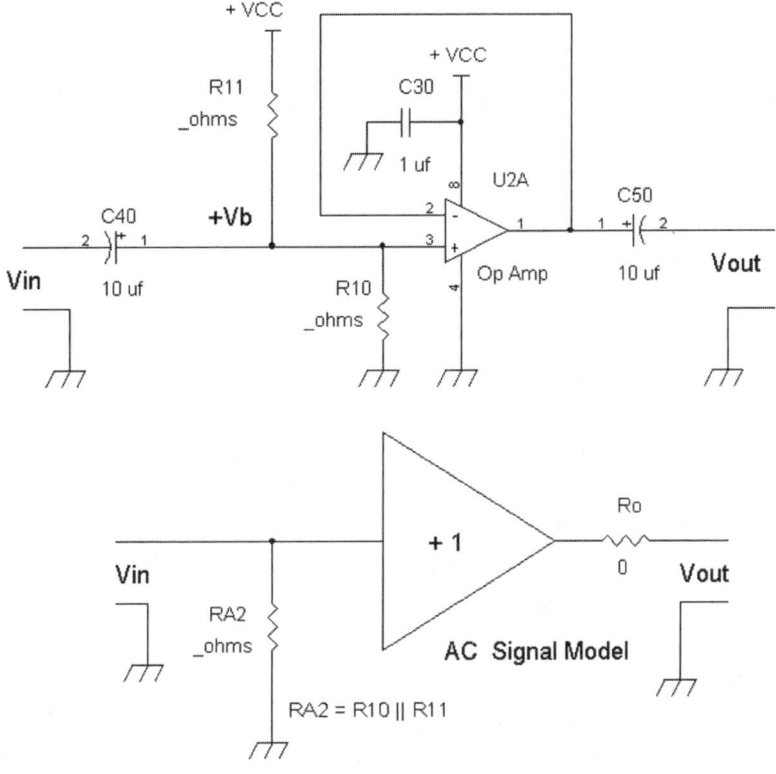

FIGURE 9-7   Block diagram of a single supply voltage follower circuit.

We will now look into some very important specifications to keep in mind when working with op amps. The voltage follower configuration will be used as an example.

Maximum Safe Power Supply Voltage

Older op amps such as the LM741, TL081, LF351 and LM318 have a 36-volt maximum across their V$^+$ and V$^-$ terminals. Some of the newer single supply Texas Instruments LinCMOS™ amplifiers such as the TLC251or TLC271 have about a 16-volt maximum. Yet some of the newer op amps (e.g., LMV358 and NJM2100) only have a 6- or 7-volt supply limit.

See Table 9-1 on dual op amps.

TABLE 9-1 Various Op Amps with Maximum Voltage Across Their V⁺ and V⁻ Supply Terminals

Op amp	Max Supply Voltage ±	Max Supply voltage + V	Pins V⁺ and V⁻
LMV358	2.5 volts	5.0 volts	8 and 4
TLC272	8.0 volts	16.0 volts	8 and 4
LM358	15.0 volts	30.0 volts	8 and 4
TL082	16.0 volts	32.0 volts	8 and 4
ISL28218	20.0 volts	40.0 volts	8 and 4
NE5532	21.0 volts	42.0 volts	8 and 4

Note: Max Supply voltage +V refers to a single supply

Although Table 9-1 shows the maximum supply voltages, to be on the safe side, the op amps are operated at ≤ 90 percent of maximum rated supply voltages. For example, the TLC272 would be operating with a +12-volt supply that is 4 volts below its maximum rating of 16 volts. Because not all volt meters (DVMs) are absolutely accurate, it's always good to be on the safe side by running the op amps with a good safety margin below the maximum rated voltage.

Minimum Power Supply Voltage

Not all op amp will work correctly with low-voltage power supplies, such as a single +5-volt supply or at ± 3 volts.

For example, the LM1458 or TL082 op amp will not work properly or with very poor performance (e.g., very low output swing < 2 volts p-p) when used with a +2.5-volt and −2.5-volt supply. A new op amp such as the LMV358 will work at these voltages, but a higher supply voltage such as ± 5 volts will damage it. See Table 9-2 for the manufacturer's minimum recommended supply voltages.

TABLE 9-2 Minimum Supply Voltages for Various Op Amps Via Data Sheets

Op amp	Minimum Supply ± Voltage	Minimum Single Supply Voltage
TLC272	1.5 volts	3.0 volts
LM358	1.5 volts	3.0 volts
ISL28218	1.5 volts	3.0 volts
OPA2134	2.5 volts	5.0 volts
OPA2604	4.5 volts	9.0 volts
LM1458	5.0 volts	10.0 volts

As an experiment using a voltage follower configuration (see Figure 9-6, U2A) Table 9-3 shows a summary of minimum supply voltages and their outcomes. The op amps are loaded into 10MΩ (essentially a no-load condition), and note that the results will be worse when the load resistance is 1KΩ.

TABLE 9-3 Experimental Results of Various Op Amps in Voltage Follower Configuration and with "No Load" at the Output

Op amp	Supply ± Voltage	Output Voltage
MC1458	± 2.5 volts	0.6 volt peak to peak
TL082	± 2.5 volts	1.4 volts peak to peak
LM4558	± 2.5 volts	1.8 volts peak to peak
LF353	± 2.5 volts	2.5 volts peak to peak
OPA2134	± 2.5 volts	3.0 volts peak to peak
ISL28218	± 2.5 volts	3.0 volts peak to peak
TLC272	± 2.5 volts	3.5 volts peak to peak
OPA2604	± 3.5 volts	2.0 volts peak to peak
LT1632 (Rail to Rail in/out)*	± 2.5 volts	5.0 volts peak to peak

Note that the LT1632 op amp can still amplify properly over the entire supply range for the input and output. This type of op amp is called a "rail to rail" op amp for both input signal range and output voltage swing.

There are basically two types of rail to rail op amps.

- An op amp that can swing the full supply voltage range, but a smaller input range that does not match the supply voltage(s). For example, for the ISL28218 op amp, if the power supply voltage is +12 volts, the output voltage swing is 0 volts to 12 volts, but the input voltage range would be 0 volts to about 10 volts.
- An op amp that can swing the full supply voltage range that also has an input voltage range that matches the full power supply. For example, if the supply voltage is 5 volts, then the LT1632 has an input range from 0 volts to 5 volts while the output can swing from 0 volts to 5 volts also.

Caution on Providing Supply Voltages

We have to choose carefully the op amp to suit the power supply voltage or the op amp may be damaged due to overvoltage. But we must be aware of the minimum operating voltage for the op amps, or the op amp will not amplify the signal properly.

If the incoming power supply has voltage fluctuations, then we need to ensure that there is no overvoltage to the op amps. One solution is to preprocess the raw voltage into a regulated voltage using a voltage regulator integrated circuit.

For example, if you buy an unregulated wall charger DC supply that states it is a 12-volt DC 1 amp supply, usually it will provide a higher voltage (e.g., 17 volts) under a low-current load, while delivering the specified 12 volts DC at 1 amp. For safe operation, you can choose an op amp with a higher maximum voltage rating like an NE5532 that can operate up to 42 volts. Or you can use a lower-voltage op amp such as the TLC272 and regulate the "raw" voltage (e.g., a 12-volt to 17-volt range) by using an 8-volt regulator chip such as an LM7808 to provide a constant 8.0 volts to the TLC272 op amp whose maximum supply voltage is 16 volts.

Alternatively, buy a regulated wall supply, which is usually a switching power supply.

To reiterate, when powering an op amp, it is best to use some type of voltage regulation. This way, if there are spikes in the AC power line, your op amps and other circuits (e.g., logic chips) will stay protected from damage.

Voltage regulator circuits will be discussed further at the end of this chapter.

Maximum Output Current

In general, most op amps have push pull class AB output stages that provide about 10 mA of positive and negative current output. With this type of output current capability, the op amp can drive resistive load $\geq 1000\Omega$ with ≥ 10 volts output. But this 10-mA output current can be smaller if lower supply voltages such as a single 5-volt or 7-volt supply are used.

For example, a TLC272 amplifier with a $+12$ volts supply in a voltage follower configuration loaded into a 220Ω resistor outputs about 5.6 volts peak to peak, whereas with a $+6$-volt supply, the output (before distortion) drops to 2.25 volts peak to peak.

NOTE: Trying to pull more than the maximum current results in clipping or limiting to the output voltage.

If these types of op amps that have typically a 10-mA maximum output current are used to drive low-impedance headphones such as a 16Ω or 32Ω version, signal clipping will occur.

For example, a 32Ω load is driven directly with a TLC272 with $+12$-volt supply with a voltage follower circuit, as shown in Figure 9-8.

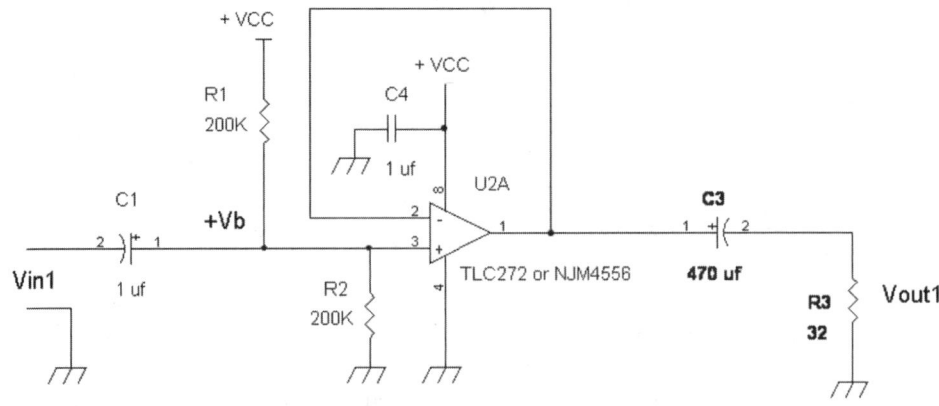

FIGURE 9-8 Voltage follower circuit.

The voltage follower circuit in Figure 9-8 having a +12 volt supply and a 32Ω resistor connected to the output produces a clipped (distorted) sine wave signal as shown in Figure 9-9.

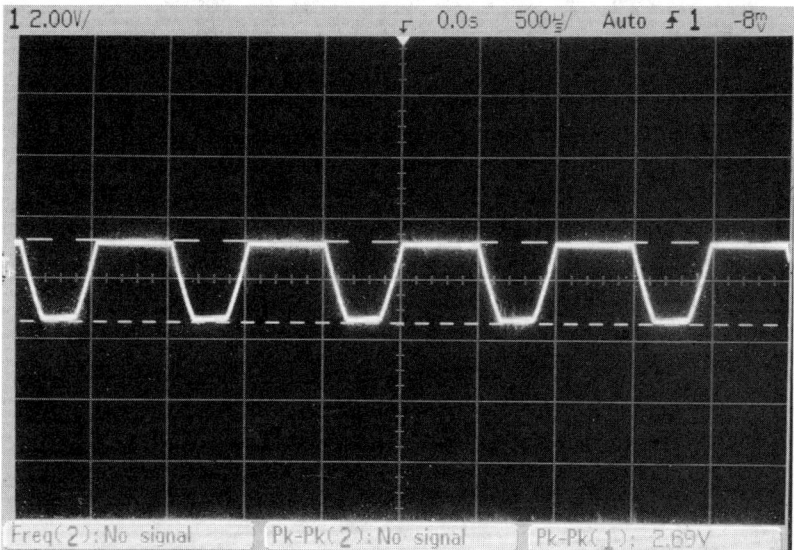

FIGURE 9-9 Waveform output with TLC272 with 5 volts peak to peak input at 1000Hz, Vin1, and 2.69 volts peak to peak output at Vout1 as measured by the dashed horizontal lines/cursors.

There are amplifiers that can supply 20 mA to 30 mA such as LM833 or NE5532 dual op amps. These can drive lower-impedance loads such as 600Ω.

A high-current op amp such as the NJM4556 can output up to ± 70 mA. This allows driving low-impedance headphones (e.g., 16Ω to 64Ω stereo headphones), or even drive small loudspeakers with a 32Ω or 50Ω impedance. See Figure 9-10, where a 32Ω load is driven with an NJM4556 using 12-volt supply with a voltage follower circuit like the one shown in Figure 9-10. The NJM4556 peak output current into 32Ω is then the peak voltage, 2.5 volts peak/32Ω or 78 mA peak.

Note that 5 volts peak to peak is equal to 2.5 volts peak.

But there is a word of warning on some op amps with very limited output current such as the LM358 and LM324 that are commonly used in DIY (do-it-yourself) projects. These op amps have very poor output drive capability. Figure 9-11 shows an LM358 with a single 5-volt supply circuit driving a 10MΩ load (see Figure 9-8, where U2A = LM358 and VCC = 5 volts, and R3 = 10MΩ or 2000Ω). With a 3000 Hz 0.5-volt peak to peak signal at Vin, see Figure 9-11 for the output waveform.

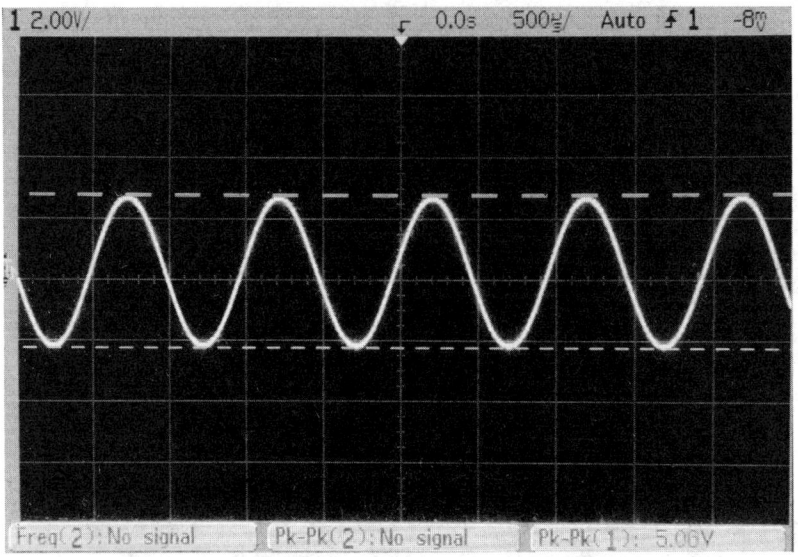

FIGURE 9-10   Waveform from NJM4556 delivers a clean sine wave with 5 volts peak to peak into 32Ω. Note that 5 volts peak to peak is the same as 2.5 volts peak.

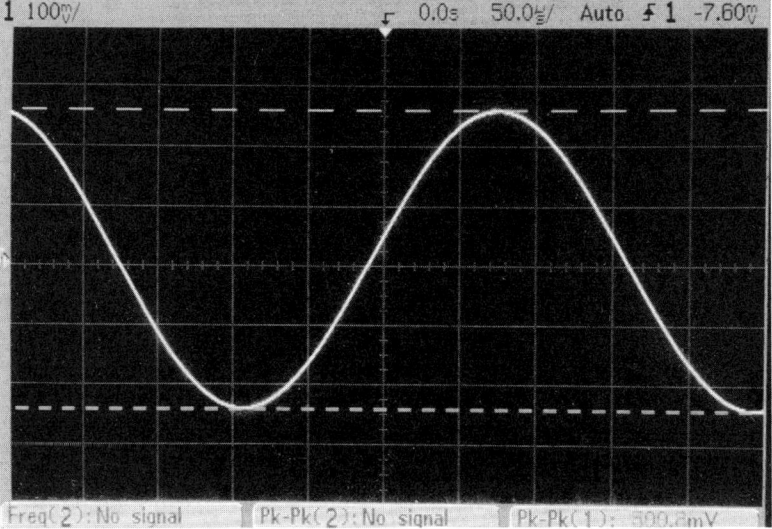

FIGURE 9-11   An LM358 voltage follower circuit's output signal into a 10MΩ resistor with a clean 500-mV peak to peak sine wave.

However, if the LM358 voltage follower drives a 2000Ω (for R3 in Figure 9-8) instead of the previous 10MΩ, the waveform distorts quite badly. See Figure 9-12.

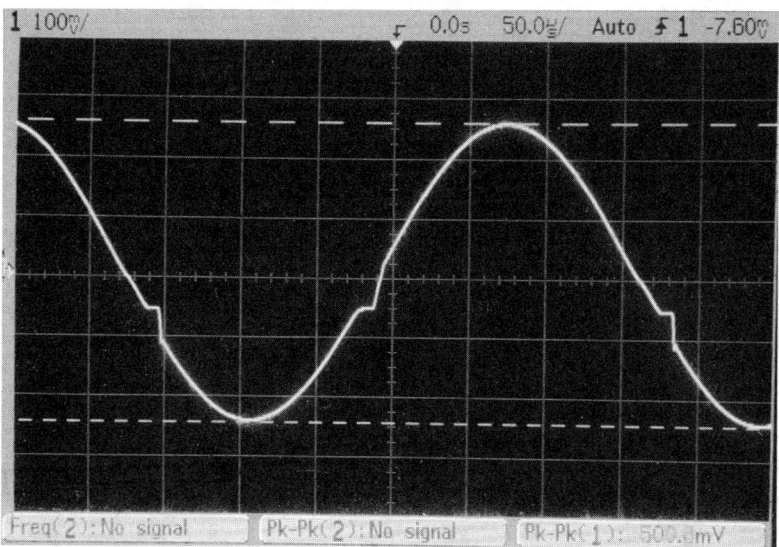

FIGURE 9-12   The LM358 voltage follower's distorted output waveform (~ 500 mV peak to peak) when driving a 2000Ω resistor (R3). Note the discontinuities near the "zero" crossing of the sine wave. This is known as crossover distortion.

One way to reduce or eliminate the crossover distortion in an LM358 or LM324 is to connect a pull-up or pull-down resistor (e.g., 1KΩ or 1.5KΩ) from the output terminal to either the +V or –V terminal or ground terminal. See Figures 9-13 and 9-14.

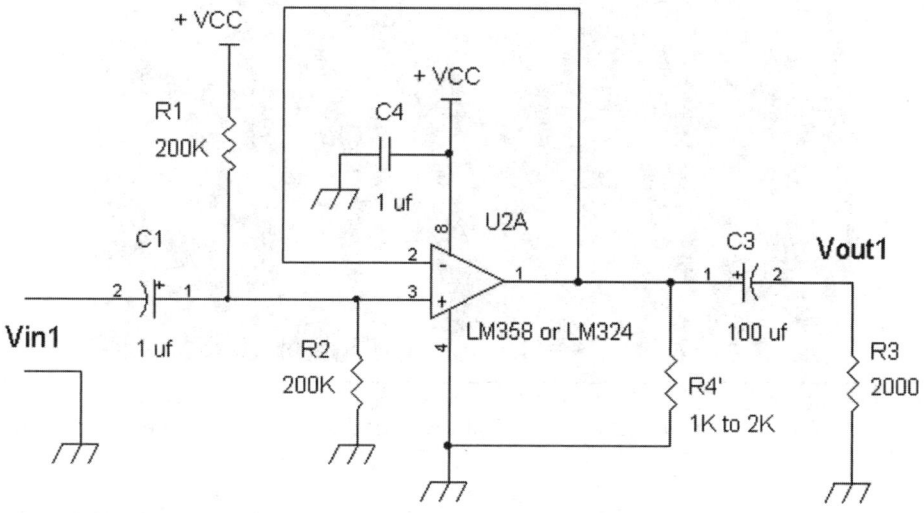

FIGURE 9-13   A pull-down resistor R4' is connected from the output pin 1 to the –V or ground terminal pin 4 of the LM358 op amp to reduce or eliminate crossover distortion. Alternatively, you can use a pull-up resistor tied to pin 8 or +VCC as shown in Figure 9-14.

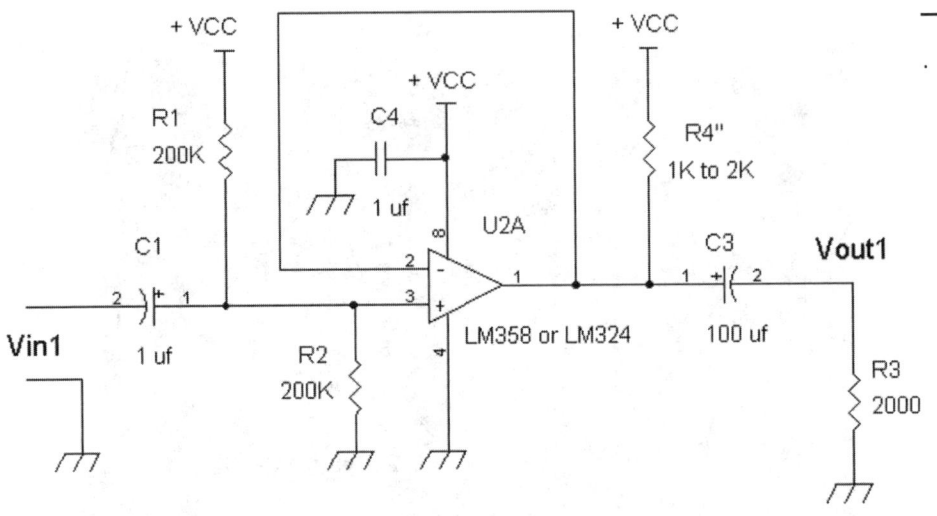

FIGURE 9-14 A pull-up resistor, R4", is connected to the LM358's output pin 1 and pin 8 (+VCC) to reduce or eliminate crossover distortion.

The modifications via R4' and R4" in Figures 9-13 and 9-14 result in a cleaner looking sine wave as shown in Figure 9-15.

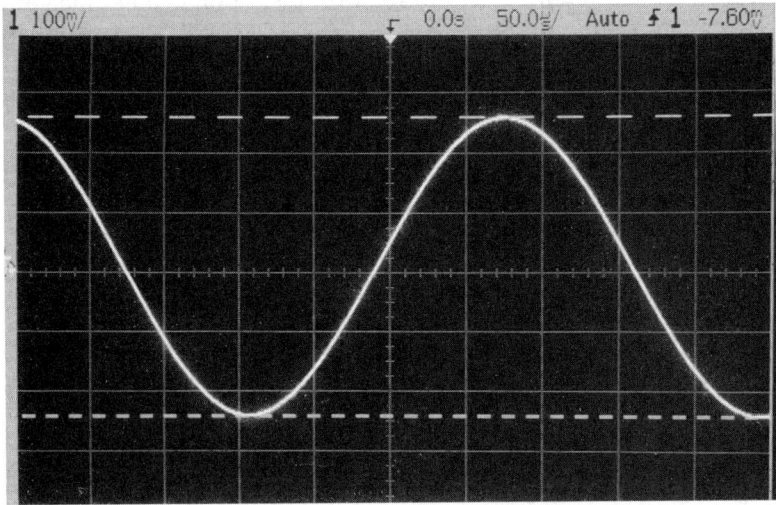

FIGURE 9-15 LM358 with a pull-down or a pull-up resistor that eliminates crossover distortion on the sine wave. The cursors with dashed lines across are also displayed for 500 mV peak to peak.

Output Voltage Range

When we supply an op amp with ± supply voltages such as ± 5 volts, we would expect that the output can swing approximately ± 5 volts. In practice, the op amp generally has a more restricted range. For example, in a TLC272 op amp, there is about a 1-volt loss from the V^+ supply, but the op amp can swing almost to V^-. If we use ± 5 volts, the output range is about (5 volts − 1 volt) to −5 volts, or +4 volts to −5 volts. Of course, the output range will further reduce when loaded into a resistor that draws at least 1 mA of current.

Other op amps can have more loss such as the LM 741 that loses about 2 volts on both V^+ and V^- supplies. For example, with a ± 9-volt power supply, the output swing is about +7 volt to −7 volts.

Special op amps that are called "rail to rail" usually provide just that—output swings that match the supply voltages. For example, an ISL28208 will provide nearly the same output swing voltages as the power supply within about 100 mV. However, because the ISL28218 op amp does not have rail to rail performance at the input terminals, when it is configured as a voltage follower the output swing will be restricted to the input range, which loses about a volt or two. For example, in Table 9-3 the ISL28218 has a swing of 3 volts with a 5-volt power supply when it is configured as a voltage follower. However, we will see later that we can take full advantage of the rail to rail output voltage range even if the input voltage range is restricted by using the amplifier as an inverting gain amplifier or as a non-inverting gain amplifier where the gain is generally ≥ 2.

Input Signal Range

Even though an op amp can swing rail to rail, that does not ensure that the input range is rail to rail. That is, if the power supply voltages are ± 5 volts, most op amps have a *common mode* input range that is less than or within the power supply voltages. For example, the CA3130 (Renesas) op amp can swing rail to rail at its output, but the input has a range of V^- to V^+ −2 volts. For a 5-volt supply where V^- = ground or 0 volts and V^+ = +5 volts, the input range is 0 volts to V^+ −2 or 0 volts to 3 volts. So, in a voltage follower circuit even if the input signal is from 0 to 5 volts, the output will be 0 to 3 volts.

See Table 9-4 for some op amps with various output swing and input ranges for a single positive voltage supply, +V, that is referenced to ground.

TABLE 9-4   Some Op Amps' Input and Output Voltage Ranges for a Single Supply

Op Amp	Input Voltage Range	Output Voltage Range
LM358 or LM324	−0.2 v to +V − 1.5 v	+0.6 v to +V − 1.5 v
LM1458 or LM741	+3.0 v to +V − 0.5 v	+1.7 v to +V − 1.0 v
NE5532 or NE5534	+2.0 v to +V − 3.0 v	+1.5 v to +V − 1.5 v
TLC272	0 v to +V − 1.0 v	0 v to +V − 1.0 v
ISL28218	0 v to +V − 2.0 v	0 v to +V
LT1632	0 v to +V	0 v to +V

Let's take examples where $+V = 6.0$ volts or four 1.5-volt cells connected in series.

See Table 9-5.

TABLE 9-5   Examples of Voltage Ranges Where the Single Supply Voltage Is + 6.0 Volts

Op Amp	Input Voltage Range	Output Voltage Range
LM358 or LM324	−0.2 v to +4.5 v	+0.6 v to +4.5 v
LM1458 or LM741	+3.0 v to +5.5 v	+1.7 v to +5.0 v
NE5532 or NE5534	+2.0 v to +3.0 v	+1.5 v to +4.5 v
TLC272	0 v to + 5.0 v	0 v to + 5.0 v
ISL28218	0 v to + 4.0 v	0 to + 6.0 v
LT1632	0 v to + 6.0 v	0 v to + 6.0 v

A main takeaway from Tables 9-4 and 9-5 is that there are only certain types of op amps that will work well with a single 5 volt or 6 volt supply such as the LM358, ISL28218, TLC272, and LT1632. The traditional op amps used with higher supply voltages such as the LM741, LM1458, and NE5532 do not generally work at low voltages.

Also, if you notice carefully, op amps that state single supply capability have an input voltage range of at least down to ground, 0 volts, or the −V supply pin of the op amps. See the input ranges of the LM358, ISL28218, TLC272, and LT1632. The ability to "sense" input signal near ground or the −V supply of the op amp can be very handy. See Figure 9-16, where we can make a constant current circuit out of an LM358.

Q1's collector provides a constant DC current to light-emitting diode LED1 by regulating the emitter voltage, VE to 0.1 volt DC. By having emitter voltage stable at 0.1 volt the emitter current, IE, then is 0.1 volt/5Ω or IE = 20 mA. Since IE = IC, the collector current, IC = 20 mA. Since the collector is supplying current to the LED, the LED current is also 20 mA.

So, the question is how do we get a regulated 0.1 volt at the emitter? The answer is by negative feedback, which is done in the following manner:

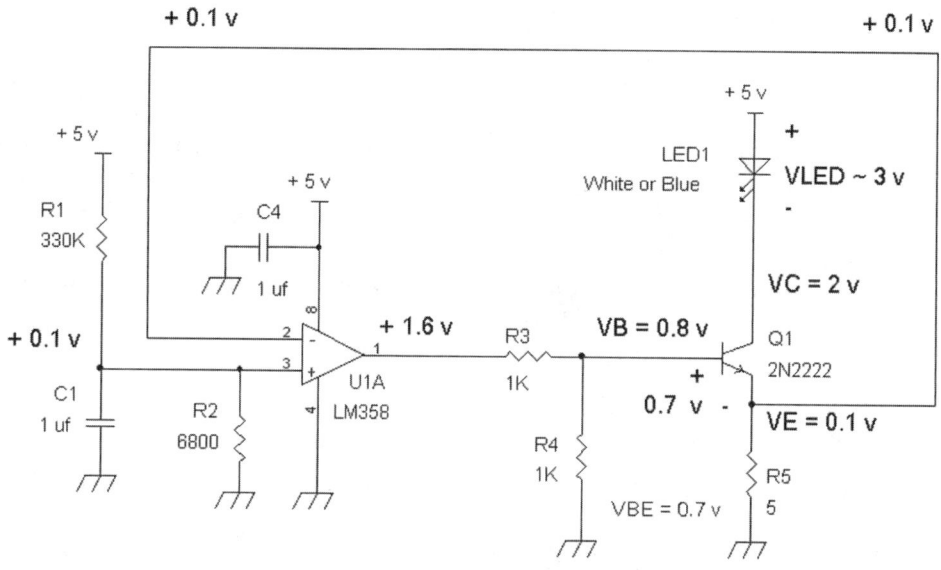

FIGURE 9-16   A feedback current source circuit with nearly 0 volt sensing at the input terminals of the LM358 op amp at pins 3 and 2.

Via voltage divider R1 and R2 with the 5-volt supply voltage, we have 0.1 volt at the (+) input of the op amp U1A. Because there is negative feedback, the (+) input terminal's voltage at pin 3 forces the (−) input terminal's voltage at pin 2 to be the same voltage. Thus, pin 2 has 0.1 volt. If we follow where pin 2 is connected to, pin 2 is connected to Q1's emitter. To force the 0.1 volt at the emitter, the base voltage will have to be high enough to cause conduction in the transistor via forward biasing the base-emitter junction. The base-emitter voltage is VBE ~ 0.7 volt. So, the base voltage referenced to ground has to be VB = VBE + VE or VB = 0.7 volt + 0.1 volt. Thus, VB ~ 0.8 volt. By observing that if we make the approximation that the base current is small, then the divide-by-two voltage divider via R3 and R4 will cause the drive voltage into R3 to be twice that of VB. Therefore, the output pin 1 has to be twice VB or 2 × 0.8 volt. Thus, we get 1.6 volts at pin 1, which is within a "comfortable voltage output range" for the op amp. Because this is a negative feedback circuit, if the transistor's VBE changes with temperature to a lower or higher voltage than 0.7 volts, the output pin 1 will readjust automatically to ensure that there is always 0.1 volt at Q1's emitter terminal.

NOTE: The voltage divider via R1 and R2 provides the 0.1 volt via 5 volts [R2/(R1 + R2)] or 5 volts [6800/(330,000 + 6800)] = 5 volts [0.02019] = 0.10 volt.

Non-Inverting Gain Amplifiers

Previously, we discussed about voltage follower circuits that have a voltage gain of 1. They have very high input resistance set by the input bias resistor to the (+) input, and a very low output resistance that is capable of driving loads typically in the 2KΩ range or lower. The voltage follower preserves the signal's phase from input to output, whereas in Chapter 8, we saw that a common emitter amplifier's output inverts the phase of the input signal.

Op amps can be configured to have the advantages of the voltage follower while providing voltage gains ≥ 1. See Figure 9-17.

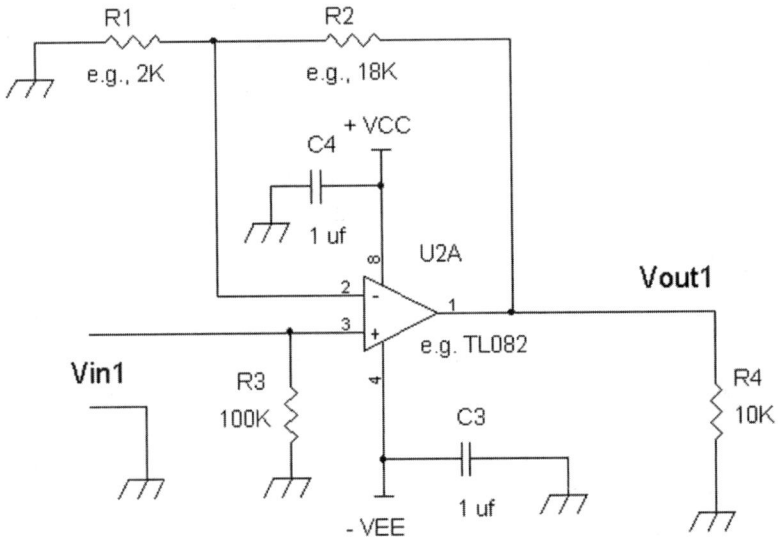

FIGURE 9-17   An example non-inverting amplifier with plus and minus power supplies.

The amplifier shown in Figure 9-17 is directly coupled to the input signal source, Vin1, and its output is directly coupled to a load resistor, R4. Load resistor R4 is just shown as an illustration and can be omitted. In essence this amplifier will amplify not only AC signals, but also DC signals.

Note that an input bias resistor, R3, is required so that even if the input signal Vin1 is disconnected, the op amp will be properly biased at the input terminal. Although R3 is shown as 100KΩ, it can be almost any value within reason. That is, it does not make sense to have R3 = 1Ω because that will load down the input signal source, Vin1, and drain excessive current, or worse, cause the input signal source to be damaged. Typical input bias resistances for R3 are in the range of 100Ω to 10KΩ when using op amps with bipolar input stages such as the NE5532 or LM741, etc. For op amps that have field effect transistor (FET, MOSFET, or JFET) input stages such as the TLC272, TL082, and OPA2134, the input bias resistor R3 can be as high as 10MΩ.

If in doubt, you can find a variety of field effect transistor input op amps by a web search or by searching your op amp at www.mouser.com or www.digikey.com and specifying an input bias current < 2 nA (< 0.002 μA or < 2000 pA) at the 25-degree Celsius/centigrade rating. For example, the TL082 has a maximum input bias current 200 pA rating at 25 degrees C.

A general rule of thumb is if the input bias current multiplied by the input resistance (R3) or the feedback resistor (R2) is less than 10 mV, then you are OK. For example, if we used an LM1458 op amp, with a worst case 500 nA input bias current, then we want 500 nA × R3 < 10 mV or R3 <10 mV/500 nA, R3 < 20KΩ. This is just a general rule of thumb, and if you are working in measuring or instrumentation circuits, then it may be best to choose a low offset voltage (V_{os}) field effect transistor op amp such as the OPA2134 or LF412.

For a non-inverting gain amplifier, the gain is:

$$A_v = Vout1/Vin1 = (R1 + R2)/R1$$

For example, in the circuit, R1 = 2KΩ and R2 = 18KΩ:

$$A_v = (R1 + R2)/R1 = (2KΩ + 18KΩ)/2KΩ = 20KΩ/2KΩ = 10$$

$$A_v = 10$$

For non-inverting gain op amp circuits, the gain is $A_v = (R1 + R2)/R1$ and the output resistance at Vout1 → 0Ω. That is, we normally do not have to worry about having the gain reduced if we load the output of the op amp with any range of resistance values as long as the op amp's output stage can deliver sufficient current. But be aware that normally the output current limit is in the ± 10 mA range.

Now let's intuitively understand how gain equation:

$$A_v = Vout1/Vin1 = (R1 + R2)/R1 \text{ is derived. (See Figure 9-17.)}$$

Recall the voltage at the (+) input terminal and the (−) input terminal is the same. In a negative feedback system, the (+) input terminal (e.g., U2A pin 3) determines the voltages at the output and (−) input terminal.

As a reminder note that you have to ensure that the (+) input always has a DC path to a voltage source. If you leave the (+) input unconnected or coupled to a capacitor, your circuit will not work. Now let's go back to how this circuit works.

We see that R1 and R2 form a voltage divider for the output voltage, Vout1. That is: Vout1[R1/(R1 + R2)] = voltage at the (−) input terminal (e.g., U2A pin 2 in Figure 9-17). But we know that the (−) input terminal's voltage = (+) input terminal's volts, which is Vin1.

Therefore, Vout1[R1/(R1 + R2)] = Vin1, which leads to:

$$[Vout1/Vin1][R1/(R1 + R2)] = 1$$

and if we divide by [R1/(R1 + R2)] on both sides of the equation:

$$[Vout1/Vin1] = 1/[R1/(R1 + R2)] = (R1 + R2)/R1$$

$$Vout1/Vin1 = (R1 + R2)/R1$$

This confirms the previously stated gain equation for Av for a non-inverting gain amplifier. What's happening inside the op amp is that it is providing a voltage at Vout1 such that when the voltage is attenuated via R1 and R2, the voltage at the (−) input terminal matches the voltage at the (+) input terminal, which is Vin1. This is how a negative feedback system works. So even if you try to load down the op amp by lowering the resistance of R4, the op amp will compensate to deliver a voltage at Vout1 that ensures that the voltage at the (−) input terminal matches the voltage at the (+) input terminal.

But also note that the op amp has output current limitations, and we have to be mindful how we set the gain of the amplifier while working within the output current capability of the op amp. The feedback resistors R1 and R2 also load the op amp at Vout1. For example, we will get into trouble if we set R1 = 2Ω and R2 = 18Ω. The op amp's output terminal will see the (R1 + R2) as an "extra" load resistor, which will be 2Ω + 18Ω or 20Ω. If the op amp has a 10-mA maximum output current, then the maximum output voltage (e.g., at U2A pin1) will be ± 20Ω × 10 mA or ± 200 mV, which is a very limited output swing.

Often it is better to protect your input terminals from external DC bias voltages riding on top of AC signals. There are some devices that will provide an output signal with a DC bias voltage, which will cause a directly coupled amplifier such as in Figure 9-17 to overload. For example, if +VCC = +6 volts and −VEE = −6 volts and Vin1 = (1 volt DC + V_{AC_signal}), and A_v = 10, then:

$$Vout1 = 10 \times Vin1$$

$$Vout1 = 10 \times (1 \text{ volt DC} + V_{AC\_signal}) \text{ or}$$

$$Vout1 = 10 \text{ volts DC} + 10 \ V_{AC\_signal}$$

However, since the power supplies are ± 6 volts, there is no way that the op amp U2A can provide 10 volts DC at Vout1. Therefore, the amplifier will be clipped to the positive rail at Vout1 and there will be no AC voltage at Vout1. We can fix this "problem" by AC coupling via an input capacitor. See Figure 9-18.

By using an input coupling capacitor, C2, any DC bias voltage riding on top of the AC signal from Vin1 is eliminated at the (+) input terminal (e.g., U2A pin 3). Thus, the DC voltage at pin 3 is 0 volts, which leads pin 2, the (−) input, to have also 0 volts DC, and this further leads to the DC voltage at pin 1 is also 0 volts. Capacitor C1 at 1μf is used to provide an AC short circuit to ground such that the AC gain is still (R1 + R2)/R1 or 10 in this example with R1 = 20KΩ and R2 = 180KΩ. If you notice carefully, the resistors R1 and R2 have been scaled up tenfold from Figure 9-17. Why? The reason is if we want the same low-frequency response with R1 = 2KΩ and R2 = 18KΩ, C1 would have to be 10 μf. In practice, it is harder to find a 10 μf non-polarized capacitor when compared to obtaining a 1 μf capacitor.

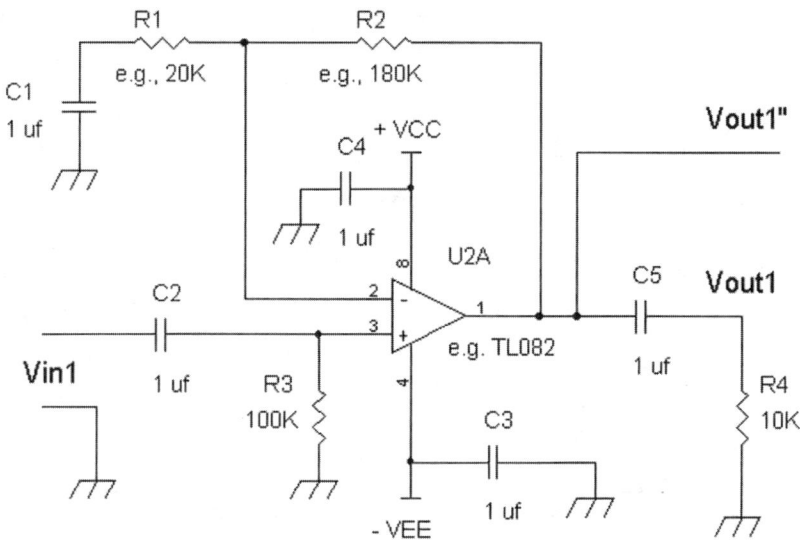

FIGURE 9-18   Using an input capacitor, C2, to block any DC voltage from Vin1 while passing AC signals.

Capacitor C1 also is used to prevent amplifying the op amp's internal offset voltage, V_{os}, which is usually in the 5 mV to 10 mV range for general purpose amplifiers. If C1 were replaced with a 0Ω resistor, then we have a DC gain of 10, which will result in a DC voltage at the output of $10 \times V_{os}$. For example, if $V_{os} = \pm 10$ mV, then the output DC voltage can be $\pm 10 \times 10$ mV or ± 100 mV DC. With C1, the offset voltage is just voltage "followed" and the DC gain with C1 is 1. So, with C1 installed, Vout1's DC voltage is just Vos or typically ≤ 10 mV DC.

Note that the built-in offset voltage in the op amp can be a positive or negative DC voltage. An ideal op amp will have $V_{os} = 0$ volt. You can buy precision op amps with $V_{os} \leq 1.0$ mV such as the LF412**A** (note the LF412A has a tighter V_{os} specification ≤ 1.0 mV than the LF412 without the "A" designation, which has $V_{os} \leq 3.0$ mV).

Figure 9-18 also shows that you can AC couple the output signal via C5 with Vout1 or have the output signal DC coupled via Vout1". An AC coupled output is useful for blocking any DC voltage that may go into the output terminal of U2A. This is important in some cases where the output terminal of an amplifier is driving another circuit that has a DC bias voltage. If you use the Vout1" terminal, the DC bias voltage from the other circuit will be "shorted" DC wise to ground or zero volts that may cause this other circuit to have the wrong DC bias and not operate correctly. Thus, to play it safe, use a DC blocking output capacitor, C5, and also remove the load resistor R4.

If you are using a single power supply non-inverting amplifier, see Figure 9-19.

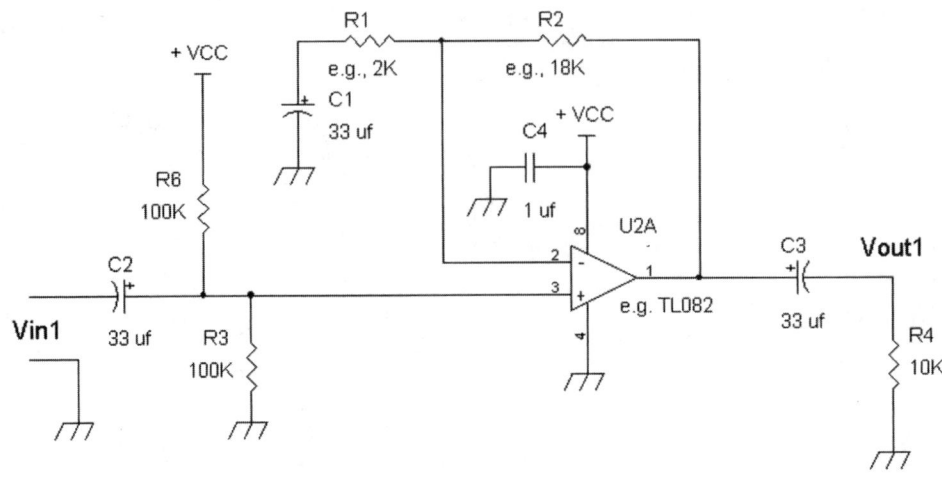

FIGURE 9-19 A single power supply non-inverting gain amplifier.

The AC gain of this amplifier is again $A_v = (R1 + R2)/R1$. To maximize on voltage swing, the DC bias voltage at the (+) input is set at +VCC/2. For example, if +VCC = 9 volts, the DC bias voltage is set to 4.5 volts at pin 3 of U2A via voltage divider R3 and R6, where R3 = R6. The AC input resistance is R3 ‖ R6, which is 100KΩ ‖ 100KΩ or 50KΩ. Note that resistor R6 is connected to +VCC, which is an AC ground.

Because we are using a single supply, we can use polarized electrolytic capacitors (C1, C2, and C3) that allow for higher capacitance values at low cost.

To analyze this circuit DC-wise, see Figure 9-20 where we remove the capacitors C1, C2, and C3. Since these capacitors block DC currents when charged to their final value, they are open circuits in terms of DC analysis.

By removing the capacitors, a simplified circuit is shown in Figure 9-20 on the right. On the left side circuit, since R1 and R4 are disconnected on one end, they can be removed, which is shown on the right side circuit.

Here we see only a voltage divider circuit with R3 and R6, along with R2, the feedback resistor. However, because the op amp's (−) input terminal has a high input resistance, there is no current flowing through R2. This implies there is no voltage across R2 since the voltage across R2 is equal to VR2 = I2 × R2, where I2 = current flowing into the (−) input terminal. Since I2 = 0 amp flowing into the (−) input:

$$VR2 = 0 \text{ amp} \times R2 = 0 \text{ volt}$$

Therefore, we can simplify further this right side circuit. See Figure 9-21.

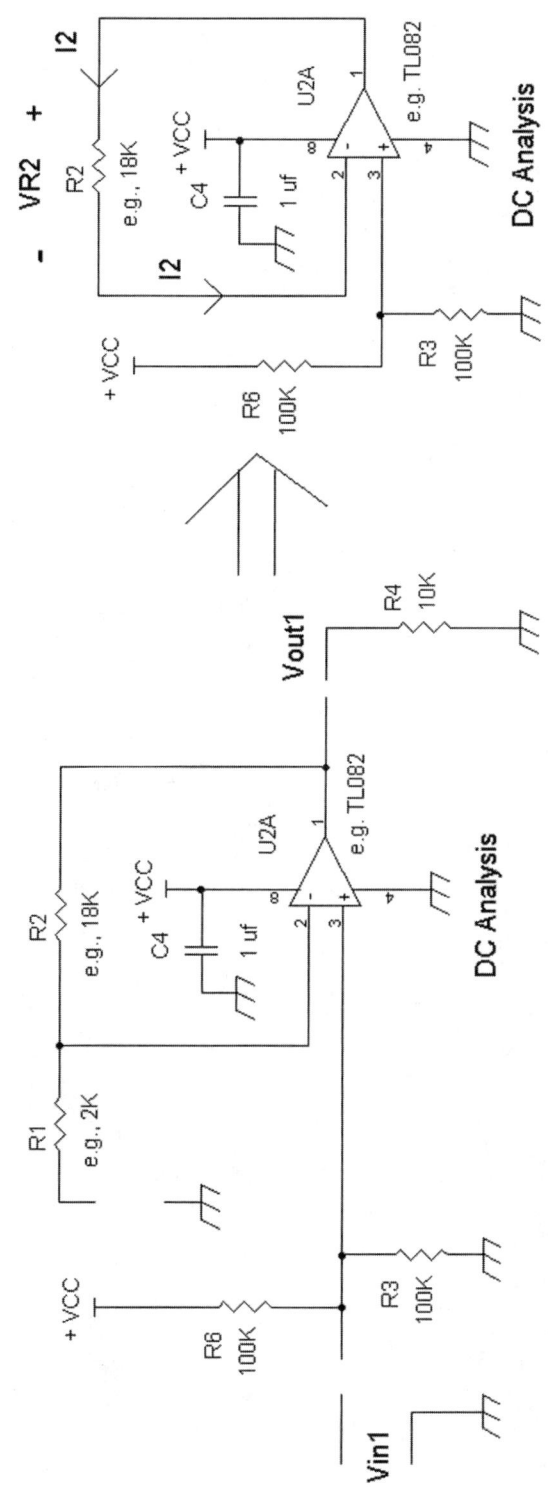

FIGURE 9-20 Removing C1, C2, and C3 simplifies the circuit for DC analysis.

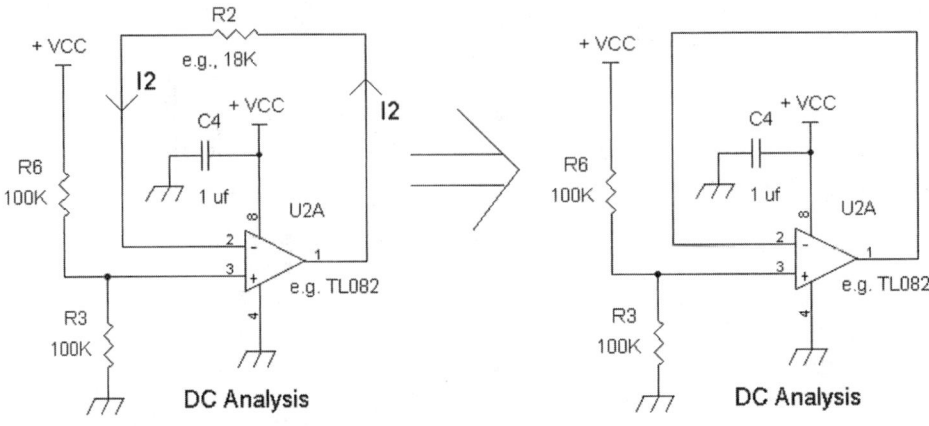

FIGURE 9-21 With zero current I2 flowing through feedback resistor R2, the circuit reduces to a simple voltage follower amplifier for DC analysis.

Because there is 0 volts across R2, this means the voltage at output pin 1 is the same voltage at pin 2, the (–) input. This then means we can replace R2 with 0Ω or a wire since we will get the same result, the voltage at pin 1 is the same as the voltage at pin 2. Thus, the DC voltage at pin1 is just +VCC/2 since the voltage at the (+) input pin 3 is the same as the voltage at (–) input pin 2 when negative feedback is applied.

Figure 9-22 shows a model of a non-inverting gain amplifier for AC signals.

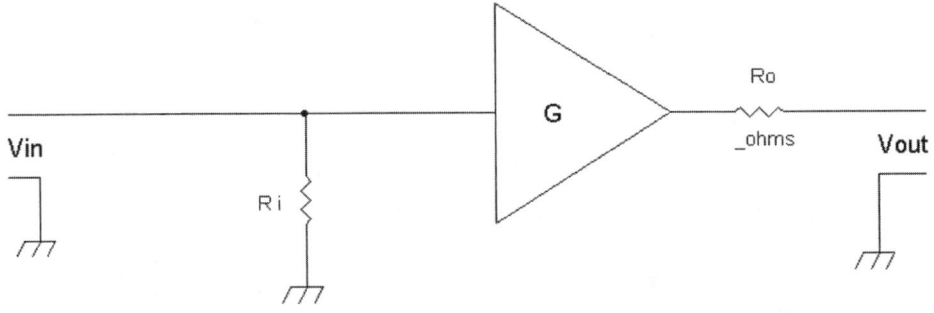

FIGURE 9-22 A generalized model for the non-inverting amplifiers, with $G = A_v$.

From Figure 9-22 we have:

Ro ~ 0Ω

NOTE: Keep in mind that op amps generally deliver about ± 10 mA into a load.

Ri = R3 in Figures 9-17 and 9-18, Ri = R3 ‖ R6 in Figures 9-19 and 9-20. Typically, Ri ≥ 10KΩ.

$G = A_v = [(R1 + R2)/R1]$ for Figures 9-17, 9-18, 9-19, and 9-20.

Also, be sure to choose resistance values for the feedback network R1 and R2 such that it does not load the op amp's output current excessively. Determine the maximum signal output and divide by (R1 + R2) to determine the current draw from the op amp. Generally, this current draw should be less than 50 percent of the op amp's maximum output current.

For example, in Figure 9-7, if the maximum current output is 10 mA and you are using ± 12 supplies, for $A_v = 5$ you can use R2 = 4020Ω and R1 = 1000Ω. Note $V_{supply} = 12$ volts.

Worst case, the load current due to the feedback network will be:

$$V_{supply}/(R1 + R2)$$

For this example, it will be 12 volts/5020Ω. This translates to about 2.4 mA, which is less than the 10-mA output limit. So, this is OK.

On the other hand if you choose R2 = 120Ω and R1 = 30Ω, the current draw will "want to be" 12 volts/(120Ω + 30Ω) = 80 mA.

This will cause the amplifier to clip at the output, and the waveform will clip at $V_{output\ max} \le (R1 + R2) \times 10$ mA or ≤ 150Ω × 10 mA or ≤ 1.5 volts peak.

If a non-inverting gain amplifier circuit is generally set for a gain or $A_v \ge 2$ with an op amp that has rail to rail output voltage but does not have rail to rail performance at the input terminal, then this circuit can deliver rail to rail at the output while keeping within the input voltage range for proper operation. The reason is that by having a gain of ≥ 2, the input voltage range no longer has to go from rail to rail, but instead ≤ 50 percent of the output voltage range. For example, suppose a single supply amplifier shown in Figure 9-17 has an input range of +VCC – 1.2 volts to –VEE, and the power supplies are +VCC = +2.5 volts and –2.5 volts. This means the input terminal must be in the range of –2.5 volts to (+2.5 volts – 1.2 volts), which is between –2.5 volts to +1.3 volts. Now suppose the gain is +2 by having R1 = R2 = 10KΩ. At the zero-signal condition, the output voltage is 0 volts.

If the input signal is between –1.25 volts and +1.25 volts (e.g., a sine wave input signal whose peak amplitude is 1.25 volts), then the output will be twice that due to setting $A_v = 2$. Therefore, the output voltage will swing from –2.5 volts to +2.5 volts. And note that the input voltage signal range of –1.25 volts and +1.25 volts is within the input range of the op amp at –2.5 volts to +1.3 volts.

Troubleshooting Non-Inverting Op Amp Circuits via an Oscilloscope, Signal Generator, and DVM

Here are some tips on basic debugging techniques:

1. Make sure that the power supply terminals are connected correctly and use a voltmeter to confirm proper power supply voltages at the power pins. Be sure to have the op amp's data or spec sheet handy.

2.   Make sure there is at least a ≥ 0.1 µf decoupling capacitor wired or soldered closely to each power pin (e.g., plus supply pin 8 and negative supply pin 4 of an LM1458) and ground. For example, see the 1 µf power supply decoupling capacitors C4 and C3 in Figure 9-17. For a single supply circuit, you need only one decoupling capacitor since the other power pin is usually grounded. See Figure 9-19 where positive supply pin 8 of U2A is decoupled with 1 µf capacitor C4 to ground, while its negative power pin 4 is connected to ground. Without power supply decoupling capacitors, op amps can oscillate or give noisy output signals on their own.

3.   Confirm that the input DC voltages are within the range as specified by the data sheet. For example, an LM358 at +5 volts power supply has an input voltage range from about 0 volts to 3.5 volts. Use a voltmeter to confirm that the non-inverting input terminal (e.g., pin 3 or 5 in an LM358) is within the input voltage range.

4.   Measure the voltage with a DVM across the non-inverting input pin and the inverting input pin of the op amp. Confirm that the voltage is ~ 0 volt. If the voltage is not zero such as 0.5 volts DC, then the op amp circuit is not operating properly.

5.   For op amp circuits with positive and negative supplies, verify that the DC output voltage is close to zero. Verify that the non-inverting input terminal is connected via a wire or resistor (or a coil) to a DC potential such as ground or a bias voltage. With an ohm meter (DVM) and with the power supplies turned off, confirm that there is some resistance from the non-inverting input terminal to ground to a DC voltage line. For example, in Figure 9-18, the non-inverting input pin 3 is connected to ground via a 100K ohm resistor (R3) to establish a 0-volt DC bias.

6.   For op amp circuits with positive and negative supplies, if you are getting unexpected DC voltages at the output pin, then with the input signal source disconnected, try temporarily connecting the non-inverting input terminal to ground via a resistor (e.g., 10KΩ or lower) and re-measure the DC output voltage. If it is now correct, look for a bad bias resistor or bad connection in the non-inverting input terminal area. If connecting the non-inverting input terminal to ground does not help, then temporarily connect a wire or low-value resistor (e.g., 100Ω) from the output to inverting input terminal to form a voltage follower. If you now get a DC voltage ~ 0 volts, then recheck the feedback resistor network for the correct values. But it may be easier to just use a DVM and measure the resistance of the feedback network (e.g., R1 and R2 in Figure 9-7) with the power turned off.

7.   If after #1 to #6, you are not getting anywhere, replace the op amp, but make sure the original op amp could operate at the power supply voltages. For example, if the power supply provides ± 12 volts DC and the circuit does not seem to work, verify that the op amp is not a lower voltage version such as an

LMV358 (e.g., +5.5 volts maximum) or a TLC272 where a supply voltage of more than ± 8 volts will cause damage.

8. If the DC bias conditions are correct, but the AC signals are distorted or do not have correct amplitudes given the gain setting [(R1 + R2)/R1], then first check the resistance values of the feedback resistors. Also make sure the feedback network resistance value of (R1 + R2) $\geq$ 500Ω is not low enough to cause current limiting at the output that will cause clipped waveforms.

9. If the low-frequency (e.g., 20 Hz) response is poor—such as less than 50 percent of amplitude when compared to the signal amplitude at a mid-band frequency (e.g., 1000 Hz)—check the input and output coupling capacitors. If they measure correctly, check the frequency response at the non-inverting input pin (e.g., pin 3 or 5 in a dual op amp). If that's OK, check the frequency response at the output pin of the op amp such as pin 1 or 7 of a dual op amp. If there is a loss in low-frequency response, see if there is an AC coupling capacitor in the feedback network, such as C1 in Figure 9-18 or C1 in Figure 9-19. If you suspect there is a problem like the capacitance is not high enough, parallel a larger capacitance across C1, such as 2.2 µf capacitor. If the capacitor is ceramic, check the maximum working voltage. For example, if you use a 1-µf, 6.3-volt capacitor and the DC voltage across it is +5 volts, chances are that the true capacitance will be half of that, or 0.5 µf, which will cause a poorer low-frequency response.

10. Also, if electrolytic capacitors are used, measure with a volt meter (DVM) that the DC voltage across the capacitor matches the polarity markings of the capacitor. *If you measure a reverse bias voltage across an electrolytic capacitor, turn off the power immediately, replace the capacitor and install a new capacitor with the correct polarity.*

11. Another troubleshooting technique is to confirm that the feedback decoupling capacitor is an AC short circuit. Apply a sine wave signal between 200 Hz and 500 Hz at the input and measure the AC voltage across the C1/R1 junction (e.g., of Figure 9-19). There should be a very small AC voltage there when compared to the AC signal at the non-inverting input pin (e.g., pin 3 or 5 in a dual op amp).

12. Also check if the circuit was for a dual op amp but a single op amp was installed instead, or vice versa. Both single and dual op amps have 8 pins total and some pins are common, such as the –V supply pin 4, non-inverting input pin 3, and inverting input pin 2; however, the other pins 1, 5, 6, 7, and 8 are not the same for a single op amp and dual op amp.

This concludes an introduction to non-inverting gain op amp circuits. Next, let's take a look at inverting gain amplifiers.

Inverting Gain Amplifiers

We now turn to inverting gain op amp circuits where they produce an inverted waveform at the output when compared to the input signal. See Figure 9-23.

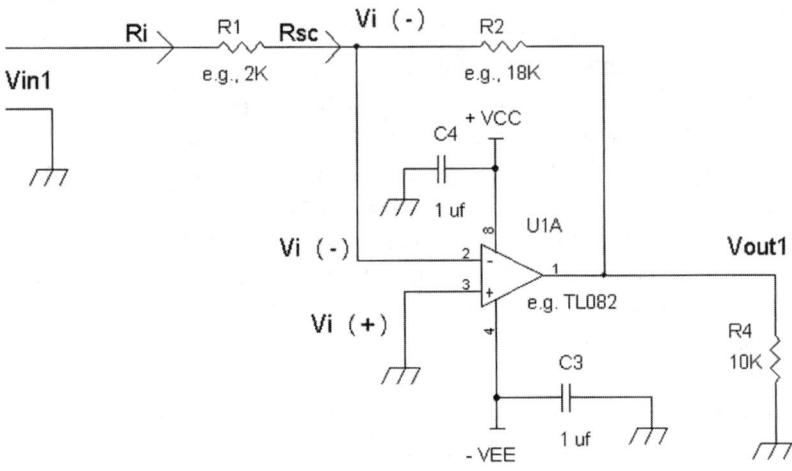

FIGURE 9-23　An inverting gain op amp circuit with plus and minus power supplies.

> ### Side Note: How Does the Inverting Gain Amplifier Work, Really?
>
> An inverting gain op amp circuit also still relies on the voltage applied to the (+) input terminal that defines how the output signal will react to match the (–) input's voltage to be the same as the (+) input's voltage. For a simple example, let R1 = R2 in Figure 9-23. If Vin1 = +1 volt DC, what voltage must Vout1 be such that at the junction of R1 and R2 that is connected to the (–) input matches the voltage at the (+) input terminal? We see that the (+) input terminal is grounded or at 0 volts. R1 and R2 form a voltage summing circuit for Vin1 and Vout1. If the resistors R1 and R2 are equal in value (e.g., R1 = R2 = 2KΩ) and we have + 1 volt DC into R1 via Vin1, it makes sense that the output voltage at Vout1 should be – 1 volt DC. This way at the junction of R1 and R2 we will get 0 volts. The output terminal (e.g., pin 1) is adjusting its voltage such that when summed via R2, the voltage at pin 2, the (–) input terminal, is zero volts. This output voltage adjustment via R2 provides the "virtual" short circuit voltage across the (–) input and (+) input terminals, with a voltage applied to input resistor R1.

The gain A_v = Vout1/Vin1 = –R2/R1. So, if we keep the same feedback resistors as in the circuit of Figure 9-17, that is, R1 = 2KΩ and R2 = 18KΩ, we find that A_v = –18K/2K or A_v = –9. Note the magnitude of the gain for the inverting amplifier: –R2/R1 is different from the non-inverting gain amplifier's formula of

(R1 + R2)/R1 = 1 + R2/R1. The difference between the magnitudes of the two gains of non-inverting amplifier and inverting amplifier is just 1. For example, the non-inverting gain amplifier will have a gain of 10 with R1 = 20KΩ and R2 = 180KΩ, while the inverting gain amplifier will have a gain of –9.

The input resistance is just R1. So why is this? The non-inverting input is connected to ground; its voltage is 0 volts or Vi (+) = 0 volts. Since in a negative feedback amplifier the voltage at the inverting input terminal is the "same" as the voltage in the non-inverting input terminal, we can deduce that the voltage at the inverting input terminal Vi (–) is also 0 volts. Thus, R1 on the right side connection to pin 2 of U1A is "virtually grounded." Therefore, referenced to ground, the input resistance is R1. In other words, if we look at an equivalent resistor referenced to ground from the inverting input node Vi (–), its resistance is Rsc, where Rsc is a short circuit to ground, equivalently. Rsc is a very low equivalent resistor because there is a large magnitude open loop gain inside the op amp. It is very similar to the feedback resistor in a one-transistor amplifier from Chapter 8 where we found that with a high-gain amplifier the input resistance looking into the base with the collector-to-base resistor is very low.

Because of R2, the feedback resistor, Rsc = R2/(1 + a), where "a" is the open loop gain. For example, if we have an NE5532 amplifier whose open loop gain is ~ 20,000 at 2 kHz, then the Rsc = R2/(1 + 20,000) ~ R2/20,000. Thus, if R2 = 180KΩ, Rsc ~ 180KΩ/20,000 or Rsc ~ 9Ω.

Since R1 = 20KΩ and Rsc = 9Ω << 20KΩ, the 9Ω Rsc resistor equivalently shorts to ground the right side lead of R1 at Vi (–).

In terms of the output resistance at Vout1, the op amp with the negative feedback resistor R2 has an output resistance close to zero ohms. The amplifier in Figure 9-23 amplifies DC and AC voltages at Vin1. For example, if Vin1 = 1.0 v DC, then Vout1 = –(R2/R1) + 1 volt DC. For example, if R2 = 180KΩ and R1 = 20KΩ, the output voltage will be: Vout1 = –(180K/20K) × 1 volt DC or Vout = –9 volts DC. Note that the supply voltage should be at least a few volts higher than 9 volts, such as ± 12-volt supplies for +VCC and – VEE.

To amplify only AC signals, use AC coupling capacitors C1 for the input, and C5 for the output (see Figure 9-24). Alternatively, you can directly couple your output signal via Vout1". Note the low frequency cut-off is f_c = 1/(2πR1C1) at the input when driven by the low source resistance, Rs, signal such that Rs << R1. For example, many generators have 50Ω source resistances that are << R1 = 20KΩ. For output Vout1, the cut-off frequency is f_c = 1/(2πR4C5).

NOTE: Because the inverting gain op amp circuit has the (+) input tied to a fixed DC voltage within its common mode voltage range, rail to rail output is achieved with any op amp specified with rail to rail output. That is the input range of the op amp does not have to include rail to rail performance. (See Figure 9-24.)

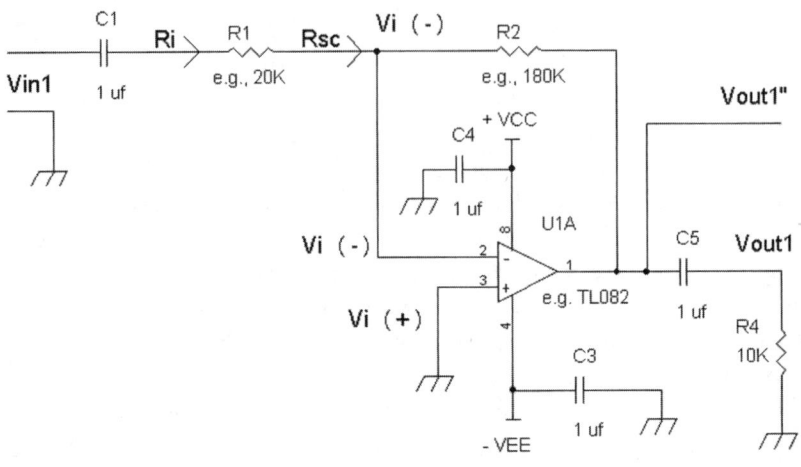

FIGURE 9-24 An inverting gain op amp circuit with AC signal coupling capacitors C1 and C5. Input resistance ~ R1 when C1 is an AC short circuit.

Note that Figures 9-24 and 9-25 look similar with the input resistors R1 and R1" with R2 and R_BC. The difference is the op amp U1A has very high gain via many transistors inside, and a very low output resistance compared to the amplifier in Figure 9-25.

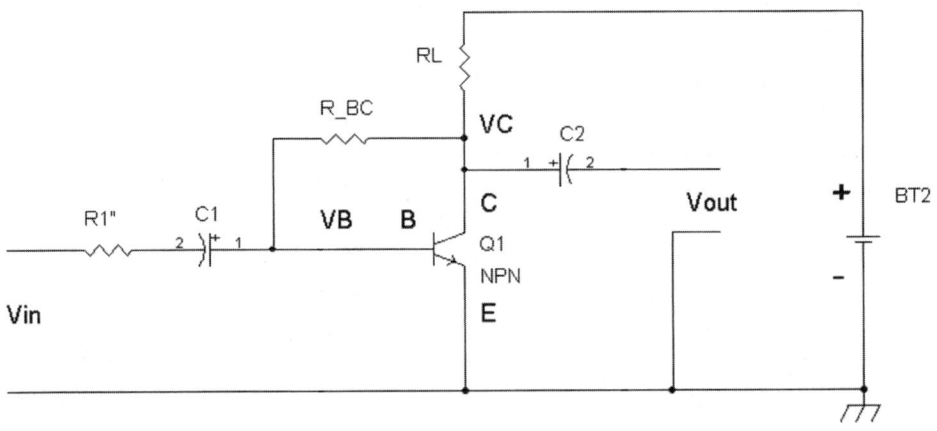

FIGURE 9-25 A one-transistor amplifier from Chapter 8 that resembles an inverting gain op amp circuit for small voltage gains A_v ~ –R_BC/R1", where R_BC is the feedback resistor. Input resistance ~ R1" when C1 is an AC short circuit.

One can think of Figure 9-25 as simplified inverting gain amplifier. It does have one or two advantages over an op amp. For one advantage, even with typical transistors such as the 2N3904, it can deliver very wide bandwidth. For example, if you want to make a cheap inverting gain of –2 amplifier at 1.5 volts supply, you can set RL = 470Ω, R_BC = 2KΩ and R1" = 1KΩ. The –3 dB bandwidth will be in the order of 5 MHz.

For another advantage, not too many inexpensive op amps work at +1.5 volts supply and have a bandwidth of about 5 MHz for a gain of –2.

In Figure 9-25 the amplifier's gain is: Vout/Vin ~ –R_BC/R1" where R_BC/R1" < 3. However, this simple one-transistor amplifier cannot drive low-resistance loads, and will lose performance in terms of output swing. Also, the maximum gain from this amplifier is less than 3 for distortion less than 1 percent.

But for the flexibility to provide large voltage gain, low distortion, and the ability to load into low-resistance loads with minimal board space area, op amps are the way to go. They are available in even smaller packages such as surface mount versions. These can include SO-8 or SO-14 dual inline surface mount packages, but op amps are also available in even smaller sizes, such as a single amplifier in about the size of an SOT-23 transistor with five leads (e.g., two for power, two for inputs, and 1 for output).

Now let's take a look at single supply inverting gain op amp circuits. See Figure 9-26.

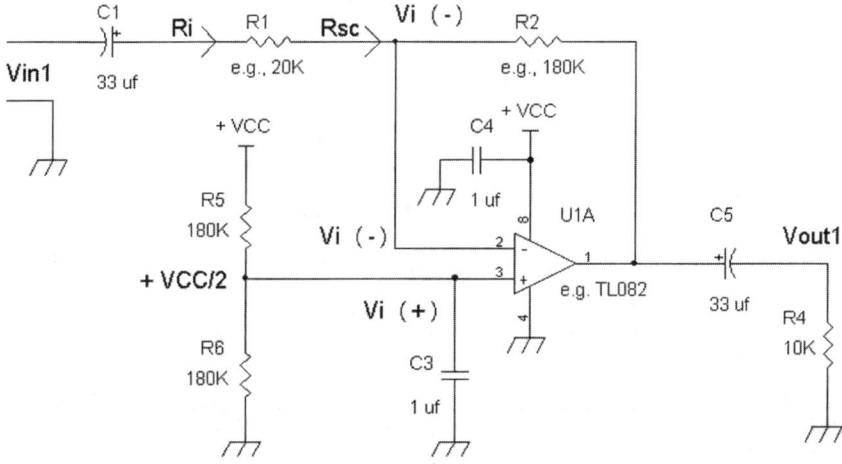

FIGURE 9-26 An inverting amplifier op amp circuit with a single power supply.

We want to maximize voltage swing by setting the output DC voltage to one-half the supply voltage. This is done by defining a DC voltage at the non-inverting input terminal (e.g., pin 3 for one section of a dual op amp). By negative feedback the voltages at the (–) and (+) input terminals pins 2 and 3 are equal, so Vi (+) = Vi (–) = +VCC/2. Also note that there is a voltage reference filter capacitor C3 that removes noise from +VCC such as power supply ripple. If there is power supply noise embedded into the +VCC/2 voltage it will be amplified by a factor of [1 + (R2/R1)]. In this example where R1 = 20KΩ and R2 = 180KΩ, and without C3 any noise from +VCC/2 will be amplified by [1 + (180K/20K)] or 10. Thus we want to filter any noise of power line frequency ripple signal that may be present in +VCC. In general, a roll-off frequency < 20 Hz is a good start. This is because C3 with R5 and R6 form a low-pass filter.

In Figure 9-26, the low-pass filter roll-off frequency is $1/[2\pi(R5 \| R6) C3]$ which is: $1/[2\pi(180K\Omega \| 180K\Omega) 1\ \mu f]$ or $1/[2\pi (90K\Omega \times 1\ \mu f)] = 1.77$ Hz, which is less than 20 Hz. This means that for any ripple voltage or noise whose frequency is above 1.77 Hz, attenuation will occur. If you need more filtering to remove noise from +VCC, you can increase C3's capacitance by using an electrolytic capacitor with its (–) terminal connected to ground. For example, a 33 μf 25-volt capacitor will reduce the noise about 33 times more when compared with the 1 μf for C3.

Since no DC currents flow into R1, and no DC current flows into R2, the DC voltage at output pin 1 is +VCC/2. See Figures 9-27 and 9-28. Also, if you look carefully in terms of DC analysis, Figure 9-26 resembles the non-inverting gain amplifier in Figure 9-19.

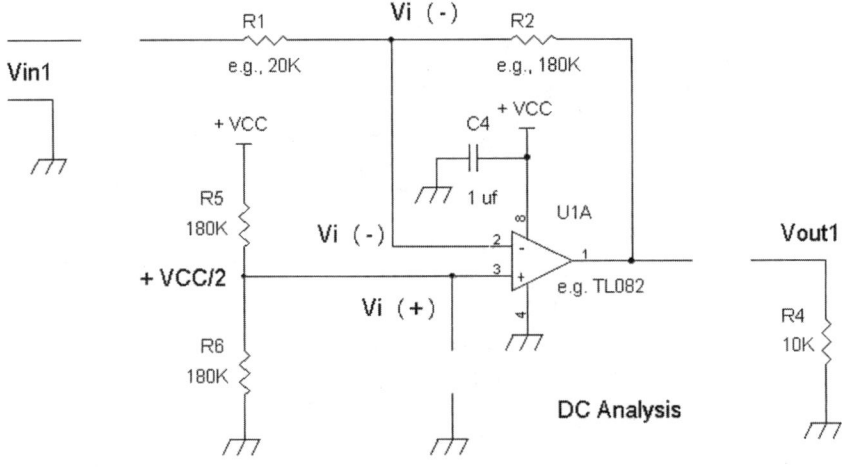

FIGURE 9-27   Capacitors C1, C3, and C5 are removed for DC analysis.

By removing the capacitors, we see on Figure 9-28 that the circuit is simplified for DC analysis and unconnected components such as R1 and R4 are removed.

Further simplification can be done as shown on the right side drawing of Figure 9-28 where R2 is replaced with a wire such that there is no current flowing through R2, which results in no voltage across R2. And, of course, a wire or 0Ω resistor has no voltage across it. Thus, as we can see on the schematic on Figure 9-28's right hand side, all we are left with is a voltage follower with +VCC/2 into the (+) input of the op amp that results in +VCC/2 at the output pin 1 of U1A.

As a reminder when using a single supply, make sure you have sufficient voltage at the power pin. If you are using general purpose op amps such as LM741, LM4558, TL082, or higher performance ones like the NE5532 or LM4562, usually you can get by with a +VCC range of 9 volts ≤ +VCC ≤ 28 volts. However, should you

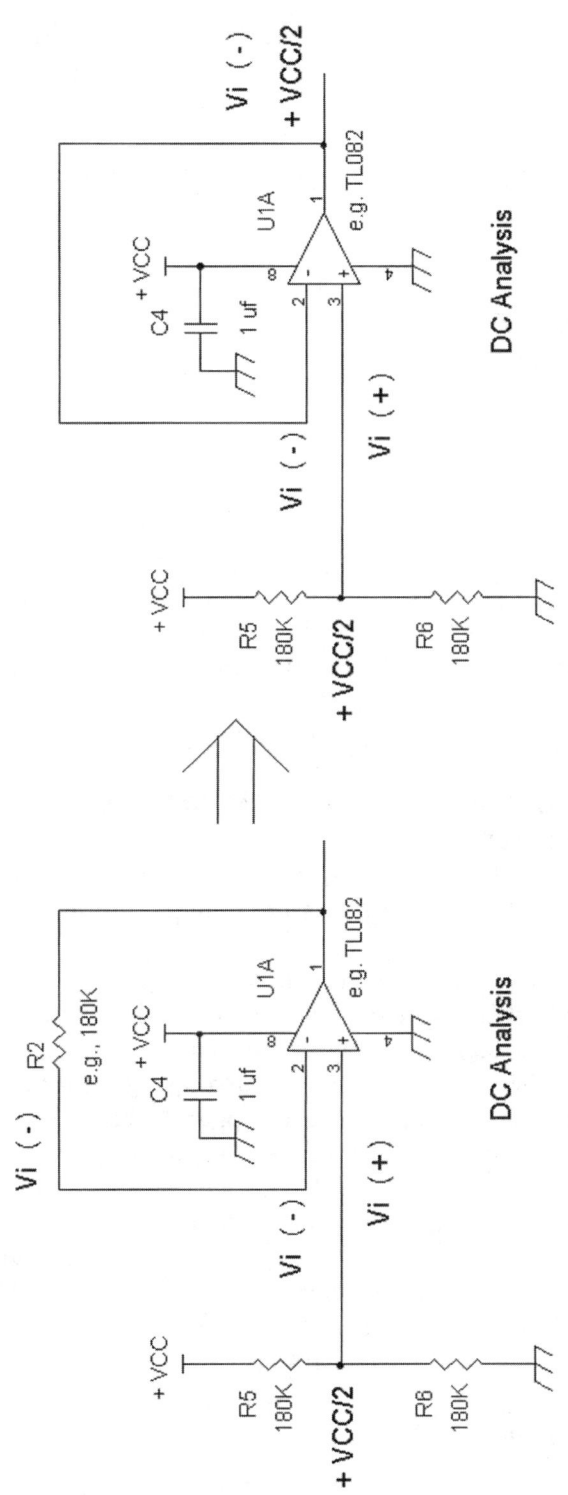

FIGURE 9-28 With R1 and R4 removed the circuit is simplified for DC analysis.

decide to use lower-voltage op amps, most will work at +5 volts, but be aware that some of these low-voltage op amps have maximum supply voltages of +5.5 volts (e.g., LMV358) and of +12 volts (e.g., TL972 low-noise op amp has rail to rail output voltage).

We can model an inverting gain op amp circuit for **AC input and output signals** as shown in Figure 9-29.

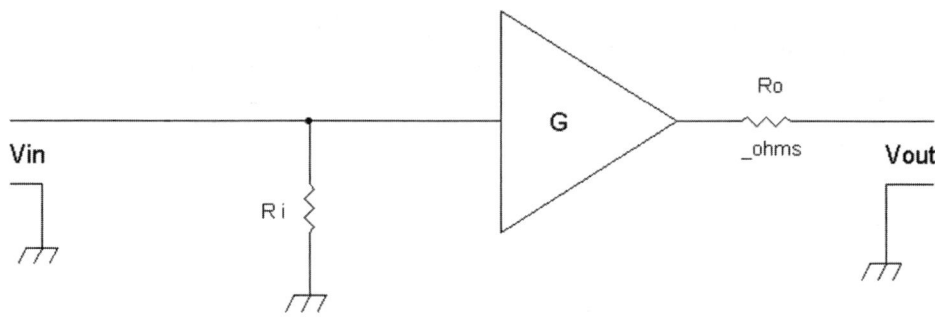

FIGURE 9-29   A general model for an inverting gain amplifier.

For an inverting gain amplifier the AC gain $G = A_v = -R2/R1$ and the AC input resistance $Ri = R1$, with R1 and R2 shown in Figures 9-23, 9-24, and 9-26. In all cases the AC output resistance, $Ro \sim 0\Omega$ for the inverting gain op amp circuits.

Troubleshooting Inverting Op Amp Circuits via an Oscilloscope, Signal Generator, and DVM

Here are some tips on basic debugging techniques for inverting gain op amp circuits, which are similar to the non-inverting gain circuits:

1. Make sure the power supply terminals are connected correctly and use a voltmeter to confirm proper power supply voltages at the power pins. Be sure to have the op amp's data or spec sheet handy.

2. Make sure there is at least a ≥ 0.1 μf decoupling capacitor wired or soldered closely to each power pin (e.g., plus supply pin 8 and negative supply pin 4 of an LM1458) and ground. For example, see the 1 μf power supply decoupling capacitors C4 and C3 in Figure 9-23. For a single supply circuit you need only one decoupling capacitor since the other power pin is usually grounded. Without power supply decoupling capacitors, the op amps can oscillate or give noisy output signals on their own.

3. Measure the voltage with **only a DVM** across the non-inverting input pin and the inverting input pin of the op amp. Confirm that the voltage is ~ 0 volt. If the voltage is not zero across the two input terminals such as 0.5 volts DC, then the op amp is NOT operating properly.

NOTE: Do not try measuring the voltage across the input pins (e.g., pins 3 and 2 of a dual op amp) with an oscilloscope. You can short out one of the input pins. The ground lead of the oscilloscope probe or wire must always be connected to the ground connection of the circuit. In general, by an indirect connection, the ground lead of an oscilloscope is at ground potential. That is, the ground lead from an oscilloscope is not "floating" like the negative (and positive) lead of a DVM.

4. For op amp circuits with positive and negative supplies, verify that the DC output at the op amp voltage is close to zero. If it is not, verify that the non-inverting input terminal is connected to a DC potential such as ground.

5. If after #1 to #4, you are not getting anywhere, replace the op amp, but make sure the original op amp could operate at the power supply voltages. For example, if the power supply provides ± 12 volts DC and the circuit does not seem to work, verify that op amp is not a lower-voltage version, such as an LMV358 (e.g., +5.5 volts maximum) or a TLC272, where a supply voltage of more than ± 8 volts will cause damage.

6. If the DC bias conditions are correct, but the AC signals are distorted or do not have correct amplitudes given the gain setting [–R2/R1], then first check the resistance values of the feedback resistors. Also make sure the feedback network resistance value of R2 ≥ 500Ω is not low enough to cause current limiting at the output that will cause clipped waveforms.

7. If the low-frequency (e.g., 20 Hz) response is poor, such as less than 50 percent of amplitude when compared to the signal amplitude at a mid-band frequency (e.g., 1000 Hz), check the input and output coupling capacitors. If they are OK, check the frequency response at the output pin of the op amp such as pin 1 or 7 of a dual op amp. If there is a loss in low-frequency response, see if there is an AC coupling capacitor at the input, such as C1 in Figure 9-24 or C1 in Figure 9-26. If you suspect there is a problem like the capacitance is not high enough, parallel a larger capacitance across C1, such as a 2.2-μf capacitor. If the capacitor is ceramic, check the maximum working voltage. For example, if you use a 1-μf 6.3-volt capacitor and the DC voltage across it is +5 volts, chances are that the true capacitance will be half of that, or 0.5 μf, which will cause a poorer low-frequency response.

8. Also, if electrolytic capacitors are used, measure with a voltage meter that the voltage across the capacitor is correct with the polarity markings of the capacitor. *If you measure a reverse bias voltage across an electrolytic capacitor, turn off the power immediately, replace the capacitor, and install a new capacitor with the correct polarity.*

9. Check to make sure the phase at the output is inverted relative to the phase of the input signal. If you are getting the same phase signal at input and output, then the op amp may be damaged or open circuited from the feedback and input resistors (e.g., R2 and R1). As a result of the op amp being open circuited,

the signal will travel through R1, then to R2 to the output. If you have no power to the op amp, a feed-through condition will also occur and the signal from R1 to R2 may travel through and there will be no inversion of the input signal at the output terminal.

A telltale sign that something is wrong is if you probe the inverting input of the op amp (e.g., pin 2 or pin 6 of a dual op amp) and see a signal, then something is wrong because you should see no signal at all. This is because the inverting input in an inverting gain amplifier is a virtual ground, or a virtual AC ground (e.g., in a single supply circuit).

10.   Also check if the circuit was for a dual op amp but a single op amp was installed instead, and vice versa. Both single and dual op amps have 8 pins total and some pins are common, such as the $-V$ supply pin 4, non-inverting input pin 3, and inverting input pin 2; however, the other pins 1, 5, 6, 7, and 8 are not the same for a single op amp and dual op amp.

11.   In an inverting gain amplifier you cannot make the feedback resistor R2 = 0Ω, since this will result in a zero output. Also, making the feedback resistor (e.g., R2) low resistance (e.g., < 500Ω) can cause the op amp's output waveform to distort due to current limiting. Recall that the feedback resistor is also connected to the inverting input terminal, which is a virtual ground. So, the feedback resistor is a "direct" load resistor to the output terminal.

Brief Notes on Both Non-inverting and Inverting Gain Amplifier Circuits

- If you are providing signals to the outside world, such as to a connector that will later be used with a cable or wire, please add a 47Ω series resistor between the output of the op amp and the connector. Or if you are driving any shielded cable or coaxial cable with the op amp, also add a series resistor between the output of the op amp and the cable. The reason for adding a series resistor is to prevent the op amp from oscillating, and to provide some short circuit protection. This is especially true when a person can inadvertently short out a connector while trying to connect a cable to it.
- For both types of amplifier circuits, choose an op amp that is "unity gain compensated." This will ensure that there will be no parasitic oscillations. Most op amps are unity gain compensated such as the ones listed in this chapter.
- The high-frequency bandwidth is determined by the op amp's unity gain bandwidth divided by [1 + (R2/R1)]. Usually, there is a typical or minimum unity gain bandwidth specification on the data sheet. For example, if you are using a TLO82 op amp that has about a 3-MHz unity gain bandwidth, the high-frequency bandwidth of either inverting or non-inverting amplifiers is determined by the ratio of R2 over R1. In a previous example R1 = 2KΩ and R2 = 18KΩ. Therefore, with a TL082 the high-frequency bandwidth is

3 MHz/[1 + (18KΩ/2KΩ)] = 3 MHz/10 or 300 kHz. This 300-kHz bandwidth is for a non-inverting gain of +10 or an inverting gain of –9.

- The op amps also have a power bandwidth specification or slew rate specification. Generally, for instrumentation such as monitoring slow changing DC voltages, a slew rate of ≥ 0.2 volts per microsecond will work. For audio circuits with ≤ 5 volts peak to peak output, ≥ 0.5 volts per microsecond is fine. But for a large audio signal voltage output of ~ 24 volts peak to peak, ≥ 1 volt per microsecond will do. Once you have to work with signals beyond 20 kHz, it's best to find an op amp ≥ 20 volts per microsecond. For example, for even standard resolution TV (SDTV) analog signals such as 480i, you will need ≥ 50 volts per microsecond with a high-frequency bandwidth ≥ 5 MHz.
- Op amps with negative feedback, such the non-inverting and inverting amplifiers, have power supply noise rejection at their power pins. This means if there is a power supply ripple from a raw supply, the op amp will keep the ripple from showing up at the output. However, it is usually better to use a voltage regulator for powering op amps. See the next section in this chapter.

A Short Look at Linear Voltage Regulators

Voltage regulators provide a constant voltage to an electronic circuit, even though the raw power supply is varying in DC voltages. For linear voltage regulators to operate correctly, the incoming voltage must be larger than the regulated voltage at the output. For example, if you have a power supply that varies between 9 volts and 12 volts due to power supply ripple or changes in the AC power line voltage, you can provide a regulated +6 volts or +5 volts via an LM7806 or LM7805 three-terminal regulator. If you want to adjust to different regulated voltage, you can use an LM317T integrated circuit and two resistors to set the voltage, which will be greater than two volts difference between the incoming voltage and the regulated output voltage.

Another reason for using linear voltage regulators is to remove high-frequency noise from switching power supplies. For example, an efficient wall power supply may have switching noise in the 1-MHz region added to its regulated DC voltage. A voltage regulator can remove the switching noise.

Voltage regulators are available in through-hole and surface mount packages. We will be looking at through-hole versions, which are easier to use for DIY projects, plus they can be mounted onto heat sinks.

Types of Linear Regulators Starting with "Standard" Fixed Voltage Versions

What is a voltage regulator? It is a device that takes in an incoming raw DC voltage and provides a constant lower voltage. The incoming raw DC voltage may vary in DC amplitude with noise or ripple voltages riding on top of a DC voltage. The voltage regulator then removes the noise or ripple voltages and provides a lower constant voltage.

For example, if the incoming DC voltage has a range of +7 volts to +10 volts, that is, a minimum of +7 volts and a maximum of +10 volts, and this incoming DC voltage is connected to a +5 volt regulator, the output of the regulator will be a constant +5 volts.

A voltage regulator requires a DC input voltage that does not dip below its minimum input voltage rating. This means that the DC input voltage must not be a pulsating DC voltage such as a rectified AC voltage without a filter capacitor, which means the DC voltage can dip to 0 volts. If such an instance happens, the regulator will just pass through the input voltage that includes the pulsations. Figure 9-30 shows an example of a pulsating DC voltage waveform.

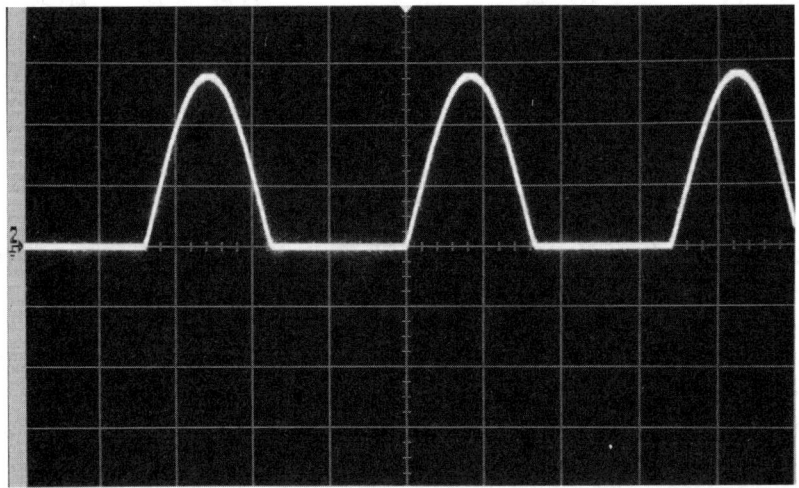

FIGURE 9-30 A half wave rectified voltage without a filter capacitor that is NOT the type of DC input voltage required for a voltage regulator.

The waveform at times goes to 0 volts or ground. Instead, see Figure 9-31, a half wave rectified voltage with a filter capacitor that is more suitable for use with a voltage regulator.

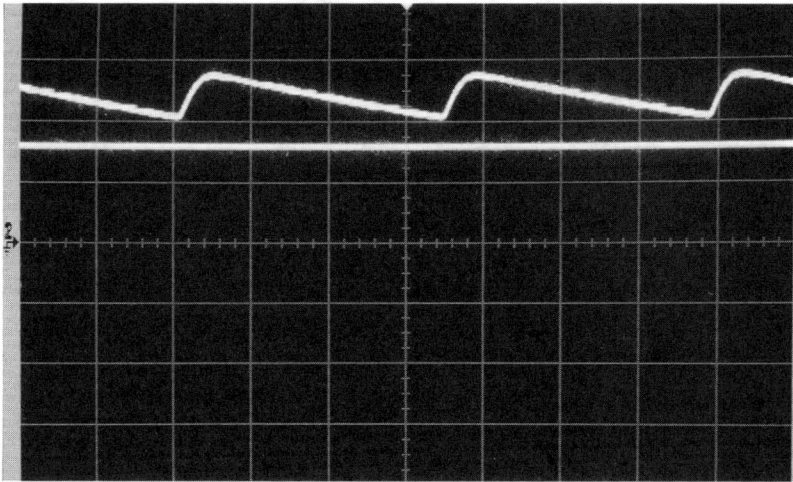

FIGURE 9-31   Top trace: A half wave rectified voltage that provides a DC voltage with a "saw-tooth" ripple voltage riding on top of the average DC voltage for the input voltage into a voltage regulator. Bottom trace: Regulated voltage output. The center horizontal line is at 0 volts.

Note the wave does not go to ground and has a minimum DC voltage at the lowest point of the sawtooth. This lowest point voltage must be equal to or greater than the "drop-out" voltage specification of the voltage regulator. For example, if Figure 9-31 is at 2 volts per division, the DC voltage has a minimum voltage of about +4 volts and a maximum voltage of about 2.8 div × 2 volts/div, or about 5.6 volts peak. If a standard voltage 3.3-volt voltage regulator is used, generally, the drop-out voltage is about 2 volts to be on the safe side, which would mean a minimum input voltage = 3.3 volts + $V_{drop-out}$ is needed. This would be 3.3 v + 2 v = 5.3 volts, which would mean when the waveform dips below 5.3 volts, the voltage regulator will stop working properly since the example has the waveform at a peak of 5.6 volts only briefly and dipping to down 4.0 volts. To "fix" this, you can use a low–drop-out regulator that usually has a drop-out voltage of about 0.5 volts. This means the minimum input voltage = 3.3 volts + 0.5 volt or 3.8 volts. Since the example waveform has the minimum voltage at 4.0 volts, the low–drop-out voltage regulator will work properly, but just barely.

Drop-Out Voltage Summary

Again, most standard linear voltage regulators (e.g., with fixed and adjustable regulated output voltages) require that the "dip" or minimum of the DC input voltage is at least 2 volts above the regulated output voltage. The drop-out voltage is specified as the minimum voltage between the input and output of the regulator to ensure that voltage regulation takes place. In the example of having 2 volts above the regulated voltage, the drop-out voltage is 2 volts.

However, there are low–drop-out voltage linear voltage regulators that can regulate the voltage when the "dip" or minimum of the DC input voltage is at least 0.50 volt above the regulated output voltage. In this case, the drop-out voltage is 0.5 volts, which when compared to the 2-volt drop output voltage for standard voltage regulators, we can see why the term "low–drop-out voltage" is used.

Voltage Selections, Packages, Pin Outs and Schematics

For standard fixed-voltage output regulators in TO-220 (1 amp) and TO-92 (100 mA) packages, see Figure 9-32. Notice that the pin outs are not the same for positive and negative voltage regulators.

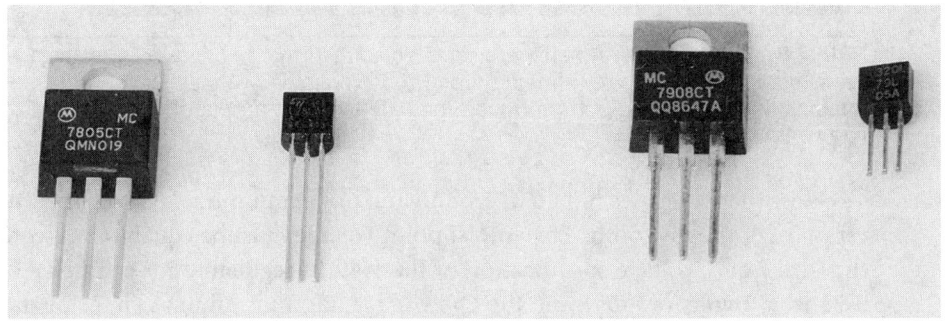

FIGURE 9-32   Fixed-output voltage regulators shown with 1 amp TO-220 and their mounting tabs, and the smaller 100 mA TO-92 packages that look like transistors (e.g., 2N3904), but they are actually integrated circuits.

This is important to notice that in a ***positive regulator***, reading from the side where the part number is printed, the sequence is **input, ground, and output**, whereas in a ***negative voltage regulator***, the first two terminals are reversed have the sequence, ***ground, input, and output***. For TO-220 packages, the metal tab is always connected to the middle terminal. That is in a TO-220 fixed-voltage regulator, the metal (mounting) tab is ground for a positive version, while for a negative fixed-voltage regulator the tab is the input voltage terminal. It is very important to never ground the negative voltage regulator's tab, or you will short out the raw input DC supply.

If you are using a low-power TO-92 linear voltage regulator, such as the positive voltage 78Lxx or negative voltage type 79Lxx, be aware of the pin out because they are different. Again, from the flat side where you are reading the part number, the sequence for the positive TO-92 package is output, ground, and input, which is a reverse sequence from its bigger TO-220 version. For the negative voltage regulator in TO-92 package, the pin out sequence as read from the flat side is ground, input, and output, which is the same as its TO-220 higher current version. Keeping track of all these packages in your head can be confusing. If in doubt, look up the data sheet.

See a schematic diagram for fixed-output voltage regulators in Figure 9-33 and their associated decoupling capacitors.

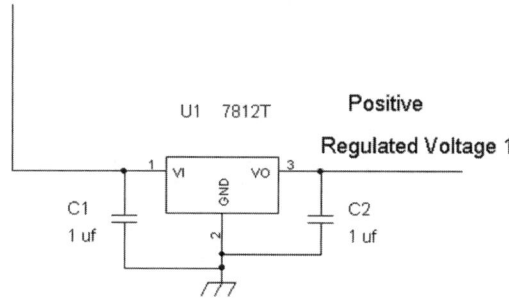

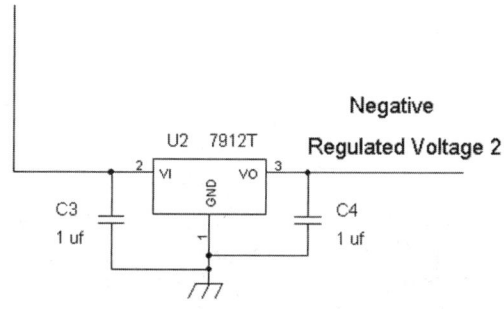

FIGURE 9-33   Schematic diagrams for standard positive and negative voltage TO-220 regulators with decoupling capacitors C1, C2, C3, and C4. Notice that the first two pins U1 are reversed from the pins in U2 (Vi and Gnd).

Knowing the Pin Out Sequence Is Important

See Figure 9-34 for the pin out sequence for both TO-220 and TO-92 fixed (e.g., not adjustable) voltage regulators.

Common fixed output voltages include $+3.3$ volts (78Mxx for 500 mA or 78Lxx for 100 mA), with 78xx (positive) for 1 amp at $+5$ volts, $+6$ volts, $+8$ volts, $+10$ volts, $+12$ volts, $+15$ volts, $+18$ volts, and $+24$ volts. In the 78xx part number, you replace the "xx" with the voltage you require. For example, if you need a $+12$-volt regulator 78xx $\rightarrow$ 7812, or if you need a $+5$-volt regulator, the correct part number is 7805.

Equivalently, there are LM340 series regulators equivalent to the 78xx series. Here, the numbering system requires a dash such as a $+5$-volt regulator is named as LM340-5. Voltage selection on the LM340 series is the same for the 78xx series when using parts made by Texas Instruments (TI).

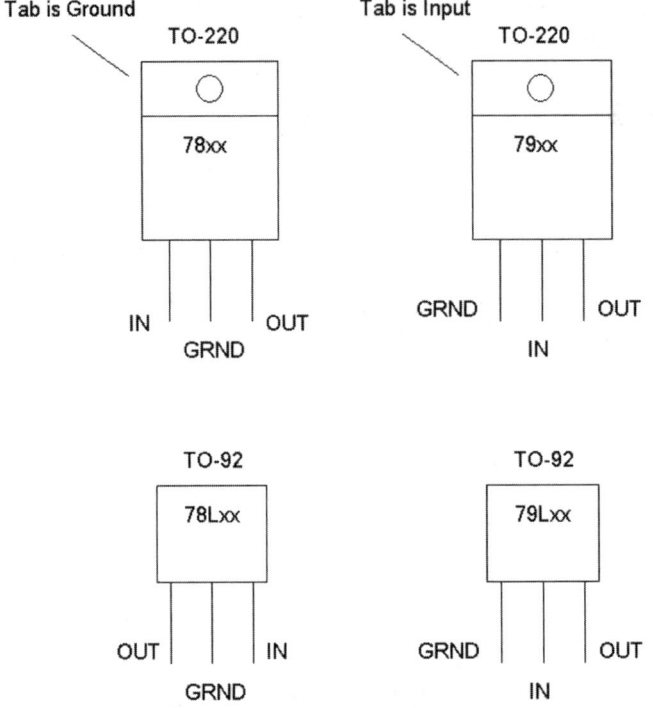

FIGURE 9-34   Pin out sequences as read from the front side of the voltage regulators that have their printed markings. The TO-92 illustrations are drawn larger for clarity of pin outs. See Figure 9-32 for a comparison of the real sizes of the different packages. (7800 series = Positive regulator, 7900 series = Negative regulator).

For negative voltage regulators, the 79xx series can output regulated negative voltages of –5 volts, –6 volts, –12 volts, –15 volts, –18 volts, and –24 volts. Again, as an example, if you need a –6-volt regulator, the part number will be 7906.

NOTE: The raw input DC volts should not exceed any voltage higher than 30 volts at its peak for positive voltage regulators (78xx) and no lower than – 30 volts peak for negative voltage regulators (79xx); otherwise, they can be damaged.

Low–Drop-Out Voltage Regulators

There are low–drop-out (LDO) voltage regulators that require that the input voltage is only ≥ 0.5 volt above the regulated output voltage. Again, this means the drop-out voltage is 0.5 volt. Also, these LDO regulators have the same pin out sequence as the previously mentioned standard fixed regulators. That is, the pin outs for the low–drop-out positive voltage regulator LM2941-x are the same as the LM78xx TO-220 regulators. The low–drop-out negative voltage regulator, LM2990-xx, will have the same pin out for the LM79xx TO-220 packages. See Figure 9-35.

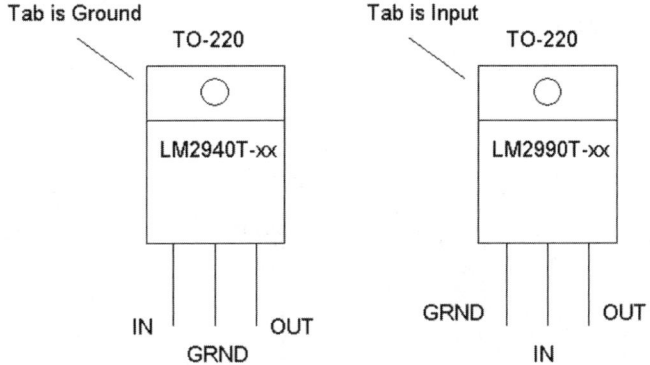

FIGURE 9-35 Low–drop-out voltage regulators and their pin outs. (LM2940 series = Positive regulator, LM2990 series = Negative regulator).

However, there are a couple of noteworthy differences between the low–drop-out regulators compared to the standard ones.

- The maximum positive input voltage is +25 volts for the LDO regulators versus +35 volts absolute maximum for the standard versions (LM78xx).
- The decoupling capacitors have to be larger for the output capacitors in low–drop-out regulators. Typically, you can use electrolytic capacitors of 33 µf at 35 volts for both the input and output decoupling capacitors, and most of all observe the polarity. For the positive LDO regulators, a 1 µf (e.g., ceramic) at the input terminal will suffice while using a 33 µf at the output terminal. And with the LM2990-x series, the negative terminals of the electrolytic decoupling capacitors are connected to the input and output terminals of the regulator. See Figure 9-36 and observe the electrolytic capacitors' (C2, C3, and C4) polarity markings.

The available fixed-output voltages for the positive voltage LDO regulators LM2940T-xx are: 5 volts, 8 volts, 9 volts, 10 volts, 12 volts, and 15 volts. For example, a positive 5-volt version would be listed as LM2941T-5.0, whereas a 9-volt version would be LM2940T-9.0, and a 12-volt part number will be LM2940T-12.

For the negative voltage LDO regulator LM2990T-xx, the available fixed-output voltages are: –5 volts, –12 volts, and –15 volts. An example part number for a –5-volt version would be listed as LM2990T-5.0, whereas a negative 12-volt version would be LM2990T-12.

Positive Unregulated Voltage 1

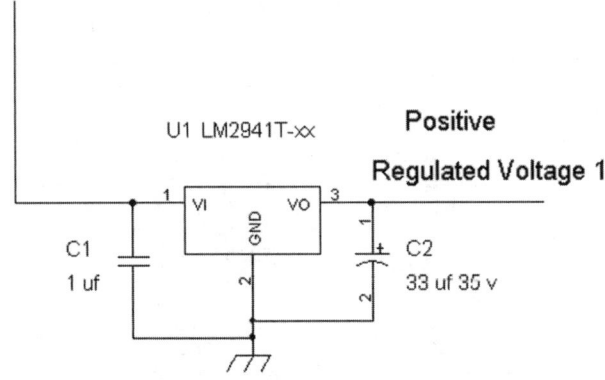

Negative Unregulated Voltage 2

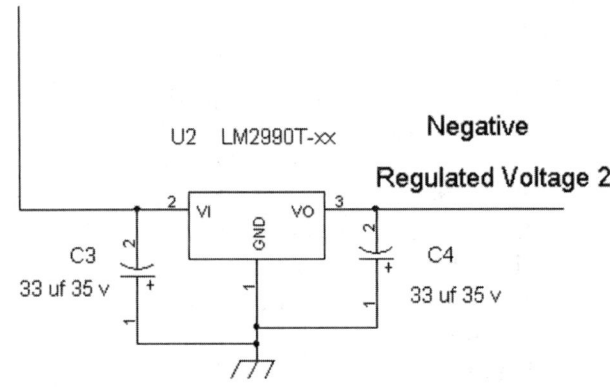

FIGURE 9-36 LDO regulators with decoupling capacitors larger in capacitance for C2, C3, and C4 when compared to standard voltage regulators.

Things That Can Go Wrong with 7800 and 7900 Series Voltage Regulators

- If the input voltage to a positive voltage regulator (e.g., 7805) is a negative voltage that is the wrong polarity, the positive voltage regulator will drain excessive current and the output will be about −0.9 volts. You will need to shut off power immediately and replace the regulator with a new positive voltage regulator. You then connect with the correct polarity input voltage (e.g., a positive voltage) to the input of the new positive voltage regulator.
- If the input voltage is correctly positive, but the ground lead (e.g., middle lead) is left unconnected (e.g., floating), the output voltage will be roughly the same voltage as the input. This can lead to damaging ICs or circuits that require a regulated voltage. For example, suppose you have a TLC272 op

amp that is connected to a +12-volt regulator's output, 7812, but the 7812's middle ground lead is not connected to ground. With an input voltage of +24 volts, the output voltage of the 7812 will be about 23 volts, which will damage the TLC272 op amp because it has a maximum operating voltage of 16 volts.

- If you reverse the pin outs by swapping the input and output leads of a positive voltage regulator (e.g., 7805) while still grounding the middle lead, then you will be applying the unregulated voltage to the output pin and taking the output of the regulator via its input pin. This will cause **excessive current** and also pass the unregulated voltage from the regulator's output pin to the regulator's input pin. In other words, you will be applying an unregulated high voltage to your circuits, which can damage them. Swapping the positive voltage regulators' two outer leads can be the result of not paying attention to the silkscreen on a printed circuit board that denotes which way the TO-220 package is installed. Power should be turned off immediately and the regulator replaced and installed properly.
- If the input voltage to a negative voltage regulator (e.g., 7905) is an incorrect positive voltage of the wrong polarity, the negative voltage regulator will drain excessive current and the output will be about +2 volts. You will need to shut off power immediately and replace the regulator. Then fix the wiring of the input voltage source such that it provides a negative voltage to the input pin of the negative voltage regulator.
- If the input voltage is correctly a negative voltage, but the ground lead (e.g., first lead on the left side) is unconnected (e.g., floating), the output voltage will be roughly the same negative voltage as the input's negative voltage. Hence the raw input voltage is passed onto the circuits requiring a well-defined or limited negative voltage. This can lead to damaging ICs or circuits that require a regulated negative voltage.
- If you reverse the pin outs by swapping the pin sequence from 1, 2, 3 to 3, 2, 1 of a negative voltage regulator (e.g., 7905), the output lead will be grounded, the input lead stays the same since it is the middle pin, and output is now provided via the ground pin. This will cause **excessive current** and output of about –0.7 volt. You will need to shut off power immediately and replace the regulator with the correct pin out sequence.

Accidents Involving the 7800 Series Positive and 7900 Series Negative Voltage Regulators

If you load onto a printed circuit board a positive 78xx voltage regulator in the place of a negative 79xx voltage regulator and 79xx regulator in the place of a 78xx voltage regulator, you will have fed through the raw voltage with a voltage drop of about the rated regulated voltage.

For example, if 7808 is installed in the place of a negative voltage regulator, 79xx, and the raw supply is minus 20 volts, then the output voltage will be –(raw voltage – rated regulated voltage), which is then: –(20 volts – 8 volts) = –12 volts.

And if a negative voltage regulator such as a 7906 is installed where a positive voltage 78xx regulator should be, the output voltage will be (raw voltage – rated regulated voltage) = (20 volts – 6 volts) = +14 volts.

If the raw positive voltage is equal to or less than the regulated voltage rating, the output will be close to zero for the 7900 negative series regulator that is loaded into a board that is meant for a 7800 series regulator.

Some Mishaps with Low–Drop-Out LM2940 and LM2990 Series Regulators

The low–drop-out (LDO) regulators have different internal circuitry as the standard regulator circuits. Thus, there will be some differences when LDO regulators are connected incorrectly.

- When a positive voltage LDO regulator (e.g., LM2940-xx) is connected with the input and output leads swapped, and a correct raw positive voltage is applied, the regulator will pass through the raw positive voltage. For example, if the raw voltage is +20 volts DC, and the input and output leads of the LDO positive voltage regulator is swapped, the output voltage will be about +20 volts. Therefore, there will be no regulation, and subsequent circuits requiring a regulated voltage may be damaged. You should turn off the raw power supply and replace the LDO and install a new one with the correct pin out sequence.
- If you apply a negative input voltage to a positive voltage LDO's input terminal, the output will be close to zero volts.
- If you forget to connect the middle terminal of the positive voltage LDO to ground and leave the middle pin unconnected or floating, again the output will be essentially the unregulated input voltage that can cause damage to other circuits.
- If you have inadvertently installed a negative LDO (e.g., LM2990-xx) in place of a positive voltage regulator, and turned on the positive raw supply, the negative voltage LDO will pass to its output the raw input supply voltage. This raw input voltage will be higher than the expected regulated voltage and can damage other circuits.
- For a negative LDO voltage regulator (e.g., LM2990-xx), if you apply an incorrect positive voltage at the input, the output will be about +1 volt from the negative LDO regulator, and excessive current will draw. The raw power supply must be turned off immediately. Then the correct negative supply must be wired to the input terminal.
- If you reverse the pin sequence of the negative LDO voltage regulator, excessive current will be drained and about –0.7 volt will be at the output. Again, turn off the supply immediately and install a new negative LDO voltage regulator with the correct pin out sequence.

- If the circuit requires a negative LDO voltage regulator and everything is connected correctly, but a positive LDO voltage regulator is mistakenly installed, the raw supply voltage will pass to the output and possibly cause overvoltage damage to other circuits.

Some Troubleshooting Hints When Working with Regulators

- One thing to remember about voltage regulator chips is that they can only deliver current in one direction. In a positive regulator, the current flows out. But if you try to have the positive voltage regulator absorb or sink current, you may turn off the voltage regulator and as a result, the output voltage will no longer be the expected regulated voltage. An example is if you have a +5-volt regulator (7805) and if one end of a resistor is connected to the output of the 7805 and the other end of the resistor is connected to +20-volt source, then current will be flowing the wrong way into the output terminal. This can cause the 7805 to shut off because it can only absorb a small amount of current.
- If the voltage regulators get too hot, then they will shut down and output close to zero volts. You will need to first check if you have the correct voltage regulator installed accordingly. Then you need to monitor the current into the voltage regulator's input terminal via an ammeter. See if it's within the range of the output current capability of the voltage regulator. If it is, then the voltage regulator may need to be mounted on a heat sink.
- If the output is lower than the regulated voltage, use a DVM and measure the voltage at the input and output terminals. For a positive voltage regulator, the input voltage should be greater than 2 volts above the rated output voltage. For example, if you are using a 7808 for +8 volts output, the minimum input voltage is 2 volts above that, which means that the input voltage is greater than or equal to 10 volts.
- If you are monitoring with an oscilloscope on the input and output terminals and there is a high-frequency oscillation signal, then make sure you have the decoupling capacitors soldered or wired within one-half inch of the input, ground, and output terminals. Also make sure the capacitances are large enough (≥ 1 μf).

If you are using ceramic capacitors, use a voltage rating at least twice the voltage at the input and output terminals. This is because some ceramic capacitors will derate when there is a DC voltage close to its maximum rating.

Should an electrolytic capacitor be used, observe its polarity. If it is installed incorrectly, shut off the power immediately and install a new one with the correct polarity.

This completes Chapter 9, and we will be looking into audio circuits in the next chapter.

CHAPTER 10

Audio Circuits

We will explore two types of audio amplifiers—preamplifiers, commonly known as preamps, and power amplifiers.

The examples chosen will be from commercially manufactured designs. What's different in this chapter is that in some cases, we will see associated circuits such as a power supply and a voltage regulator. Although integrated circuit (IC) amplifiers could have been chosen for this chapter, it would be good to know something more about amplifiers on the transistor level. This way you can use this knowledge to augment IC designs with discrete circuits.

NOTE: Audio signals are usually measured in RMS voltages. For a sine wave signal, the peak to peak voltage is 2.82 x RMS voltage. For example, a 1-volt RMS sine wave is 2.82 volts peak to peak. However, when you are measuring the audio signal on an oscilloscope, you will often measure the peak to peak amplitude instead of the RMS value.

Preamps and Power Amps

For transducers such as microphones and phonograph pickup cartridges that generate small amplitude signals in the 1 mV to 10 mV range, preamps are needed to provide large voltage gains for these devices. For example, a microphone preamp has a voltage gain of 30 to 1000.

A dynamic microphone needs a preamp gain of about 300 or 50 dB. There are higher output microphones such as electret types. These require power, and they generally have about 10 times more output signal than a dynamic microphone; so they only require gains of 30 or about 30 dB. Low-level microphone preamps have voltage gains from about 30 to 1000 (30 dB to 60 dB) with flat frequency response from 20 Hz to 20,000 Hz.

NOTE: Decibels or dB = 20 log$_{10}$(IVout/VinI) where the voltage gain = IVout/VinI. For example if you have a 2 volt RMS input signal and the output is 10 volts RMS, then the gain = I10 volts RMS/2 volts RMSI = 5 and in decibels we have 20log$_{10}$(5) dB = 14 dB since log$_{10}$(5) = 0.7. Also log$_{10}$(x) is commonly known as just log(x), where x is a non-negative number. For example, $x \geq 0$.

Magnetic phonograph pick up cartridges (a.k.a., phono cartridges such as Shure M97xE or Audio Technica AT95E) for playing (LP) vinyl records generally require a gain of 50 to 100 at 1 kHz. The phono cartridge preamp requires a frequency equalization to provide flat frequency response from the vinyl record to the output of the preamp. That is, the vinyl recording is cut with a non-flat frequency response. This is done to maximize playing time by essentially attenuating low frequencies while boosting higher frequencies.

For higher (amplitude) level signals, we use *line-level* preamps that have a gain of 1 to about 10 or 0 dB to 20 dB. Line-level signals are in the order of 70 mV RMS to about 1 volt RMS. A typical portable digital audio device may output a maximum of about 1 volt RMS or about 2.82 volts peak to peak. Preamplifiers, whether low-level or line-level types, do not provide enough current output to drive a loudspeaker. However, some preamps can drive headphones.

The ultimate goal for the preamp is to provide enough signal level to a power amplifier. Typically, an audio power amplifier will be specified for an input voltage for full power output.

For example, many audio power amps have a gain of 26 dB or 20. If the amplifier is specified for 2 volts RMS input for full output into an 8Ω speaker load, then the power output will be output RMS voltage squared, divided by the load resistance. In this example, power output will then be P = [(2 volts RMS × 20)^2/8 Ω]. Thus, P = [40^2/8] watts = [1600/8] watts, or P = 200 watts output into the 8Ω load.

For this chapter we will look into a Dynaco PAT-5 low-level preamp and two different power amplifiers from the Pioneer SX-626 and SX-636 stereo receivers.

First, we need to discuss differencing amplifiers and emitter follower amplifiers and how they will relate to preamps, power amps, or voltage regulators.

Now let's look at some building blocks to preamp and power amps.

A Basic Difference Amplifier

In most modern amplifiers, negative feedback is used. This requires an inverting input terminal, and a non-inverting input terminal. For example, the voltage follower circuit using an op amp has the output terminal connected to the inverting input terminal while the input signal is applied to the non-inverting input terminal (e.g., Chapter 9, Figure 9-3).

A difference amplifier can be made with a common emitter amplifier and a common base amplifier. What we want is for the difference amplifier to provide an output signal that somehow subtracts two input signals. First let's take a look at each amplifier in Figure 10-1.

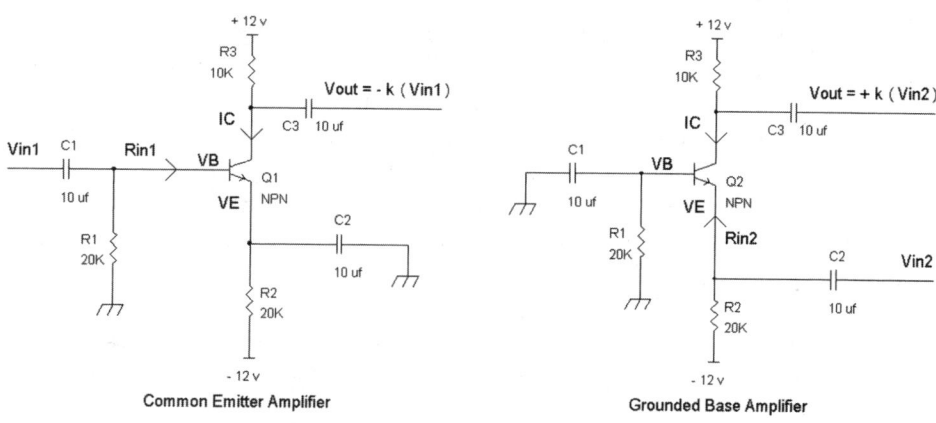

FIGURE 10-1   A common emitter amplifier with input signal Vin1 and a common base amplifier with a second input signal Vin2.

The common emitter amplifier in Figure 10-1 provides an inverting gain referenced to input signal Vin1. In both figures, VE = emitter voltage, VB = base voltage, IC = collector current, Rin1 = input AC resistance Q1's base, and Rin2 = input AC resistance into Q1's emitter.

Generally for the common emitter amplifier, the gain – k = $-R3/r_e$, where r_e = 0.026 v/IC, where in this case, IC = (VE – –12 v)/R2, and where VE ~ –0.7 volt since Q1's DC base voltage ~ 0 volt.

Therefore, IC = (–0.7 v – –12 v)/20KΩ or IC = 11.3 v/20KΩ = 0.565 mA or IC = 0.565 mA = 0.000565 A, and r_e = 0.026 v/0.000565 A or r_e = 46Ω.

With r_e = 46Ω, the gain, Vout/Vin1 = –k is then $-R3/r_e$ or –k = –10KΩ/46Ω **Vout/Vin1 = –k = –217.**

If –k = –217, then by multiplying by –1 on both sides, **k = 217.** Knowing +k will be important for the common base amplifier shown in Figure 10-1.

The input resistance looking into Q1's base, Rin1 is β r_e, and if β = 100 (as an example only because β can typically be anywhere from 50 to 300), then Rin1 = 100 (46Ω) or Rin1 = 4600Ω.

The actual input resistance as seen by Vin1 is then R1 || Rin1, and in this case, R1 || Rin1 = 20KΩ || 4600Ω = 3740Ω.

NOTE: R1 || Rin1 = (R1 x Rin1)/(R1 + Rin1).

Now let's look again at Figure 10-1's common base amplifier on the right side. The base of the Q2 is AC grounded via C1. And we call it a common base amplifier because the base is AC grounded and both input and output signals have ground connections. Here we see that the common base amplifier has no phase inversion at the output with the collector output voltage being Vout = +k (Vin2).

It turns out a common base amplifier, when the input signal Vin2 is a voltage source with 0Ω source resistance, has the same voltage gain as the common emitter amplifier except there is no phase inversion at the output. That is, the common base amplifier's output waveform has the same phase as the input signal. Given the same collector load and emitter bias resistors R3 and R2 with the same power supplies at ± 12 volts, the collector current of the common base amplifier for Q2 is also 11.3 volts/ R3 or 11.3 volts/20KΩ = 0.565 mA = IC. Thus r_e = 0.026 volt/0.000565 A, or r_e = 46Ω is the same for the common emitter amplifier in Figure 10-1.

The gain of the common base amplifier Vout/Vin2 = + k = R2/r_e = 10KΩ/46Ω = +217, which is the same gain of the common emitter amplifier except that there is no phase inversion in a common base amplifier between the input and output signals.

When we look into the input resistance via Rin2 for a common base amplifier, we find that there is no current gain in terms of input emitter current and output collector current since they are essentially the same. However, in the common emitter amplifier, the base current is very small and is 1/β of the collector current. Thus, we can surmise that since there is no current gain in the common base amplifier, the input resistance must be very low. In fact, the input resistance into the emitter of Q2 is Rin2 = r_e = 46Ω, which smaller than Rin1 (input resistance into Q1's base) by a factor of 1/β since Rin1 = β r_e. Recall that if β = 100, then Rin1 = 4600Ω for the common emitter amplifier. But Rin2 for the common base amplifier is 46Ω. The actual input resistance from Vin2's viewpoint is R2 || Rin2 = 20KΩ || 46Ω ~ 46Ω.

NOTE: Anytime you parallel a first resistor with second resistor of at least 35 times more resistance, the paralleled resistance is essentially the value of the first resistor within 3 percent accuracy. For example, a 1000Ω resistor in parallel with a 39KΩ is about 975Ω, which for most practical purposes is still 1000Ω.

In Figure 10-1, we now have shown that a one-transistor circuit can have two inputs, one at the base and another at the emitter. By utilizing two inputs with a single transistor, we can make a difference amplifier as shown in Figure 10-2.

Since the amplification factor, k = R3/r_e, is the same for both common emitter and common base configurations, we see in Figure 10-2, a difference amplifier is possible with a single transistor Q1. The output signal Vout = k (Vin2 – Vin1).

Or put another way, for the common emitter amplifier's input signal Vin1, Vout = –k (Vin1); and for the common base amplifier, Vout = k (Vin2).

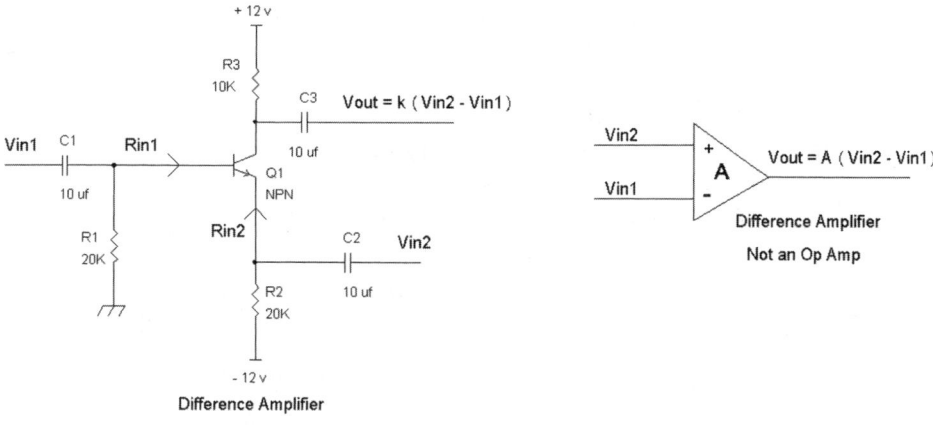

FIGURE 10-2 A difference amplifier having two input signals, Vin1 and Vin2, and a block diagram of the difference amplifier on the right side.

We can use this difference amplifier with another common emitter amplifier to make a feedback amplifier as shown in Figure 10-3.

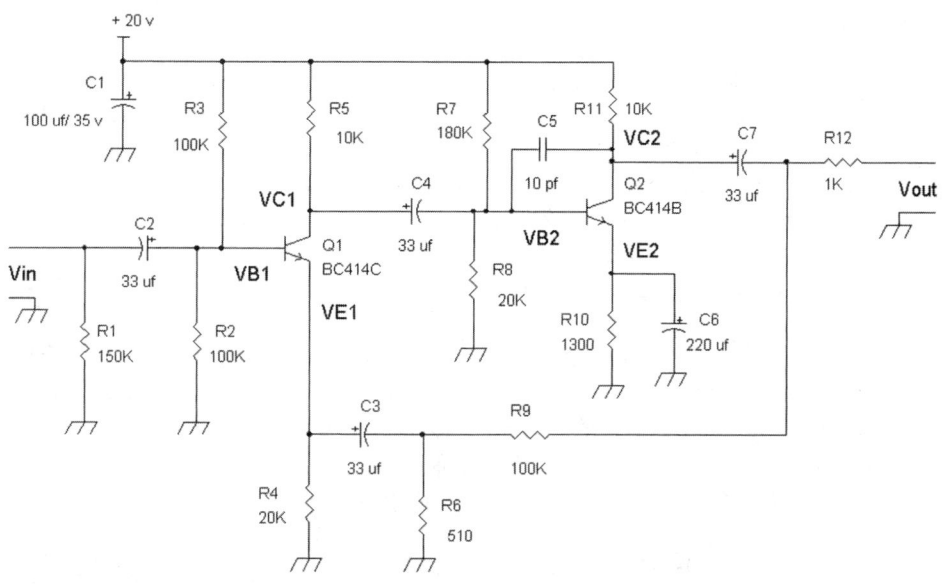

FIGURE 10-3 An example two-stage feedback amplifier with the first stage transistor with Q1 similar to the difference amplifier shown in Figure 10-2.

Let's first find the DC bias voltages. VB1 = 10 volts DC since R3 and R2 form a divide by two voltage divider circuit for +20 volts. For an estimate as long as R2 || R3 ≤ 10 × R4, we can ignore base currents as long as $\beta \gg 10$ (e.g., the $\beta > 300$ for both Q1 and Q2).

With VBE_{Q1} = VB1 – VE1, VE1 = VB1 – VBE_{Q1} or VE1 = 10 v –0.7 v or VE1 = 9.3 volts. From this, Q1's DC collector current IC_{Q1} = IE_{Q1} = 9.3 v/20KΩ = 0.465 mA. Thus, VC1 = 20 v – IC_{Q1} × R5 = 20 v – 0.465 mA × 10KΩ or VC1 = 20 v – 4.65 v, which leads to VC1 = 15.35 volts.

Note that although we are not optimized for maximum swing, it will be fine because the voltage gain of the next stage is very high and less than 1 volt peak to peak swing from Q1's collector will do. Also note VCE_{Q1} = VC1 – VE1 = 15.35 v – 9.3 v = +6.05 volts, which is good since the collector to emitter voltage should typically > 0.5 volt to ensure that Q1 is operating as an amplifying device.

For Q2, the base voltage is VB2 = 20 v [R8/(R8 + R7)], and thus:

VB2 = 20 v [20K/(20K + 180K)] = 20 v [20K/200K] = 20 v /10

VB2 = 2 volts, and since VBE_{Q2} = 0.7 volt

VE2 = VB2 – VBE_{Q2} = VE2 – 0.7 volt = 1.3 volts or VE2 = 1.3 volts

This means IC_{Q2} = IE_{Q2} = VE2/R10 = 1.3 v/1300Ω or IC_{Q2} = 1 mA.

VC2 = 20 v – IC_{Q2} R11 = 20 v – 1 mA (10KΩ) = 20 v – 10 v

VC2 = +10 volts

A single-supply two-transistor amplifier utilizes a difference amplifier Q1 where the emitter acts as a second input terminal for the feedback resistor network, R9 and (R6 ‖ R4). Q1's base can be thought of as a first input terminal. The difference voltage across the base-emitter junction of Q1 provides an output signal at the collector, VC1. This output signal is then sent to the second-stage amplifier, Q2, which is another inverting gain amplifier.

To understand this amplifier in terms of a **feedback system**, we can find which terminal of Q1 is the non-inverting input and which one is the inverting input. One way is to look at Q1 and Q2. We know that Q1 is a common emitter amplifier with phase inversion from the base to the collector in terms of AC signals (e.g., see Figures 10-1 left side and 10-3). Since the second stage with Q2 is also a common emitter amplifier, it inverts the phase from Q1's collector signal. Thus, from the base of Q1 to the collector of Q2, we get two 180-degree phase inversions so that the output signal at Vout is in phase with the input signal Vin.

For Figure 10-3's amplifier, we can say that Q1's base terminal is a non-inverting input. By process of elimination, is the remaining input at the emitter the inverting input for this two-transistor amplifier? Yes, that is the case. If we couple an input emitter terminal AC signal at the emitter of Q1 via C3, the collector voltage VC1 will have an AC signal in the same phase with this emitter terminal input signal. Since Q1's collector signal is sent to Q2, which is a common emitter amplifier, there will be a phase inversion from the collector signal of Q1 to the collector signal of Q2.

In other words, since a signal coupled to the emitter terminal at Q1 has the same phase as the collector signal at Q1, the phase is thus inverted from the Q1 emitter terminal signal to the collector of Q2, which is coupled to Vout. Thus, the emitter terminal of Q1 is an inverting input terminal for this amplifier.

We can estimate the non-inverting gain by just observing that the two-resistor feedback network R9 and R6 looks like the feedback network used in non-inverting gain op amp circuits. For a more accurate estimate we can replace R6's value with R6 || R4.

However, since R4 (20KΩ) >> R6 (510Ω) we can say that (R4 || R6) ~ R6 = 510Ω.

The estimated (closed loop) gain is then [1 + (R9/R6)] ~ Vout/Vin, *providing the total (open loop) gain from the first and second stages is* >> [1 + (R9/R6)].

For this example, R9 = 100KΩ and R6 = 510Ω, so our estimated gain is Vout/Vin ~ [1 + (100K/510)] or Vout / Vin ~ 197.

The quick analysis of the gain of the first stage involving Q1 is about:

(R5 || R7 || R8 || Rin2) / R6, Given R5 = 10KΩ, R7 = 180KΩ, R8 = 20KΩ, and Rin2 ~ 10KΩ when Q2's β ~ 400 at IC_{Q2} = 1 mA collector current

To calculate Rin2, with IC_{Q2} = 1 mA, r_{e2} = 0.026 v/IC_{Q2} = 0.026 v/0.001 A = 26Ω. Or r_{e2} = 26Ω. Rin2 = β r_{e2} = 400 $\times$ 26Ω or Rin2 = 10.4kΩ or Rin2 ~ 10kΩ.

The gain of the first stage is about (R5 || R7 || R8 || Rin2)/R6 ~ 4K/510 ~ 8.

Note VB2 ~ 2 volts, so VE2 ~ 1.3 volts and with R10 = 1300Ω, the emitter and collector currents of Q2 is ~ 1.3 volts/1300Ω or ~ 1 mA = IC2. The second stage's gain from the base of Q2 to its collector is about R11/r_{e2}, where **R11 = 10kΩ** with r_{e2} = 0.026 volt/IC2 r_{e2} = (0.026 volt) / (0.001 A) or **r_{e2} = 26Ω**.

Gain of the second stage = ~ 10KΩ/26Ω or about 380.

Total **open loop gain** is then = **gain of 1st stage** $\times$ **gain of 2nd stage** = 8 $\times$ 380 ~3040. The estimated "closed loop" gain is ~ 197 and so we should be OK since 3040 >> 197. Should the open loop gain be in the order of [1 + (100K/510)], such as the gain of 1st stage $\times$ gain of 2nd stage = 197, the actual gain Vout/Vin will be less than than [1 + (100K/510)] = 197.

Also note that the gain calculations are estimates, by which we mean if the measured gain is within 20 percent of the calculated value, that's good enough. Estimates within 20 percent are good enough for troubleshooting. If more accurate analysis is needed probably the best thing to do is to run a simulation via a program such as LT Spice if you are not into deep engineering analysis via hand calculations. See link to LTSPICE from Analog Devices (http://www.analog.com/en/design-center/design -tools-and-calculators/ltspice-simulator.html).

Before we look at the specific Dynaco PAT-5's voltage regulator and low-level preamp circuit, we should examine the emitter follower circuit. See Figure 10-4.

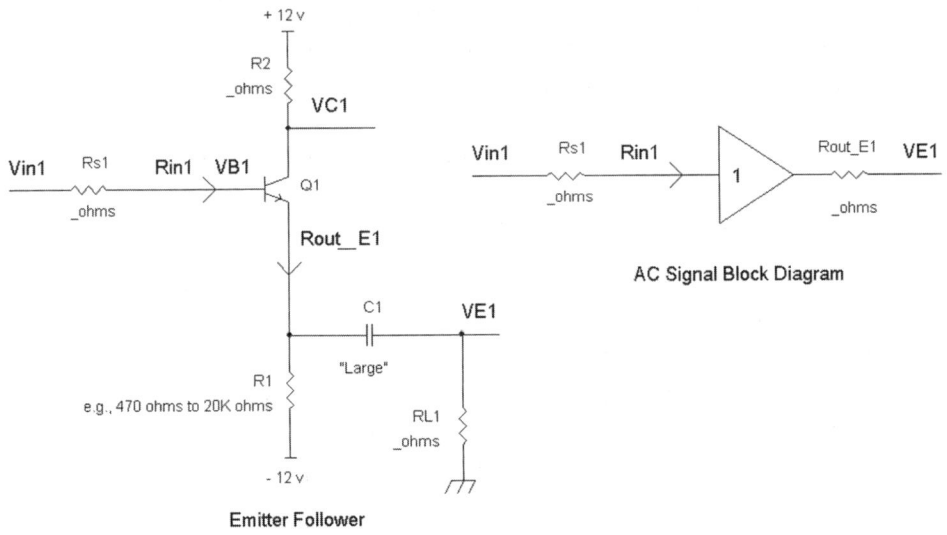

FIGURE 10-4 An emitter follower circuit on the left side, and on the right side a block diagram model of the emitter follower circuit for AC signals.

An emitter follower circuit is often used as a gain of 1 buffer amplifier to provide a high-resistance input (Rin1) and a low-resistance output (Rout_E1).

The input resistance Rin1 ~ β r_e + (β + 1)(R1 || RL1), where r_e = 0.026 v/IC wherein IC is the DC collector current.

And the output resistance from Q1's emitter terminal before being connected to R1 and RL1 is Rout_E1 = [Rs1/(β + 1)] + r_e.

For example, suppose Vin is an audio signal generator with a 600Ω as its source resistance, **Rs1 = 600Ω**, and **R1 = 1130Ω** so that **IC = 10 mA**. Then r_e = 0.026 v/0.010 A or r_e = **2.6Ω**. Let Q1 = PN2222 (TO-92 EBC pin out) with β = 75, then we have the following with **R1 || RL1 = 530Ω given RL1 = 1000Ω**.

Rin1 ~ β r_e + (β + 1)(R1 || RL1) = 75 ×26Ω + (75 + 1)(1130Ω || 1000Ω)

Rin1 ~ 1950Ω + 40.28KΩ or **Rin1 = 42.23KΩ**

Rout_E1 = [Rs1/(β + 1)] + r_e = 600Ω/(75 + 1) + 2.6Ω = 7.89Ω + 2.6Ω or **Rout_E1 = 10.49Ω**

As we can see, indeed the input resistance, Rin1 = 42.23KΩ, is reasonably high and the output resistance, Rout_E1 = 10.49Ω, is fairly low.

One other characteristic of the emitter follower is that it has power supply rejection. That is, if there is noise or ripple from the power supplies, little of it shows up at the output, Vout_E1. For example, if there is even large amounts of noise (e.g., 1-volt peak to peak ripple) on the +12-volt supply in Figure 10-4, the emitter follower will reject essentially all of the positive supply ripple voltage at Vout_E1. If there is ripple on the negative supply (–12 v), it will be reduced via a voltage divider as seen from the –12-volt supply via R1 and Rout_E1. The resulting voltage dividing factor is **Rout_E1/(Rout_E1 + R1)**, which is 10.49/(10.49 + 1130) ~ 0.01 ~ 1%. For example, if there is 100 mV of ripple on the negative supply for this example, then Vout_E1 will have only about 1 percent of 100 mV or 1 mV of noise at the output of the emitter follower.

NOTE: C1 has large capacitance to provide an AC short circuit (e.g., C1 is ~ 0Ω for AC signals when compared to RL1's resistance value).

Figure 10-4 shows an example of this circuit with an extra resistor, R2, which does not affect Rin1 and Rout_E1. Collector series resistor R2 can be used as a current-limiting resistor to prevent Q1 from sourcing excessive current that may damage it. Typically, R2 has a resistance value in the order of 2.2Ω to about 100Ω to provide current limiting just in case the Q1's emitter is shorted to ground. Also, R2 can be chosen with generally a higher-resistance value such as 510Ω to 10KΩ to act as a load for a common emitter amplifier for Q1. Care must be taken such that there is sufficient DC voltage across the collector-emitter terminals to avoid having Q1 in the saturation region, which will cause Q1 to not amplify.

Dynaco PAT-5 Low-Level Preamp Section and Power Supply

In the mid-1970s, the Dynaco corporation designed one of their last high-fidelity preamps, the PAT-5. We will look at a couple of its circuits for estimating performance and troubleshooting. See Figures 10-5 to 10-7. The design included a phono preamp section via an equalization network and a microphone channel. These figures have had their original Dynaco component reference designations renumbered for this book.

Let's split Figure 10-5 into two parts—the power supply–voltage regulator and the low-level preamp section. See Figure 10-6 for the power supply–regulator section.

The power supply uses a 90-volt RMS center tapped (CT) transformer, T1, that delivers 45 VAC RMS into each diode's anode for D1 and D2. The peak DC voltage into C1 is then 1.414 × 45 volts ~ 65 volts peak DC voltage. The cathodes of D1 and D1 provide an unregulated + 65 volts DC into capacitor C1 via the full-wave rectifier circuit of D1 and D2. R1 and C2 form a low-pass filter that removes most of the ripple (e.g., a 120 Hz sawtooth waveform for the 60 Hz transformer T1). A quick approximation for the cut-off frequency requires having R2 in parallel with R1 because ZD1, a 43-volt Zener diode, is turned on and for practical purposes the ripple voltage at ZD1's cathode is nonexistent. Thus, ZD1's cathode looks like an AC ground for R2 connected to C3. To find the cut-off frequency in a low-pass filter, we have to look at not only the driving resistor into C2 but also any resistor connected to C2 whose other end goes to AC ground. When we have this, the actual RC

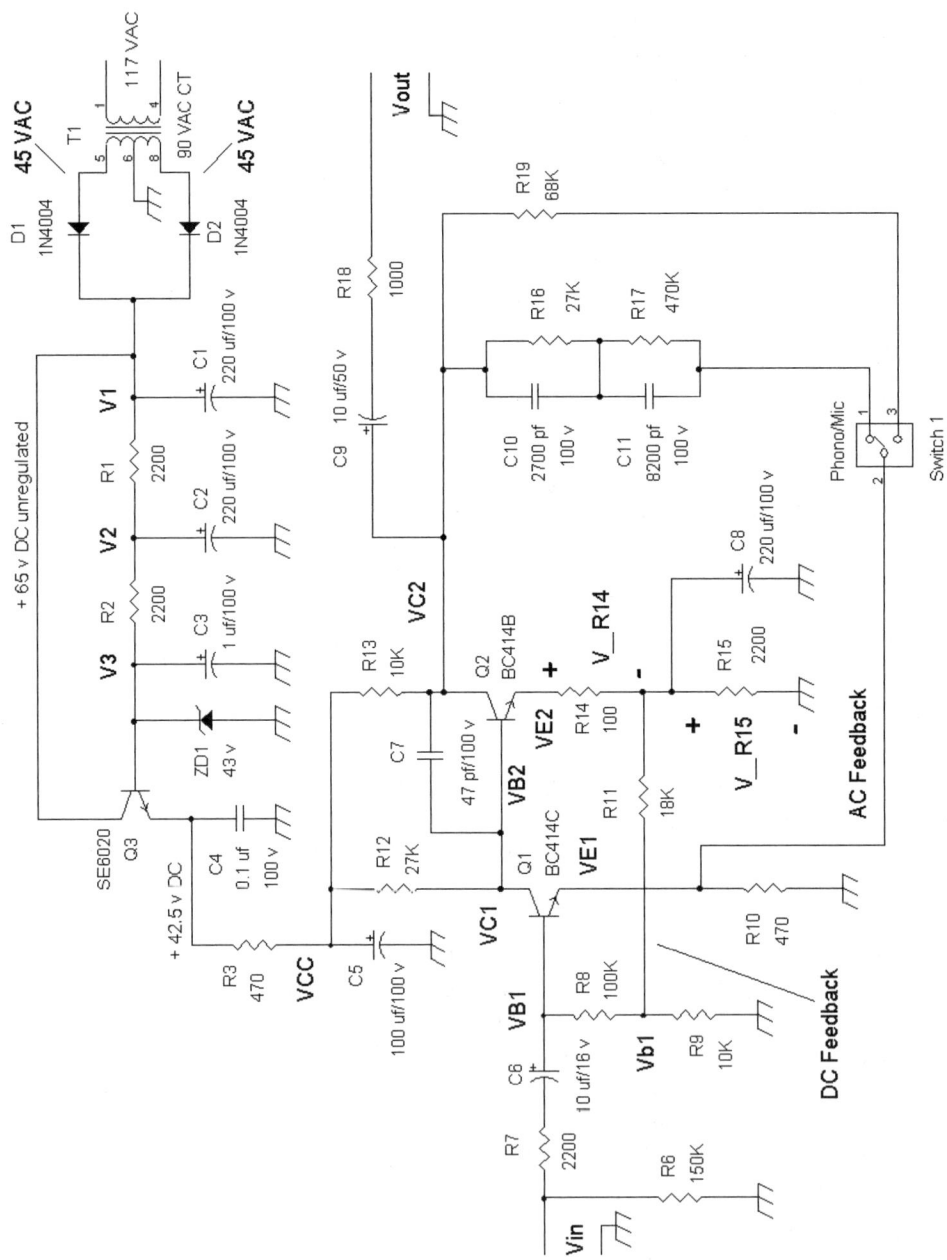

FIGURE 10-5 Dynaco PAT-5 low-level preamp section and power supply with regulator.

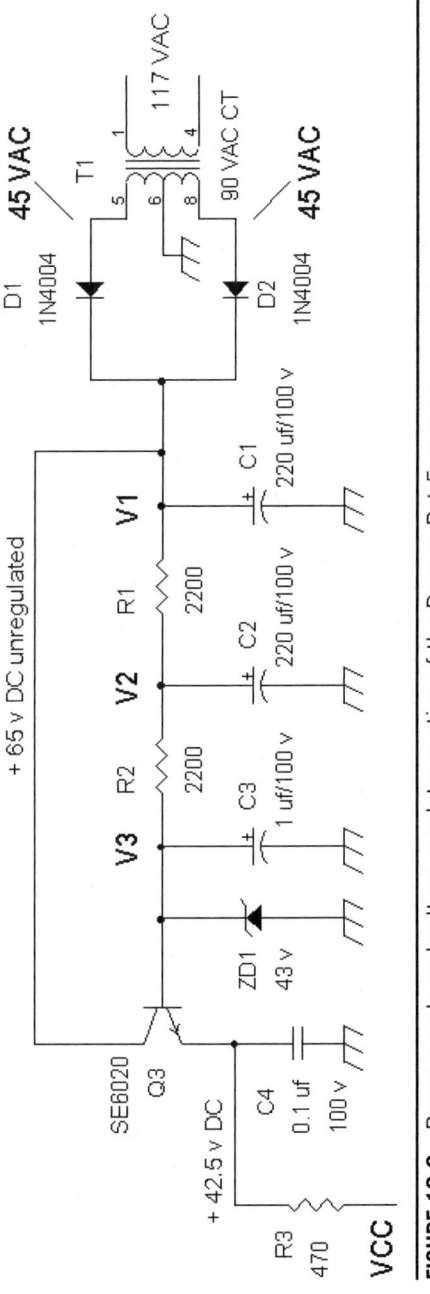

FIGURE 10-6 Power supply and voltage regulator section of the Dynaco Pat-5 preamp.

time-constant of the circuit is τ = (R1 ‖ R2) C2 and the cut-off frequency f_c = 1/[2 π τ] = 1/[2 π (R1 ‖ R2) C2] = 1/[2 π (1100Ω) 220 μf] or f_c = 0.66 Hz. For a rough estimate of how much the 120 Hz is attenuated we just take the ratio of the two frequencies, f_c / 120 Hz = 0.66 Hz/120 Hz = 0.0055 or 0.55 percent of the ripple voltage from the +65-volt unregulated supply. For example, if there is a 1-volt peak to peak 120 Hz sawtooth waveform at +65 volts, then we can expect about 0.55 percent of 1 volt or about 0.55 mV of ripple on the + terminal of C2. Because R1 and R2 have equal resistances, they form an averaging voltage summing circuit. Whenever you have two equal resistors and there are two different voltages, V1 and V2, driving each resistor, the voltage at the junction of the two resistors are (V1 + V2)/2. We have 65 volts at C1 and R1, and 43 volts at R2 and C3. Thus, the estimated voltage at C2 is (65 v + 43 v)/2 or + 54 volts DC at C2's positive terminal. Put it another way, the voltage at C2 should be the halfway point of the two voltages on each side of C1 and C3.

Transistor Q3 forms an emitter follower circuit with 43 volts at its base terminal from ZD1. We will lose about 0.7 volts due to the VBE or base emitter turn-on voltage. So, at Q1's emitter we would expect something like 42.3 volts. Q1's emitter can then provide about 42.3 volts at high currents while allowing ZD1 to be turned on. For example, the Zener diode current is approximately the voltage between R1 and R2 divided by (R1 + R2). This current is then (65 v – 43 v)/(R1 + R2) = 22 v/(2200Ω + 2200Ω) or ZD1's current = 5 mA. If we have 1 mA of base current from Q3, the Zener current will be reduced to 4 mA, which is still good for turning on ZD1. However, with 1 mA of drive into the base of Q3, and with Q3 having a β ≥ 50, at least 50 mA can be provided via Q3's emitter output terminal to the preamp circuits.

Capacitor C4 (0.1 μf) does not reduce any residual 120 Hz ripple voltage because the capacitance is too small. However, it does get rid of any higher-frequency noise, and C4 also "stomps" out any chance for Q3 to parasitically oscillate at very high frequencies (e.g., > 10 MHz). Further low-frequency filtering is provided by R3, which is coupled to C5, 100 μf at the preamp. (See Figure 10-5.)

Troubleshooting the Power Supply and Regulator

1. Confirm that the Zener diode, diodes, transistor, and capacitors are installed correctly. For example, if a 43-volt Zener diode is put in backwards, you will get a "fake" Zener voltage of 0.7 volts instead of the correct 43 volts. Remember in a Zener diode, the positive voltage via a resistor or current source will flows into the **cathode**. A transistor installed backwards, such as having its collector and emitter terminals reversed, may turn into a 7-volt Zener diode and it may have excessive base current due to running the transistor in reverse β mode. For example, the reverse β mode or reverse current gain mode in a transistor results in the current gain dropping to about 1 to 5, whereas a transistor normally has a current gain of over 30.

2. For your safety, be sure to avoid areas where the AC power line voltage comes in such as the power switch and or the primary windings of T1, the power transformer.

3. With the power turned on, confirm with a volt meter (negative or black lead at ground) that on the anode of D1 and D2 that there is about 40 AC volts RMS to 50 AC volts RMS. Most DVMs will measure AC in RMS voltages.

4. Then with the volt meter set to DC measurements, confirm that there is about 65 volts DC on C1's positive terminal. If the DC voltage is low, the capacitor may be old and have lost capacitance. Parallel a $\geq$ 220 μf 100-volt (extra) external capacitor with C1 and see if the DC voltage rises to about 65 volts. If the voltage is restored to about 65 volts, replace C1 with the external capacitor. Note that C1 can actually be 220 μf to 470 μf at 100 volts or more. Be careful to discharge the extra external capacitor by connecting across a 47KΩ to 220KΩ resistor for about a minute. If you short out the extra capacitor to discharge it, you will encounter a large spark, which may be unsafe.

5. Measure the DC voltage at C2's positive terminal. It should be higher than 43 volts (e.g., $\geq$ 50 volts DC) if there is about 65 volts at C1's positive terminal. Then confirm that there is about 43 volts at C3, anode of ZD1, and the base of Q3. If this is confirmed, Q3's emitter should have about 42.3 volts. You should also confirm that Q3 has a base-to-emitter DC voltage of about 0.7 volt. If the voltages are lower than expected at the base-to-emitter of Q3, then confirm there is about 65 volts at the collector. If Q3 does not have collector voltage > 43 volts, then the base-emitter junction acts like a diode, and excessive base current will be drawn because Q3 is not acting like an emitter follower circuit having a high-input resistance at the base. Also, if the base-to-emitter voltage is greater than 0.7 volts, such as VBE > 1 volt, then Q1's base-to-emitter junction may be blown, which results in an open circuit. You will need to replace Q3 and a suitable NTE replacement part is found via http://www.nteinc.com/index.php.

6. If the voltage at Q1's emitter is too high, that is, greater than the Zener voltage, 43 volts, then there could be a short across the collector-to-emitter junction of Q3. You will need to replace Q3.

7. Generally, if the Zener diode is damaged, it can be shorted out, which can result in 0 volts at the base and emitter of Q3. If the Zener diode is an open circuit, Q1's base voltage will be higher than 43 volts.

8. Measure the DC voltage across R3 (470Ω), which should be a few volts DC at most. If the voltage across R3 has over 6 volts, then the preamp may be drawing excessive current via C5 or other circuits.

Now let's look at the preamp section in Figure 10-7.

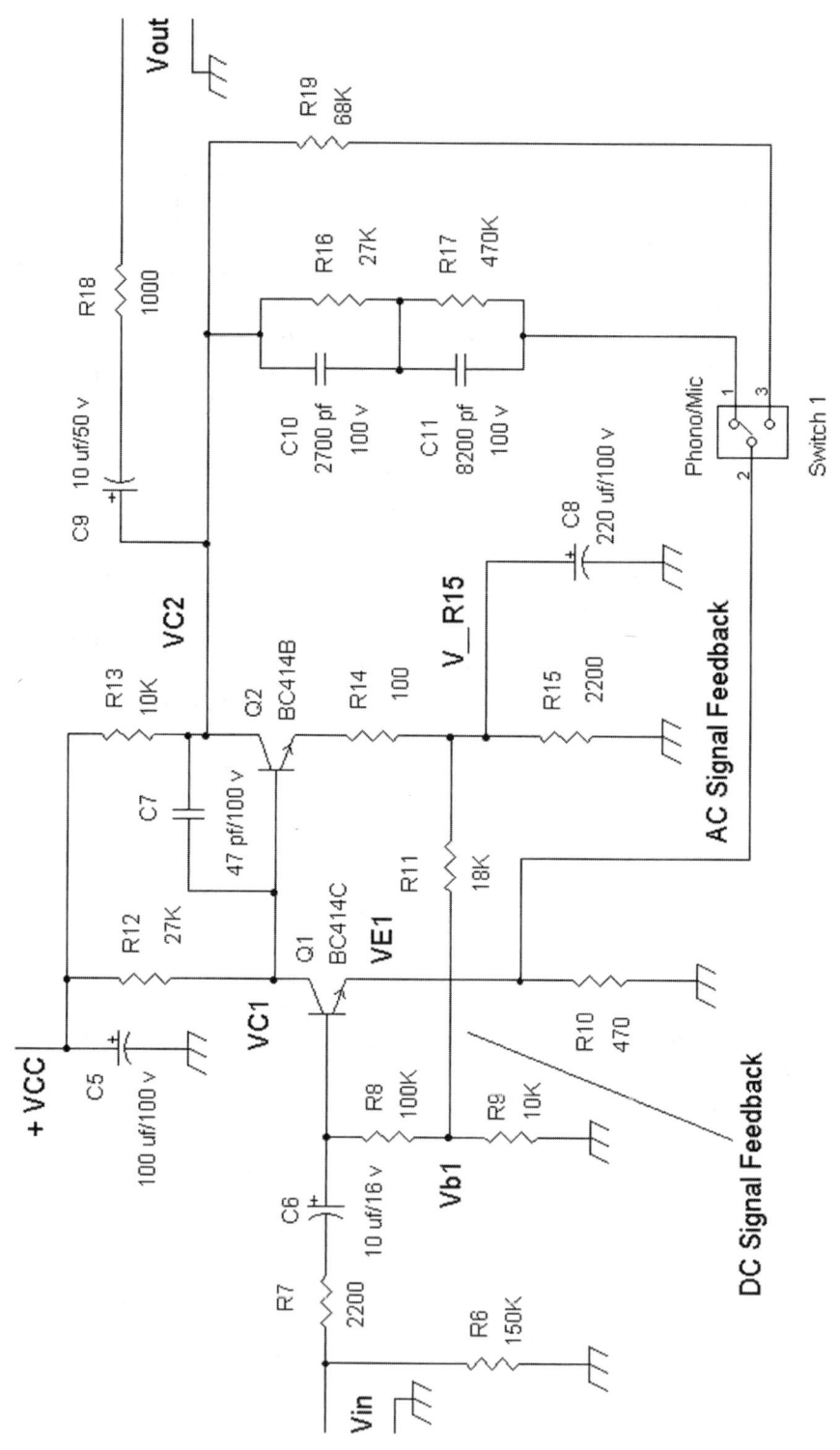

FIGURE 10-7 Preamp section of a modified PAT-5 preamp.

This preamp has two feedback loops. One is a DC bias feedback loop that includes R15, R11, R9, and R8. The second is an AC signal feedback loop via R19 and R10 when Switch 1 is at position "3" (microphone), and also with C10, C11, R16, R17, and R10 when Switch 1 is at position "1" (phono cartridge).

Preamp's DC Bias Point Estimates

We will look at the DC feedback loop first. Q1 by itself is a common emitter amplifier, which means the collector of Q1 has an inverted phase relative to the input signal into its base. In Figure 8-19 of Chapter 8 we saw a common emitter amplifier with a feedback resistor (R_BC) from collector to base that sets up the collector current of the transistor (along with base to ground resistor R_BE in Figure 8-19). When we look at Figure 10-7 we notice that Q1's collector is coupled indirectly back to its Q1 base by a "circuitous" route through Q2. This route starts from Q1's collector connected to the base of Q2, and then via the emitter of Q2 by way of R14 and R15, a feedback network of R11 and R9 feeds a DC signal back to the base of Q1 via R8.

In terms of DC biasing, Q2 is an emitter follower amplifier for Q1's collector output. Since an emitter follower has no phase inversion from base to emitter of Q2, we can connect a feedback resistor R11 back to the base of Q1 (via R8). Resistor R9 forms a voltage divider with feedback resistor R11. Notice that R11 is connected to C8, a large capacitance capacitor to "completely" filter out any AC signal via Vin such that only a DC signal voltage is fed to R11 to work properly for the DC feedback loop. In Figure 10-6 we can estimate VCC ~ 40 volts for a few volts drop across R3.

Let's look at estimating Q1's collector current. It is important to note that Q1's collector (VC1) is connected to Q2's base. We first want Q1's collector voltage, VC1, to be much less than one-half of VCC ~ +40 volts. The reason for this is to maximize voltage swing at Q2's collector output. Ideally, we want the base voltage of Q2 to be sitting at a low voltage such as less than 6 volts, but high enough to bias Q2's base-emitter junction and provide sufficient voltage at R15 to bias Q1's circuitry via base resistor R8. An example of a "bad" design would have the collector DC voltage at Q1 at +35 volts. Given that we have a 43-volt supply, this would mean Q2's collector can swing only from 35 volts to 43 volts, which is about 8 volts peak to peak. However, a better design having Q1's DC collector voltage at 6 volts instead means that Q2's output can swing from 6 volts to about 43 volt or about 37 volts peak to peak, which is much better than having an 8-volt peak to peak swing.

Let's assume that VC1 is in the order of 5 volts to 10 volts so that the DC voltage at the base of Q2 will allow good output voltage swing at Q2's collector. Recall that the collector voltage should always be above the base voltage. The collector current $IC_{Q1} = (VCC - VC1)/R12$. If we pick a guess of VC1 = 6 volts, then:

$$IC_{Q1} = (40 \text{ v} - 6 \text{ v})/27K\Omega \text{ or } IC_{Q1} \sim 1.26 \text{ mA}$$

NOTE: If we had guessed that VC1 = 10 volts instead, then $IC_{Q1} \sim 1.11$ mA, which is not much of a difference from 1.26 mA.

Now we can estimate Q1's DC emitter voltage as:

$$VE1 = IC_{Q1} \times R10 \sim 1.26 \text{ mA} \times 470\Omega \text{ or } VE1 = +0.59 \text{ v DC}$$

We can estimate that the VBE turn on voltage of Q1 and Q2 is about 0.7 volt.

Because the DC current gain for Q1 is very high, we can "ignore" base currents flowing through R8, so there is almost no voltage drop across R8. Thus, Vb1 ~ Q1's base voltage. In practice if Q1's β ~ 400 and Q1's collector current ~ 1.26 mA, its base current is about 1.26 mA/β or about 3.15 μA. The voltage drop across R8 is then 100kΩ × 3.15 μA or about 0.315 volt.

Q1's DC base volts ~ Vb1 = VBE_{Q1} + VE1 ~ 0.7 v + 0.59 v or Vb1 ~ 1.29 volts.
We can now find V_R15 via a voltage divider formula:

$$V\_15 [R9/(R11 + R9)] = Vb1$$

$$\text{Or } V\_R15 = Vb1/\{[R9/(R11 + R9)]\} = Vb1[(R11 + R9)/R9]$$

$$V\_R15 = Vb1 [(R11/R9) + 1]$$

With Vb1 = 1.29 volt, R9 = 10KΩ, and R11 = 18KΩ, we have for V_R15:

$$V\_R15 = 1.29 \text{ v} [(18K/10K) + 1] = 1.29 \text{ v} [1.8 + 1] = 1.29 \text{ v} [2.8]$$
or V_R15 = 3.6 volts

Thus, IC_{Q2} = V_R15/R15 = 3.6 volts/2200Ω or IC_{Q2} = 1.64 mA.
And this means VC2 = VCC – IC_{Q2} × R13 ~ (40 v – 1.64 mA × 10KΩ) VC2 = (40 v– 16.4 v) or VC2 = 25.6 volts DC.

The estimates for the bias points can be given a ± 20 percent tolerance. We are not after the exact values but something within the "ballpark." In some cases, you can find the schematic or service manual to high-fidelity product on the web. One such site, www.hifiengine.com, requires login information including email address and password. See Table 10-1 from the calculated values versus information from the Dynaco PAT-5 Preamplifier's construction and service manual.

TABLE 10-1   Estimated Calculations Versus Service Manual's Information, Which Are Within About 20 Percent

Voltage Node	Estimated Calculation	Service Manual's Information
Q1 emitter voltage (VE1)	0.59 volt	0.65 volt
Q1 base voltage (Vb1)	1.29 volts	1.25 volts
Q1 collector voltage (VC1)	6.0 volts	5.0 volts
Q2 emitter voltage (V_R15)	3.6 volts	4.4 volts
Q2 base voltage (VC1)	6.0 volts	5.0 volts
Q2 collector voltage (VC2)	25 volts	20 volts

Making estimates of the DC bias points can be handy because the service manual or schematic may not always be available. You may have to "manually" draw out a schematic by tracing the circuit you are troubleshooting. From that, many discrete amplifiers can be analyzed quickly either by making estimates as shown here, or by entering the schematic into a simulation program such as SPICE. You can download a free version from Analog Devices via the following link: http://www.analog.com/en/design-center/design-tools-and-calculators/ltspice-simulator.html.

AC Analysis

For looking into the small signal gain of Figure 10-7, we notice that there is an AC feedback path via Q2's collector to R10 at the Q1's emitter either through C10, C11, R16, R17, or through R19. In the larger scheme of things, Q1's base is like the non-inverting input terminal of an op amp or feedback amplifier, while Q1's emitter terminal is like the inverting input terminal. In the simplest case where Switch 1 selects position 3 (for microphones) to utilize R19 and R10 as the feedback elements, the gain of the preamplifier is Vout/Vin ~ [1 + (R19/R10)]. If this equation looks familiar, it is the same equation for calculating voltage gain for an op amp circuit for a non-inverting amplifier. With R19 = 68KΩ and R10 = 470Ω, Vout/Vin ~ [1 + (68K/470)] or Vout/Vin ~ 145.

With Switch 1 set for phono preamp, the feedback network C10, C11, R16, and R17 works like a frequency dependent "resistance." This feedback network complies to the ***RIAA phono equalization curve***. This network is more properly termed as an impedance network. At high audio frequencies the two capacitors will dominate the impedance characteristic and act like a low-value impedance. We can model this network to give an idea of what we would expect the gain to be as a function of frequency. Also, with a signal generator and oscilloscope you can confirm the following:

Vout/Vin ~ [1 + ($Z_{network}$/R10)]

where $Z_{network}$ is a combination of C10, C11, R16, and R17.

At a low frequency such as 20Hz, the capacitors have very high impedances and the resistors dominate so $Z_{network}$ ~ R16 + R17 = 27KΩ + 470KΩ = 497KΩ. Thus, at around 20 Hz, the expected gain within about 20 percent is Vout/Vin ~ [1 + (497K/470)] ~ 1000. Try setting the generator at 10 mV peak to peak at 20 Hz for Vin, and Vout should be about 10 volts peak to peak within 20 percent.

To find what we may expect at 1000 Hz, we can look at the RIAA feedback network and pick out R16 at 27KΩ as the resistor value we will use to find the gain. This means at 1000 Hz the gain will be [1 + (27K/470)] = 58.4 within 20 percent. The actual gain may be slightly higher, like 70. Again, you can set the generator at 1000 Hz with 50 mV peak to peak at Vin and Vout should be about 2.9 volts to 3.5 volts peak to peak.

Finally, at 20 kHz, the RIAA equalization curve requires that the gain is dropped to 10 percent of the 1000 Hz gain. So, at 20 kHz we will expect Vout/Vin ~ 5.84 within 20 percent. Again, set the generator to about 100 mV peak to peak at 20 kHz for Vin, and Vout should be about 0.6 volt peak to peak within 20 percent. If Vout is around 0.75 volts peak to peak, that's fine too.

Another way to test this is: Set the generator to about 10 mV RMS or about 28 mV peak to peak at Vin, and test at 20 Hz, 1000 Hz, and 20 kHz and note Vout's amplitude. At 20 Hz, the amplitude should be about 10x of the output at 1000 Hz. And at 20 kHz, its amplitude should be about 10 percent or one-tenth of the amplitude of the 1000 Hz signal.

Troubleshooting the Preamp in Figure 10-7

- Make sure there is about +40 volts DC from the power supply to C5 and then check for the DC bias points listed on Table 10-1. If the DC voltages are within about 20 percent, then the preamp is ready to be tested with AC signals.
- If there are DC bias problems, with a DVM, check the VBE (base to emitter) turn-on voltage for transistors Q1 and Q2. Typically, this should be between 0.6 volt and 0.7 volt. If you still have problems and the DC voltages are incorrect per Table 10-1, inspect each of the transistors to make sure the pin outs for emitter, base, and collector are connected correctly. For example, if any of the transistors are connected in reverse, where the collector and emitter terminals are reversed, the β will drop from about 400 or 500 to 1 to 5, which will cause excessive base currents to be drawn. If there is excessive base current from Q1, the DC voltage across resistor R8 will be large, since typically with a β ≥ 400, the voltage across R8 will be < 0.5 volt. If Q1 is wired with collector and emitter reversed, the voltage across R8 will be >> 0.5 volt.
- Check that all electrolytic capacitors are installed correctly polarity-wise. An incorrect installation in electrolytic capacitors will cause capacitor leakage currents, which can "throw off" the DC bias voltages.
- With a signal generator at 1000 Hz sine wave and 50 mV peak to peak for Vin, select Switch 1 for phono operation, and confirm that Vout has about 60 to 75 times the input amplitude or 3 volts to 3.75 volts peak to peak. You can probe Q2's base to verify that you get about 100 times less amplitude of the 1000 Hz signal than at the collector. The gain of the second stage Q2 is ~ R13/R14 = 10K/100 ~ 100 within 25 percent. So expect Q2's base to have about 30 mV of signal. Note that R15 is AC shorted to ground via C8. Confirm that there is no 1000 Hz sinewave signal C8 (220 μf) with an oscilloscope set at 10 mV per division sensitivity and 500 μsec per division. You should get close to a flat line at this oscilloscope setting. If there's an AC signal at C8's plus terminal, replace it with a new 220 μf capacitor of at least

50 working volts. Also confirm that Q1's emitter terminal has almost the same AC amplitude voltage as Vin. Note that Q1's base and emitter terminals are like the (+) and (−) inputs to an op amp, which have the same AC voltage (e.g., virtual short circuit across the inputs) when negative feedback is applied.

- If Vout seems to have some high-frequency parasitic oscillation, check or replace C7 (47 pf) since this capacitor is used to ensure that this preamplifier is stable with feedback components (R16, R17, C10, and C11).
- If the Vout's signal is clipped, check the load resistance value, which typically should be ≥ 10KΩ. A typical audio power amplifier has at input load resistance in the range of 10kΩ to 250kΩ. If the preamp's output is loading into a headphone or resistor in the 1KΩ range, then Vout will clip at voltages < 3 volts peak to peak. Again, this preamp is not meant to drive earphones or headphones. The Q2 output stage can only deliver about a milliamp or two.

A High-Fidelity Audio Power Amplifier

Now let's look at a SX-636 Pioneer stereo receiver power amplifier. Although this is a 1970s amplifier design, the circuit topology still held up into the early 2000s. Many of the newer designs still include its basic features, a differential pair input stage and a complementary push pull output stage. We will explore some of the sub-circuits in this amplifier including the input stage, the second voltage gain stage, and its output stage.

As engineers learned how to design with transistors, they migrated from earlier circuit topologies that included audio signal transformers to output transformer-less designs using output capacitors (e.g., 2200 μf) to direct DC coupled output stages. The SX-636 stereo receiver's design used positive and negative power supplies, and in essence, it is almost like a high-power output op amp. See Figure 10-8.

Most power amplifiers include two-stage preamplifiers, and this circuit is no exception. In place of a single transistor difference amplifier from Figure 10-7 Q1 (BC414C) where Q1's base is the (+) input terminal and its emitter is the (−) input terminal, we have in Figure 10-9 transistors Q1 and Q3 where the Q1's base is the (+) input terminal and Q3's base is the (−) input terminal.

In Figure 10-9, Q1 and Q3 form a "true" differential amplifier where the input resistance input into each base is the same. *Recall that in a single transistor difference amplifier, the input resistance into the base is much greater than the input resistance into its emitter by a factor of β.* **Note: Start with confirming all supply voltages.**

We can observe that the emitter-to-base voltages of Q1 and Q3 are both approximately 0.7 volts. From this we can also surmise that the DC voltage across VB1 and VB3 is ~ 0 volt since $VB1 = VBE_{Q1} + VEB_{Q3} + VB3$, or $VB1 - VB3 = VBE_{Q1} + VEB_{Q3}$.

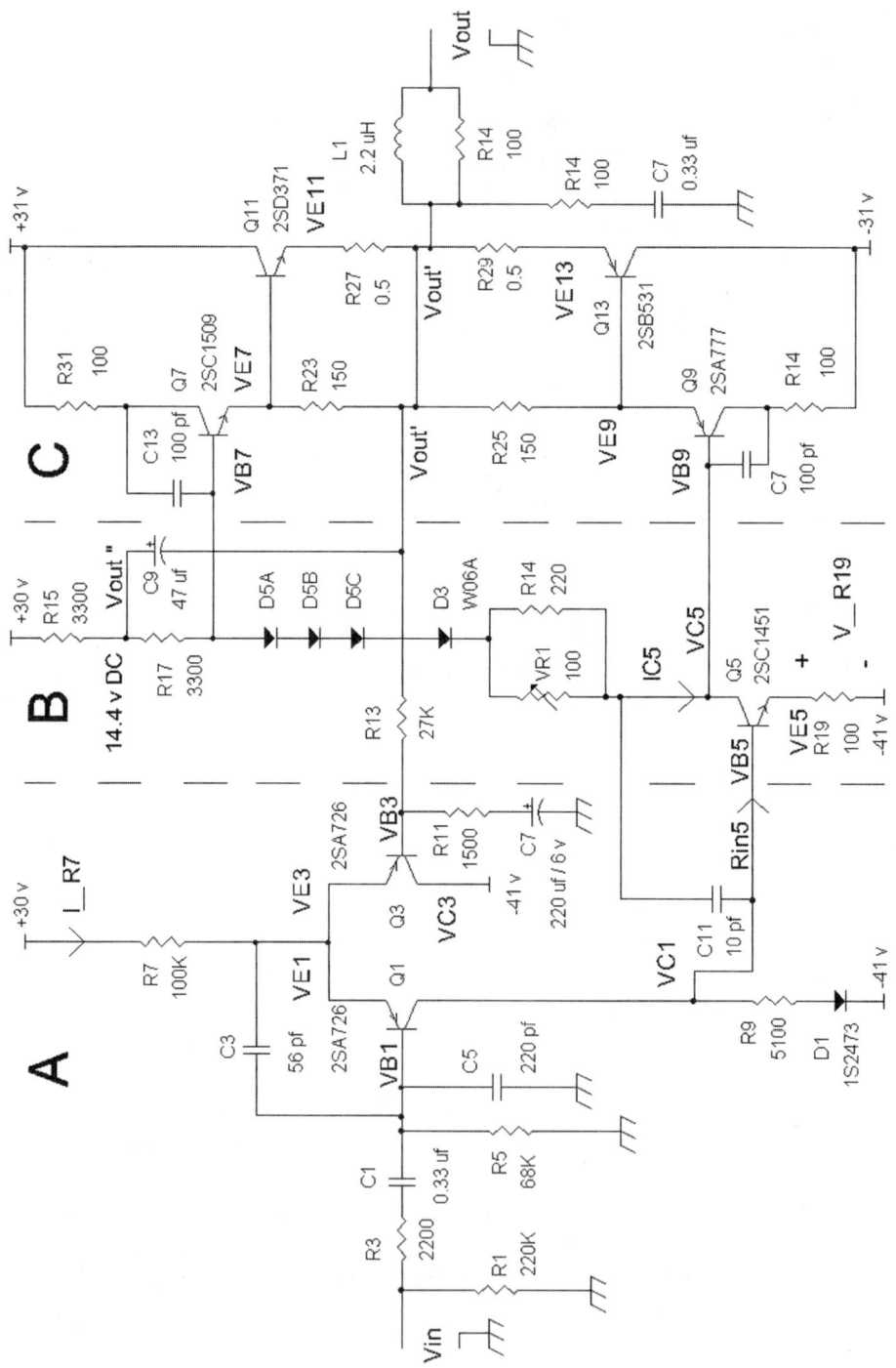

FIGURE 10-8 A redrawn full-page schematic of the Pioneer SX-636 audio power amplifier. Sections A = first stage differential pair, B = second stage voltage gain amplifier, and C = output stage to supply large currents to a loudspeaker.

232

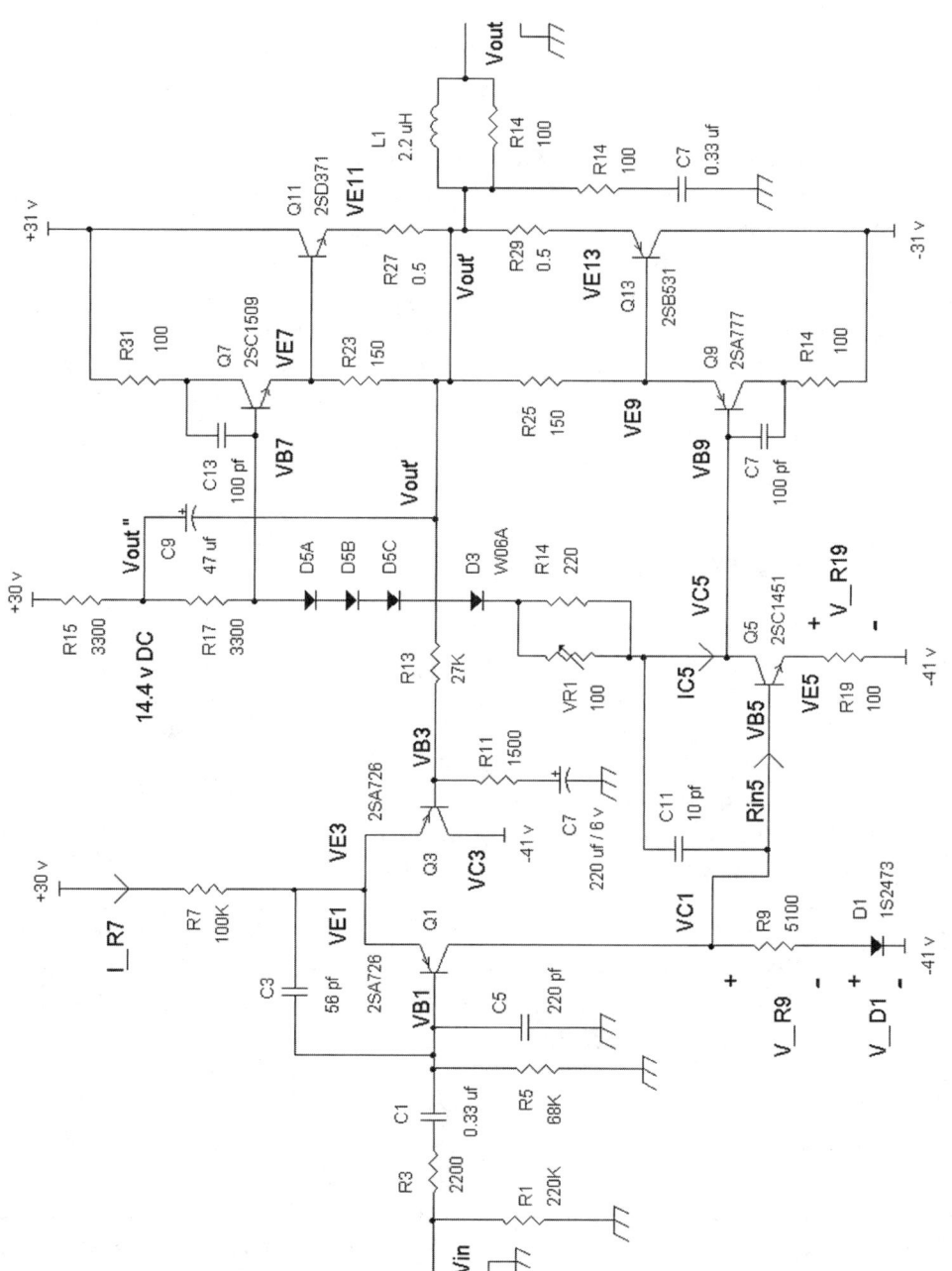

FIGURE 10-9 SX-636 power amplifier schematic.

However, **VBEQ1** = –**VEBQ1**, thus VB1 – VB3 = (–**VEBQ1** + VEBQ3). Since the emitter base turn-on voltages of PNP transistors Q1 and Q3 are both ~ 0.7 volt, so **VEBQ1** = VEBQ3 = 0.7 v. And this means the difference voltage between the two bases is VB1 – VB3 = –**0.7 v** + 0.7 v, or **VB1 – VB3 = 0 v**, which also leads to **VB1 = VB3**, which is the same approximation for an operational amplifier circuit with negative feedback where the (+) and (–) inputs are about the same voltage. The output of the Q1 and Q3 amplifier is VC1 that feeds the amplified signal to the second stage, Q5, a common emitter amplifier. The output signal from Q5's collector (VC5) is then fed to a two-stage complementary transistor emitter follower with the first emitter follower stage formed by Q7 and Q9, and whose second stage emitter follower circuit includes Q11 and Q13, which are high-current output transistors that are mounted on a heat sink. Q7, Q11, Q13, and Q9 biasing diodes are D5A-D5C and D3. These diodes are also on the heat sink with output transistors Q11 and Q13. Diodes D5A-D5C and D3 then track the temperature of the output devices and prevent Q11 and Q13 from heat damage. For example, if the output transistors heat up, diodes D5A-D5C and D3 will also heat up and provide a lower biasing voltage to the output transistor, therefore reducing the output transistors' DC quiescent collector current and thus reducing power dissipated (e.g., which lowers the temperature) in the output devices Q11 and Q13.

DC Biasing Conditions in Figure 10-9

Q1 and Q3 form a differential pair amplifier where it is the difference in voltage between their two bases that provides an amplified output at Q1's collector, VC1. To bias the differential amplifier, resistor R7 provides an emitter tail current to Q1's and Q3's emitter. Ideally, when with no signal, Vin → 0 volt AC, *we want the tail current I_R7 to split evenly between Q1 and Q3*. Reminder: Confirm all supply voltages.

When Vin → 0 volt AC, Q1's base voltage, VB1 ~ 0 volt DC. If VB1 ~ 0 volt, then VE1 ~ 0.7 volt since in a PNP transistor the turn-on voltage is 0.7 volt from emitter to base (VEB ~ 0.7 volt).

The emitter tail current is then the current flowing through R7, I_R7 = [(30 v – 0.7v)/R7] = I_R7. With R7 = 100KΩ, we have I_R7 = (29.3 v)/100KΩ or I_R7 = 0.293 mA. With the tail current 0.293 mA, we have split evenly this current between the collector currents of Q1 and Q3. This means that Q1's collector current IC1 is one-half of the emitter tail current I_R7, so IC1 = 0.1465 mA.

With Q1's collector current known, we can take a look at the second stage, Q5's DC biasing conditions. Q5's collector needs to draw enough current via IC5 such that at the base of Q7, VB7, we have about +1.4 volts, which is the sum of the VBE turn-on voltages for Q7 and Q11. The reason for having this +1.4 v = VB7 is so that by the time we measure Vout, we should get close to 0 volts DC. For example, a large DC voltage of 10 volts DC at Vout can burn out an 8Ω loudspeaker with about 1.25 amps DC flowing through the speaker's voice coil.

Q5's collector current, IC5 = (30 v – 1.4 v)/(R15 + R17). With R15 = R17 = 3300Ω, IC5 = 28.6 v/(3300Ω + 3300Ω) or IC5 = 4.33 mA. Now that we know that IC5 ~ 4.33 mA, we would deduce that the voltage across R19, V_R19, which is Q5's emitter resistor, should be IC5 × R19 or about 0.00433 A × 100Ω = 0.433 volt across R19, a 100Ω resistor.

The total voltage across Q5's base emitter junction and R19 = VBE_{Q5} + V_R19 ~ 0.7 v + 0.433 v = 1.133 volts. This total voltage having **VBEQ5 + V_R19 = 1.133 volts** is also equal to the voltage across R9, V_R9 plus the voltage across D1, V_D1. Because IC1 ~ 0.1465 mA, the forward voltage across D1, V_D1 ~ 0.4 v. You can download the spec sheet for the 1S2473 diode for D1.

This means V_R9 + V_D1 = VBE_{Q5} + V_R19 = 1.133 volts.
Or V_R9 = 1.133 v – V_D1 = 1.133 v – 0.4 v, or V_R9 ~ 0.733 volt.
So, does IC1 × R9 = 0.1465 mA × 5100Ω = 0.733 volt? Let's see.

0.1465 mA × 5100Ω = 0.747 volt

which is pretty close to 0.733 volt.

So, based on working "backwards" from Q5's collector current for 0 volts at Vout, we see that the voltage driving into the base of Q5 via Q1's collector voltage VC1 is close but not "exact." As we will see, this amplifier is part of a negative feedback system that will adjust VC1 "precisely" to the correct DC voltage such that Vout = 0 volt DC.

So, how does the amplifier circuit "guarantee" that Vout will have 0 volts DC when Vin = 0 volt? What if the transistors, Q5, Q7, and Q11, really do not have 0.7-volt VBE base-to-emitter voltages? Wouldn't we have a DC offset problem at Vout? Fortunately, the amplifier is set up as a unity gain feedback amplifier or voltage follower for DC signals. That is, for DC voltages, you can remove C7 for DC analysis. Now the amplifier looks very much like a voltage follower op amp circuit with R13 as the feedback resistor. So, the amplifier achieves near 0-volt DC at its output due to Vout' being connected back to the inverting input terminal (Q3's base) via R13.

Since the differential amplifier Q1 and Q3 and second-stage amplifier Q5 provide "high" DC gain, Figure 10-9's circuit looks like one "big" discrete op amp of sorts. Thus, by negative feedback, the DC output voltage is very close to 0 volt, because the (+) input (Q1's base) is set to ~ 0-volt DC, which then drives via feedback resistor R13 to cause the (–) input voltage at Q3's base to go to → 0 volt. Thus, Vout' and Vout both go to 0-volt DC. (At Vout, inductor L1 = 0Ω for DC currents.)

One of the most important DC conditions to look for is the bias voltages for the driver transistors Q7 and Q9 and the output transistors Q11 and Q13. Biasing diodes D5A – D5C in series with another biasing diode D3 should provide about +2.5 volts from the base of Q7 (VB7) to the cathode of D3. VR1 and R14 supply additional voltage in series with the biasing diodes such that you can set the output transistors' quiescent bias current. To do this, Vin is removed. Then measure with a volt meter across either output transistor emitter resistors R27 or R29 and adjust VR1 for about

0.025 volt to 0.038 volt so that the output devices are idling at 50 mA for "normal biasing" or at 76 mA for lower crossover distortion.

NOTE: The total DC biasing voltage for the driver and output transistors is $VBE_{Q7} + VBE_{Q11} + VEB_{Q13} + VEB_{Q9} \sim 0.7$ v $+ 0.7$ v $+ 0.7$ v $+ 0.7$ v $= 2.8$ volts.

AC Signal Conditions

For signal tracing this amplifier, we will need an oscilloscope and sine wave generator. Also, because this is indeed a power amplifier, an 8Ω load resistor rated at 40 watts or more will be connected to the output, Vout. You can improvise by connecting eight 1Ω 5-watt resistors in series or parallel six 47Ω 10-watt resistors. Although 8Ω loads are commonly used for evaluating power amplifiers, you can also use resistive loads in the 4Ω to 16Ω range at the appropriate wattage ratings of at least 40 watts. Alternatively, a 100-watt incandescent light bulb can be used and it will have about 10Ω (cold) resistance, but the resistance will go up as the lamp glows brighter.

If the DC conditions are correct in that Vout ~ 0-volt DC, you can first test the amplifier without any load resistor. Set the sine wave generator for 1 volt peak to peak at 400 Hz. The AC gain should be Vout/Vin = [1 + (R13/R11)] or [1 + (27K/1500)] = **Vout/Vin = 19**. This means Vout = 19 × 1 volt peak to peak or Vout = 19 volts peak to peak 400 Hz sine wave.

Now with the oscilloscope, you should be able to also measure about 19 volts peak to peak sine wave at VB7, VB9, and at C9's positive terminal, Vout".

The gain of the second stage, Q5, may look "obvious" because the collector load resistors, R15, in series with R17 may look like the total load resistance of R15 + R17 = 3300Ω + 3300Ω = 6600Ω. The DC load resistance is 6600Ω for Q5 and the DC gain from VB5 to VC5 ~ 6600Ω/R19 = 6600/100 = 66. However, the AC load resistance is much higher due to AC coupling capacitor C9, which is also known as a "bootstrap" capacitor. Capacitor C19 has two functions. First, it provides a same phase signal from Vout back to R17 for providing extra signal during the positive cycle of the sine wave such that there is increased current drive into Q7 and Q11.

Without C9, the amplifier will clip prematurely at the positive cycle of the sine wave. The second function of C9 is to boost the AC load resistance for Q5's collector. We can estimate the collector load's AC resistance due to C9 and R17, which will result in something >> R17. For a very rough approximation, the AC signal is identical at both terminals of R17, which are at Vout" and VB7. With "equal" AC voltages at both ends of R17 no AC current flows through R17, and this results in an infinite AC resistance for R17. However, there is a slight loss at Vout and Vout" compared to the AC signal at Q5's collector. The two-stage emitter followers from Q7 and Q11 form a gain of about 0.8 to 0.9 gain. That is Vout/VB7 ~ 0.8 to 0.9. The bootstrapping effect from C9 that couples the output signal Vout back to R17 will raise the effective collector load resistance for Q5 in the following manner: Q5's load resistance ~ R17/[1 – (Vout/VB7)]. So, if Vout/VB7 ~ 0.9, then Q5's load resistance ~ 3300Ω/[1 – (0.9)] = 3300Ω/0.1 = 33,000Ω, and the approximate AC gain from Q5 is about 33,000Ω/R19 = 33,000/100 = 330. Note that R19 = 100Ω.

Compared to the DC gain of 66, the AC gain of 330 is 5 times higher due to the bootstrapping capacitor. The actual AC gain from VB5 to VC5 will be at least twice the gain of 66 but likely less than 500 when loaded with an 8Ω load at Vout. If the gain is around 66 or less, then C9, the bootstrapping capacitor, has lost capacitance and needs to be replaced. For example, if the output voltage is 19 volts peak to peak at 400 Hz sine wave, then VC5 should also have about 19 volts peak to peak; and the VB5 should have < (19/66) volt peak to peak, which is < 287 mV volt peak to peak at Q5's base. Note that in this example, VB5 ~ 19 volts/AC gain, and *if* the AC gain is estimated to be at 330, then VB5 ~ 19 volts/330, or VB5 ~ 57.5 mV peak to peak (which is less than 287 mV peak to peak).

NOTE: The diodes D5A to D5C and D3 form a low AC resistance ≤ 50Ω because they are being used as a constant voltage source, which by definition has ~ 0Ω AC resistance. The maximum resistance by VR1 in parallel with R14 is about 70Ω.

Troubleshooting the Power Amplifier

1. With the power turned off, check the resistances from collector to emitter of the output transistors Q11 and Q13. You will need to measure the collector-to-emitter resistances a second time with the probe leads reversed. Typically, if the resistances are > 10Ω, the transistors are fine. But if the collector-to-emitter resistances fall in the 2Ω or less category, chances are that there is a shorted transistor. If you find any one of the output transistors shorted, remove both output transistors Q11 and Q13. See Figure 10-10.

    Now, measure the collector-to-emitter resistances of the driver transistors, Q7 and Q9, the same way as before, twice with the second time having the ohm meter's leads swapped. If there are no short circuits from collector to emitter of the driver transistors, you can now power up the amplifier without the output transistors, because the driver transistors will keep the negative feedback loop intact via R23 and R25 connected to feedback resistor R13. Confirm that the DC voltage at Vout is about 0 volts. Connect a 1000Ω load resistor to Vout and connect a signal generator at 1-volt peak to peak sine wave at 400 Hz to Vin. Confirm with an oscilloscope that Vout has a sine wave at about 19 volts peak to peak. If Vout does provide about a 19-volt peak to peak signal into the 1000Ω half-watt (or two 2000Ω quarter-watt resistors in parallel) load resistor, then the amplifier probably works fine up to this point and all you need is to replace the power transistors. If you cannot find the original part numbers, you can generally replace the NPN and PNP TO-3

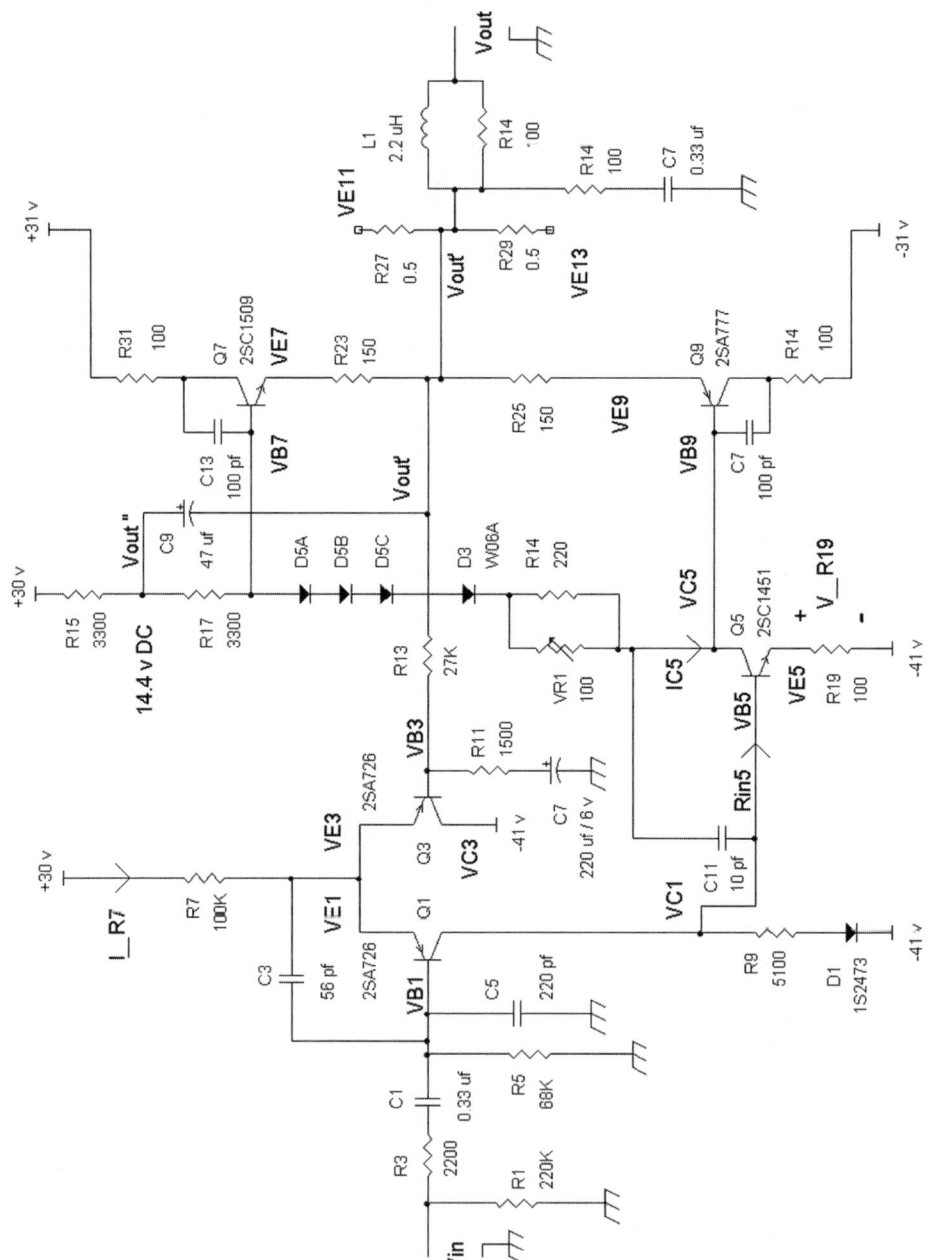

FIGURE 10-10 Schematic with output transistors removed, and driver transistors Q7 and Q9 to Vout.

238

case output transistors with parts from NTE at: http://nte01.nteinc.com/
nte%5CNTExRefSemiProd.nsf/$all/72BF14B50722ACF285257910007E5A0
D?OpenDocument.

Turn off the power or, better yet, unplug the power cord and replace the
output transistors. Keep track of which transistor goes where. The NPN transistor will have its collector connected to the +31-volt supply line, which you
can confirm with an ohm meter with the power turned off. The NPN 2SD371
can be replaced by an NTE380, MJ802, or 2N5886, and the PNP 2SB531 can
be replaced with an NTE281, MJ4502, or 2N5884.

Alternatively, you can select for TO-3 case NPN and PNP power transistors that have voltage breakdown of at least 70 volts and at least a 10A collector current rating. For example, the 70-volt transistors 2N3055G (NPN) and
MJ2955 (PNP) may serve as replacements. Although, to be on a safer side, the
higher-voltage transistors mentioned above should be used. With the output
transistors replaced, turn off the signal generator, and turn on the amplifier
and measure for about 0 volts at Vout.

2. Replace the 1000Ω load resistor with at least a 25-watt 8Ω resistor. Be sure to
keep the 8Ω load resistor away from other items because it will be very hot
once the amplifier supplies signal to it. Turn on the generator for 1 volt peak
to peak into Vin at 400 Hz. You should measure with an oscilloscope about 19
volts peak to peak across the 8Ω load resistor. Sweep the frequency from
about 50 Hz to about 10 kHz and confirm about 19 volts peak to peak at the
output. Now reduce the signal generator's input voltage by half to 0.5 volt peak
to peak, and sweep the generator from 50 Hz to 20 kHz to observe about 9.5
volts peak to peak at Vout.

CAUTION: Do not try to measure the amplifier at 20 kHz at full power. This test can
eventually destroy the output transistors due to excessive heating. At the top end of
the audio frequencies, the output power transistors become more lossy, and extra
heat builds up, which can damage them.

3. In some cases, it may be difficult to find exact replacements for all the parts.
Generally, it is better to replace "up" with higher power dissipation, voltage, or
collector current devices. This amplifier has supply voltages of +31 volts to –41
volts. Thus, to be on the safe side, we can specify the transistors' voltages as >
(+31 v – –41 v = +31 v +41 v = 72 v) or > 72 volts. Given that the AC power
line can increase as much as 10 percent, the transistors' minimum voltage rating
would be about 80 volts. For example, you can replace the input transistors,
2SA726 (BCE pin out) with careful installation using 2N5401 (EBC pin out)
because the pins outs are not the same. Similarly, the pre-driver transistor,
2SC1451 (EBC), 150 volts, TO-5 case, may be replaced with TO-5 case
transistors 2N3440 (EBC), 250 volts or 2N3439 (EBC) 300 volts. If there is a

recommended substitution with a lower power dissipation or plastic TO-92 case transistor, it's better to find another transistor with higher power ratings.

For transistors that begin with 2SA, 2SB, 2SC, and 2SD, sometimes the "2S" is dropped. For example, a **2SD371** may be also named **D371**. Or a **2SA726** and an **A726** transistor are really the same electronic part.

However, it is always best to replace any broken transistor with the original part or with an NTE version. Replacing with other components can sometimes lead to parasitic oscillations. Make sure to retest the amplifier with an oscilloscope probing the output terminal, Vout, and look for any small or large high-frequency signals. If they are there, you need to go back and look for original parts.

Also, sometimes you may be able to find the exact part number replacements on eBay, but beware of counterfeit items.

This wraps up Chapter 10. Obviously, there are many more ways to examine both the PAT-5 preamp and the SX-636 power amplifier. Troubleshooting takes experience, but make sure to take safety precautions.

Chapter 11 will look into analog integrated circuits.

CHAPTER 11
Troubleshooting Analog Integrated Circuits

We will be exploring various types of circuits and systems that use analog integrated circuits as examples of troubleshooting, debugging, and improving. Many of the circuits shown in this chapter will have come from the web (world wide web) or from magazine articles or books. When we build hobby circuits, or fix commercially manufactured products, generally one would assume that the circuits are reliable in terms of performance/design with replaceable parts. From my experience building hobby circuits, working in repair shops or radio stations, and teaching students, I have found that troubleshooting circuits can be broken down into the following categories:

- There are some circuits that will not work or will not work very well as shown in the schematic. These circuits will require some redesigning.
- There are some circuits that will work as a "one of a kind" but may not be repeatable due to the design. This type of design sometimes works by luck. For example, a photodiode circuit seems to work if a minimum amount of light falls on it because the op amp may not have the capability to output down to 0 volts but have to start at about $+1$ volt (e.g., see Figure 11-18).
- Many circuits published really do work. And these circuits will perform fine as shown in the schematics or instructions. However, sometimes it may be "easy" to mis-wire a circuit or put in the wrong part. Other times, the circuit may work in "normal" consumer everyday use but will break down under 24/7 schedules due to overheating, for example. Using a higher-power device can fix this problem.

We will concentrate on the first two categories—circuits that are redesigned for more reliable operation.

Circuits That Need Fixing or Redesigning

Let's take a look at a problem I had to solve just recently when a young student showed me a preamp circuit lifted off the web for amplifying electric guitar pickups. See Figure 11-1.

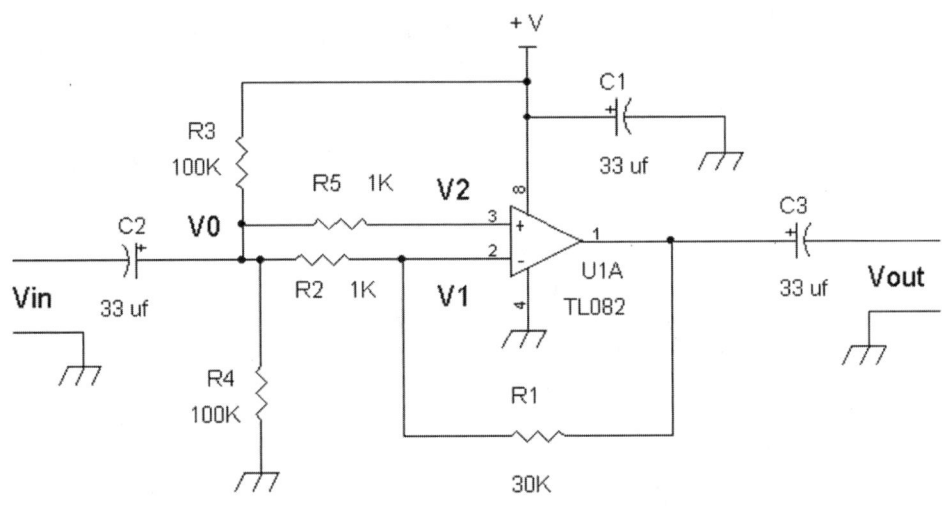

FIGURE 11-1   A preamp meant to have a gain ~ 30 but instead has a gain of 1.

As we can see, Figure 11-1 shows some good points in that the DC bias voltage is correct to provide one-half the DC supply voltage, V0, from R3 and R4. Via V0 connected to R5 and R2, the op amp's output pin will be biased to one-half the DC supply voltage for maximum AC voltage swing. With a 30KΩ R1 feedback resistor and a 1KΩ R2 input resistor, at first glance, the (inverting) voltage gain should be $-R2/R1 - 30K/1K = -30$, correct? No. Because the input signal also is coupled to the non-inverting input pin 3, there is a gain cancellation effect going on. The reason is any input signal at the (+) input of op amp with the same input voltage at R2 results in a non-inverting amplifier's gain summed with the inverting gain.

Note that the input resistance into the (+) input terminal, V2, is very high so there is "no" current flowing through R5, which means there is no voltage across R5, or R5 may as well be a short circuit or 0Ω wire. The reason is because V0 = V2, and since V2 at the (+) input terminal acts as a reference voltage for the feedback amplifier, the (−) input voltage V1 must follow the same voltage as V2. Thus V1 = V2, and since V0 = V2, V0 = V1. This means the voltage across R2 is 0 voltage and the input resistance to R2 (e.g., at V0) is infinite resistance since 0 volts across R2 implies there is no current flowing through R2.

When the input resistor R2 (left side) is connected to the (+) input (via R5) we have the following:

The non-inverting gain is $[1 + (R1/R2)]$; and the inverting gain is $-(R1/R2)$.

The total gain (AC or DC signal) is then $[1 + (R1/R2)] + -(R1/R2)$, which leads to: $1 + (R1/R2) - (R1/R2) = 1 =$ the total gain $=$ non-inverting gain plus inverting gain.

This total gain makes sense in terms of the DC gain as well. Since R3 and R4 form a voltage divider of half the supply voltage at R5 and R2, with a gain of 1, the DC output voltage at pin 1 of the op amp has to be one-half the supply voltage. For example, if $+v = 9$ volts, pin 1 $= 9$ volts/2 or 4.5 volts DC.

So how do we fix this circuit to have gain? See Figure 11-2, where we add C5.

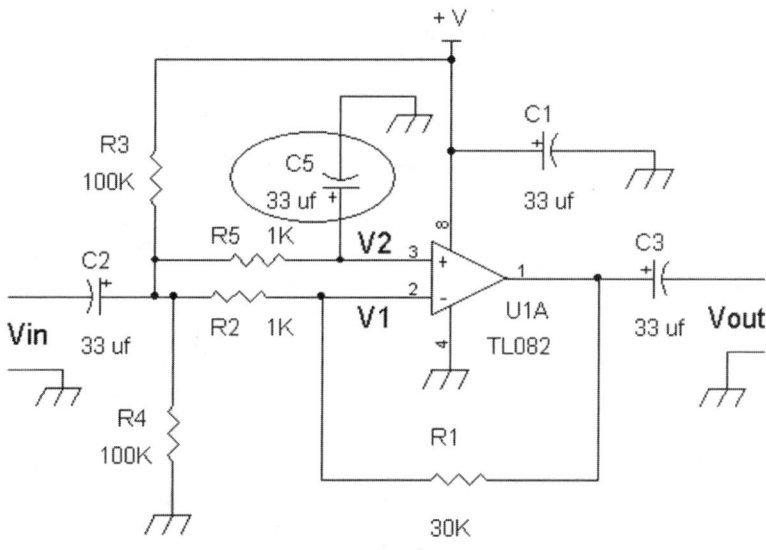

FIGURE 11-2 By adding C5, we provide an AC short to ground at the non-inverting input pin.

By AC shorting to ground the non-inverting input (pin 3) via adding C5, we now have a "standard" inverting gain amplifier that has a gain of Vout/Vin $= -R1/R2 = -30K/1K$ or Vout/Vin $= -30$. The preamp is restored to a reasonable gain for amplifying the guitar pickup's signal. R1 may be increased further to increase the gain, such as having R1 $= 100K\Omega$ for a gain of -100.

If we look at the AC input resistance, Rin, of this preamp in terms of Vin "looking" into C2, we see that C2 and C5 are AC short circuits, so we have the following pertaining to Figure 11-2.

R3's top connection is connected to $+V$, an AC ground, so AC-wise R3 is in parallel with R4. Because C5 acts as an AC short to ground, R5, which is connected to C5, is grounded as well. This means the AC input resistance has at least R3 in parallel with R4 and R5. Because the (+) input is AC grounded via C5, we know that the input resistance of an inverting op amp circuit is just the input resistance, which is R2. Thus, the AC input resistance is then R3 in parallel with R4, R5, and R2. Or, Rin = R3||R4 || R5 || R2. We can group the equal value resistors together since R3 = R4 = 100kΩ, and R5 = R2 = 1kΩ.

Rin = (R3||R4) || (R5 || R2)

Since R3||R4 = 100kΩ||100kΩ or R3||R4 = 50KΩ, and R5||R2 = 1kΩ||1kΩ or R5||R2 = 500Ω, Rin = 50kΩ||500Ω. However, Rin ~ 500Ω, since 50KΩ >> 500Ω.

In this first fix in Figure 11-2, the gain was restored, but R5 added to the loading input source, Vin. We can rescale the values for R5 and C5 to increase the input load resistance at Vin. See Figure 11-3.

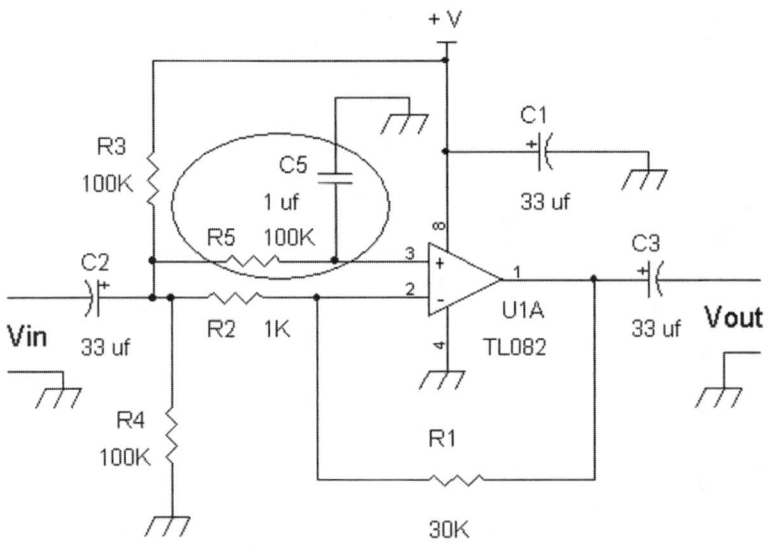

FIGURE 11-3   We can scale R5 up 100x and also use a smaller value C5 to provide an AC short circuit to ground.

By rescaling the values of R5 and C5, we mean having the time constant R5 × C5 of Figure 11-3 equal to or greater than R5 × C5 of Figure 11-2. The reason for this is so that C5 in both Figures 11-2 and 11-3 will be an AC short circuit to ground relative to R5.

In Figure 11-2, the resistor-capacitor time constant R5 × C5 = 1KΩ × 33 μf = 0.033 second; and in Figure 11-3, R5 × C5 = 100KΩ × 1 μf = 0.10 second > 0.033 second, so we are OK with this rescaling. A larger time constant means we have more filtering from noisy power supplies at +V. In general, C5 should be at least 0.1 μf. That is, if you scale R5 to 10MΩ and C5 to 0.01 μf, the preamp will work providing the circuit is enclosed in a shielded and ground metal box because pin 3 of the op amp will not have a low enough impedance at 50 Hz or 60 Hz and may pick up hum from power lines.

Given the rescaled values for R5 and C5, the input resistance now is:

R3||R4||R5||R2 = (R3||R4||R5) || R2

With R3 = R4 = R5 = 100KΩ, (R3||R4||R5) = 100KΩ/3 = 33.3KΩ, which is >> R2 = 1KΩ. Thus, Rin ~ 1KΩ in Figure 11-3 versus Rin = 500Ω in Figure 11-2.

Can we improve this circuit further and why? If the power supply is from a battery, there will not be a problem in terms of noise induced to the input via R3 and R4. However, if +V is from a power supply with some noise such as power supply ripple, the power supply noise may be coupled to the output of the op amp. See Figure 11-4 for a further improved circuit to reject power supply noise and maintain maximum input resistance for an inverting gain op amp circuit.

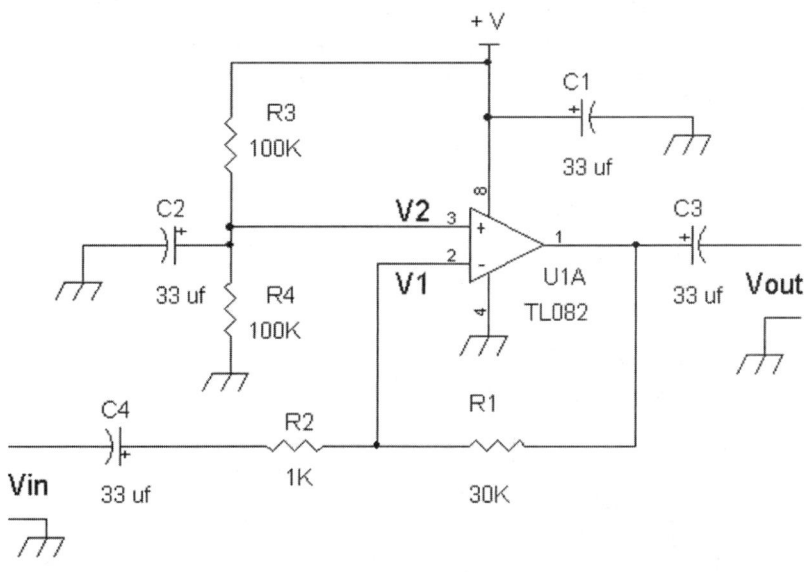

FIGURE 11-4   An improved circuit to further reject power supply noise.

By connecting the DC biasing circuit R3 and R4 with C2, a low-noise DC bias voltage is established at the non-inverting input terminal of the op amp. Since R3 and R4 are equal in value, the DC voltage is one-half +V. Thus, we get one-half the supply voltage at the (+) input terminal of the op amp and because of R1, the DC bias voltage at the (+) input terminal "forces" the same voltage at the (–) input terminal. And because there is no DC current flow through R1 (recall input bias current → 0), the voltage across R1 = 0 volts. Thus, the DC output voltage at pin 1 is also one-half the supply voltage, which is (+V/2). Again, DC-wise, R2 is "disconnected" since C4 is an open circuit at DC.

The values of R3, R4, and C2 were chosen to form a low-pass filter with a very low-frequency cut-off as to substantially reduce any power supply ripple from +V. With R3 = R4 = 100KΩ and C2 = 33 μf the cut-off frequency, $f_{cut-off} = f_c$:

$f_c = 1/[2\pi(R3||R4)C2] = 1/[2\pi(100K\Omega||100K\Omega)33\mu f]$ or $f_c = 0.0965$ Hz ~ 0.10 Hz

To find the attenuation factor of a noise signal at a frequency f_{noise}, we have:

attenuation factor $= f_c/f_{noise} = 0.1$ Hz$/f_{noise}$

For example, if we have a 60-Hz signal at 10 mV peak to peak riding on top of $+V$, then the AC signal at the $(+)$ input will be 10 mV peak to peak $\times$ (f_c/f_{noise}), which is then 10 mV peak to peak $\times$ (0.1 Hz/60 Hz) $= 0.0166$ mV peak to peak. If more filtering is required, you can increase the values of the resistors R3 and R4 to about 1MΩ with C2 $= 33$ μf, but it will take about 60 seconds or so for the bias voltage to come up to one-half $+V$.

Note that even though there is the ripple voltage at the op amp's power pin 8 with $+V$, the op amp itself has sufficient power supply ripple rejection to prevent this ripple voltage from feeding through to the op amp's output pin 1.

If we are concerned with low input resistance, Rin, and power supply noise, one of the best moves is to use a non-inverting gain configuration. See Figure 11-5, which is configured for a higher input resistance at C2.

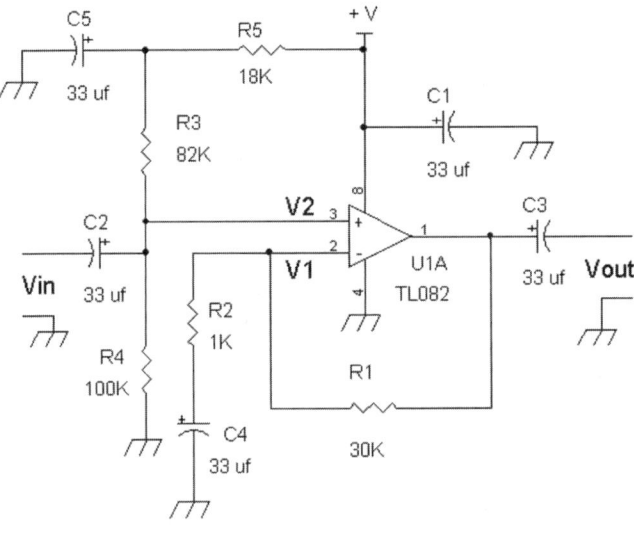

FIGURE 11-5 A non-inverting gain amplifier with extra power supply filtering via R5 and C5.

By using a non-inverting amplifier for a preamp, we have the advantage of larger input resistance, Rin, which is equal to R4||R3 $= 100K\Omega||82K\Omega$ or Rin $= 45K\Omega$, due to C5 being an AC short circuit to ground. Again, if more filtering from noise via the power supply, $+V$, is required, C5 can have a larger capacitance such as 330 μf. Generally, a TL082 op amp will be "good" enough low noise-wise for some low-level sources such as magnetic pickups or microphones.

However, you can make a lower noise preamp if the low-level source includes a low impedance device such as a dynamic microphone. In Figure 11-5, you can replace U1A with a lower noise op amp such as an OPA2134, LM833, or NE5532, and rescale the feedback network, R1, R2 for lower resistances such as:

R1 = 3000Ω, R2 = 100Ω, and C4

to a larger capacitance value to maintain good low-frequency response such as C4 = 330 μf.

We now will look at a circuit found off the web or a hobby book. See Figure 11-6, which shows a high-gain op amp circuit that drives a speaker amplifier.

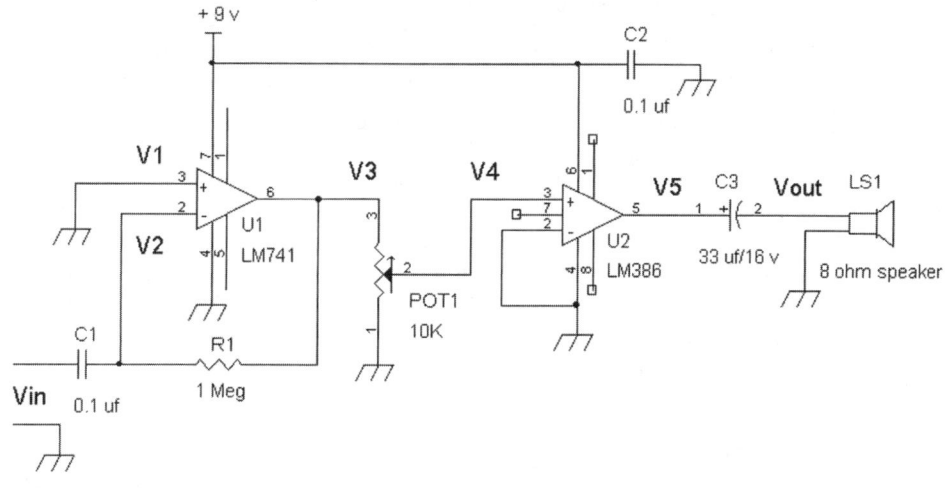

FIGURE 11-6   An amplifier circuit that kind of works, which produces distorted signals.

This amplifier has a basic flaw in that U1 LM741 really requires a negative power supply to pin 4. Also, the other problem with this amplifier is that a DC voltage from the LM741's output pin 6 couples a DC voltage to the input terminal of U2, LM386. In general, the LM386's input terminal wants to see an AC coupled signal with no DC component.

As a result of pin 4 grounded for U1, the output signal resembles an amplified half-wave rectifier. See Figure 11-7.

We can show a way to slightly redesign the circuit in Figure 11-6 to solve the problems previously mentioned.

See Figure 11-8 where C5 along with R2 and R3 are added. Please refer to the redesigned circuit in Figure 11-8.

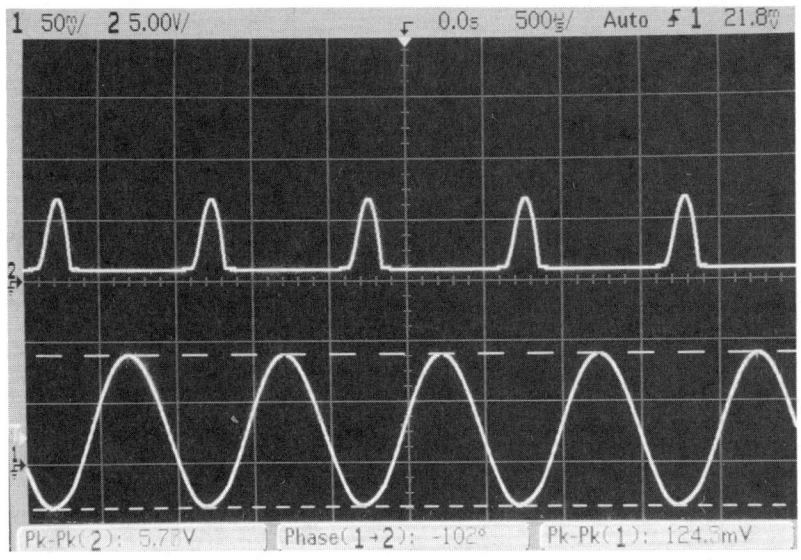

1 50⍟/ 2 5.00V/	⌐ 0.0ₛ 500⍟/ Auto ƒ 1 21.8⍟

Pk-Pk(2): 5.7ᵊV   Phase(1→2): -10ᵊ°   Pk-Pk(1): 124.ᵊmV

FIGURE 11-7   Top trace is the output signal at about 5.7 volts peak to peak; bottom trace is the input signal at about 124 mV peak to peak.

We can add a minus power supply (e.g., –9 volt) to pin 4, however, a little re-biasing of the LM741's input stage should fix our problem. See Figure 11-8.

To debug Figure 11-6, we "un-ground" the non-inverting input pin 3 of U1 and provide it a one-half supply voltage at 4.5 volts. C4 is added to reduce noise from the +9-volt supply. Input capacitor C1 is now connected to a series resistor R4 to provide an inverting gain op amp circuit, where the gain of U1 is – R1/R4. If more or less gain is required, the resistance value of R4 may be varied, such as from 1000Ω to 100KΩ. If a better low-frequency response is needed, C1 can be replaced with a 33-μf (25-volt) capacitor with its positive terminal connected to R4 (Figure 11-8).

In Figure 11-6 the signal was coupled directly via C1 to the inverting input terminal, which made the op amp work in an "open" loop configuration. When an op amp is operating in an open loop configuration, it is behaving more like a comparator and the output signal is more like either on or off. Now see Figure 11-8.

To ensure that no DC voltage appears at the input of U2, an extra coupling capacitor C5 is added in series with POT1. Finally, a snubbing network, R5 and C6, is added at the output of the LM386 pin 5 to ensure that no parasitic high-frequency oscillation occurs as recommended in the LM386's data sheet.

Other improvements to this circuit can include changing output speaker capacitor C3 to a larger capacitance value, such as 100 μf to 220 μf (16 volt).

If a "faster" op amp is needed in terms of gain bandwidth product (GBWP) for U1, the LM741 (1 MHz GBWP) may be replaced with a TL081 (3 MHz GBWP), LF351(4 MHz GBWP), or OPA134 (8 MHz GBWP).

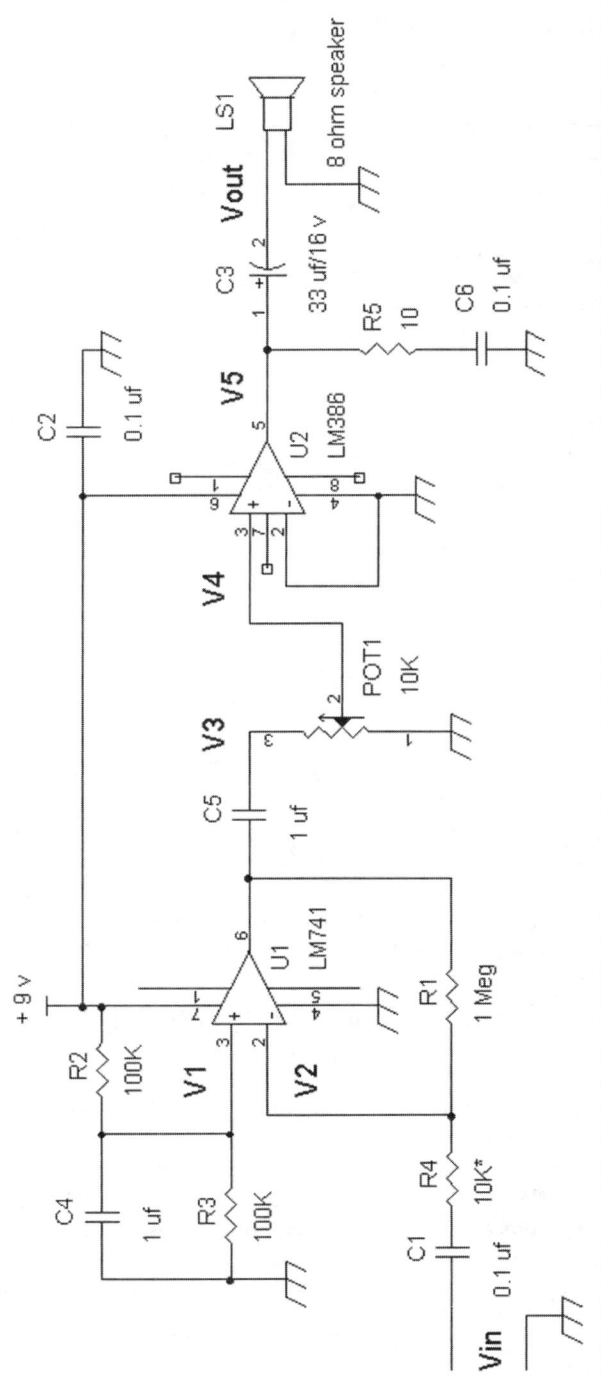

FIGURE 11-8 A "fixed" version with an extra bias circuit, R2 and R3, which allows the circuit to work off a single supply.

Our next circuit that needs debugging is an active band pass filter that can be used for audio signal applications such as an equalizer, or for applications where you want to pass a signal around a center frequency while attenuating signal whose frequencies are outside the center frequency.

For example, if you like to "isolate" a 19-kHz signal that has been added to a high-fidelity audio whose frequency ranges from 50 Hz to 15 kHz, an active band pass filter can be used. It will pass the 19-kHz signal essentially without attenuation, while removing most of the signal from 50 Hz to 15 kHz.

In Figure 11-9 there are two graphs of a lower Q band pass filter and a higher Q band pass filter. Notice that the higher Q filter is narrower in shape, which allows for better rejection of signals whose frequency is outside of the center frequency fc. A good approximation for Q is the center frequency divided by the −3 dB or 70.7 percent amplitude frequencies. That is $Q = fc/BW_{-3\,dB}$.

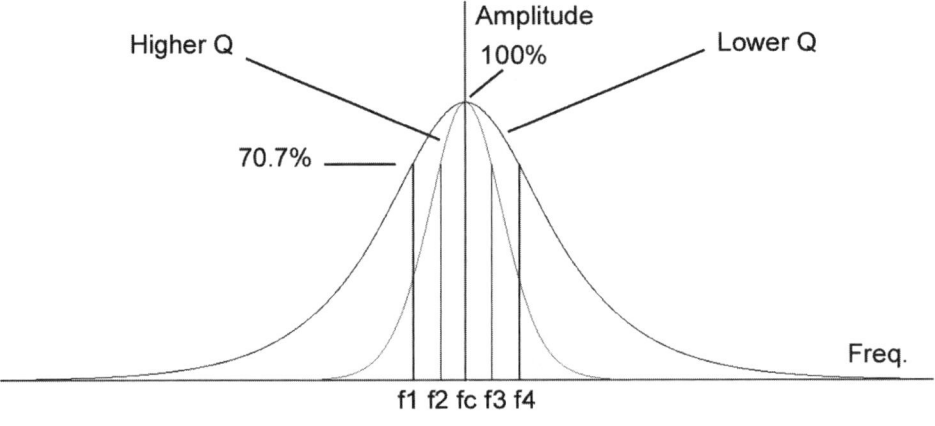

FIGURE 11-9   Graphs of band pass filters relating to −3 dB (at 70.7 percent) points and Q.

For the lower Q band pass filter: Q = fc/(f4 − f1) where (f4 − f1) = −3 dB bandwidth; and for the higher Q band pass filter: Q = fc/(f3 − f2) where (f3 − f2) = −3 dB bandwidth.

For example, if the center frequency, fc = 10 kHz and f3 = 10.5 kHz with f2 = 9.5 kHz, then the higher Q band pass filter has:

Q = 10 kHz/(10.5 kHz − 9.5 kHz) = 10 kHz/1kHz or Q = 10

If the lower Q band pass filter has f4 = 11 kHz and f1 = 9 kHz, and recall that fc = 10 kHz, then its Q = 10 kHz/(11 kHz − 9 kHz) = 10 kHz/2 kHz or Q = 5.

The graph in Figure 11-9 overlays the "normalized" curves of the lower Q and higher Q band pass filters. In reality, often the higher Q band pass filter will have more gain at the center frequency when compared to the lower Q band pass filter. For example, if the higher Q band pass filter provides 2 volts AC at fc, then the lower Q band pass filter may provide 1 volt AC at fc. Note: On the next page f_{center} = fc.

Now let's take a look a band pass filter that does not work with AC signals. See Figure 11-10.

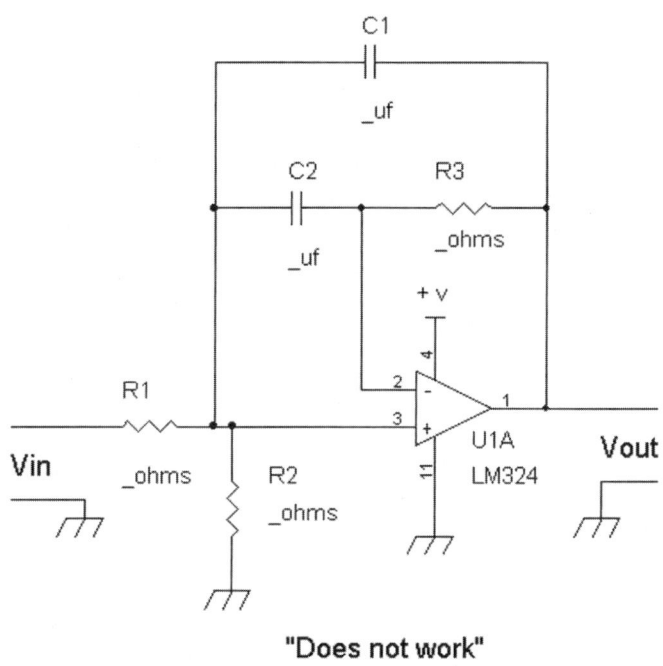

"Does not work"

FIGURE 11-10 Example wrong active band pass circuit that has DC biasing problems at the Vin input terminal with R1.

The band pass filter in Figure 11-10 would work if the minus supply pin 11 that is connected to ground would be instead connected to a negative power supply such as –9 volts. As a result of having the negative supply pin 11 connected to ground, the op amp is not capable of outputting a signal that swings below ground. Also, the non-inverting input at pin 3 cannot tolerate a signal below –0.5 volts. So again, an AC signal connected to Vin will produce a distorted Vout. Also note that the LM324 is known for crossover distortion, so using that op amp is not the best choice along with its low gain bandwidth product of 1 MHz. This bandpass filter can run the limit of an op amp's gain bandwidth product because it is "gain hungry." For example, suppose we want to build a 19-kHz bandpass filter with a –3 dB bandwidth of 2 kHz. This means at 1 kHz below and above the center frequency, 19 kHz, the response is 70.7 percent of the peak amplitude at 19 kHz. This means at 18 kHz and 20 kHz these are the –3 dB frequencies. The Q of a bandpass filter is defined as $Q = f_{center}/BW$, where f_{center} = center frequency of the band pass filter.

With this example, $Q = 19$ kHz/2 kHz or $Q = 9.5$.

The gain at the center frequency is $A_v = 2 Q^2$.

For this example, $A_v = 2 (9.5)^2$ or $A_v = 180$.

To determine the op amp's open loop gain at 19 kHz, an approximation is GBWP/ f_{center}.

For the LM324 that has a 1 MHz GPWP, the open loop gain at 19 kHz ~ 1 MHz/19 kHz ~ 52.

Note that we need at least 180, and generally we want the open loop gain to be about 4 to 10 times of $A_v = 2\,Q^2$. That is, we want the op amp's open loop gain at 19 kHz to be 720 to 1800.

For example, an LT1359 Quad op amp with 25 MHz GBWP would work, having an approximate gain of $25 \times 52 = 1300$ at 19 kHz.

Sometimes it's better to use dual op amps rather than quad op amps. The reasons are a wider selection, lower cost, and a greater chance to find a faster op amp. Now let's take a look at the debugged version in Figure 11-11.

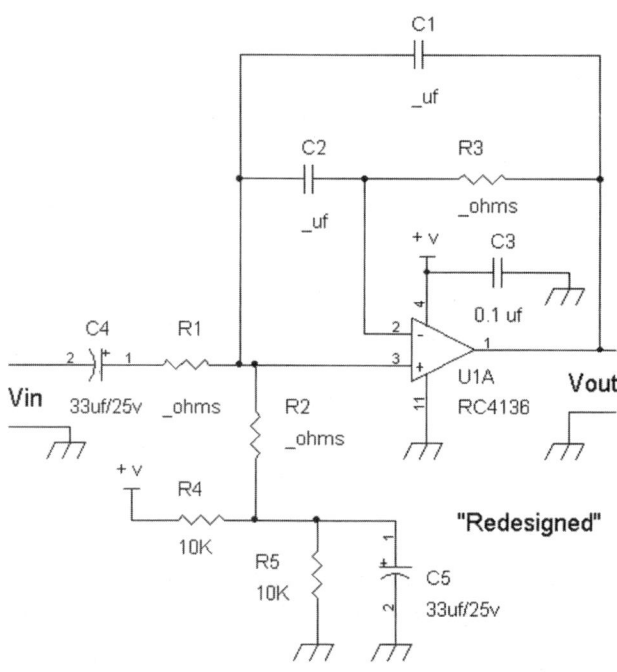

FIGURE 11-11   Corrected band pass filter circuit with a DC bias circuit, R4, R5, and C5.

By adding a bias circuit to form a one-half supply voltage with a large capacitance C5, the circuit is debugged. Also, a different quad op amp, the RC4136, which has a 3 MHz GBWP, is used instead of the LM324. The RC4136 has a very good output stage that does not exhibit crossover distortion.

Because the RC4136 has a 3-MHz gain bandwidth product (GBWP), the band pass filter's center frequency is limited to Q < 10 and f_{center} < 3.75 kHz. A worst case calculation was made by the following:

$2 Q^2 = 2 (10)^2 = 2 (100) = 200$

This gain number of $2 Q^2$ is when R2 is removed from the circuit and R1 drives the band pass filter.

We want an open loop gain of at least 4 times the $2 Q^2$ at the band pass filter's center frequency.

The open loop gain required is then $4 \times 2 Q^2 = 4 \times 200 = 800$. Now we need to find the band pass filter's maximum center frequency, $f_{center-max}$:

$f_{center-max} = GBWP/(4 \times 2 Q^2) = 3 \text{ MHz}/800 = 3.75 \text{ kHz}$

To determine the values of the resistors R1, R2, and R3, and the capacitors C1 and 2 for a given band pass filter with a center frequency, f_{center}, and Q based on the desired –3 dB bandwidth, see the following:

–3 dB bandwidth $= f_{center}/Q$

$R1 \| R2 = R3/(4 Q^2)$, where also $R1 \| R2 = (R1 \times R2)/(R1 + R2)$

$R3 = Q/(\pi C f_{center})$

$C = C1 = C2$

The gain of the overall band pass filter is $[R2/(R1 + R2)] \times 2 Q^2$.

When the value of Q may be in the order of 3 to 10, this gives a gain range, $2 Q^2$, from 18 to 200. This means that the band pass filter has quite a bit of gain at the center frequency. To ensure that the band pass filter does not overload too easily, the input resistors R1 and R2 form a voltage divider that will reduce the overall gain. A good starting point for R1 and R2 is to make a voltage divider of about 10:1 or 11:1. For example, in an 11:1 voltage divider, if $R1 = 10R2$, then $[R2/(R1 + R2)] = 1/11$; the gain range from Vin to Vout is now reduced to 1.6 to 18 instead of 18 to 200.

One way to start designing the band pass filter is to determine C as a function of $R1 \| R2$, Q, and f_{center}. We can substitute $R3 = Q/(\pi C f_{center})$ into $R1 \| R2 = R3/(4 Q^2)$ to express $R1 \| R2$ as a function of Q, C, and f_{center}, which leads to the following:

$R1 \| R2 = 1/[\pi C 4Q f_{center}]$

and now we can find C as a function $R1 \| R2$, Q, and f_{center} as:

$C = 1/[\pi (R1 \| R2) 4Q f_{center}] = C1 = C2$

Once C is determined, we can find R3 via:

$R3 = Q/(\pi C f_{center})$

For example: Suppose we want to build a band pass filter having $f_{center} = 1 \text{ kHz}$ with Q = 3, and given that R1 = 100KΩ and R2 = 10KΩ such that $R1 \| R2 = 100 \text{K}\Omega \| 10 \text{K}\Omega \sim 9.1 \text{K}$.

$C = 1/[\pi (R1||R2) 4Q f_{center}] = 1/[\pi (9.1K\Omega) (4)(3) 1kHz]$ or $C = 2916$ pf $= C1 = C2$.

$C1 = C2 \sim 2700$ pf in parallel with 220 pf or a 0.0027 µf in parallel with 220 pf.

$R3 = Q/(\pi C f_{center}) = 3/(\pi 2916$ pf 1 kHz) or $R3 = 109K\Omega \sim 110K\Omega$, and to reiterate, $R1 = 100K\Omega$ and $R2 = 10K\Omega$.

The overall gain at the center frequency is Vout/Vin $= [R2/(R1 + R2)] \times 2 Q^2$ or in this example:

$[10K/(100K + 10K)] \times 2 (3)^2 = [1/11] \times 2 \times 9 = 18/11 = 1.6 =$ gain or Vout/Vin $= 1.6$ @ 1kHz.

Photodiode Circuits

We now turn our attention to photodiode preamps. These are often used with sensor or communication systems. Here we will explore some mistakes as posted on the web or in some hobbyist books.

Photodiode sensors with light sources such as LEDs can be used to sense the presence or distance of an object. Also, photodiodes with a focused light beam can be used to read black and white lines such as bar codes. In this section, we will look at photodiode "receiver" circuits that convert the light sensed into a signal voltage. Let's first look at a typical photodiode that has a cathode lead and an anode lead along with its associated schematic symbol. See Figure 11-12.

FIGURE 11-12 Schematic symbol of a photodiode (note arrows pointing into the diode whereas in an LED the arrows are pointing away from the diode), and identification via the lead length where the anode lead is slightly longer than that cathode's lead.

If the leads are the same length, you can use an ohm meter to measure the forward voltage from anode to cathode, which is about 0.7 volts or so. Measure again with a standard diode (1N914) where the cathode is clearly marked with a band, and you should be able to identify the cathode of the photodiode. Now, let's take a look at what is really a photodiode. See Figure 11-13.

The photodiode acts like a light intensity current source, with the positive current flowing from cathode to anode. In a sense it's like an NPN transistor current source where the current flows from collector to emitter. Hence, the photodiode is

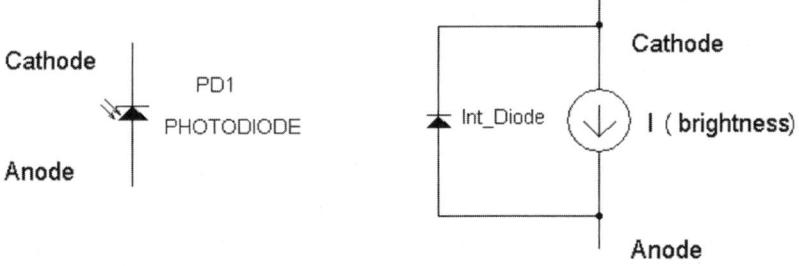

FIGURE 11-13 A photodiode schematic diagram is on the left side, and a current generator representation of it shown on the right hand side with the internal diode, Int_Diode.

polarity dependent. The photodiode current is modeled as "I (brightness)" to depict that it is proportional to light intensity. That is, the more light there is, the higher the current is generated from the photodiode. Typically, the photodiode will generate currents as low as I (brightness) < 0.001 μA when it is covered or in the dark, or with intense lighting such as in sunlight, it will generate I (brightness) >> 1 μA.

In this book, we can also sometimes "shorthand" I (brightness) with I (lum), where "lum" is the luminance level that is the same as brightness. For example, when you buy an LED bulb for a ceiling lamp, it is rated in **lum**ens (e.g., a 9-watt LED lamp equivalent to a 60-watt incandescent bulb gives out about 800 lumens).

Note let's take a look at some working circuits before we examine non-working ones. A photodiode can be connected to load resistor RL to develop a voltage across it as shown in Figure 11-14.

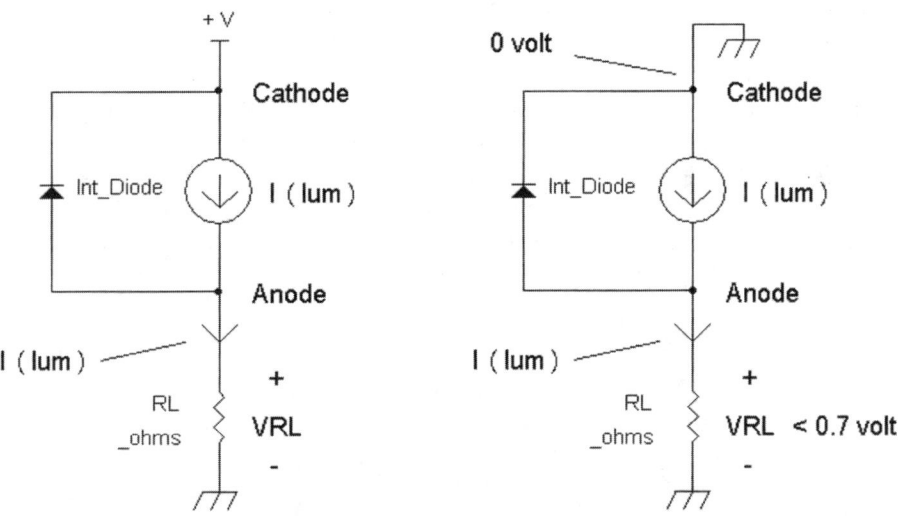

FIGURE 11-14 Two photodiode circuits, one with a +V bias (left side), and the other on the right side with 0-volt bias (via the cathode being connected to ground).

To "extract" the photodiode's signal, generally it is reverse biased or zero biased to prevent the photodiode's current from diverting into its internal diode instead of flowing out into a load resistor. As shown in Figure 11-14, we see on the left side's circuit that +V is forming a reverse bias to the internal diode, Int_Diode. The maximum voltage across RL is VRL = (+V) +0. 7 volt.

For example, if we are using a 5-volt supply, then +V = +5 volts and the maximum voltage across load resistor RL = +5 volts + 0.7 volt = 5.7 volts.

The voltage across the load resistor is VRL = I (lum) × RL and equivalently, I (lum) = VRL/RL.

If, for example, RL = 1MΩ and +V = +5 volts, then the maximum photodiode current, I (lum)$_{max}$ = 5.7 volts/1MΩ or I (lum)$_{max}$ = 5.7 µA.

In Figure 11-14, the circuit on the right shows what happens when +V $\rightarrow$ 0 volts that essentially grounds the cathode of the internal diode. This means that Int_Diode is now in parallel with RL, and also means it "clips" or limits the maximum voltage across RL by the turn-on voltage of a silicon diode or about 0.7 volt. As an example, if RL = 1 MΩ with +V = 0 volts, I (lum)$_{max}$ = 0.7 volts/1MΩ or I (lum)$_{max}$ = 0.7 µA = 700 nA.

If we want to buffer (amplify) the signal output of Figure 11-14, we can use a voltage follower or non-inverting gain amplifier that can sense at ground and provide an output voltage at near ground, such as a TLC272 op amp. However, there can be a disadvantage by merely amplifying VRL with a non-inverting gain amplifier. Because the RL is typically 100kΩ to 1 Meg ohm, even small stray capacitances can cause a frequency response problem.

For example, if the photodiode has a 10-pf internal capacitance, Cpd, across the cathode and anode, the high-frequency response with RL = 1MΩ will be about 16 kHz where the cut-off frequency, f_c = 1/[2π (RL × Cpd)] which is: f_c = 1/[2π (1MΩ × 10 pf)] = 15.924 kHz = f_c.

We can improve on high-frequency response by using a "trans-resistance" amplifier, which means current in and voltage out. But before we show you some trans-resistance amplifier circuits, let's take a look at testing the photodiode itself with a digital voltmeter (DVM). See Figure 11-15.

Even a typical low-cost DVM can test photodiode via its 200 mV or 2000 mV DC full-scale setting because it has a built-in 1MΩ resistance across the positive and negative leads. You can use a flashlight and shine it directly into the photodiode and read the voltage output with the voltmeter setting. Note that the voltage will be limited to about 700 mV DC. As shown in Figure 11-15, the voltmeter shows 206 mV across the input 1-MΩ input resistance of this particular DVM. This means the photodiode is 206 mV/1 MΩ or 206 nA (nano amps) or equivalently 0.206 µA. If you use a higher input resistance DVM that normally has a 10MΩ input resistance at the voltmeter setting, you can measure the photodiode current with 10x sensitivity.

Also shown in Figure 11-15 is that you can measure the photodiode's current directly by using the 200 µA full scale setting as shown with a flashlight shining directly into the photodiode measuring 54.1 µA.

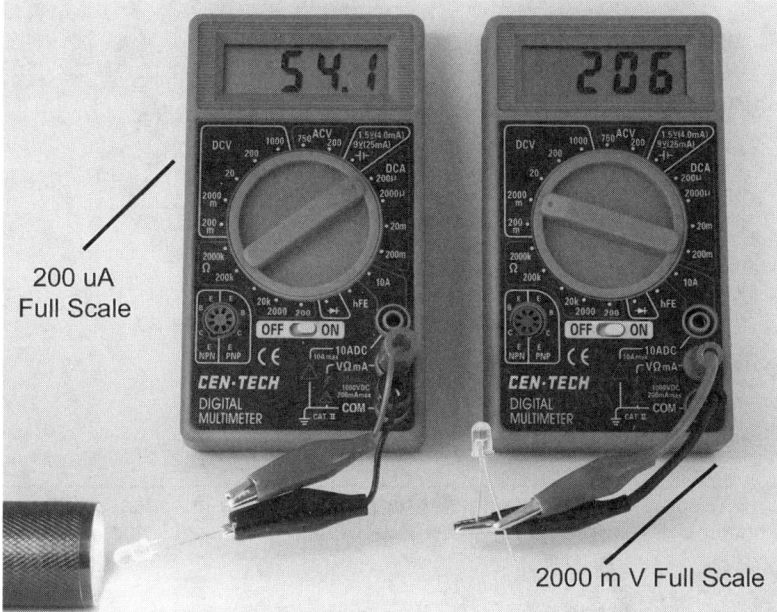

200 uA
Full Scale

2000 m V Full Scale

FIGURE 11-15 Two ways to measure photodiodes via the voltmeter (on the right side) and via the current (micro amp meter) settings (on the left side).

If you need to measure lower photodiode currents, such as using a lower sensitivity photodiode or an LED, use a slightly higher-performance DVM with a 10MΩ resistance across the positive and negative leads. With 10MΩ load resistance the sensitivity is increased 10x over the lower priced DVM. For example, most auto-ranging and >$20 DVMs have 10MΩ resistance in the voltmeter mode.

Note if you use an LED as a photodiode, the equivalent internal diode, Int_Diode, as depicted in Figure 11-14 will have a maximum voltage of the LED's turn-on voltage, V_{LED}. For example, in a red, green, and yellow standard LED, the turn-on voltage is about 1.7 volts. For a blue or white LED, expect the turn-on voltage to be about 2.5 volts to 3.0 volts. This means if you shine light into the LED, the maximum voltage across RL will be VRL ≤ LED's turn on voltage, which can be in the order of 1.7 volts or more for VRL depending on what type of LED you are using. Also, the pin out will be same for the LED as with the photodiode; the shorter lead will be the cathode.

Trans-resistance Amplifiers

Figure 11-16 shows a trans-resistance amplifier with a photodiode connected to the inverting input of an op amp. The input signal is a current and the output provides a voltage. In terms of "gain" we have a voltage output divided by a current input or V/I that has the unit of resistance.

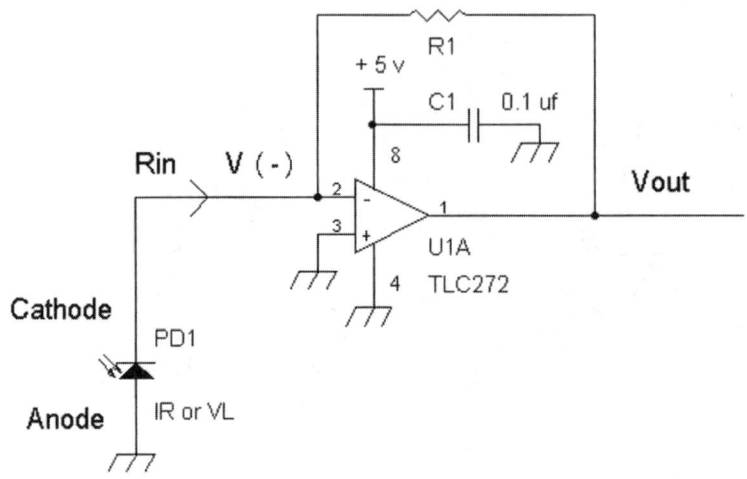

FIGURE 11-16 A trans-resistance amplifier for providing a low-impedance voltage output related to photocurrent from PD1, a photodiode. Note: IR = Infrared and VL = Visible Light.

The photodiode, PD1 in Figure 11-16, has its cathode connected to the (−) input terminal of U1A, which serves as a low resistance point referenced to ground. That is, the input resistance, Rin, looking into the combination of feedback resistor R1 and pin 2, the inverting input of U1A, is very low for two reasons.

The first reason is that since the pin 3 (+) input terminal is grounded (or AC grounded if pin 3 is tied to a +V positive bias voltage), the (−) input terminal must have the "same" voltage as the (+) input voltage due to negative feedback via R1. Thus, Rin is looking into a virtual short circuit to ground via the (+) input being grounded. In actuality, Vout is adjusting the voltage such that the current through R1 matches the photodiode's current from PD1. When this happens, the voltage at V(−) is "adjusted" to ground or 0 volts via Vout.

The second reason why Rin is close to 0Ω is that in an inverting gain op amp amplifier, the input resistance looking into the inverting input terminal is $Rin \sim R1/[A_0(f)]$, where $A_0(f)$ = open loop gain of the op amp as a function of frequency. You can also estimate $A_0(f)$ by knowing the gain bandwidth product (GBWP) for frequencies beyond the open loop frequency response.

For example, if U1A = TLC272, the gain bandwidth product, GBWP, is about **1.7 MHz** for a 5-volt power supply. If we want to know the approximate Rin at f = 100Hz with R1 = 1MΩ, we can estimate $A_0(f) \sim (GBWP/f)$, with the understanding that in most cases the **maximum** $A_0(f)$ is between 100,000 and 1,000,000.

$Rin = R1/ A_0(f) \sim R1/(GBWP/f) = 1MΩ/(1.7MHz/100Hz) = 1MΩ/17,000$ or $Rin \sim 59Ω$. When we consider that PD1 outputs something in the order of 10 μA maximum, the voltage at V(−) will be 10 μA x Rin or V(−) ~ 0.59 mV, which is ~ 0 volts.

Note at DC or slowly changing signals with time, $A_0(f)$ is $\geq$ 100,000. So, given that the open loop gain, $A_0(f)$ is $\geq$ 100,000 at DC or low frequencies, Rin $\leq$ 1MΩ/100,000 or Rin $\leq$ 10Ω, which pretty much looks like a short circuit to ground considering PD1's photocurrents are typically $\leq$ 10 μA.

You can download op amp spec sheets from various manufacturers (e.g., Texas Instruments, Analog Device, New Japan Radio, ST Electronics, etc.) for open loop gain, $A_0(f)$, data.

Now let's take a look at Figure 11-17. The output voltage Vout = R1 × I (lum), where I (lum) is the photodiode's current flowing downward from cathode to anode of PD1.

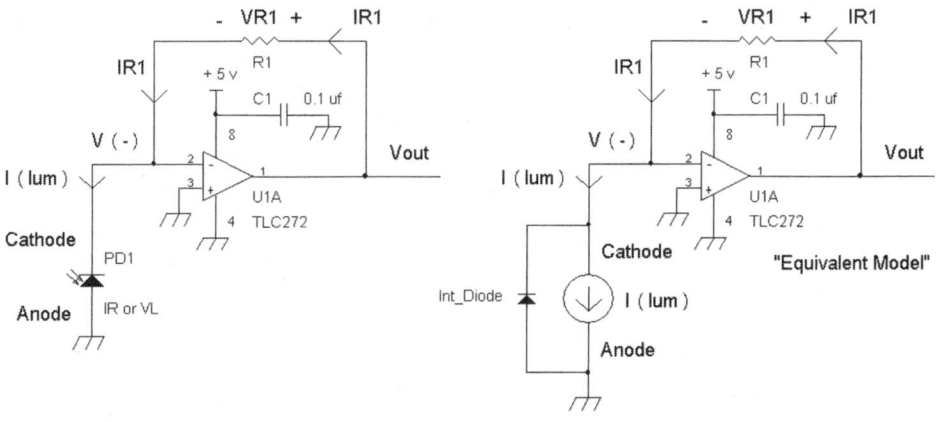

FIGURE 11-17 A trans-resistance amplifier with photodiode PD1 in the left circuit with its "Equivalent Model" on the right side showing a light intensity controlled current source generator, I (lum).

Notice in Figure 11-17 the TLC272 op amp is chosen specifically to work with +5 volts for sensing at 0 volts at the input terminals pins 2 and 3, and also for the ability for the output at pin 3 to swing all the way down to ground. Typically, the feedback resistor R1 is in the range of 100KΩ to 4.7MΩ. For example, if R1 = 470KΩ, then Vout = I (lum) × 470KΩ, which is a positive voltage.

Essentially, how we get the positive voltage for Vout goes like this: We know that no current flows into the (–) input at pin 2 of U1, so this means R1's current IR1 = I (lum) , the photodiode current. When I (lum) starts generating current based on light intensity, it will "pull" current through R1. Since the (+) input terminal at pin 3 is connected to ground or zero volts, the negative feedback system has to correct itself to ensure the (–) input voltage matches the (+) input voltage, which is ground. Thus, the (–) input is "servoed" to ground via Vout and feedback resistor R1. When the light shines in the photodiode, the photodiode's current, I(lum), forces Vout, then rises to a positive voltage just enough such that Vout/R1 = I (lum). This forms a balanced condition whereby the voltage at R1 connected to the (–) input is now

0 volts, which is what the negative feedback system wants. Essentially, it is saying that it wants the following condition at the (–) input terminal Vout – IR1 × R1 = 0 volts = voltage at (–) input, or put in another way, Vout = VR1 = IR1 × R1.

Since IR1 = I (lum), then we have Vout – I (lum) × R1 = 0 volts at the (–) input terminal.

So Vout is always doing a balancing act to ensure that whatever photodiode current is pulling from its cathode, the voltage at the (–) input is "servoed" to zero volts. For example, suppose the following in Figure 11-7 (or Figure 11-18 below left side): R1 = 1MΩ and the photodiode current I (lum) = 0.5 μA. If Vout were to be grounded temporarily, then right side terminal of R1 would be grounded; and the left side of R1 connected to the (–) input terminal would be –0.5 volts which is = –I (lum) × R1 = –0.5 μa × 1MΩ by having the photodiode current "sinking" current instead of sourcing current (e.g., flowing into R1). If Vout rises from 0 volts to +0.5 volts to counter exactly the –0.5 volts, then the voltage at the (–) input terminal will be 0 volts. Thus, Vout = I (lum) × R1 or in this example 0. 5 μA × 1MΩ = 0.5 volt. Note that Vout = VR1 + V(–) or V(–) = Vout – VR1, and we want V(–) → 0 volts, which leads to 0 = Vout – VR1 or Vout = VR1 that leads to **Vout = IR1 × I (lum)**, thus Vout = I (lum) × R1.

Now let's take a look at some other photodiode amplifier circuits with possible problems and their fixes. See Figure 11-18.

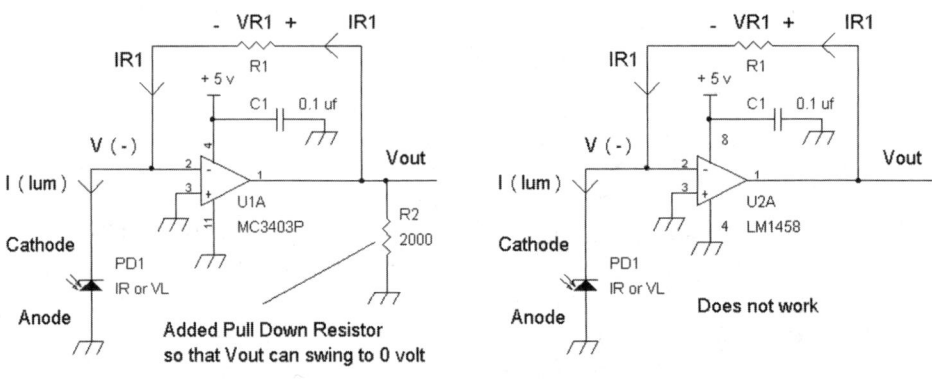

FIGURE 11-18 MC3403P (left side) and LM1458 (right side) photodiode preamps.

In Figure 11-18, the left circuit uses an MC3403P op amp that has an NPN-PNP push pull emitter follower output stage. Because the PNP output transistor's emitter cannot deliver a voltage to 0 volts, there will be a problem with sensing small photocurrents. For example, if the photodiode current is 0 μA with R2 removed from the preamp circuit, and R1 = 100kΩ, Vout will be in the +50 mV to +200 mV range when measured with a DVM. To re-establish a truer linear relationship between light intensity and output voltage, pull-down resistor R2 is connected from pin 1 to ground. This ensures that the MC3403P's output stage works as an NPN emitter follower capable of operating correctly all the way down to 0 volts. Typically, R2 can be a resistor in the range of 3300Ω to about 1000Ω. Thus, with R2 installed, when the photodiode current is 0 μA, Vout = 0 volt. Also note that the MC3403P's input terminals will operate properly down to 0 volt.

The LM1458 circuit (right side) in Figure 11-18 will not work because the input circuit of an LM1458 or similar op amp cannot operate correctly when the (−) input terminal is at the same potential as its minus supply pin 4. The (−) input of an LM1458 or similar op amp requires at least 1.5 volts to 2 volts above the negative supply pin (e.g., pin 4).

A fix then is to replace the LM1458 op amp with op amps such as the TLC272 (16-volt max supply), MAX407 (10-volt max supply), NJU7002 (16-volt max supply) or CA3240 (32-volt max supply) whose input terminals can sense at ground and whose outputs can swing to ground.

Suppose we have the correct op amp in terms of input and output voltage ranges, can there be other problems? Yes, see Figure 11-19.

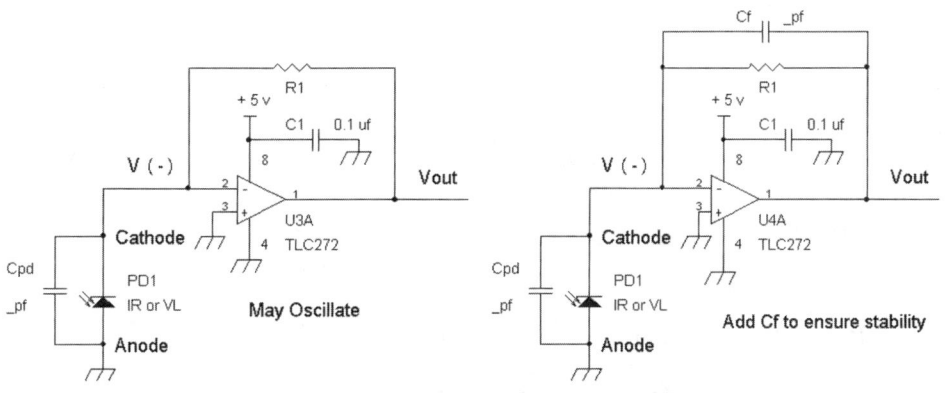

FIGURE 11-19 The photodiode's internal capacitor, Cpd, can cause oscillations by added extra lagging phase shift in the feedback network via R1 and Cpd.

All semiconductors including diodes and transistors have internal capacitances across their terminals. In Figure 11-19 on the left side, the photodiode, PD1, has a built-in capacitor, Cpd. Typically, Cpd is in the range of 1 pf to as much as 100 pf, depending on the surface area of the photodiode. Since R1 typically has a value in the range of ≥ 100KΩ, the low-pass filter effect via R1 from Vout to V(−) with Cpd adds phase shift that can cause the circuit to oscillate.

For example, if you replace U3A with a faster op amp and or connect Vout to a capacitance load, it is possible to cause the op amp to oscillate. The oscillation frequency is generally > 1kHz such as 10kHz or more. To fix this problem, the circuit on the right side of Figure 11-19 shows a feedback capacitor, Cf, in parallel with R1. Cf with R1 forms a positive phase or leading phase network with respect to diode capacitance Cpd.

This positive phase or leading phase offsets/cancels the lagging phase caused by Cpd. At very high frequencies, the impedance of Cf is much smaller than the resistance of R1 such that we get a capacitive feedback network of Cf and Cpd. Since a capacitive voltage divider from Vout to V(−) does not exhibit phase shift, the circuit has less chance of oscillation. Generally, we choose Cf for minimum capacitance to maintain good high-frequency response or bandwidth. Typically, Cf is in the order of 1 pf to 10 pf. However, if we need to filter out noise while maintaining a lower but sufficient bandwidth, Cf can be increased.

The −3 dB frequency for bandwidth is: $f_c = 1/[2\pi (R1)(Cf)]$.

For example, if R1 = 1MΩ and Cf = 10 pf, then $f_c = 1/[2\pi (1M\Omega)(10 \text{ pf})]$ or $f_c = 15.9$ kHz. If you want twice the bandwidth, reduce the capacitance of Cf to half by setting Cf to 5 pf so that $f_c = 31.8$ kHz.

We now turn to examples from the web (world wide web) or from some publications that show mistakes, but these circuits are easily remedied. Let's start with Figure 11-20.

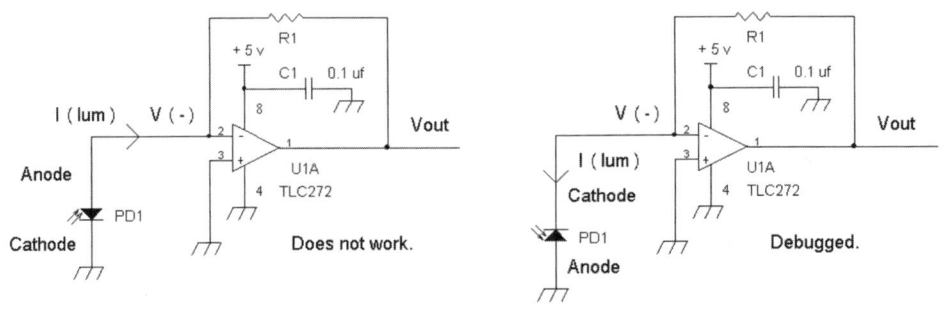

FIGURE 11-20   The left side circuit shows photodiode PD1 connected in reverse.

In one of the classes I taught concerning photo emitters and sensors, a basic single supply photodiode preamp was built. The output Vout seems to be stuck at 0 volts no matter how much light was shined into the photodiode. As can be seen in Figure 11-20's left circuit, the photodiode's anode is connected to the inverting input of U1A. Because the photocurrent positive current is flowing **in**to the inverting input, we can expect to have a negative voltage at the output. This negative voltage output would be true if the negative supply pin 4 of U1 would be connected to a negative supply such as −5 volts DC. However, since pin 4 is grounded at 0 volts, this means the output voltage at pin 1 cannot go lower than 0 volts. Therefore, the output voltage is "clipped" at 0 volts. To fix this problem, just reverse the photodiode's connections as shown in the right side circuit of Figure 11-20. This then has the photodiode current flowing **out** from the inverting input that provides a positive Vout. Also, note the op amp that was chosen allows the input terminals to operate correctly at ground, and that allows the output voltage to swing as low as ground. The output voltage for the right side circuit is Vout = I (lum) × R1. In *Star Trek* (TOS, the original series), the science officer usually saves the day by telling the chief engineer to invert phase. And this is one time it is correct to invert the phase of the photodiode to get the circuit working correctly. Now let's look at Figure 11-21.

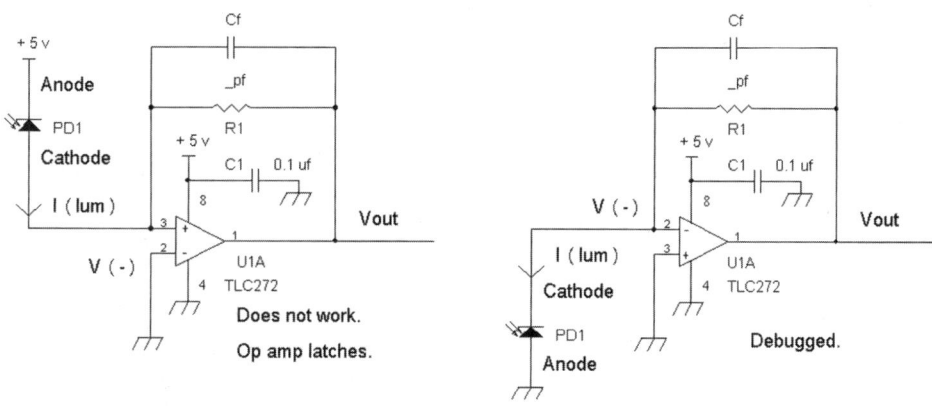

FIGURE 11-21   A example of inverting the phase of the photodiode and the inputs at pins 2 and 3 fixes the problems.

In this example, we have to invert phase twice. See the left circuit in Figure 11-21. And the mistakes we shall see are somewhat devoid of proper design.

For the left side circuit in Figure 11-21, the photodiode looks similar to the correct orientation for the photodiode's current flow via I (lum); the configuration, however, is incorrect. Although the cathode of PD1 is properly biased to a positive +5 volts for reverse bias on the photo diode, there are two problems. First there is **positive feedback** via R1 via the (+) input being connected to the feedback resistor R1. When an op amp is connected in a **positive feedback** configuration like this, the concept of V(+) = V(–) at the input terminals **does not apply.** Generally, the op amp's output signal at Vout may show oscillation or may latch to a DC voltage. With positive feedback it does not matter which way the photodiode is connected because the output signal will not be related to photodiode current. To fix the two problems, we have to first, reverse the input pins of the op amp and second, connect the photodiode as shown in the right side circuit of Figure11-21. Although the photodiode now sits at 0 volts DC bias, which is fine for most cases, there may be a reason for biasing the cathode of PD1 to a positive voltage. By applying a positive voltage that puts the photodiode into reverse bias, the internal photodiode capacitance reduces. We can take the anode of PD1 in the right circuit of Figure 11-21 and connect it to an available negative power supply. But suppose we only have the positive power supply. Can we still apply a positive reverse bias voltage to the cathode of PD1? The answer is yes. See Figure 11-22, which incorporates a second amplifier (e.g., you can use a dual op amp).

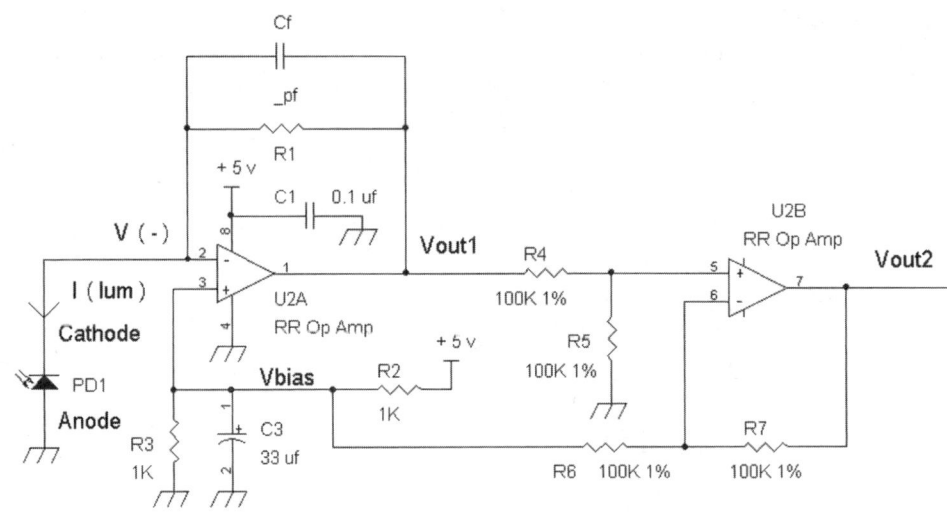

FIGURE 11-22 Biasing the photodiode's cathode via Vbias and using a differential amplifier circuit via U2B. Note: RR Op Amp = Rail to Rail op amp such as an ISL28218.

We can set a positive voltage on the cathode of PD1 by having the (+) input of U2A biased with a voltage. In this case it is about one-half the supply voltage, or 2.5 volts in this case with a positive 5-volt power supply. Because we have a negative feedback circuit via R1 from Vout1 to the (−) input of U2A, the voltage at pin 2, V(−) is the same as Vbias. That is, under a negative feedback op amp circuit, the voltages at the input terminals are the "same." Therefore with V(−) = Vbias, and Vbias = 2.5 volts, the cathode of PD1 is sitting at 2.5 volts DC.

The resistors R2 and R3 are chosen such that R2||R3 << R6. In general, we want R2||R3 ≤ 1% of R6. In this circuit, R2 = R3 = 1KΩ so R2||R3 = 500Ω and 500Ω is 0.5 percent of 100KΩ.

With R2 = R3, the current draw into R2 is (5 v)/(R2 + R3) or 5 v/2KΩ = 2.5 mA.

We can increase R2 and R3 to 2KΩ to save power if needed. Decoupling capacitor C3 with R2 and R3 forms a low-pass filter to reduce noise from the +5v supply line. The cut-off frequency of this low-pass filter is $f_c ~ 1/[2\pi(R2||R3)C3]$. In this circuit we have:

$$f_c = 1/[2\pi(1000\Omega||1000\Omega)33\mu f] = 1/[2\pi(500\Omega)33\mu f] \text{ or } f_c = 9.65 \text{ Hz}$$

And generally, we want $f_c << 60$ Hz for good filtering. If needed, we can increase C3 from 33μf to 330 μf. A ≥ 10-volt capacitor should work fine since the supply voltage is +5 volts and with R2 and R3, nominally, there are only 2.5 volts across C3.

Vout1 is the summation of the bias voltage, Vbias and the photodiode current times R1 or Vout1 = Vbias + I (lum) × R1.

Essentially, this circuit is a voltage follower for Vbias since the photodiode, PD1, is a current source that has "infinite" DC resistance. So, there is no DC resistance value from cathode to anode of PD1. Again, see Figure 11-22.

The end game is to provide an output that is in the form of only I (lum) × R1, without any DC bias voltage. We then need to subtract out "exactly" Vbias from Vout1 = Vbias + I (lum) × R1. By using a differential amplifier U2B, the output of it is Vout2 = Vout1 – Vbias. Or put in another way (for the circuit in Figure 11-22), Vout2 = **Vbias** + I (lum) × R1 – **Vbias** = I (lum) × R1 = Vout2.

Differential amplifier U2B works as a unity gain subtractor when all four of the resistors, R4, R5, R6, and R7, are the same value. Note that they are 1 percent tolerance resistors.

Ideally, op amps U2A and U2B are from a dual package rail to rail (RR) output for maximum voltage swing such as NJU7062 or NJM2762. However, if you bias the cathode of PD1 to a slightly lower voltage such as 1.67 volts by having R2 = 2KΩ and R3 = 1KΩ, you can use other op amps such as TLC272 or CA3240 and still get good output voltage swing. Of course, if you just need a slight positive bias voltage at the photodiode cathode, such as a few hundred millivolts, you can set R2 = 10KΩ to have Vbias ~ +0.45 volts.

We now turn to a circuit found in a publication and on the web that is devoid of reliable operation. In a sense it is "bankrupt" in good electrical engineering practice.

(After all, this is Chapter 11 and we should cover at least one bankrupted circuit . . . just kidding!)

See Figure 11-23.

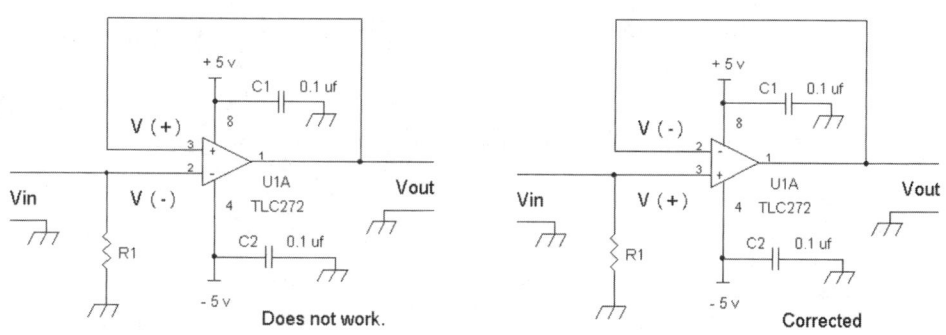

FIGURE 11-23   The left circuit is sometimes called an inverting voltage follower, which is incorrect.

In Figure 11-23, the left side circuit shows positive feedback because the output signal from Vout is connected back to the non-inverting input. This will cause the output signal to oscillate or latch to either a positive or negative voltage. With positive feedback we will find that the voltages measured at the non-inverting and inverting input terminals are not equal. That is, for the left circuit with positive feedback V(+) ≠ V(–).

To fix this circuit, you need to reverse the input leads as shown on the right side circuit of Figure 11-23. With this done we now have a standard unity gain non-inverting voltage follower where Vout = Vin, and V(+) = V(−) .

Summary

In this chapter, we have been shown some examples of "bad" circuits, and there is a reason for showing these. As a hobbyist or even an engineer, oftentimes we assume what's published in books or magazines or posted on the web (world wide web) is correct. In my experience, some of these non-working circuits have "typos," others include misunderstandings of electronic principles such as the left side circuit in Figure 11-23, and yet other circuits that do not work can be attributed to an incorrect part or not knowing limitations of components (e.g., rail to rail op amps versus standard op amps). So, when we build such "bad" circuits exactly and check them over and over again and find nothing wrong construction-wise, it's time to examine the circuit itself, and not assume that it actually works as advertised. As we get out of Chapter 11, which we did in a sense showing how to fix erroneous or "bankrupted" circuits, we will find even more types of these circuits ahead. Stay tuned to the next couple of chapters—more examples to come.

Reference Books

1. Arthur B. Williams, *Electronic Filter Design Handbook, Second Edition.* McGraw-Hill, 1988.
2. Paul R. Gray and Robert G. Meyer, *Analysis and Design of Analog Integrated Circuits, Second Edition.* Wiley, 1984.

CHAPTER 12

Some Ham Radio Circuits Related to SDR

There are many amateur radio circuits available today such as those used in software defined radios (SDR). In this chapter we will look into some of these circuits including building blocks that include crystal oscillator circuits, mixers, and simple RF matching circuits.

We will explore circuits published in magazines and books, and also those posted on the web. Because these are DIY (do-it-yourself) type circuits, there may be room for improving performance and reliability, or room to simplify these circuits.

This chapter will cover mainly amateur radio receiving circuits so that anyone can build and explore these without having to obtain a license (e.g., Technician, General, or Extra).

However, if you are transmitting ham radio signals you will require an FCC (Federal Communications Commission) license in the United States or an equivalent amateur radio license elsewhere.

Software Defined Radio Circuits

A software defined radio uses a computer to handle "middle and back" portions (intermediate frequency, IF, filter, demodulator/detector, audio filtering, and audio output circuits) of a superheterodyne (superhet) radio. Essentially, a basic superhet radio includes a tuned circuit for the received signal, a mixer circuit to shift down the received signal's frequency, an intermediate frequency filter and amplifier, and finally a detector (e.g., AM) and audio amplifier. See Figure 12-1.

NOTE: RF mixing means we are in some way multiplying two signals to get a third signal.

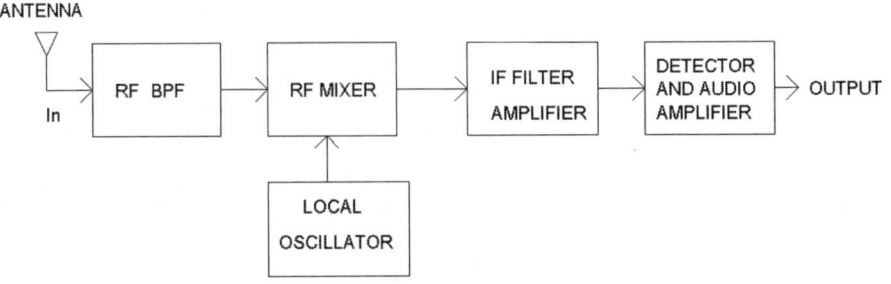

FIGURE 12-1   A block diagram of a basic conventional superhet radio.

As shown in Figure 12-1 the basic superhet radio requires an RF band-pass filter (BPF) to ensure that the RF mixer with the local oscillator provides an IF signal indicative of the tuned RF signal from the antenna. For example, in a standard AM radio, the IF is 455 kHz. This means that the local oscillator's frequency is 455 kHz above the incoming RF signal. But if there is an RF signal that is 455 kHz above the local oscillator's frequency, the IF signal will include an undesirable image signal interfering with the desired signal. For example, if the desired station is at 600 kHz, then the local oscillator's frequency is 600 kHz + 455 kHz = 1055 kHz. The mixer will output a signal at |1055 kHz – 600 kHz| = 455 kHz. However, if the antenna receives a radio station at 1510 kHz, and if the BPF tuned to 600 kHz does not sufficiently attenuate signals at 1510 kHz, then the output of the mixer will have another signal at |1055 kHz – 1510 kHz| = 455 kHz. Thus, the RF band-pass filter should be tuned to the desired station and remove substantially all undesirable RF signals that would be inadvertently mixed down to the intermediate frequency. Fortunately, in most AM radios, there is a sufficiently high Q (e.g., narrow bandwidth) tuned circuit that passes the desired RF signal while substantially attenuating the image signal (e.g., 1510 kHz) that is 455 kHz above the local oscillator frequency (e.g., 1055 kHz). The IF filter and amplifier is usually a ceramic filter or an LC (inductor capacitor) band-pass filter with one or more transistors, or a dedicated IC (e.g., TA2003). The output of the IF filter-amplifier is then connected to a demodulator such as an AM (amplitude modulation) detector (e.g., half-wave rectifier circuit). If the system is an FM radio, the detector will employ a frequency modulation detector (e.g., a quadrature detector, a ratio detector, or a frequency discriminator). The detector is coupled to an audio amplifier that provides an audio signal to an earphone or loudspeaker.

One disadvantage of the traditional superhet radio is that the system blocks are fixed to one type of function. For example, if the incoming RF signals also include Morse Code CW (continuous wave) and single sideband signal (SSB) signals, then the detector system will be complex with multiple demodulation circuits (e.g., product detector with CW filter, product detector with SSB filter, etc.). Although these different types of detectors can be made in hardware, they can be mimicked via a software program in a computer, once signals are digitized through the soundcard. See Figure 12-2.

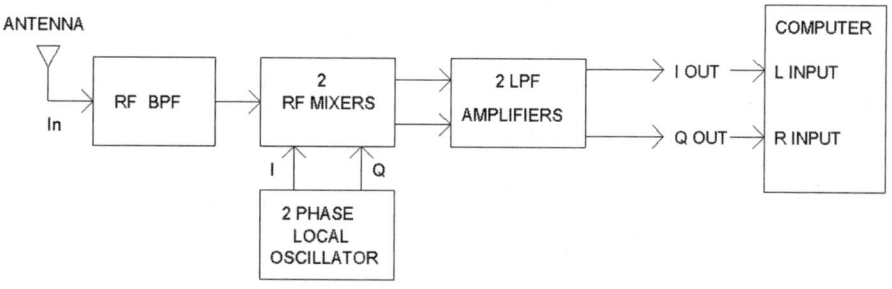

FIGURE 12-2   A block diagram of an SDR front-end system with low-pass filters for the IF.

As we can see in Figure 12-2, there are some similarities with the standard super-het radio. There's still an RF band-pass filter, but now we have twice as many RF mixers, twice as many signals for the local oscillator, and twice as many IF amplifiers and filters. Fortunately, these types of mixers and IF filters (low-pass filters, LPFs) can be implemented with op amps and simple resistor capacitor filters. The reason for having twice as many parts after the RF band-pass filter is so the computer can take the two signals, I OUT and Q OUT, and process them such that any image signal will be cancelled out via the software program installed in the computer (e.g., Winrad). Basically, the local oscillator has an I phase signal (defined as 0 degrees phase reference) and a Q phase signal (90 degrees phase shift from the I phase signal). The outputs of each of these I and Q oscillator signals are fed to separate I and Q RF mixers. From the outputs of the I and Q mixer are frequency translated signals of two low-frequency IF signals that are 90 degrees phase shifted from each other in terms of their carrier signals. Figures 12-3 and 12-4 illustrate this. The two low-pass

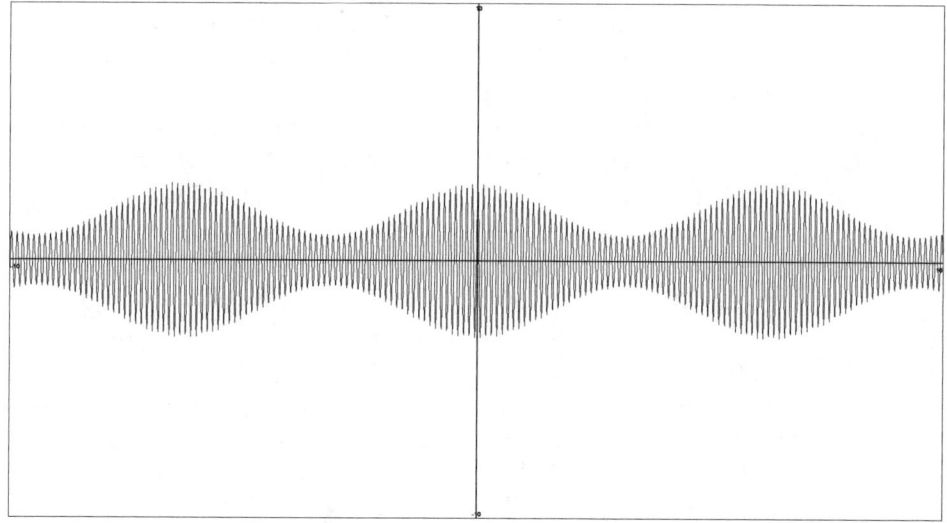

FIGURE 12-3   An example RF amplitude modulation signal at a high frequency such as > 1 MHz or more.

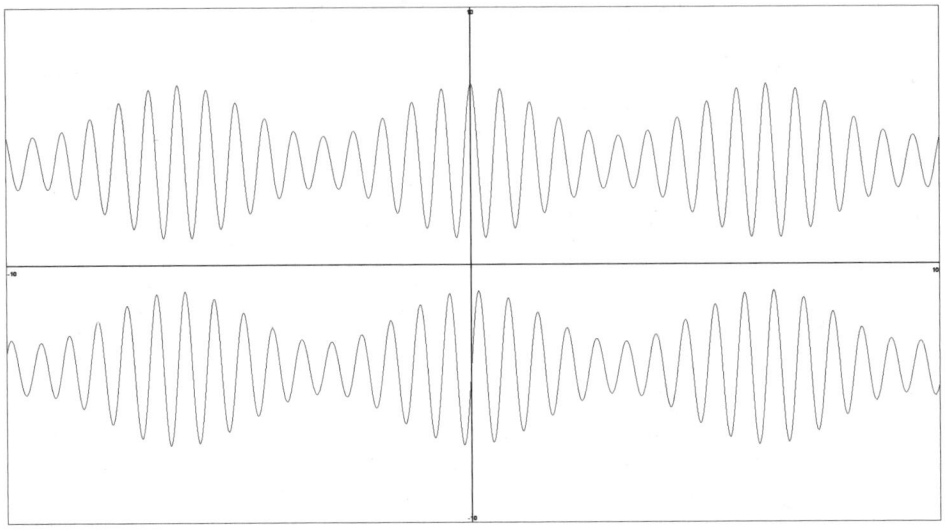

FIGURE 12-4 Examples of two lower intermediate frequency (IF) signals where the carrier signal's phase of the bottom AM signal is 90 degrees shifted with respect to the top waveform. See center vertical axis as a reference.

filters then provide signals whose intermediate frequency is generally within a 20-kHz, 40-kHz, or 96-kHz bandwidth for a computer soundcard.

Now let's examine a simple mixer SDR front-end circuit that includes the RF band-pass filter, I and Q analog switch mixers, and low-pass filter amplifiers. We will first start with Figure 12-5, the local oscillator that generates I phase and Q phase carrier signals.

A basic crystal oscillator via IC1E utilizes a logic inverter gate configured as an inverting gain amplifier via negative feedback resistor R1 to establish a DC operating point such that the logic inverter gate IC1E is self-biased in an amplifying region. The value of R1 is chosen to be high resistance, with $R1 \geq 1M\Omega$ so that it will not load down the signal from the crystal Y1 at C5. If R1's resistance is too small (e.g., $< 1000\Omega$), the circuit may not oscillate due to excess attenuation at pin 11. The oscillator circuit requires a phase lag of about 60 degrees or more from RC filter R8 and C4 so that the crystal Y1 and capacitor C5 form another phase lag circuit of at least 90 degrees (e.g., 90 degrees to 135 degrees). To ensure oscillation, the phase lag from R8 and C4 combined with the phase lag from Y1 and C5 has to be in the order of 180 degrees since the inverter gate "automatically" gives another 180 degrees of shift. With a total phase shift of 360 degrees via the phase shift networks (R8, C4, Y1, C5), an oscillation occurs at IC1E pin 10. The oscillator signal is coupled to a buffer inverter IC1A that drives a binary divider, IC3A, which provides a divide by 2 signal at pin 3 and a divide by 4 signal at pin 4 of the 74HC393 IC. A D flip flop is clocked by the divide by 2 signal (e.g., 7.16MHz) at pin 3 IC3A, which provides one-bit sample-and-hold signals of the divide by 4 signal at pin 2 of the 74HC74, which in

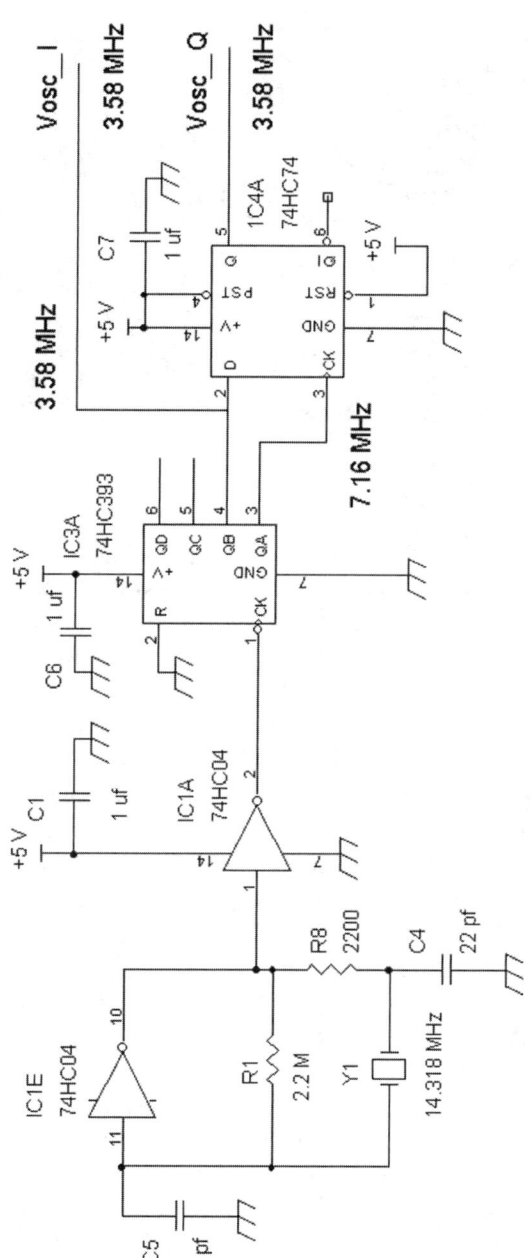

FIGURE 12-5 A crystal oscillator circuit with flip flops to generate I and Q carrier signals, where Vosc_I is defined as the 0 degrees signal and Vosc_Q is 90 degrees shifted in reference to the Vosc_I signal.

turn provides a quarter-cycle or 90 degrees delayed signal of the original divide by 4 signal. See Figure12-6.

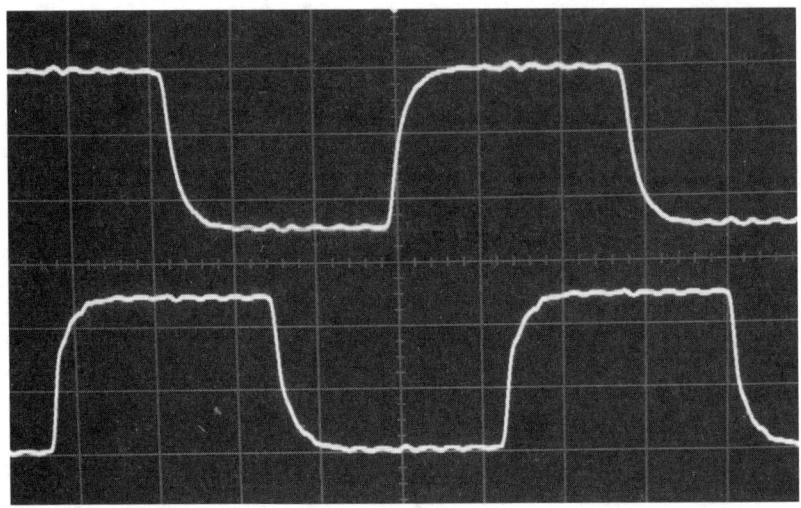

FIGURE 12-6 Top trace is Vosc_I (0 degrees reference) square-wave signal and the bottom trace is Vout_Q (90 degrees), a delayed square-wave signal.

The two oscillator signals, Vosc_I and Vosc_Q, will be used for the I and Q mixers as shown in Figure 12-7 where **IC5A and IC5B are square-wave RF mixers**.

Vin is from a 50Ω antenna tuned to ~ 3.60 MHz for the 80-meter ham radio band. RF matching network Cvar1, L1, and R13 form an RF matching network for Rin = 50Ω. This RF matching network is a high Q high-pass filter that includes some band-pass filtering around the resonant frequency of about 3.60 MHz. Generally, we want to step up the RF input voltage via Cvar1 and L1 many times. This step-up factor is Q and generally we want Q ≥ 3 so that the subsequent calculation will be accurate for practical purposes (e.g., within 10 percent).

The RF voltage across L1 is basically Q × Vin. So, this RF matching network provides an equivalent step-up transformer effect. For example: Q = XL1/50Ω, for 3.60 MHz = f_{res} and L1 ~ 22 uH, X_{L1} = $2\pi f_{res}$(L1) so Q = 2π(3.60MHz)(22 uH)/50Ω or **Q ~ 10**. We can find Cvar1 from f_{res} = 3.60 MHz = $1/[2\pi\sqrt{(L1)(Cvar1)}]$, which leads to Cvar1 = 1/[L1(3.60 MHz × 2π)²] Cvar1 = 1/[22 uH (3.58MHz × 2π)²] Cvar1 = $[1/(1.112)^{10}]$ farad or Cvar1 = 90 pf. Cvar1 can be a polyvaricon variable capacitor used in AM radios (e.g., two sections, 0–60 pf and 0–140 pf).

To ensure that Rin = 50Ω, R13 is required to "terminate" at L1 so at the resonant frequency of 3.60 MHz, Rin is resistive and at 50Ω. Having Rin = 50 Ω is important so that the antenna's coaxial cable is properly loaded/terminated to ensure low SWR (standing wave ratio), which means maximum power from the antenna is transferred to the front in circuit at Vin via Cvar1. To calculate R13, you can use the following formulas for Q ≥ 3.

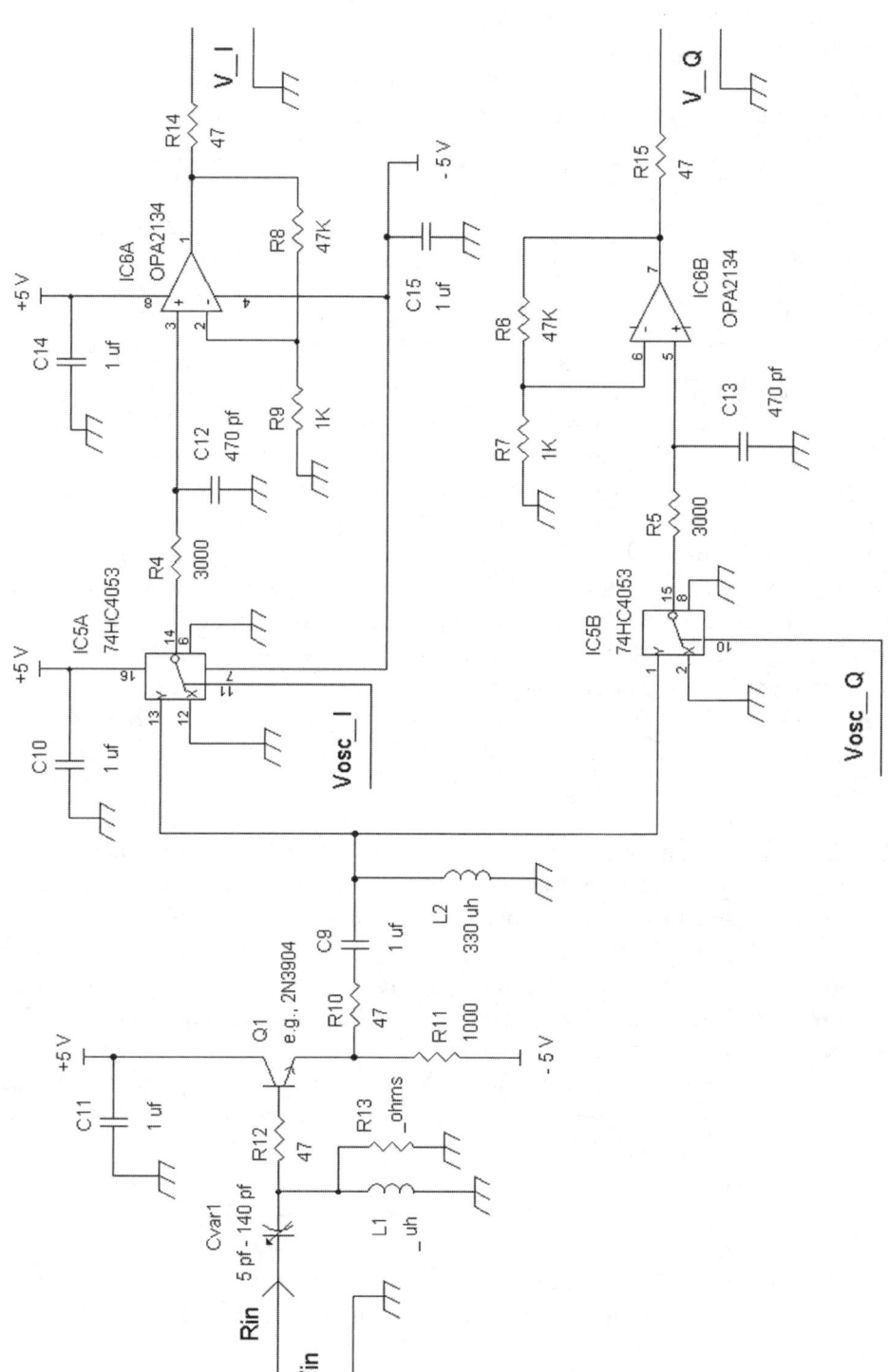

FIGURE 12-7 An SDR front-end circuit with input band-pass filter (Cvar1 and L1) amplifier Q1, I and Q mixers (IC5A and IC5B) and low-pass filter amplifiers (R4, C12, IC6A; R5, C13, IC6B).

Note that Cvar1 is adjusted for maximum amplitude at transistor Q1's emitter since L1's inductance is accurate to about 10 percent to 20 percent.

$R13 = Q \times X_{L1} = 10 \times 2\pi f_{res}(L1) = 10 \times 2\pi(3.60 \text{MHz})(22 \text{ uH})$, or **R13 ~ 5000Ω**. When you know Q, another faster way of calculating R13 is to use the following:

$Q = \sqrt{\frac{R13}{RS}}$, where Rs = input RF signal's source resistance, which is generally 50Ω.

For example, **Rs = 50Ω.**
This leads to: $R13 = Q^2 \times Rs. \ Q = \sqrt{(R13/Rs)} = \sqrt{(5000Ω/50Ω)} = \sqrt{100} = 10$
Since Q = 10:

$R13 = 10^2 \times 50Ω = 100 \times 50Ω$ or $R13 = 5000Ω$

If higher Q is desired, we can raise the L1's inductance such as L1 = 47 uH, then Q = 23.7. This will lead to having Cvar1 = 42 pf and $R13 = (23.7)^2 \times 50Ω$ or R13 = 28KΩ. (Note: You can use 27KΩ, which is close enough.) Note that Q1 does not have infinite input resistance to its base terminal. This input resistance at Q1's base is roughly $\beta \times (R11 \parallel Z_{L2})$, which will be in parallel with R13. For an approximation given R11 = 1KΩ and $|Z_{L2}| = 2\pi(3.6 \text{ MHz})L2 = 2\pi(3.6 \text{ MHz})330 \text{ uH} = 7.46KΩ \gg 1KΩ$. So $\beta \times (R11 \parallel Z_{L2}) \sim \beta(R11) = \beta(1KΩ)$. Since β ranges between 50 to 100, the input resistance is about 50KΩ to 100KΩ, which is in parallel with R13. Given R13 in these two examples can be 5KΩ or 28KΩ, a good approximation is just to increase R13 by 20 percent from the calculated value. Thus 5KΩ → 6KΩ (6.2KΩ for a standard resistor value), and 28KΩ → 33.6KΩ (33KΩ or 36KΩ). Thus, slightly increasing R13's calculated resistance compensates for Q1's input load resistance.

NOTE: In many instances, the antenna may just be a long wire connected to Vin or Cvar1 directly. Here, the source resistance of the long wire antenna may not be 50Ω, and it is possible that R13 can be removed to maximize gain at the resonant frequency.

Emitter follower Q1 acts as a unity gain amplifier with high input resistance at the base of Q1. With L1 providing a 0-volt Q1 DC base bias voltage, Q1's emitter DC voltage is –0.7 volt since VBEQ1 = | 0.7 volt. *If you are not sure how the DC voltages for this Q1 emitter follower are calculated, you can build it with L1, Q1, R12, R11 with ± 5-volt supplies and measure the emitter voltage with a voltmeter with the negative test lead connected to ground.*

The DC collector current is set by R11, $IC_{Q1} \sim (-0.7 \text{ v} - -5 \text{ v})/R11 \sim 4.3 \text{ v}/1000Ω$ or $IC_{Q1} = 4.3$ mA. However, the DC collector current can be lowered to as low as ~ 2 mA or R11 → 2200Ω, because the input signal will be very small (< 100 mV) and does not require large signal swings at Q1's emitter. The output of Q1 is fed through a coupling capacitor C9 into a large value inductor that establishes 0 volt DC at the

"Y" inputs of the 74HC4053 analog switches at pin 13 and pin 1 of IC5A and IC5B. Note that series resistor R10 = 47Ω provides some isolation from the "kickback" spikes or glitches that will emanate to the input terminals at pins 13 and 1 due to switching characteristics of the 74HC4053. In general, when using an analog switch as an RF mixer, it is always good to add a small value series resistor to the input so that the amplifier driving it (e.g., Q1 in this case) does not exhibit instability due to the dynamic loading caused by the 74HC4053 input terminals' glitch signals. Note that there is an alternative to lowering the amount of glitch kickback by choosing a lower capacitance analog switch such as the SD5000 series FET (Note that the SD5000 or SD5200 quad switch has a different pin out than the 74HC4053).

Now let's take a look at the mixers, IC5A and IC5B, that gate through the RF input signal via Q1's emitter at a rate of the local oscillator's frequency. The output of each of the mixers at pins 14 and 15 will contain signals related to RF input signal, an IF signal related to the difference of the frequency of the RF signal and the frequency of the local oscillator, plus other high-frequency signals beyond the frequency of the RF input signal. See Figures 12-8 and 12-9.

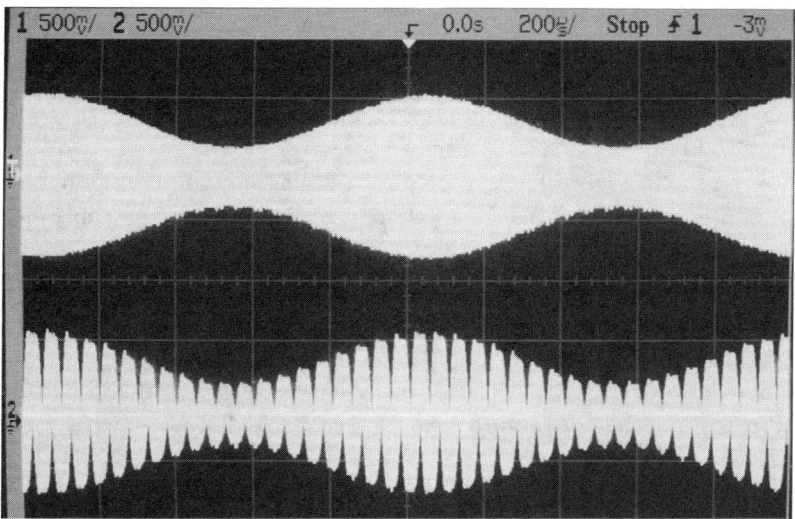

FIGURE 12- 8   Top trace is the RF input AM signal at 3.6 MHz carrier at 1 kHz modulation; bottom trace is a mixer output at IC5A pin 14. Note the two amplitudes are the same.

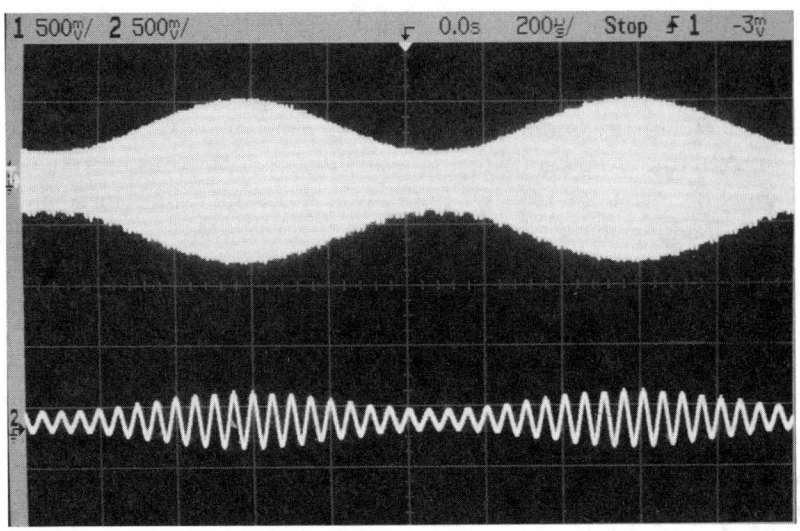

FIGURE 12-9   Top trace is the same RF input signal; bottom trace is the low-pass filtered version at R4/C12. Note that the filtered version is about one-third the amplitude of the RF input signal.

With the local oscillator providing 3.58 MHz and with a 3.60 MHz RF input signal, the IF (intermediate frequency) will be |3.60 MHz – 3.58 MHz| = 0.02 MHz or the IF = 20 kHz.

As we can see in Figure 12-9, when all the high-frequency components are filtered out of the amplitude of the IF signal, we see a signal that still retains the type of amplitude modulation envelope of the RF input signal. However, the amplitude is reduced to about one-third. That is the conversion gain = amplitude of the IF signal after filtering (e.g., at R4/C12) divided by the amplitude of the RF input signal at the input of the mixer (e.g., pin 13 of IC5A).

Theoretically, for the IC5A/B mixer using *Fourier Series* (which the reader does not need to know), the **conversion gain = (1/π) = 31.847%**, which is close to 33 percent or one-third.

Amplifiers U6A and U6B are non-inverting gain amplifiers each with a gain of (1 + 47K/1K) or a gain of 48. The total conversion gain from the input of the mixer at L2 or pins 13 and 1 of IC5A/B to the output of either U6A or U6B is then **(1/π) × 48 = 15.2**.

NOTE: The op amp chosen in this example is the OPA2134, which has an 8 MHz gain bandwidth product. So, for a gain of 48, the resulting bandwidth will be 8 MHz/48 ~ 166 kHz, which is sufficient for an audio sound card that accepts a 96 kHz analog input bandwidth using a 192-kHz sampling rate.

Some Troubleshooting Tips Concerning Figure 12-5 and Figure 12-7

1. Check wiring on the pin outs for all the ICs. If an IC is loaded in reverse, you may burn it out because the power and ground pins are at pins 14 and 7 or at pins 16 and 8. Reversing the IC will supply positive voltage to the ground pin and ground the + power pin, which will damage the chip. If this happens, shut off the power and carefully remove the chip and put in a new one with the correct orientation. It generally helps to use IC sockets instead of soldering the chip directly into the circuit board.

2. Confirm that the power supplies +5 volts and –5 volts are at the appropriate pins for IC5, IC6, the collector of Q1, and emitter resistor R11. And confirm +5 volts on the power pins of IC1, IC3, and IC4. Also, with an ohm meter, confirm the ground terminal pins for IC1, IC3, IC4, and IC5. Also, make sure that the power supply decoupling capacitors C10, C14, C15, and C16 are each located near (within half an inch of) the power pins to the ICs. Confirm that these decoupling caps are making continuity to the ICs' power pins (e.g., pin 14 for a 14-pin chip and pin 16 for a 16-pin chip), and that the other sides of the decoupling caps are connected to ground with short leads. The ground connection of a decoupling capacitor can be made via a ground plane on a printed circuit board or to pin 7 of a 14-pin chip or for a 16-pin chip to pin 8.

3. For the oscillator board in Figure 12-5, confirm that the crystal oscillator is working with an oscilloscope. Make sure that the oscilloscope is set to a bandwidth of at least 50 MHz bandwidth, otherwise a 14.318-MHz square wave will look too sinusoidal. Confirm a 14.318-MHz sine wave of at least 1 volt peak to peak at pin 11 of IC1E, and an approximate square wave at IC1E pin 10 **and IC1A pin 2** of about 5 volts peak to peak also at 14.318MHz. If you see noise at IC1E pin 10, check if R1 (2.2MΩ) is the correct value and that it is connected as shown in Figure 12-5; or try another crystal. Also, check R8 (2200Ω) for the correct resistance, and confirm that C4 and C5 are 22 pf.

4. Check for a 5-volt peak to peak square wave at IC3A pin 3 that should be one-half the frequency (7.16 MHz) from IC1A pin 2, and confirm that pin 4 of IC3A has a one-quarter frequency (3.58 MHz) square wave at about 5 volts peak to peak. If you see no output, confirm that pin 2 of IC3A is grounded.

5. With an oscilloscope, check for 90 degrees difference in waveform in Vosc_I at pin 2 of IC4A and Vosc_Q at pin 5 of IC4A. See Figure 12-5. If signal Vosc_Q is not a 90-degree phase-shifted square-wave signal relative to Vosc_I, then make sure pin 1, pin 4, and pin 14 of IC4A have +5 volts on all of them, and that pin 7 of IC4A is connected to ground.

6. At this point if you see very small amplitude waveforms, check to see if any of the output pins such as IC1E pin 10 and IC1A pin 2, IC3A pins 3 and 4, or IC4A pin 5 are shorted to ground or shorted to +5 volts.

7. For the SDR front-end board in Figure 12-7, we can confirm that the oscillator signals, Vosc_I and Vosc_Q, are at IC5A pin 11 and at IC5A pin 10, respectively.

8. Connect a 50Ω source resistance generator at 3.60 MHz and 200 mV peak to peak into the Vin terminal at Cvar1. With an oscilloscope probe at the emitter of Q1, confirm at least 1.5 volts peak to peak when Cvar1 is adjusted for maximum amplitude. The gain should be close to 10 ± 20 percent from Vin to the emitter of Q1. You may need to set your scope to "Bandwidth Limit" and AC coupling. Bandwidth Limit reduces the analog signal bandwidth to 20 MHz in most oscilloscopes but allows for viewing cleaner waveforms. Since the frequency is 3.6 MHz, the 20-MHz bandwidth limitation does not interfere in observing waveforms accurately for the SDR front-end circuit. *Usually, you can obtain an accurate frequency (< 0.05 percent error) digitally synthesized waveform generator for less than $100.*

9. Confirm sine waveforms at pins 13 and 1 of IC5 that are at least 1.4 volts peak to peak at 3.60 MHz.

10. Confirm sine waveforms at pin 3 and pin 5 of U6, which should be about one-third the amplitude of the waveform at pins 13 and 1 of IC5. For example, we should see at least 1.4 volts/3 or at least 0.46 volt peak to peak at about 22 kHz at pins 3 and 5 of U6.

11. The gain of U6 is about 48, so we will get clipping at the outputs pin 1 and pin 7 of U6. Thus, turn down the RF generator from 200 mV peak to peak to 40 mV peak to peak at 3.60 MHz. With a reduction by fivefold, we should get about 0.092 volt peak to peak at pins 3 and 5 of U6. The output of U6 pins 1 and 7 should now have 48 × 0.092 volt peak to peak or 4.4 volts peak to peak sine wave at about 22 kHz. Confirm that V_I and V_Q have a phase difference of about 90 degrees.

12. If there are problems with getting waveforms in the "ball park," confirm the component values of R4, R5, C12, C13, and R8, R9, R6, and R7. For example, here's a test for the bandwidth of the low-pass filters R4/C12 and R5/C13. You can sweep the RF frequency, f_{RF}, of the input RF signal from 3.59 MHz to 3.69 MHz while probing low-pass filter output at R4/C12 and low-pass filter output at R5/C13 to confirm that the frequency response is "flat" from 10 kHz to about 40 kHz and has a −3 dB amplitude point (70.7 percent reference) at 100 kHz. Remember that the resulting frequencies from the down-converted IF (intermediate frequency) signal at the low-pass filters is $|f_{RF} - f_{LO}|$, where f_{LO} = local oscillator frequency, which is 3.58 MHz = f_{LO}. Actually, the precise local oscillator frequency is 3.579545 MHz ~ 3.58 MHz, so more accurately, the IF = $|f_{RF} - 3.579545 \text{ MHz}|$.

A Common Sample-and-Hold RF Mixer Circuit

The switch mode RF mixer shown in Figure 12-7 has the advantage that it does not load the input signal much because of the series resistors in the low-pass filter circuits, R4 and R5 (both 3000Ω), which can be made larger for less loading if needed. For example, if the same bandwidth is required, we can just keep the same RC time constant, which is R4 × C12 or 3000Ω × 470 pf = 1.41 μsec. So, for instance, we can have R4 = R5 → 10KΩ and C12 = C13 → 140 pf. This way, the loading at the input terminals pins 13 and 1 of IC5A/B will be 10KΩ, worst case. Again, see Figure 12-7. The IC5A/B mixer can be made to switch very fast if other analog switches are used such as the SD5000 quad FET (Field Effect Transistor) or if two FST3253 ICs are used. Note that neither SD5000 nor the FST3253 devices are pin for pin compatible with the 74HC4053 chip.

However, the conversion gain of the IC5A/B in Figure 12-7 is "lossy" because it has about a 32 percent conversion gain (see Figures 12-8 and 12-9). If the RF mixer is a sample-and-hold circuit where the sampling time is relatively small, such as $\leq 25\% \times (1/f_{LO})$, then the conversion gain goes up to ≥ 80 percent. See Figure 12-10. In this section we will explore the Tayloe RF Mixer, which is based on sample-and-hold circuits.

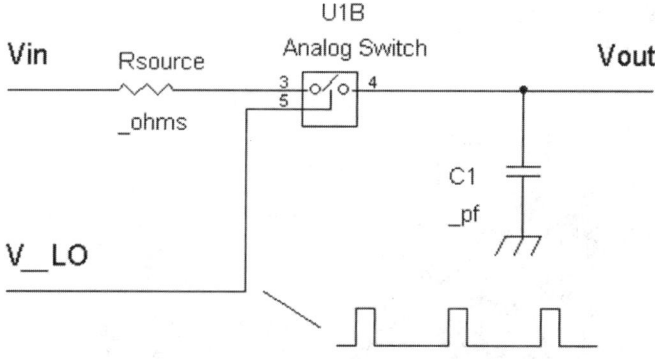

FIGURE 12-10   A simple sample-and-hold circuit; note that the sampling pulse < 50 percent duty cycle.

A simple sample-and-hold circuit is modeled by a finite source resistance, Rsource, such as a 50Ω source from Vin. However, Rsource should also include in series the ON resistance of the analog switch. For example, if Vin has a 50Ω source resistance (such as from a 50Ω antenna) and U1B, has an ON resistance of 20Ω, then Rsource = 50Ω + 20Ω = 70Ω. As we will find out soon Rsource plays an important part in terms of using an analog switch as an RF mixer. The basic operation of the sample-and-hold circuit is when V_LO's signal is logic high, the analog switch closes and connects for a short duration signal from Vin through Rsource to charge up capacitor C1. When V_LO goes to logic low, the analog switch becomes an open circuit, which disconnects Vin and Rsource from C1. Then capacitor C1 retains the charge and provides a constant voltage during the logic low state from V_LO. See Figure 12-11.

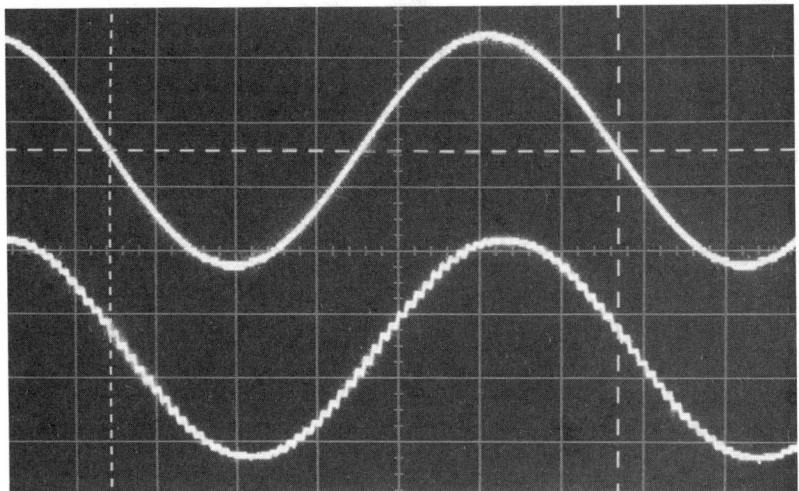

FIGURE 12-11   An example output signal from a sample-and-hold circuit on the bottom trace versus the input signal on the top trace. The stair-step waveform (bottom trace) from the sample-and-hold circuit maintains the same voltage until the next sampling pulse comes along.

See Figure 12-12 for magnified view of the stair step waveform.

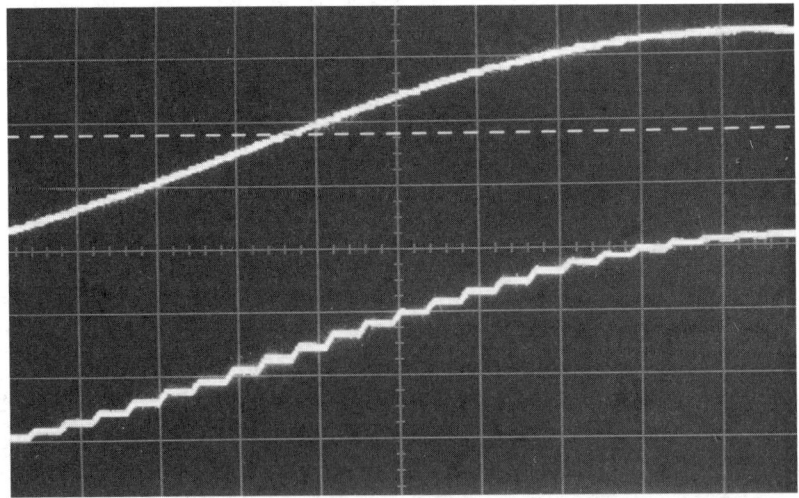

FIGURE 12-12   A closer look and the sample-and-hold waveform with input signal at the top trace and the sampled-and-held waveform on the bottom trace.

Note that frequency of the local oscillator or V_LO is close to the frequency of Vin, the RF input signal. The output at Vout is then an IF signal that has a low frequency due to $|f_{RF} - f_{LO}| = IF$ and f_{RF} almost equal to f_{LO}. For example, we want $|f_{RF} - f_{LO}| < 100$ kHz so that the IF signal at Vout can be sent to a sound card in a computer to run an SDR program. If we are trying to down-convert an 80-meter ham radio RF input signal frequency $f_{RF} = 3.800$ MHz, the local oscillator signal, V_LO, can have a frequency of $f_{LO} = 3.810$ MHz such that: $|f_{RF} - f_{LO}| = IF = |3.800$ MHz $- 3.810$ MHz$| = |-0.01$ MHz$| = 0.01$ MHz or the IF $= 10$ kHz.

In general, the sampling pulse (high logic state duration) should be << 50 percent duty cycle. That is, if you have a 1-MHz sampling frequency, the period is 1 μsecond (μsecond = μsec). The duty cycle is the "ON" time divided by the period. In this example, it will be ON time/1 μsec. Normally, you would like to have the ON time as short a duration as possible. In sampling theory the ON time duration → 0 μsec, but this is not practical because with Rsource in series with sampling capacitor C1, the capacitor C1 in Figure 12-10 cannot charge up instantaneously. So, we need to first define what the duty cycle will be. Then second, determine what we need for a good C1 capacitance value to use given that Rsource is known. To get a feel of the effects of too much capacitance in C1 see Figure 12-13, which shows examples of pulse responses with "too much capacitance" on the longer RC time constant waveform compared to the shorter RC time constant waveform that provides the final voltage with a small time interval.

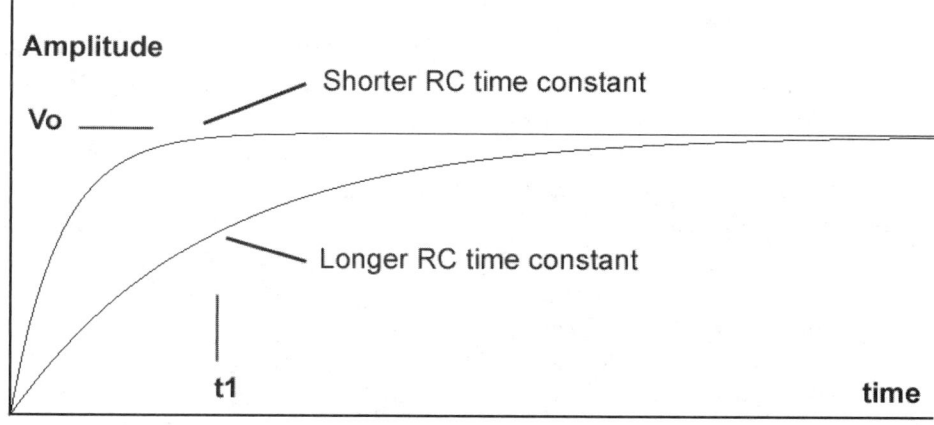

FIGURE 12 -13   With different RC time constants we see that it takes a longer sampling pulse (t) to arrive at the final voltage V_0. And ~ V_0 is achieved at t1 with a shorter RC time constant when compared to ~$V_0/2$ on the longer RC time constant waveform.

The actual equation for the voltage across the capacitor as the analog switch is closed for the duration a positive going pulse from 0 volts to V_0 volts is:

Vout = $V_0(1 - e^{-t/RC})$

where V_0 = peak to peak amplitude of the pulse and where the pulse starts at 0 volts, t = time duration of the sampling pulse (e.g., time duration of the V_LO in the logic high state), and where R = Rsource and C = C1 as shown in the circuit of Figure 12-10, and where e = 2.71828.

Ideally, we want Vout to have the same value as V_0 instantly. But this will not happen due to the finite resistance, Rsource. So, here's what the charge-up voltage at Vout looks like in terms of RC time constants where "t" is the ON time duration:

Vout = 63% of V_0 when t = RC

Vout = 86% of V_0 when t = 2RC

Vout = 95% of V_0 when t = 3RC

For example, suppose we want to build a sample-and-hold mixer for the 80-meter band around 3.600 kHz, and we want to use a 25 percent duty cycle base on a local oscillator frequency at 3.600 MHz, whose period for one cycle is 1/(3.6 MHz) = 0.277 μsec or 277 nsec (nano seconds). For a 25 percent duty cycle, the logic high "ON" duration will be 25% × (277 nsec) = 69 nsec. If we want to have Vout come up to 95 percent of input value, we have to set the RC time constant to one-third of the "ON" duration or (1/3) (69 nsec) or t' = RC = 23 nsec. This means if Rsource = 50Ω, then C1 = t'/Rsource = (23 nsec)/50Ω or C1 = 463 pf ~ 470 pf. We label this as $C1_{80meters}$ = 463 pf. Note: For 86 percent of input value set RC to (1/2) (69 nsec).

But what if we want to receive other ham radio bands up to 20 meters? Then we simply scale C1 accordingly to a smaller value in proportion to C1 for 80 meters.

That is, $C1_{20meters}$ = (20 meters/80 meters) $C1_{80meters}$.

Or for 20 meters, C1 = (1/4) 463 pf or C1 = 115 pf ~ 120 pf or 100 pf.

But what this really means also is that you can use the 20 meters C1 value for any lower frequency. For example, if C1 = 100 pf in Figure 12-10 with Rsource = 50Ω, this means for the lower frequencies (e.g., 40 meters, 80 meters, and 160 meters) the pulse duration will be longer and Vout will be greater than 95 percent (but always less than 100 percent).

Now let's look at the consequence of having too large of a C1 value when we want to measure the IF bandwidth or roll-off. That is what happens when the RF input signal's frequency starts separating more from the local oscillator frequency. Will the IF signal's amplitude have the same output voltages at 5 kHz versus 20 kHz or 100 kHz? The answer is no. The larger the C1 capacitance, the poorer the frequency response will be on the higher frequency end of the IF signal. See Table 12-1, where Rsource is 50Ω that represents a "best" case scenario. There are times when the source resistance is much higher, such as when connecting a single "untuned" long wire antenna directly into the RF mixer without an amplifier. Such resistances can be much higher, such as 300Ω or more. Table 12-1 will "stop" at the –3 dB frequency for each C1 value from Figure 12-10. The lowest frequency, 100 Hz, is defined as 100 percent with the 0.01 uf = C1.

The test setup has the local oscillator signal, V_LO, at 3.6000 MHz with a 25 percent duty cycle (69.4 nsec high logic level for a one-cycle period of 277.7 nsec). Vin is a signal at 3.6000 MHz plus the frequencies shown in Table 12-1.

For example, to provide a 1000 Hz IF signal at Vout, the input RF signal is 3.6010 MHz, such that 3.6010 MHz – 3.6000 MHz = 0.001 MHz = 1 kHz.

TABLE 12-1   Comparing Frequency Response of the IF Signal at Vout of Figure 12-10, a Single Sample-and-Hold Circuit with Different Hold Capacitor Values for C1 and with Rsource = 50Ω

Vout Freq.	% Vout @ C1 = 0.01µf	%Vout @ C1 = 0.22µf	–3 dB (70.7% of Max)
100 Hz	100%	96%	
1000 Hz	99%	95%	
5000 Hz	97%	66%	4900 Hz C1 = 0.22uf
10 kHz	96%	41%	
20 kHz	94%	24%	
30 kHz	90%	17%	
90 kHz	71%	— (too small to measure)	90 kHz C1 = 0.01uf

As we can see, there is a penalty in having a large capacitance hold capacitor C1. With the 0.22 µf for C1, the usable bandwidth is about 5 kHz, which is about one-fourth the bandwidth most minimum sampling rate sound cards can do at 44.1 kHz, which provides an analog signal bandwidth of 20 kHz. Reducing the capacitance from 0.22 µf to 0.01 µf is much better and a 90-kHz bandwidth is achieved when Rsource = 50Ω. However, if Rsource > 50Ω, such as 300Ω, then we are back to a reduced bandwidth situation. Also, if we want to listen in on higher frequency signals such as in the 40- or 20-meter ham radio bands, then the high logic level "ON" time reduces accordingly to about 35 nsecs or 17.5 nsecs. Thus, the 0.01 µf value for C1 will start providing some more bandwidth loss due to the shorter sampling intervals.

For RF sample-and-hold mixer circuits, the hold capacitor should be in the range of 47 pf to 1000 pf; this will allow for providing a good IF bandwidth up to 90 kHz or more, which can be taken advantage of with a high-quality sound card with 192 kHz sampling frequency. With higher IF bandwidth, more of the radio band can be tuned with a fixed frequency local oscillator.

In general, when using a software defined radio (SDR) program such as Winrad (or others), the span of tuning across the radio band is related to twice the IF bandwidth. Recall that in Figure 12-2, there are two mixers, one I and one Q. Both mixers will provide the same IF bandwidth, and if the hold capacitor has too much capacitance, then the tuning range within the SDR software program will be restricted.

For example, with C1 = 0.22 µf for each of the sample-and-hold mixers (I and Q), the IF bandwidth is ~ 5 kHz. This means only 2 × 5 kHz or **10 kHz of tuning** across the band is possible. If C1 = 0.01 µf for a 90 kHz IF bandwidth, then we get 2 × 90 kHz or **180 kHz of tuning** range. This then is eighteen times better than the 5 kHz IF bandwidth.

A Preferred Implementation with Sample-and-Hold Circuits

As have already seen in Table 12-1, using a large-value hold capacitor dramatically reduces the IF bandwidth that hinders tuning range. With all sample-and-hold circuits, an important principle to keep in mind is the hold capacitor such as C1 in Figure 12-10 needs to be connected to a high impedance input amplifier. That is, any (e.g., low to medium) resistive loading (e.g., ≤ 10kΩ) across the hold capacitor (e.g., C1) will discharge the capacitor's voltage before the next sample pulse and result in a loss of output signal voltage.

In some cases, the type of amplifier used to amplify the capacitance voltage at C1 of Figure 12-10 should be FET or amplifiers with extremely low input bias currents (e.g., ≤ 0.200 nA for JFET op amps). In general, almost any FET op amp will work, such as the LF353, TL082, TLC272, etc., providing you keep track of the gain and IF bandwidth.

For example, if you want a 100-kHz closed loop bandwidth from the op amp non-inverting gain amplifier and the gain bandwidth product is 3 MHz, then the maximum gain you can have is about +30 since +30 × 100 kHz = 3 MHz.

Figure 12-14 shows a preferred way of providing amplification to a sample-and-hold circuit, and Figures 12-15, 12-16, and 12-17 illustrate ways of causing problems to the sample-and-hold circuit that should not be used unless corrected or modified.

The voltage across C1, Vout, preferably should be amplified or buffered with a high input resistance amplifier (e.g., > 1 MΩ). If an op amp is used, this means that only non-inverting gain amplifier configurations will work. At first look, Figure 12-14 may seem "funny" or incorrect because there is no bias resistor to set the DC bias for the (+) input terminal of the op amp. The reason we can get away from having a bias resistor at the (+) input is that the analog switch, U1B, will transfer a sample of the input voltage to C1, which in turn biases the (+) input terminal.

In Figure 12-14, note that the input signal Vin must have a DC path to a DC voltage source or to ground. The gain of the amplifier is Vout2/Vin = [1 + (R1/R2)]. Generally, the gain is in the order of 10 to 100 in most cases. With higher bandwidths such as having a gain of 100 with an IF bandwidth of 96 kHz, you can choose an op amp (e.g., TLE082) with a gain bandwidth product of ≥ 10 MHz since 100 × 96 kHz equals 9.6 MHz. *See the end of this chapter for more on gain bandwidth product.*

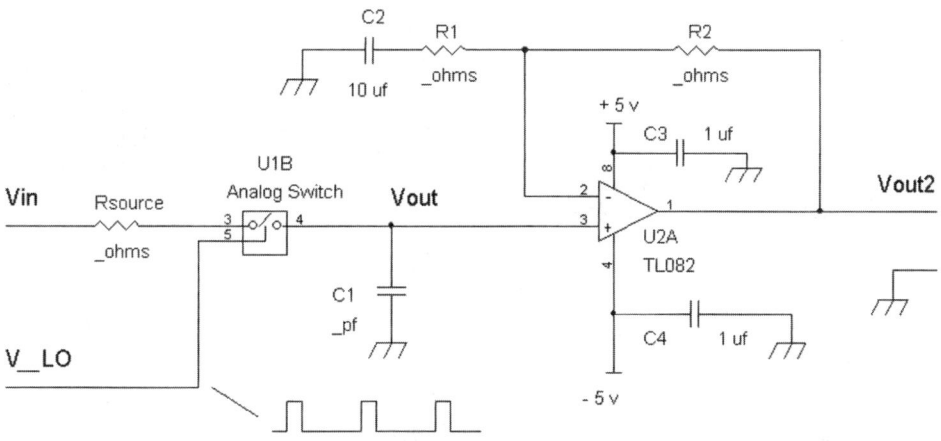

FIGURE 12-14 A preferred way of buffering or amplifying voltage across a hold capacitor to keep the hold capacitor's voltage at C1 steady until the next sampling pulse arrives via V_LO.

Alternatively, you can split the gain of 100 by cascading two amplifiers by having each non-inverting gain op amp stage provide a gain of 10. This way, you can get away with an op amp with a gain bandwidth product of about 1 MHz. However, it's best to be on the safe side and use commonly available op amps such as the TLC272, TL082, or LF353 that have gain bandwidth products of > 2MHz. The penalty for this is adding more op amps, but usually these op amps are less costly than the high-speed versions.

AC bypass capacitor, C2, can be omitted with R1 grounded on the R1/C2 side. However, C2 may be used to avoid amplifying DC voltages. Recall the input signal into the computer for the SDR program typically will be AC coupled into the sound card. If Vin has a DC bias voltage, then the op amp with C2 will just pass the same DC offset voltage to Vout2 without amplifying the DC offset voltage that can cause the op amp to saturate or clip at its output Vout2.

In Figure 12-15, there is no need to include R3 in the circuit, especially if this resistor discharges C1 too quickly. Setting a value such as < 50KΩ (e.g., R3 = 10KΩ or R3 = 1KΩ) will cause the sample-and-hold circuit to have a reduced output voltage at Vout2. If resistor R3 were a higher resistance value in the range of 1MΩ to 10MΩ, then there would generally be a minimal signal amplitude reduction at Vout2, and this change into much higher resistance values will fix the problem. However, you can just remove R3.

We now turn our attention to another error with sample-and-hold circuits that are connected to inverting gain amplifiers. One of the most "major" mistakes is to connect the hold capacitor (e.g., C1) to an inverting gain operational amplifier circuit that has a low input resistance. This low input resistance will then discharge the hold capacitor that results in an attenuated output voltage. See Figure 12-16.

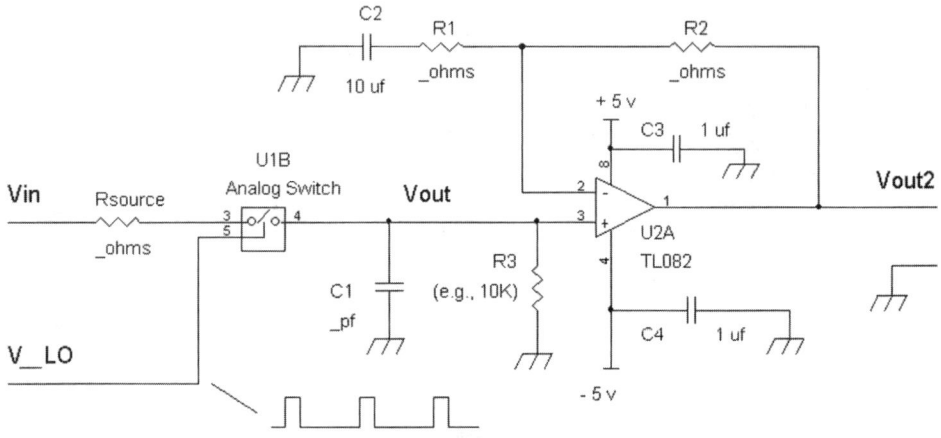

FIGURE 12-15 A non-inverting gain amplifier with a 10kΩ resistor R3 across hold capacitor C1. R3 prevents C1 from doing its job of holding a steady voltage until the next local oscillator sampling pulse arrives. To correct the situation, R3 may be removed.

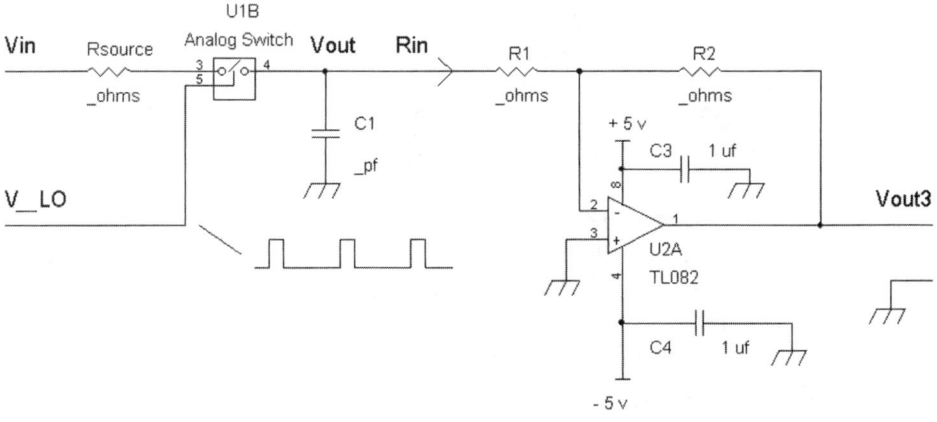

FIGURE 12-16 A sample-and-hold circuit that will definitely not work as planned because the hold capacitor is connected to an inverting gain amplifier that has a low input resistance via R1.

In Figure 12-16, the load resistance as seen by the hold capacitor, C1, is Rin, which is an equivalent resistor that is in parallel to C1. The reason for this is because the (–) or inverting input terminal of the op amp U2A is a virtual ground due to its (+) input terminal (e.g., pin 3) being connected to ground. Typically, R1 will be ≤ 1KΩ (e.g., R1 has a range of 10Ω to 220Ω in some circuits found on the web or Internet). This very low resistance will then cause the hold capacitor C1 to discharge almost fully or substantially before the next sampling pulse arrives in the V_LO signal. Because the IF signal is attenuated at C1 (Vout) due to Rin's low input resistance of R1, the gain has to be pumped back up. So typically, R1 ~ 10Ω and R2 ~ 5KΩ for a gain of 500. Because of the high gain a required IF bandwidth of up to 96 kHz, the op amp U2A has to be a high-speed op amp with a gain bandwidth product of ≥ 50 MHz. So, a 3 MHz TL082 as shown inFigure 12-13 will not work very well and result in a closed loop bandwidth of 3 MHz/500 = 6 kHz that yields a 6 kHz IF bandwidth.

NOTE: The sampling pulses as shown in Figure 12-16 show about a 25 percent duty cycle. For only Figures 12-16 and 12-17 more conversion gain may be had if V_LO were 50 percent square waves. At this point the hold capacitor, C1, does almost nothing since the low impedance of Rin diverts most of the IF signal current away from C1.

Our next circuit in Figure 12-17 shows another added problem when the input resistor R1 → 0Ω, which can cause an op amp to oscillate due to hold capacitor C1 adding extra phase shift.

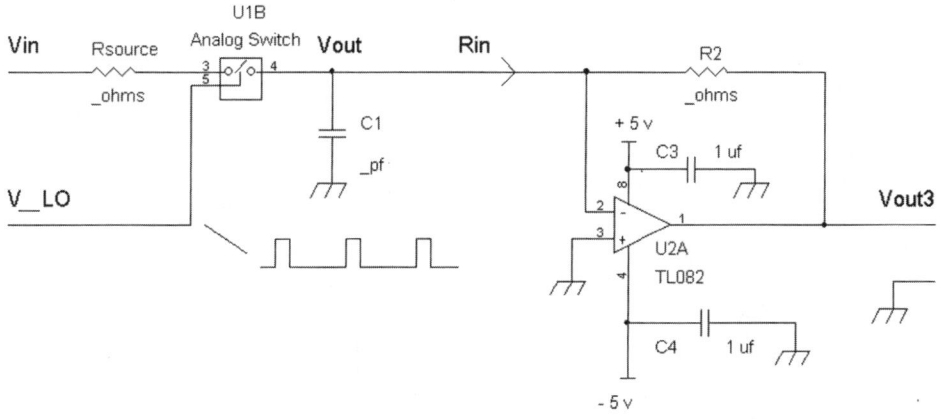

FIGURE 12-17   A sample-and-hold circuit that will not work as planned because the hold capacitor C1 is connected directly to the inverting input of the op amp that can cause parasitic oscillations from the op amp.

By having a hold capacitor, C1, connected directly to the inverting input of the op amp with feedback resistor, R2, an oscillation can occur. The reason is that R2 and C1 form a low-pass filter that causes a phase lag from Vout3 to pin 2, the (−) input terminal. This extra phase shift along with the phase shift of the op amp can then cause a net positive feedback situation, which in turn produces oscillation. Only by "luck" of picking the op amp and having an R2 × C1 time constant that does not have sufficient phase shift is there no oscillation. If other capacitance values for C1 are chosen or if a different op amp is used, then oscillation at Vout3 can occur. In summary, Figure 12-17 represents a circuit "waiting for a problem."

And of course, the input resistance, Rin, is very low. In general, the |Rin| ~ R2/[a(f)], where a(f) is the open loop gain of the op amp. So even if R2 → 5KΩ, we know that Rin < 5KΩ because a(f) is generally much greater than 1 (e.g., a 50 MHz op amp will still have an open loop gain of 10 at 5MHz, and an open loop gain of 100 at 500 kHz, and an **open loop gain of 1000 at 50kHz**). For example, if a 50 MHz op amp U2A is used such as an LM4562, and R2 = 10kΩ, and the IF or mix down frequency is 50 kHz:

Rin = R2/(open loop gain at 50 kHz) = 10kΩ/1000 or Rin ~ 10Ω

So again in Figure 12-14, the hold capacitor, C1, cannot hold a constant voltage from one sample pulse to another (e.g., see Figure 12-14 where the hold capacitor loading into infinite or high resistance (e.g. ≥ 1 MΩ) results in a constant voltage before the next sample pulse and a stair step waveform (flat steps) is produced).

A Cool Four-Phase Commutating Mixer

Figure 12-7 showed a simple I and Q mixer for software defined radios (SDRs). We will now look into the terrific Tayloe four-phase RF mixer that can be used for generating I and Q signals for SDRs. First let's look at a basic configuration. See Figure 12-18.

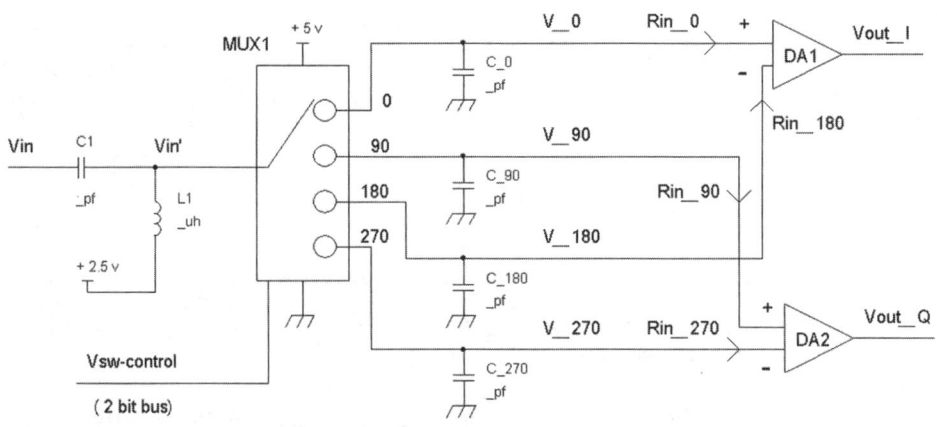

FIGURE 12-18 A Tayloe RF mixer that provides I and Q signals.

In some of the most basic configurations, input capacitor C1 (e.g., C1 = 1 μf) serves as an AC short circuit and blocks DC voltages from Vin. However, it is better for having L1 and C1 forming a basic high-pass filter that removes low-frequency interference signals. This high-pass filter may be a broadband filter such as rejecting signals below several hundred kHz, while passing RF signals above 500 kHz. Generally, as with any sampling or switch mode filter, an RF band-pass filter is desirable before the RF mixer to ensure that any out-of-band signal does not down-convert to an interfering IF signal. For example, we can also select values for L1 and C1 as a high Q (e.g., Q ≥3) high-pass filter by having [2πf$_{res}$ (L1)]/50Ω ≥3, where

$$f_{res} = 1/[(2\pi\sqrt{(L1)(C1)}]$$

and where Vin (e.g., an antenna) has a source resistance of 50Ω.

This mixer operates by sampling and holding the (four) signals between four equally spaced intervals that are one-quarter cycle apart. So, there is a four-position sequence of 0 degrees, 90 degrees, 180 degrees, 270 degrees, and then the sequence repeats over time. For example, in a 160-meter receiver at 1.8000 MHz, the switch control signal, Vsw-control, provides a sampling pulse for each phase at a 1.8000 MHz rate. With an equivalent circuit using four separate switches, Figure 12-19 shows the "ON" time when the multiplexor, MUX gates Vin' to each of the capacitors, C_0, C_90, C_180, and C_270. Figure 12-19 shows that each capacitor is charged via Vin' one at a time and just as one pulse ends, another starts via Vsw_0, Vsw_90, Vsw_180, and Vsw_270. The period of each pulse waveform will still be [1/(1.8000 MHz)] = 0.555 μsec in this example for a frequency of 1.8000 MHz.

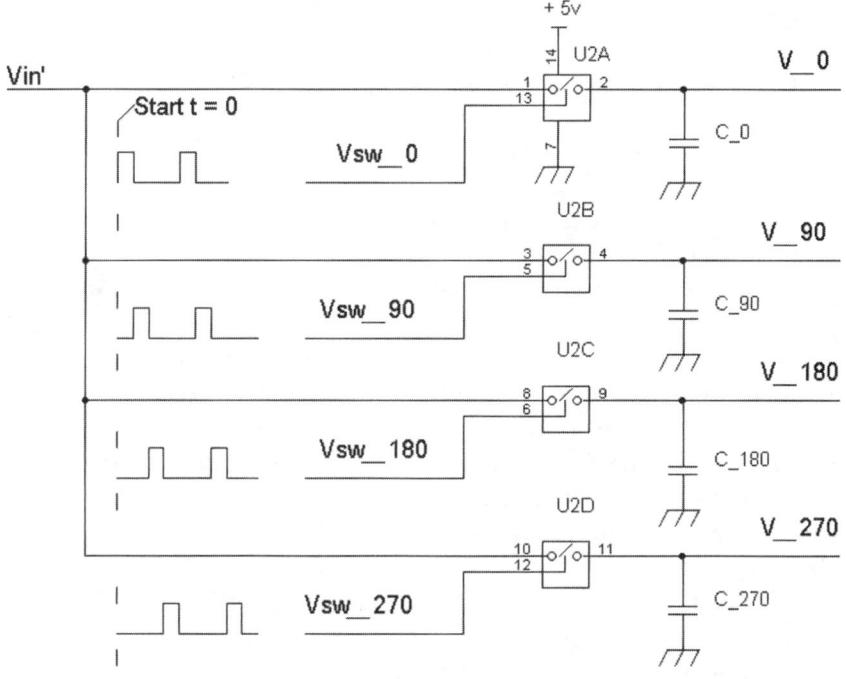

FIGURE 12-19 Timing pulses for four sampling switches. This circuit mimics MUX1 in Figure 12-18.

Now getting back to Figure 12-18, the voltage across each capacitor is connected to an input terminal of a differential amplifier (e.g., DA1 or DA2), which can be implemented with instrumentation amplifiers (e.g., INA163). Ideally, the input resistance to each input (e.g., Rin_0 to Rin_270) is infinite or in the order of ≥ 1MΩ so that the hold capacitors C_0, C_90, C_180, and C_270 do not discharge, causing a drop in voltage before the next sampling pulse comes along.

The output of DA1, Vout_I = k(V_0 − V_180), and the output of DA2, Vout2 = k(V_90 − V_270), where k is the gain of the differential amplifier (or instrumentation amplifier).

Now let's take a look at a circuit that can be improved as shown in Figure 12-20.

Some circuits posted on the web (world wide web) are similar to the one in Figure 12-20. In some cases, R7 = R8 have a range from 0Ω to 220Ω. The circuit will still provide I and Q signals at Vout_I and Vout_Q. However, the amplitude signal levels at V_180 and V_270 will be very small due to the hold capacitors, C4 and C6, being loaded to a low-resistance load via R7 and R8, whereas the two other hold capacitors, C5 and C7, are not loaded at all since these two capacitors are connected to the high resistance non-inverting inputs of U4 and U5.

The frequency bandwidth at V_0 and V_90 will be low, in the order of less than 10 kHz. This is because the hold capacitors (C5 = C7 = 0.27 μf) have large capacitances and form a low-pass filter effect. As for the voltages at V_180 and V_270, we will see that they will produce "unexpected" signals and they will not be the expected phase inversion signals of V_0 and V_90.

The V_0 and V_90 signals are on the non-inverting inputs (pin 3) of both op amps U3 and U4. Since the inverting input terminal voltages must match the non-inverting input voltages, the signal voltages at V_180 and V_270 are "overridden" by V_0 and V_90. And the voltage at the inverting input terminal is like a low-impedance voltage source since the inverting input signal matches the signal at the non-inverting input terminal. Again, see Figure 12-20.

For example, you can think of a voltage follower circuit via the inverting input terminal pin 2 of U3 is driving C4 via low-resistance resistor R7. So, if you look at the signal voltage V_0 at C5, you will see almost the same signal voltage V_180 at C4. Likewise, the V_90 voltage you see at C7 will be approximately the same as the voltage at V_270 at C6. Remember in an op amp circuit with negative feedback, it is the voltage at the non-inverting input terminal (e.g., pin 3) that determines what the voltage will be at the inverting input terminal (e.g., pin 2). That is, the voltages at pin 3 (positive input) and pin 2 (negative input) will be the same.

For example, if we look ahead at Figure 12-23, the signal voltages at C4 and C6 are "washed out" and you will not see the expected inverted phase signals there. In essence, the inverted phase signals V_180 and V_270 are "thrown away" in this circuit. But we can fix this by using an instrumentation op amp (see Figure 12-25) or by using voltage followers (see Figure 12-27) that will amplify each hold capacitor's (C4, C5, C6, and C7) voltage with high-resistance input terminals.

In terms of troubleshooting this circuit (Figure 12-20), the first place to start is at V_CLK, which is 3 volts to 5 volts peak to peak pulsed waveform such as a square wave signal. At pin3 of U2A and pin 11 of U2B, you should confirm the clock signal. To determine the frequency, you need to know which radio frequencies you want to receive via Vin. V_CLK will have a frequency 4x the incoming RF frequency. For example, if you want to listen in on the 80-meter band around 3.6000 MHz,

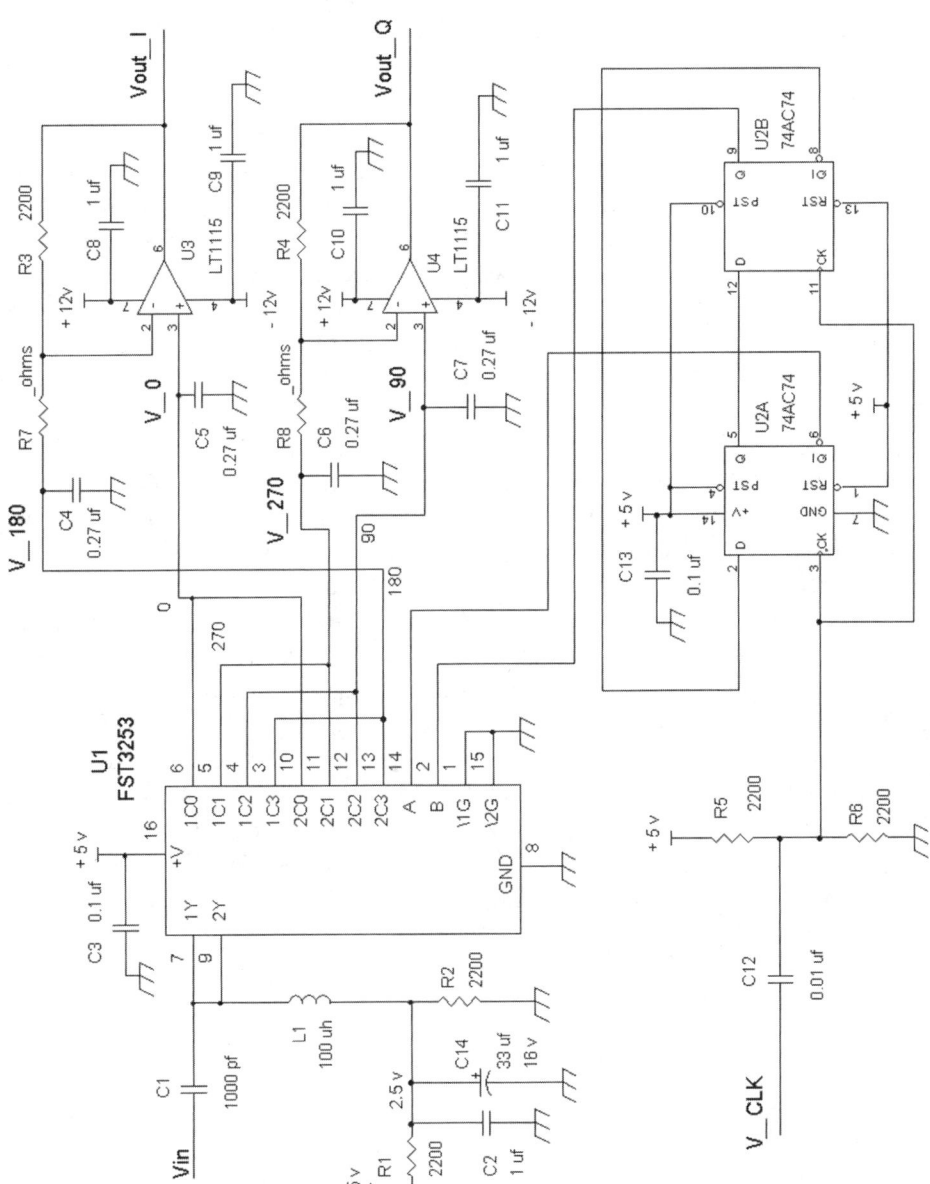

FIGURE 12-20 An example four-phase RF mixer that can be improved. Capacitors C4 and C6 can cause op amps U3 and U4 to oscillate if R7 = R8 = 0Ω.

V_CLK will have a frequency of 4 × 3.6000 MHz or 14.400 MHz. A V_CLK generator may come from a frequency synthesizer chip with crystal oscillator or from a crystal oscillator. If frequency generators are using an LC (inductor capacitor) or RC (resistor capacitor) oscillators, the frequency stability is most likely inadequate. Fortunately, today you can purchase an inexpensive crystal-controlled, two-channel variable frequency signal generator for testing these types of circuits. Again, see Figure 12-20.

For digital counter/divider chips (U2A and U2B) the first thing to check for is that there is +5 volts at the power pin 14, and pin 7 is tied to ground. Then confirm that the preset and reset pins are tied to logic high (e.g., +5 volts). We should see +5 volts at pins 4 and 10 for the preset pins and pins 1 and 13 for the reset pints. Second, confirm the (clock) CLK pins 3 and 11 U2A and U2B are tied together with a pulsed waveform on these pins, that is, in the 3-volt to 5-volt peak to peak range. Next, confirm with a two-channel oscilloscope with U2A pin 6 connected to Channel 1 and pin 9 of U2B to Channel 2 (of the scope), that you see 5-volt peak to peak waveforms that are 90 degrees phase- shifted from each other. Also, these two waveforms, when starting from both signals at 0 volts, should have a sequence of 0, 2, 3, and 1 where pin 6 of U2A is the least significant bit, and pin 9 is the most significant bit. Confirm these two waveforms are at U1 pins 14 and 2 (Figure 12-21). Also note the waveforms are not always clean pulses and you may see some small overshoot or slight ringing, which should not be a problem with logic circuits.

In Figure 12-21 the bottom waveform is one-quarter cycle or 90 degrees shifted from the top waveform. Also, note the repeating binary count sequence from the top waveform for the least significant bit and the bottom waveform for the most significant bit. Two-bit binary numbers equal the following in decimal numbers: 00 binary = 0 in decimal, 01 binary = 1 in decimal, 10 binary = 2 in decimal, and 11 binary = 3 in decimal. A low logic state is 0 in binary, and a high logic state is 1 in binary. The binary count sequence as shown in Figure 12-21 is: 00, 10, 11, 01, 00 . . . Thus, there is a four-state repeating pattern, which translates into 0, 2, 3, 1, 0, . . . When we look at the output pins of U1 for C0, C1, C2, and C3, the numeral following the "Cx" is the state of the address line. So, the sequence of 0, 2, 3, 1 translates to a sequence of C0 first, C2 second, C3 third, and C1 fourth. Because 0 degrees → C0, 90 degrees → C2, 180 degrees → C3, and 270 degrees → C1, we see that the sequence is correct in having 0, 90, 180, and 270 degrees in order.

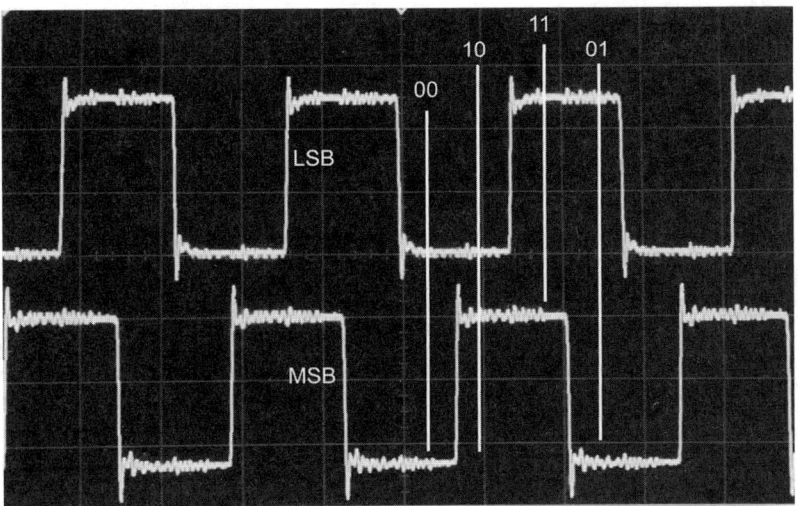

FIGURE 12-21   Top waveform at pin 14 (A input control line LSB) and bottom waveform at pin 2 (B input control line MSB) of the FST3253 analog switch with a 00, 01, 11, and 10 binary sequence or equivalently a 0, 2, 3, 1 order.

DC Bias Conditions

Confirm that pin 14 and pin 2 of U1 have the ~ 5-volt peak to peak waveforms as shown in Figure 12-21 when V_CLK is a 5-volt peak to peak square wave signal at 14.400 MHz.

Make sure the connections in U1 are connected as shown in the schematic in Figure 12-20, and that there is +5 volts at pin 16 and that pin 8 is tied to ground. Confirm that there is 2.5 volts DC at pins 7 and 9. If you do not have this, check the R1 and R2 voltage divider circuit and that there is 2.5 volts at C2 as shown in the schematic.

With Vin disconnected for now, measure the DC voltages at the hold capacitors C4, C5, C6, and C7 that should all measure about 2.5 volts within 15 percent. Also measure the DC output voltage at pin 6 of U3 and U4, which should be ~ 2.5 volts DC within 15 percent. Also confirm that the power supply decoupling capacitor, C3, has short leads (less than 0.5 inch) and is very close to pin 16 of U1. Likewise, decoupling capacitor C13 should have short leads and be close to pin 14 of U2A.

Testing Circuit with an RF or Function Generator

Again, with the generator at V_CLK set to 14.4000 MHz, with R7 = R8 = 47Ω and with a 240-mV peak to peak sine wave signal 3.600900 MHz set for 50Ω source resistance for Vin, you should see 900 Hz IF signals at V_0 and V_90 that are 90 degrees apart.

Confirm large amplitude signals (e.g., 6.0 to 8.0 volts peak to peak) at Vout_I and Vout_Q that are the same amplitude and 90 degrees apart. See Figure 12-22 where the amplitudes are ~ 7.3 volts peak to peak.

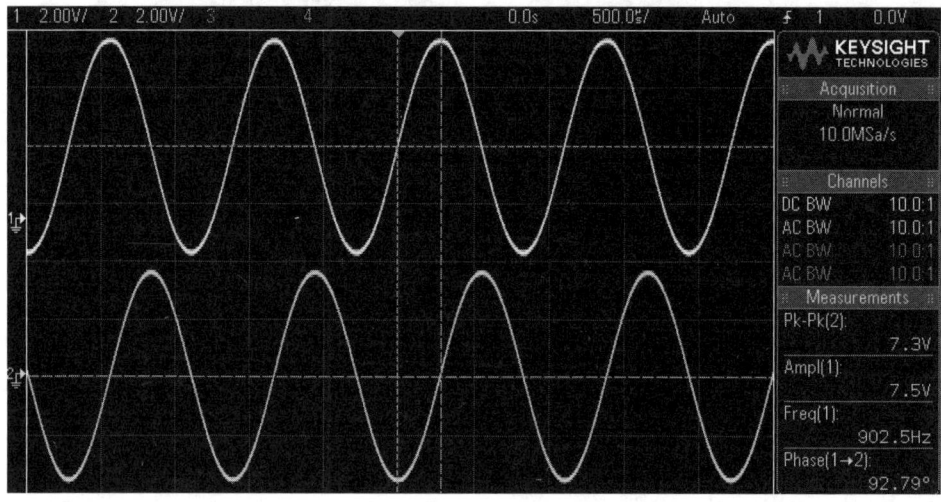

FIGURE 12-22   V_I and V_Q output signals (~ 900 Hz) from top and bottom waveforms with ~ 7.3 volts to 7.5 volts peak to peak output, and with a 92-degree phase shift between the two waveforms.

However, does Figure 12-20's circuit really work the way it should with two of its hold capacitors, C4 and C6, coupled to the low-resistance input terminals via R7 and R8? We should see signal from the hold capacitors C5, C7, C4, and C6 waveforms at 0, 90, 180, and 270 degrees. Just because we see the final result at V_I and V_Q that seems to look correct with about 90 degrees shifted apart in Figure 12-23, it does not mean everything is working correctly. We need to check further back and probe the hold capacitors. See Figure 12-23.

In Figure 12-23 the first and second waveforms should be 90 degrees apart, which is not the case. Also, the amplitudes of all waveforms should be equal, which is not the case.

Clearly, the low resistance loading of C4 and C6 in Figure 12-20 has an effect of "messing" up the amplitudes and phases of the waveforms.

The order of the waveforms can be viewed from the top as the first followed below by the second, followed again below by the third, and the fourth at the bottom.

As a matter of fact, the first and third waveforms just like the second and fourth waveforms *should be* 180 degrees apart or out of phase, but in fact they are nearly identical in phase.

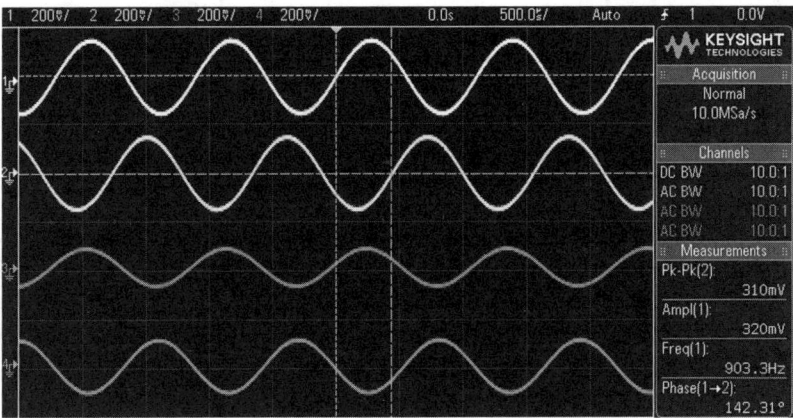

FIGURE 12-23 Signals from top to bottom at C5, C7, C4, and C6 that we expect the waveforms to be at 0, 90, 180, and 270 degrees. However, this is clearly not the case. Amplitudes are measured at 200 mV/division for all waveforms.

If we disconnect or remove R7 from C4 and R8 from C6 in Figure 12-20 so that all hold capacitors are either loading into a high-resistance input of the op amps or left open circuit we see the hold capacitors' waveforms in Figure 12-24, which looks like what we expect.

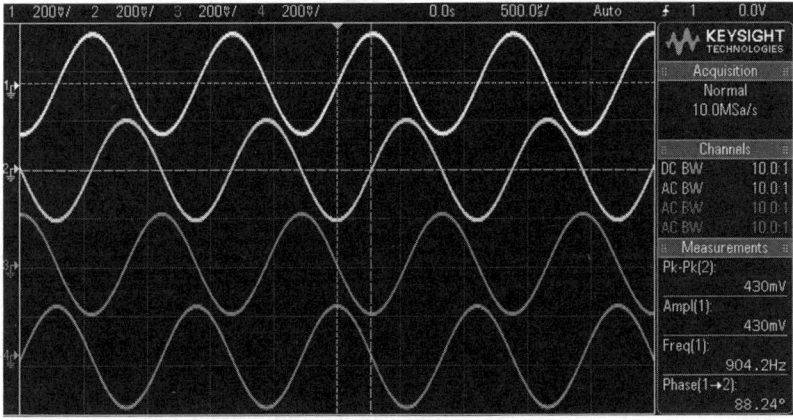

FIGURE 12-24 Waveforms from top to bottom of C5, C7, C4, and C6 that show the expected 0-, 90-, 180-, and 270-degree phase shifts, and note all waveforms have the same amplitude.

From Figure 12-24, as long as we keep all hold capacitors, C5, C7, C4, and C6, loading into a very high resistance (e.g., > 1MΩ, or infinite resistance), we will get the correct results. The waveforms shown in Figure 12-24 were done by disconnecting R7 and R8 from capacitors C4 and C6. Note now that the third waveform is 180 degrees phase shifted from the first waveform, and that the fourth waveform is an inverted version (180 degrees) of the second waveform. These inversions are what we expect as shown in the original concept via Figures 12-18 and 12-19.

Improving the "Original Design"

One way to improve the design is to use instrumentation amplifiers. This will provide high-resistance inputs to all the hold capacitors. Note that the 0.27-μf hold capacitors C4, C5, C6, and C7 will still yield only a < 10 kHz bandwidth, but they can be reduced in capacitance to as low as 1000 pf, which will provide close to 100 kHz IF bandwidth. See Figure 12-25 for one implementation that provides a high resistance load to all the hold capacitors.

Because we are using instrumentation amplifiers for U3 and U4, the DC voltages at Vout_I and Vout_Q will be close to 0 volts. This is because the DC voltages are about the same at all the hold capacitors C4, C5, C6, and C7. The instrumentation amplifiers will provide a pure subtraction of the DC voltages between V_0 and V_180 and also between V_90 and V_270, which leaves 0 volts at their output terminals (pins 8 and 9 of U3 and U4).

By replacing the LT1115 single op amps in Figure 12-20 with INA163 instrumentation op amps in Figure 12-25, this circuit now works more ideally to what a sample-and-hold RF mixer is capable of. The IF bandwidth at V_0, V_90, V_180, and V_270 are now equal and can be controlled further by changing the capacitances of the hold capacitors, C4, C5, C6, and C7. For example, C4, C5, C6, and C7 can be changed from 0.27 μf to 1000 pf to provide a wider IF bandwidth. We can now compare the phases with the four signals at V_0, V_90, V_180, and V_270. Set the frequency of V_CLK to 14.4000 MHz at 5 volts peak to peak. Set an RF signal generator or function generator to 100 mV peak to peak with a source resistance of 50Ω at 3.600100 MHz (for an IF = 100 Hz). Note that the unloaded signal from the generator will be twice the amplitude, or 200 mV peak to peak when observed connected directly to an oscilloscope oscilloscope with a 1MΩ or 10MΩ input resistance.

Observe with an oscilloscope the 100 Hz IF signals at the hold capacitors, C4, C5, C6, and C7 as shown in Figure 12-25. The voltages, V_0, V_90, V_180, and V_270 should be about 90 percent of the open circuit voltage of the generator. This will then be 90% × 200 mV peak to peak or about 180 mV peak to peak at the hold capacitors.

Note because the IF bandwidth will be low, < 10 kHz, you can use a VFO (variable frequency oscillator, crystal controlled via a synthesizer chip) to tune across the ham radio band. Another way is to use a fixed oscillator for a particular band and tune just portion of it, such as a 96 kHz span in the 80-meter band. If you have a 192 kHz analog to digital converter sound card (which is somewhat expensive), you can

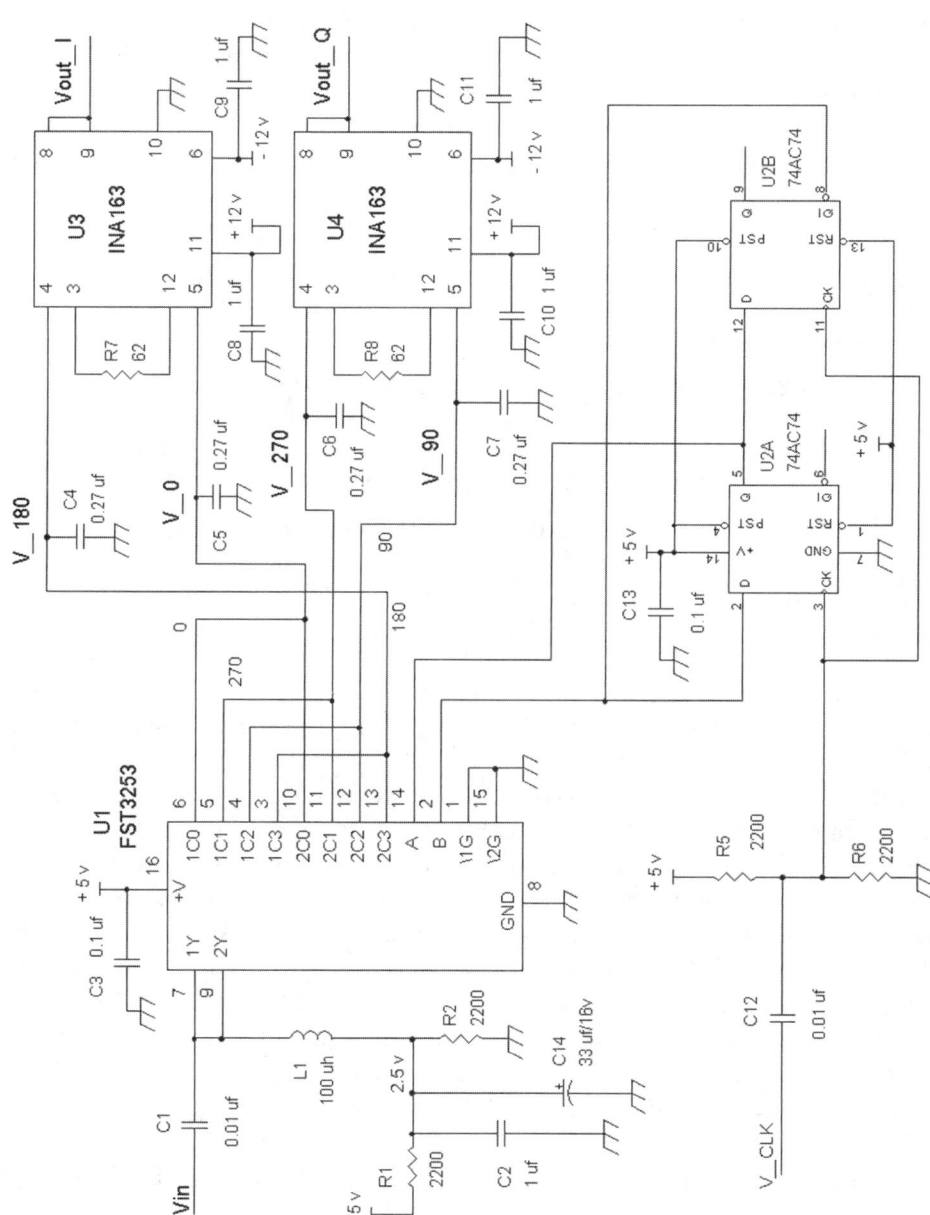

FIGURE 12-25 An improved version with instrumentation amplifiers INA163 (U3 and U4) to provide equal high resistance loading to the hold capacitors C4, C5, C6, and C7.

297

tune up to 192 kHz across a ham radio band when C4, C5, C6, and C7 are changed to a lower capacitance such as 1000 pf.

To measure the output signals from the instrumentation amplifiers, reduce the amplitude of the signal generator from 100 mV peak to peak at 50Ω source resistance to 10 mV peak to peak at 50Ω source resistance and at 3.600100 MHz. The open circuit voltage will then be 20 mV peak to peak. Each capacitor voltage will have 90 percent of 20 mV peak to peak or 18 mV peak to peak. However, since the phases of the V_180 and V_270 are inverted phases of the signals from V_0 and V_90, the difference or subtraction of V_0 and V_180 will be twice the signal.

That is V_180 = –V_0, so (V_0 – V_180) = (V_0 – –V_0) = (V_0 + V_0) or (V_0 – V_180) = 2 × V_0.

With V_0 having 18 mV peak to peak, (V_0 – V_180) = 2 × V_0 = 2 × 18 mV peak to peak or **(V_0 – V_180) = 36 mV peak to peak**.

Note that V_270 = –V_90, and with V_90 = 18 mV peak to peak we have:

(V_90 – V_270) = (V_90 – –V_90) = (V_90 + V_90) = 2 × V_90

or

(V_90 – V_270) = 2 × 18 mV peak to peak or
(V_90 – V_270) = 36 mV peak to peak

The outputs of the instrumentation amplifiers are approximately:

(1 + 6000Ω/R7)(V_0 – V_180) and (1 + 6000Ω/R8)(V_90 – V_270)

In this case, R7 = R8 = 62Ω and 6000Ω/62Ω = 6000/62, so the outputs are:

(1 + 6000/62)(V_0 – V_180) = **97.7(V_0 – V_180)**, and **97.7(V_90 – V_270)**

The expected outputs from the Vout_I = 97.7(V_0 – V_180); Vout_I = 97.7 (36 mV peak to peak) or Vout_I = 3.5 volts peak to peak. Similarly, Vout_Q = 3.5 volts peak to peak.

Because low noise instrumentation amplifiers are not always available, we can slightly modify Figure 12-20's circuit at least two ways. See Figure 12-26 where the 180- and 270-degree signals are not used so that the op amps use only their high-resistance non-inverting input terminals for amplifying the 0- and 90-degree signals. In Figure 12-26 sample-and-hold capacitors C4 and C6 are not used and can be removed. However, all sample-and-hold capacitors are reduced in capacitance to 1000 pf to provide a wider IF bandwidth out to about 100 kHz.

For Figure 12-26's circuit, the DC bias conditions should have 2.5 volts at pins 7 and 9 of U1 and also 2.5 volts DC at hold capacitors C5 and C7. Amplifiers U3 and U4 are configured as unity gain voltage followers for DC signals (e.g., for DC analysis, imagine removing the 100 uf capacitors C15 and C16). This means that the DC voltages at pin 6 of U3 and U4 will also be about 2.5 volts since the hold capacitor voltages at C5 and C7 are at 2.5 volts.

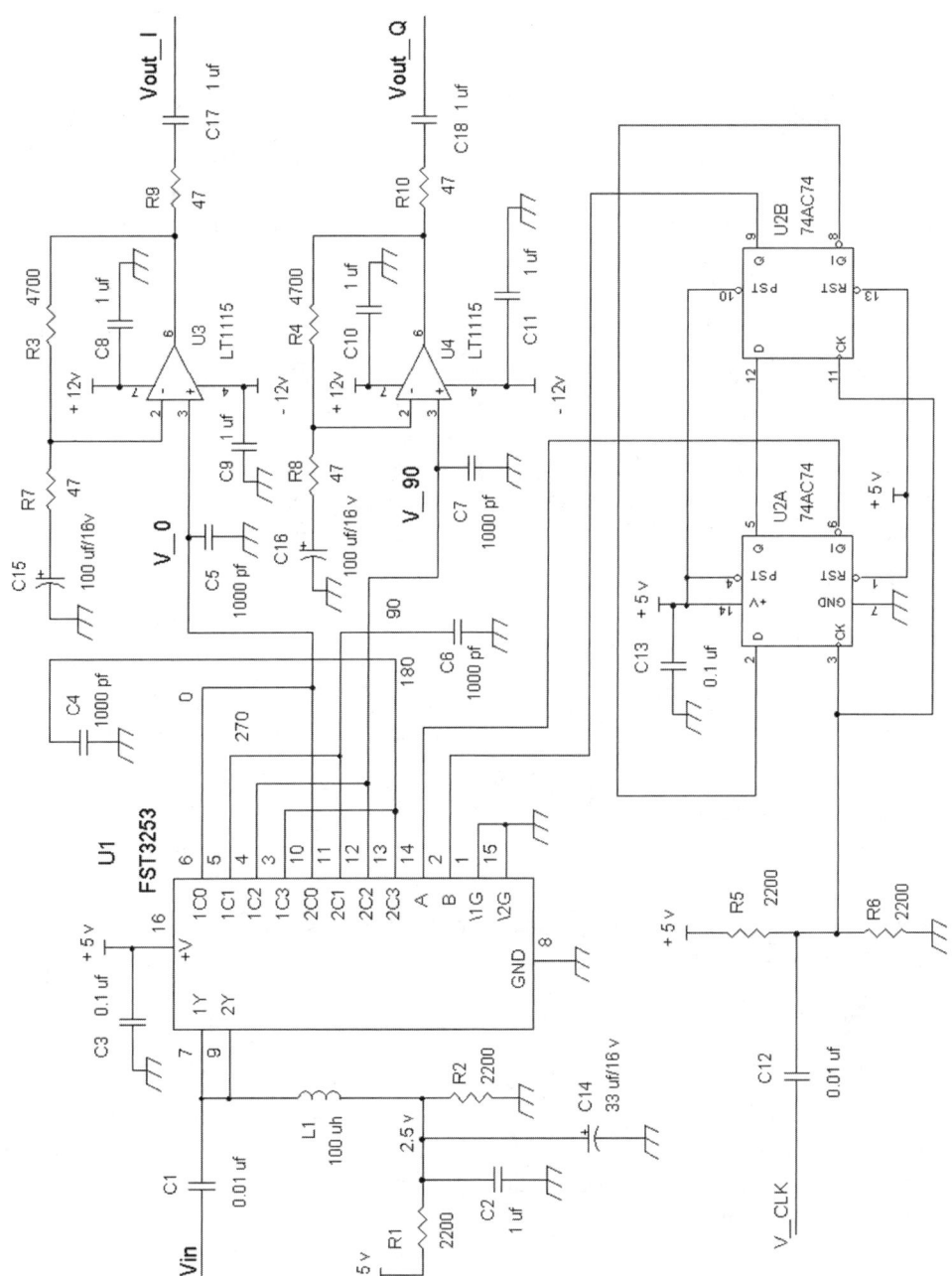

FIGURE 12-26 A circuit requiring only the 0- and 90-degree signals via V_0 and V_90.

The AC gains for U3 and U4 are (1 + R3/R7) or (1 + R4/R8), which in this case is (1 + 4700/47), which leads to (1 + 100) = 101.

With the same test conditions for Figure 12-25's circuit where V_CLK is at 14.400 MHz and 5 volts peak to peak square wave, the RF or function generator provides Vin = 10 mV peak for 50Ω source resistance with 20 mV peak to peak into a "no load" condition with frequency at 3.600100 MHz. The output amplitudes from Vout_I and Vout_Q should be 20 mV × 90% × 101 or 1.8 volts peak to peak. There should be a 90-degree phase shift on the Vout_Q referenced to Vout_I.

If we utilize all the signal voltages from the hold capacitors as shown in Figure 12-27, then we can about double the gain by adding voltage follower amplifiers U3A and U3B. These two voltage followers provide an equivalent instrumentation amplifier circuit. Via R7 and R8, the added voltage followers drive the inverting input terminals of U4A and U4B.

Not shown in Figure 12-27 are the connections to pins 7, 9, 14, and 2 of U1, which are connected to the circuits in Figure 12-26. For example, U1 pins 14 and 2 are connected to pins 6 and 9 of the U2A/B circuit in Figure 12-26. Also pins 7 and 9 of U1 in Figure 12-27 will be connected to an input circuit L1 and C1 as shown in Figure 12-26.

In this configuration, dual op amps are used. U3A and U3B are unity gain voltage followers with U4A and U4B as voltage gain amplifiers. The voltage gains for the various AC signals across the hold capacitors are as follows:

Vout_I = (1 + R3/R7) × V_0 + −(R3/R7) × V_180

However, V_180 = −V_0. This leads to:

Vout_I = (1 + R3/R7) × V_0 + −(R3/R7) × −V_0 which leads to:
(1 + R3/R7) × V_0 + (R3/R7) × V_0, or Vout_I = [1 + 2(R3/R7)] × V_0

With R3 = 7500Ω and R7 = 150Ω, Vout_I = [1 + 2(7500/150)] × V_0 or Vout_I = (1 + 100) × V_0

Vout_I = 101 × V_0

Vout_Q = (1 + R4/R8) × V_90 + −(R4/R8) × V_2700. But V_270 = −V_90. Likewise, from the same type of calculations for Vout_I:

Vout_Q = [1 + 2(R4/R8)] × V_90

Vout_Q = 101 × V_90

With a Vin at 10 mV peak to peak signal for 50Ω source resistance (20 mV peak to peak into an open circuit and at 3.600100 MHz with V_CLK being a 14.4000 MHz 5-volt peak to peak square wave, the amplitude of Vout_I = 20 mV × 90% × 101 = 1.8 volts peak to peak.

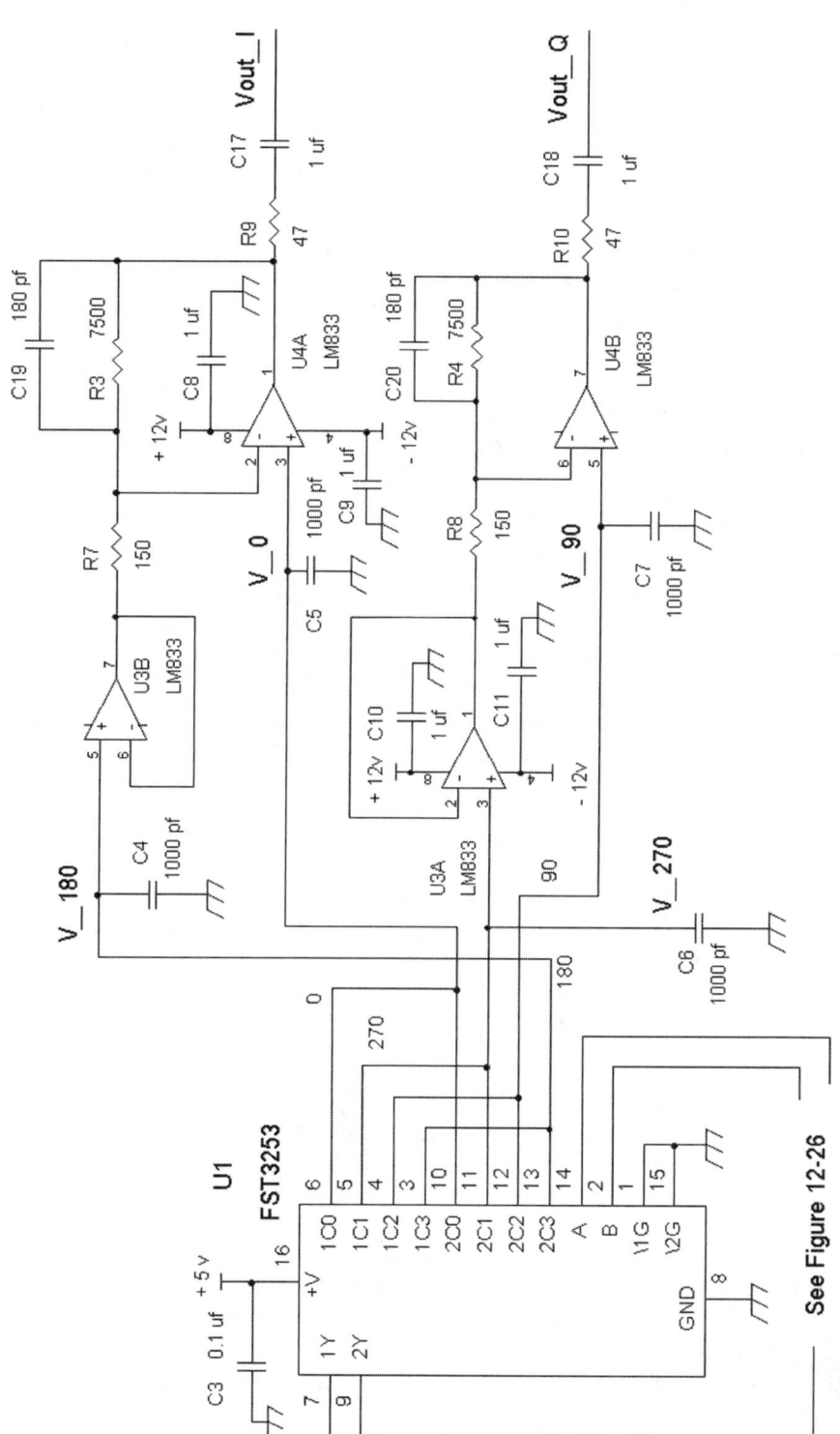

FIGURE 12-27 Partial schematic with LM833 dual op amps that have 15-MHz gain bandwidth products, which replace instrumentation amplifiers (e.g., INA163 in Figure 12-25).

301

Similarly, the amplitude of Vout_Q is also 1.8 volts peak to peak and phase shifted 90 degrees referenced to Vout_I with sine waves at 100 Hz.

In terms of the DC voltages at pins 1 and 7 of U4A/B, there are 2.5 volts DC at all the hold capacitors C4, C5, C6, and C7 so we have the following:

U4A pin 1 DC = (1 + R3/R7) × 2.5 v + –(R3/R7) × 2.5 v, which results in: U4A pin 1 DC = 2.5 v + (R3/R7) × 2.5 v + –(R3/R7) × 2.5 v U4A pin 1 DC = 2.5 volts. This is because the (R3/R7) × 2.5 v terms cancel out.

Thus, U4A pin 1 DC = 2.5 volts DC.

And similarly, U4B pin 7 DC = 2.5 volts DC.

Another View of Op Amp Circuits (Where the Inverting Input Drives a Load)

If we examine Figure 12-20's circuit with C4 and R7 and also look at the waveforms of Figure 12-23, then we would expect the first and third waveforms to be 180 degrees apart, which they are not. Then the question is why? It turns out that the (+) input to a negative feedback amplifier actually provides a low-impedance "output" impedance to the (–) input terminal. The reason is that as long as the output signal of the negative feedback amplifier is not clipping to the power supply rail, then the (–) input's voltage has to be about the same voltage as the (+) input terminal. *The (–) input voltage follows the (+) input's signal.* See Figure 12-28.

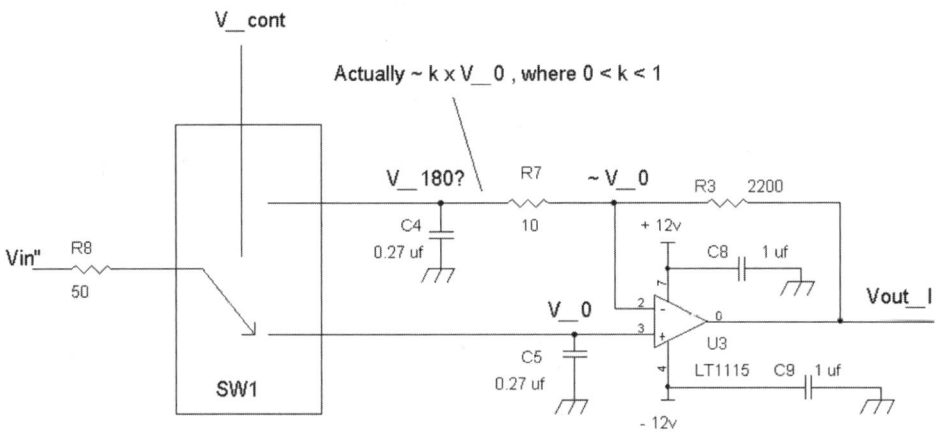

FIGURE 12-28   Examining why the voltages V_0 and V_180 are not as expected as shown in this circuit, which is similar to part of Figure 12-20.

In a non-inverting gain op amp configuration, the input voltage, V_0, is applied to the (+) input terminal. Once this happens, the (–) input terminal must be "equal" to the (+) input. The output voltage at pin 6 U3 adjusts accordingly so that the voltage at the (–) terminal matches the voltage of the (+) terminal. In essence, the signal to the (+) input terminal is the determining factor for "everything." If you try to apply another voltage (e.g., V_180) to the (–) input via R7 or a series resistor, you will change the output voltage at pin 6, but you will not be able to the change the (–) input's signal at pin 2. Thus at (–) input pin 2 we have V_0 due to the negative feedback system via feedback resistor R3. Since pin 2 represents a pure voltage source of V_0 also, it provides a voltage to the right side of R7. So, going "backwards," there is a V_0 voltage via R7 that travels into hold capacitor C4. The signal at C4 via V_0 overpowers the V_180 signal from the analog switch, so that C4 has some version of V_0 instead of V_180, where V_180 is a 180 degrees out-of-phase signal. So, the 180 degrees signal at C4 is "washed" out by the 0 degrees signal at C5. And the C4 signal looks similar in phase to the signal at C5. In fact, this is what we see when we compare the top first signal for C5 to the third signal for C4 in Figure 12-23.

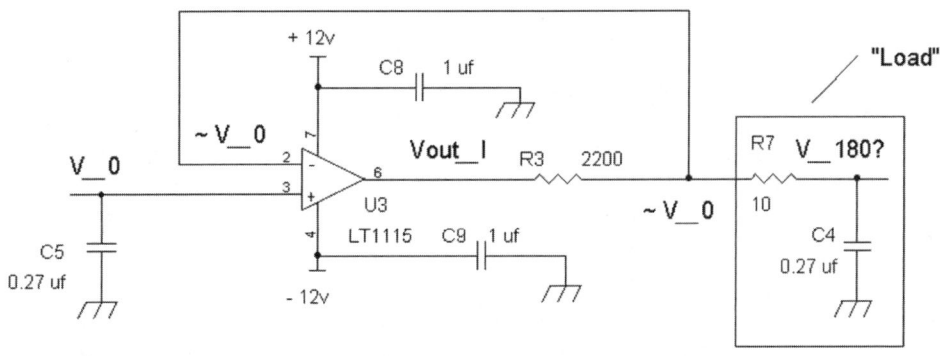

FIGURE 12- 29   Redrawn Figure 12-28 for clarity as negative feedback amplifier with a series resistance R3 within the negative feedback loop.

As shown in Figure 12-29, you can make a voltage follower with a series output resistor such as R3 (e.g., 2200Ω) and still provide a low output resistance source of V_0 into the load, R7 and C4. The reason why ~ V_0 has a low resistance in spite of R3 being in series is that the negative feedback terminal (–) input is taken at ~ V_0. The negative feedback system then turns up Vout_I to compensate for loss due to the loading effects of R7 and C4 on R3. The true output resistance at the ~V_0 is approximately $R3/a_0(f)$, where $a_0(f)$ is the frequency-dependent gain of the open loop gain from U3. For example, if the LT1115 has a 50-MHz gain bandwidth product (GBWP), and we want to know the gain at f = 1 kHz, then $a_0(f) = a_0(1kHz) = GBWP/f = $ 50 MHz/1kHz or 50,000. Then output resistance at ~V_0 is about R3/50,000 or 2200Ω/50,000 = 0.044Ω < 1Ω. We can see now that ~V_0 has low-output resistance and for all practical purposes, ~V_0 is a voltage drive signal into R7 ("Load") as shown in Figure 12-29. So that is why in Figure 12-23, C4 has a voltage that looks more like V_0 instead of C4 having a signal that is out of phase (e.g., V_180).

Suggested System Approach

In Figures 12-20, 12-25, 12-26, and 12-27, the input signal Vin (e.g., an antenna signal) goes directly to the mixer chip such as an FST3253. A better way is to take the matching network from Figure 12-7 with Cvar1, L1, R13 (e.g., R13 = 5100Ω for a Q ~ 10), R12, Q1, R10, R11, and C9. C9 from Figure 12-7 then feeds the Vin of Figures 12-20, 12-25, 12-26, and 12-27. This matching network steps up the antenna signal like a step-up transformer so that the antenna's signal level is sufficiently large such that medium low noise op amps such as NE5523 or LM833 can be used. Secondly, the RF matching network via Cvar1 and L1 forms a filter network that reduces out-of-band signals and noise that would be mixed down to the IF signal.

Crystal Oscillators

For many amateur radio circuits a very stable frequency oscillator usually points toward a crystal resonator, although sometimes a ceramic resonator can be used. The order with the best to worst frequency stability for oscillator circuits we have: crystal, ceramic, inductor-capacitor (LC), and lastly resistor–capacitor (RC).

Before we look into some oscillator circuits, let's take a look at a crystal or ceramic resonator. See Figure 12-30 of different types of crystal and ceramic resonators.

FIGURE 12-30 Cylindrical crystals or tuning forks on the left side, standard size crystals HC-49/S (shorter version) and HC-49/U (standard "longer" size) in the center, and a ceramic resonator on the right side that includes a three-lead ceramic resonator with built-in 15 pf capacitors.

The models of crystal or ceramic resonators are shown in Figure 12-31 with some circuits.

As shown in Figure 12-31(a), a crystal can be modeled as a series resonant circuit with inductor Ls, capacitor Cs, and series resistor Rs. Many oscillators are configured to use a crystal as a series resonant circuit, as shown in Figures 12-31(b), (c), and (d). Figure 12-31(a) shows the lead body capacitance, Co, which then also forms a parallel resonant circuit with Ls, Cs, and Rs. However, as we will see, most crystal oscillators run in some form of a series resonant mode.

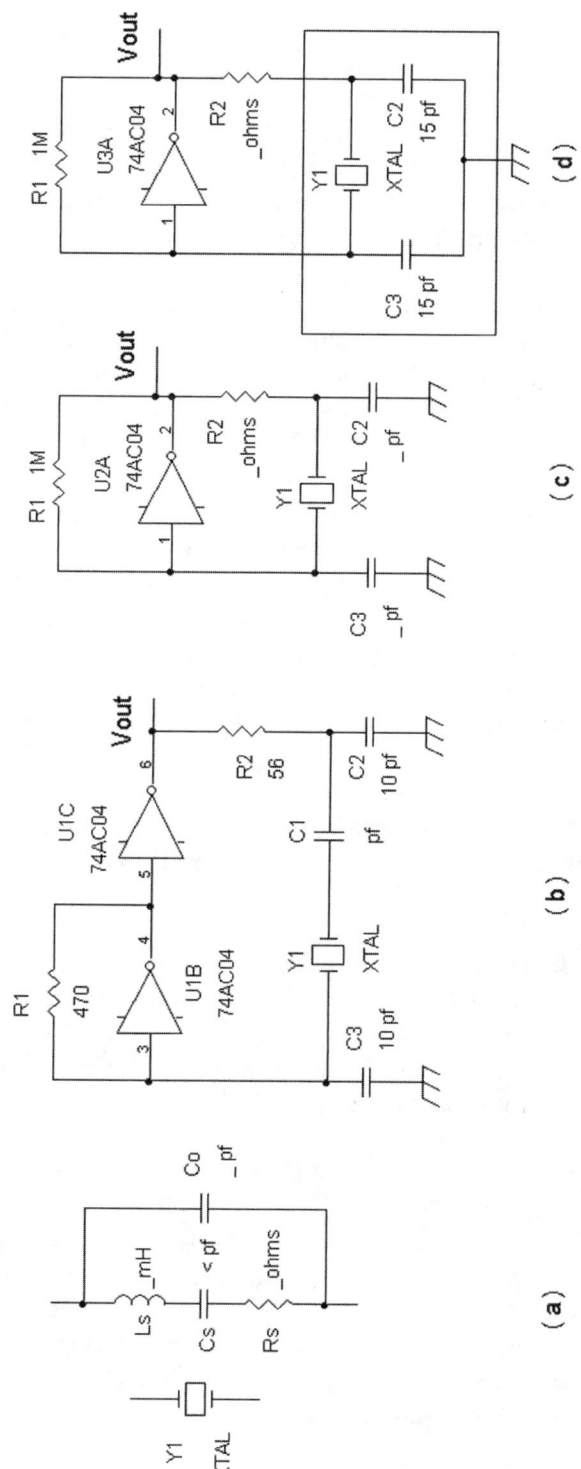

FIGURES 12-31(A) TO (D)   Schematic representation of a crystal or ceramic resonator Y1 in (a), a two-gate crystal/resonator oscillator in (b), a single-gate crystal/resonator oscillator in (c), and an oscillator where a three-terminal resonator Y1 includes internal capacitors C2 and C3 in (d).

Let's first take a look at Figure 12-31(b), a two-gate oscillator circuit. Because there are two inverters (U1B and U1C) in series, the net phase shift is 180 degrees plus 180 degrees, or 360 degrees, which is like 0 degrees. Oscillation then occurs when the crystal acts like a band-pass filter with zero degrees phase shift at the resonant or crystal frequency. With this circuit, generally, capacitor C1 has a large value such as 0.01 μf. However, C1 may be selected to a different value to adjust the oscillation frequency. C1 should be larger than 15 pf or a parasitic oscillation may occur.

In Figure 12-31(b), the first inverting gate, U1B has a feedback resistor, R1, that forms a low input resistance at pin 3. The R1 resistance value is chosen to be in the few hundreds to few thousands of ohms depending on the oscillation frequency. U1B via R1 has been configured as an analog inverting gain amplifier with low-resistance input. If the crystal, Y1, is removed, the output of the first inverter with R1 will sit at one-half the supply voltage (e.g., 2.5 volts) for 74AC (up to ~ 30 MHz), HC (up to ~ 10 MHz), and 4000 series CMOS chips (less than 1 MHz). When the crystal Y1 is put back in, it completes a positive feedback loop with the maximum gain at the crystal's resonant frequency determined by Ls and Cs in Figure 12-31(a) that form a series resonant circuit within the crystal. This series resonant circuit in Y1 provides zero phase shift and maximum gain (amplitude) at resonance. For series mode crystals, the resonant frequency is just the one imprinted on the package. And for crystals with a load capacitance, the series capacitor C1 is set to the load capacitance (e.g., 18 pf) to tune the crystal to the specified frequency. C1 allows "tweaking" the oscillation frequency further even for series mode crystals, which means C1 can have a range of 0.1 uf to about 100 pf. Note that R2 and C2 along with C3 are included in Figure 12-31(b) to prevent parasitic high frequency oscillations.

If we now look at Figure 12-31(c), we see a single inverter gate oscillator configured as an analog inverting amplifier. R1 is generally a very high resistance value ≥ 1MΩ to avoid loading down the oscillation signal at C3. The crystal Y1 is modeled as a series resistor inductor capacitor circuit with C3, the load capacitance of the crystal (e.g., 11 pf to 33 pf) that forms a low-pass filter. This low-pass filter at C3 provides 90 degrees to 135 degrees of phase shift at the resonant frequency. To achieve oscillation, we need 180 degrees of phase shift from the output pin 2 to the input pin 1 of U2A. To add more phase shift, R2 and C2 form an RC low-pass filter. Generally, we set the −3 dB cut-off frequency of the R2-C2 low-pass filter at about one-half the crystal frequency to provide about 60 degrees of phase shift, which will be sufficient for the circuit to oscillate. Typically, R2 is in the range of 470Ω and 2200Ω and C2 has a capacitance value equal to or larger than C3, the crystal's specified load capacitance. In certain cases, R2 may be increased to about 5kΩ and R1 → 4.7MΩ to 10MΩ for low-frequency (cylindrical) crystals whose frequencies ≤ 200 kHz.

Figure 12-31(d) shows essentially the same circuit as Figure 12-31(c) with a three-terminal ceramic resonator having internal C2 and C3 capacitors. This type of ceramic resonator has three leads including center ground lead. This device is symmetrical, and the two outer leads can be reversed. The center lead stays connected to ground. Typical values for R2 range from 220Ω to about 3300Ω.

NOTE: All crystals are symmetrical or non-polarized, which means that they will perform the same function with their leads reversed.

Types of Crystals

When you purchase crystals, they are separated into two groups. The most common group is crystals that are specified with loading capacitances such as 11 pf to 33 pf. Common loading capacitances are in the 18 pf to 22 pf range. The oscillator shown in Figure 12-31(b) is a series resonant circuit; the loading capacitance is C1. In series with crystal Y1 is C1 to tune the crystal to the specified frequency. For example, in Figure 12-31(b) having U1B and U1C alternatively as 74AHC04, R1 = 3300Ω, R2 = 56Ω, C2 = 10 pf, C3 = 10 pf, and with Y1 = 3.0000 MHz specified with 30 pf loading so C1 = 30pf, the oscillation frequency at pin 6, Vout measured at 2.999971 MHz or about 29 Hz off, which is within the 10-ppm tolerance of the crystal. For a 3-MHz crystal 10 ppm (10 parts per million) yields a tolerance of ±(3 MHz × 10/million), which leads to ±30 Hz.

Other oscillators that make use of the loading capacitance specification are shown in Figures 12-31(c) and 12-31(d). In these two figures, C3 is the loading capacitance value. So, if you have a crystal specified at 18 pf, then C3 = 18 pf. The other capacitor, C2, works with resistor R2 to form a low-pass filter that provides a lagging phase shift of at least 45 degrees but less than 90 degrees. Typically, the phase shift is set to about 60 degrees by having the cut-off frequency, $f_c = 1/(2\pi R2C2)$, set to one-half the crystal frequency. For example, suppose you have a 7 MHz crystal, you can set $f_c = 3.5$ MHz. Now we have to find a C2, but what's a good value for R2? R2 generally can be in the range of 470Ω to 4700Ω. So, we can just pick R2 = 1000Ω. Then this leads to C2 = $1/(2\pi R2f_c) = 1/(2\pi 1000Ω \times 3.5MHz) = 45.5$ pf, or C2 ~ 47 pf. Actually, C2 in the range of 47 pf to 68 pf will work fine. In other cases, you can use identical capacitance values for C2 and C3 that equal the specified loading capacitance such as 18 pf or 22 pf while having R2 = 470Ω to 4700Ω.

What if you want more phase shift closer to 90 degrees by lowering the cut-off frequency further to, for example, one-twentieth of the crystal frequency? This will make C2 about 10 × 47 pf or C2 → 470 pf. The oscillator may not oscillate because although the phase shift will be about 87 degrees, the amplitude is attenuated by twentyfold, so only about 5 percent of the signal will enter the crystal Y1 on the right terminal side. The inverter gate U2A configured as an inverting gain amplifier will have to have sufficient gain to make up the attenuation to ensure oscillation. A twentyfold attenuation may have too much loss for the circuit to oscillate.

The second group of crystals is the series resonant type, and there is no loading capacitance specification. Basically, it means if you were to put a series mode crystal, Y1, in Figure 12-31(b) with C1 shorted out or C1 is from 0.01 μf to 0.1 μf (e.g., an AC short-circuit at the oscillation frequency), the oscillator will oscillate at the specified frequency.

Low-Frequency Cylindrical Crystals, "Standard" Crystals, and Ceramic Resonators

In the twenty-first century, most low-frequency crystals at ≤ 200 kHz are the types shown in Figure 12-30 (left side). These provide accurate frequency references for clocks, watches, and other devices. A very popular frequency 32.768 kHz is used in almost all watches. These low frequency crystals may have about a thousand times more loss due to Rs ~ 50kΩ as shown in Figure 12-31(a), whereas the larger standard size HC-49/S and HC-49/U crystals have Rs ~ 50Ω or less. Typically, the load capacitance of the low-frequency smaller cylindrical crystals is in the 11 pf to 22 pf range. Also, they must be driven with lower AC currents when compared to the standard crystals or ceramic resonators. So, you can more easily damage a low-frequency cylindrical crystal if you are not careful.

A low-frequency oscillator such as in Figure 12-31(c) has the low-pass filter section R2 = 4700Ω, C2 = 1000 pf, Y1 = 32.768 kHz to 100 kHz, C3 = 15 pf, and R1 = 4.7MΩ. It is important to raise R1 to 4.7MΩ so that the lossy Rs = 50kΩ inside the cylindrical crystal transfers enough AC signal into the input terminal of the inverter gate, U2A that has been converted to an inverting gain amplifier. *Note: If R1= 1MΩ, this resistance value may be too low; and the circuit might not oscillate with cylindrical crystals whose frequencies are ≤ 200 kHz.* Given the low oscillation frequency, inverter U2A may be a 74C04 or CD4069 for slower rise/fall times that do not propagate output signal glitches, or you can still use the faster 74HC04 chip.

Standard HC-49 and High-Frequency Cylindrical Crystals

Standard crystals such as HC-49/S and HC-49/U have series resistance, Rs < 100Ω and with that generally as shown in Figure 12-31(c), R2 can be in the 470Ω to 2200Ω range. Again, R1 can be in the 1MΩ to 4.7MΩ range. Typically, R1 = 2.2MΩ. In general, C2 should be at least the loading capacitance, if not more. For example, suppose the crystal frequency is 20 MHz with an 18 pf load capacitance. We should choose a fast inverter gate beyond a 74HC04 such as a 74AC04, 74AHC04, or 74VHC04. C3 will equal the load capacitance of 18 pf, and if R2 = 470Ω, then C2 = 33 pf for an R2/C2 low-pass filter cut-off frequency, $f_c = 1/(2\pi R2C2)$. We set $f_c = [0.5 \times$ crystal frequency], and this will be 10MHz, half of the crystal's 20-MHz frequency to provide about 60 degrees of phase shift at 20 MHz.

Note there are fundamental frequency crystals above 200 kHz such as $\geq$ 3MHz that are available in cylindrical packages like the 32.768 kHz types. However, these cylindrical crystals have the series resistance, Rs, much lower at $\leq$ 200Ω, which allows them to be used in crystal oscillator circuits with essentially the same R2, C2, and C3 values like the HC-49/S and HC-49/U crystals.

In terms of frequency accuracy, crystals are rated in parts per million (ppm) and can range from 10 ppm to 100 ppm. For good accuracy, you can specify for better (less) than 30 ppm. But just remember the more frequency accuracy you require, the more it may cost. Fortunately, in some cases, you can replace C3 in Figure 12-31(c) with a variable trimmer capacitor (e.g., 5 pf to 35 pf) to adjust the frequency.

Ceramic Resonators

There are generally two types of ceramic resonators, low frequency and higher frequency types. The lower frequency ones resonate below 1.3 MHz with load capacitances in the 100 pf to 560pf. For the resonators above 1.3 MHz, the load capacitances are from 15 pf to 33 pf, which is similar to crystals. Their equivalent series resistance is < 500Ω, so they can be used very much like standard crystals. Ceramic resonators are generally used in situations when the required frequency accuracy is better than an LC or RC oscillator, but does not need to be as accurate as a crystal. Ceramic resonators find their way in almost all remote controls. The frequency tolerance rating is in the order of 0.3 percent to 0.5 percent or 3000 ppm to 5000 ppm. There are some ceramic resonators that have accuracies in the order of 0.1 percent or 1000 ppm, such as the surface mount package Murata CSTNR_GH5L (4.00 MHz to 7.99 MHz), CSTNE_GH5L (8.00MHz to 13.99 MHz), and CSTNE_VH3L (14.00 MHz to 20.00 MHz).

As an example, using instead a 74AHC04 inverter gate in Figure 12-31(c), set R1 = 4.7MΩ, R2 = 2200Ω, and with Y1 = 455kHz (e.g., Murata CDBLA455KCAY16-B0), set the cut-off frequency f_c = $1/(2\pi R2 C2)$ = 0.5 $\times$ 455 kHz = 227.5 kHz. This leads to:

C2 = $1/[2\pi 2200\Omega(227.5 \text{ kHz})]$ or C2 = 318 pf ~ 330 pf

Initially set C3 to 330 pf, then try different values from 100 pf to 560 pf to trim the oscillation frequency to 455 kHz using a frequency counter set to the 1MΩ input resistance mode. In this 1MΩ input resistance mode, you can generally use a $\times$10 scope probe to measure the oscillation frequency if there is sufficient output signal (e.g., > 2 volts peak to peak).

NOTE: Some frequency counters have a 50Ω input resistance mode, which will load down the oscillator's signal.

If C3 is replaced with a variable capacitor, such as 10 pf to 365 pf air dielectric type or a poly varicon (e.g., a 10 pf to 140 pf or a 10 pf to 270 pf), then this 455-kHz oscillator may be used for product detectors or beat frequency oscillators with variable pitch control.

Finally, Figure 12-31(d) shows a ceramic resonator circuit with built-in capacitors C2 and C3. For some ceramic resonators, the capacitances are 15 pf each. Other ceramic resonators may be up to 47 pf for their internal C2 and C3. As always, you can add parallel capacitor(s) across the internal C2 and/or C3 to adjust the frequency. Or a variable trimmer capacitor such as 5 pf to 35 pf can be placed between the inverter's input pin and ground. The feedback resistor R1 may be in the 1MΩ to 4.7MΩ range. Again, set the R2 resistance to provide a phase shift with the internal C2. For example, if the ceramic resonator is at 7.2 MHz and the internal C2 and C3 are 15 pf each, R2 can be set as: R2 = 1/[2π15pf(7.2MHz/2)] or R2 = 1474Ω ~ 1500Ω. Again, the cut-off frequency is set at one-half the oscillation frequency of 7.2 MHz, which results in 3.6 MHz (7.2 MHz /2).

We now will look at a one-transistor oscillator with a logic level shifting amplifier that is commonly used when a 0- to 5-volt logic signal is required to drive logic circuits including flip flops, gates, etc. See Figure 12-32.

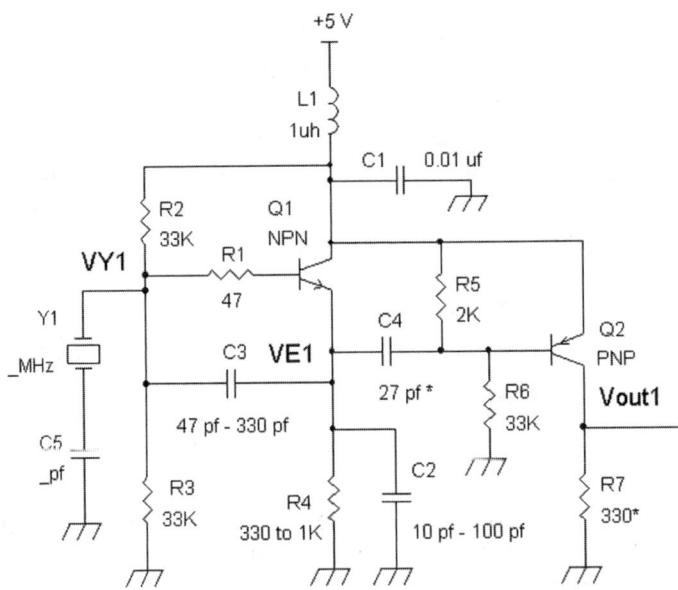

FIGURE 12-32 A "typical" one-transistor emitter follower crystal oscillator Q1 with common emitter amplifier Q2 to provide logic level signals for 5 volts peak to peak output.

C5 is generally the crystal's specified load capacitance such as 18 pf. However, C5 may be replaced with a 5 pf to 35 pf trimmer variable capacitor to adjust the frequency exactly at Vout1 using a frequency counter. If Y1 is a series resonant crystal, C5 = 0.01 uf, which is close to an AC short-circuit at the crystal's frequency.

So how does the Q1 crystal oscillator work? Q1 is an emitter follower, which means it has a voltage gain of about 1. Yet somehow this circuit must step up the signal voltage at the resonant frequency of the crystal, Y1. At the emitter, you will notice a small value capacitor C1 that forms a very slight low-pass filter effect. That is the signal at Q1's emitter has a high-frequency roll-off compared to Q1's base signal. The high-frequency roll-off at the emitter also causes a phase lag with respect to the signal at Q1's base. In order to get an oscillation going, we need **C3 and Y1 to form a high Q high-pass filter** that has a peaked amplitude response near the crystal's frequency. The signal voltage at VY1 or base of Q1 is a signal that is stepped up and has a phase lead referenced to the signal at the emitter of Q1. Crystal Y1 forms an inductive impedance at near the crystal frequency. By near the crystal frequency, we mean within less than 0.1 percent tolerance of the crystal frequency, typically within 0.01 percent or better. For example, if the crystal frequency is 10.000 MHz, the oscillator will provide signal at 10.000 MHz within 1.0 kHz. For example, 1 kHz is 0.01 percent of 10.000 MHz or 100 ppm of 10.000 MHz where ppm = parts per million, and 100/million $\times$ 10 million Hz = 1 kHz. Because there is a phase lead due to the crystal and a phase lag via C2, the net phase shift is 0 degrees. And the stepped-up signal via the high Q high-pass filter sustains an oscillation. The actual signal at VY1 (or VE1) is generally not always a sine wave, but instead it can be a distorted waveform. See Figure 12-33.

To troubleshoot this circuit, we first look at the DC conditions, which are easier to measure when the crystal Y1 is removed. The reason for this is when a circuit generates high-frequency AC signals, it causes measurement problems when using a DVM.

A first approximation for Q1's base voltage is determined by the resistive voltage divider circuit R2 and R3. Since they are equal, the Q1 base DC voltage should be about half of the 5-volt supply voltage, or 2.5 volts DC. The emitter voltage then should be about 0.7 volt below the base voltage. So, the DC emitter voltage of Q1 should be 2.5 volts – 0.7 volt or about 1.8 volts.

In general, there will be some base current from the transistors due to the current gain, β, not being "infinite." So, the emitter voltage can be as low as 80 percent of the calculated value (e.g., if the emitter voltage is calculated to be 1.8 volts, the base currents may cause the emitter voltage to drop to 1.8 volts $\times$ 0.80 or 1.44 volts).

For example, Q1 = 2N3904 and Q2 = 2N3906, where β for each transistor is generally $\geq$ 100. If by chance Q1's base voltage is much lower than 2.5 volts, such as 1 volt or less, check to see if the transistor's collector and emitter leads are installed correctly. If the collector and emitter terminals are reversed, the current gain, β, drops to < 5 and there will be excessive base currents drawn through R2 and R3 that will produce a much lower DC voltage at Q1's base terminal.

There is a base series resistor, R1 at 47Ω, to prevent parasitic high-frequency oscillations. Without it, or R1 = 0Ω, there can be a "risk" of the oscillator working incorrectly. Also note that L1 and C1 provide localized power supply decoupling so that any power supply glitches or noise does not induce noise into the +5-volt supply line that may be powering other circuits.

With the crystal Y1 reinstalled, you will need an oscilloscope next to troubleshoot this circuit. The values for C2 and C3 depend on the crystal's frequency. See Table 12-1 for suggested values for Figure 12-32.

NOTE: The crystal Y1 is a fundamental frequency crystal. Be sure to look out for that and not use an overtone frequency crystal.

TABLE 12-1   Suggested Capacitor Values Based on the Crystal's Frequency

Crystal Frequency	C2	C3	C4	R4
32 kHz to 200 kHz	Do not use	Do not use	Do not use	Do not Use
1 MHz to 5 MHZ	470 pf	470 pf	100 pf	470Ω
5 MHz to 15 MHz	47 pf	100 pf	47 pf	470Ω
15 MHz to 30 MHz	22 pf	100 pf	47 pf	470Ω
455 kHz Cer. Res.	470 pf	1000 pf	100 pf	470Ω

The signals from the oscillator section coupled to the base of the amplifier, Q2, and the final output signal at Q2's collector are shown in Figure 12-33. Notice that the output signal is not necessarily always a nice pulse signal.

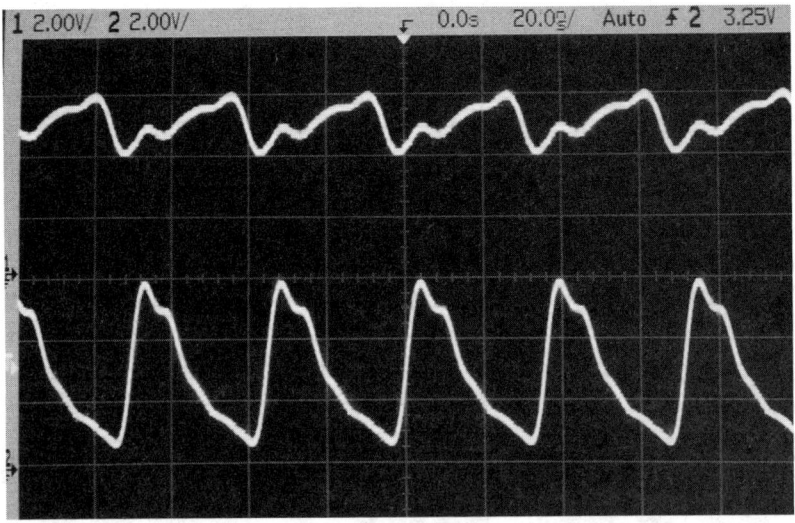

FIGURE 12-33 Top trace is a ~ 2-volt peak to peak 28 MHz signal at Q2's base or VY1; the bottom waveform shows 5 volts peak to peak 28 MHz at Vout1 for Figure 12-32's circuit.

An alternative circuit using DC feedback with a filter capacitor can also be implemented. The advantage of this circuit is that the oscillator and amplifier can work with a range of voltages without having to change value for Q2's base biasing resistors. See Figure 12-34.

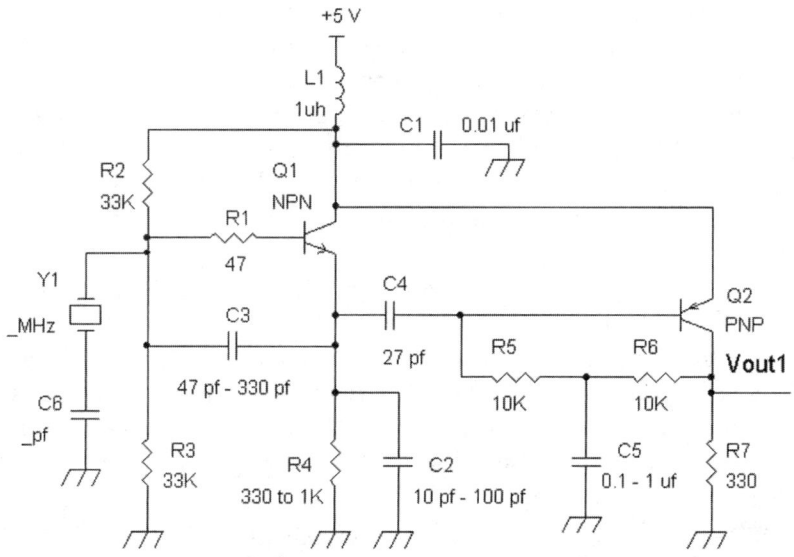

FIGURE 12-34 Automatically biasing Q2 for a range of supply voltages via resistors R5 and R6 with feedback filter capacitor C5; use the suggested C2, C3, C4, and R4 values in Table 12-1.

C6 is generally the crystal's specified load capacitance such as 18 pf. However, C6 may be replaced with a 5 pf to 35 pf trimmer variable capacitor to adjust the frequency exactly at Vout1 using a frequency counter. If Y1 is a series resonant crystal, C6 = 0.01 uf, which is close to an AC short-circuit at the crystal's frequency.

Q2's base voltage is biased to the supply voltage –0.7 volt and in this example it is 5 v – 0.7 v or 4.3 volts at Q2's base. For troubleshooting, we can safely probe the DC voltage at C5, which will have a DC voltage close to 4.3 volts, but likely to be in the range of 0.5 volt to 1.2 volts lower than Q2's base voltage.

So C5's DC voltage may be in the range of:

(4.3 volts – 0.5 volt) = **3.8 volts** to (4.3 volts – 1.2 volts) = **3.1 volts**

Again, you can have Q1 as 2N3904 and Q2 as 2N3906, or equivalent transistors. Generally, these transistors will work fine for oscillation frequencies up to 30 MHz and possibly more.

Of course, we can use a single logic gate inverter chip to make a crystal oscillator with logic-level output. We first encountered this in Figure 12-5, IC1E (74HC04 hex inverter gate) and in Figure 12-31(c). Let's take another look at these types of oscillators. In Figure 12-35(a) we see a very common single-gate inverter crystal oscilla-tor with a DC bias feedback resistor R1, which is usually between 1MΩ and 4.7MΩ. The reason for the high resistance value is to avoid loading down the signal at C5 via the crystal Y1.

NOTE: The crystal Y1 acts like an inductor and forms a two-pole, low-pass filter with C5 as a voltage divider with the internal equivalent series capacitor in Y1. This two-pole, low-pass filter provides another 90 degrees to 150 degrees of phase lag to combine with R8 and C4 so that there is a total of 180 degrees of phase shift that will enable oscillation. If you want to confirm that Y1 acts like an inductor to enable oscillation, you can substitute an inductor such as 1 μH to 10 μH in place of Y1 in Figure 12-31(c) and you will see that the circuit oscillates.

For example, if R1 has a smaller resistance value such as 470Ω, the oscillation signal at C5 will be attenuated to the point where no oscillation occurs. R1 is to bias the inverter gate to become an inverting gain amplifier with a "reasonable" high input resistance as seen by C5 looking into R1 and IC1A's input pin 1. For 74ACxx, 74HCxx, 74AHCxx, 74VHCxx, and 4000 series CMOS gates, R1 biases the input pin 1 to one-half the supply voltage such as 2.5 volts DC to provide close to 50 percent duty cycle pulses at output pin 2.

The problem with Figure 12-35(a) is that C4 is used to provide a phase lag of at least 45 degrees (via a −3 dB cut-off frequency) with the internal output resistance of IC1A. However, with faster chips such as the 74HC or 74AC logic families, the inter-nal resistance is so low that it may not provide enough phase lag, and therefore no oscillation occurs. The reason why Figure 12-35(a) works is because the crystal fre-quency is high enough (e.g., > 14 MHz) such that the gate delay of the inverter IC1A plus the output resistance with C4 has enough phase lag to ensure oscillation. If you try a lower frequency crystal in the ≤1 MHz range, chances are that the reliability of oscillation will be spotty. To examine further the phase lag requirement and role from C4, please see Figure 12-35(b).

A model of the crystal oscillator is shown in Figure 12-35(b) where Rout is the internal output resistance of the logic inverter gate. In the first generation CD4000 series CMOS (Complementary Metal Oxide Silicon) logic chips, Rout was in the order of 500Ω to 2000Ω. Thus, the phase lag network was approximated by a Rout-C4 low-pass filter with a − 3dB cut-off frequency $f_c = 1/[2\pi(\text{Rout C4})]$. If Rout = 1800Ω and C4 = 22 pf, then f_c = 4 MHz for a 5-volt supply. However, if we used a faster logic gate, such as a 74AC04 that has Rout < 10Ω, having C4 at 22 pf will not provide enough phase lag. We would have to either increase C4 to something like 4700 pf that will then cause the output pin to waste power driving C4's low impedance, or better yet we can reduce power by adding a series resistor R8 as shown in Figure 12-35(c). By having R8 in the 200Ω to 2700Ω range, we can ensure that C4 provides the necessary phase lag for oscillation. In general, the inverter's input capacitor C5

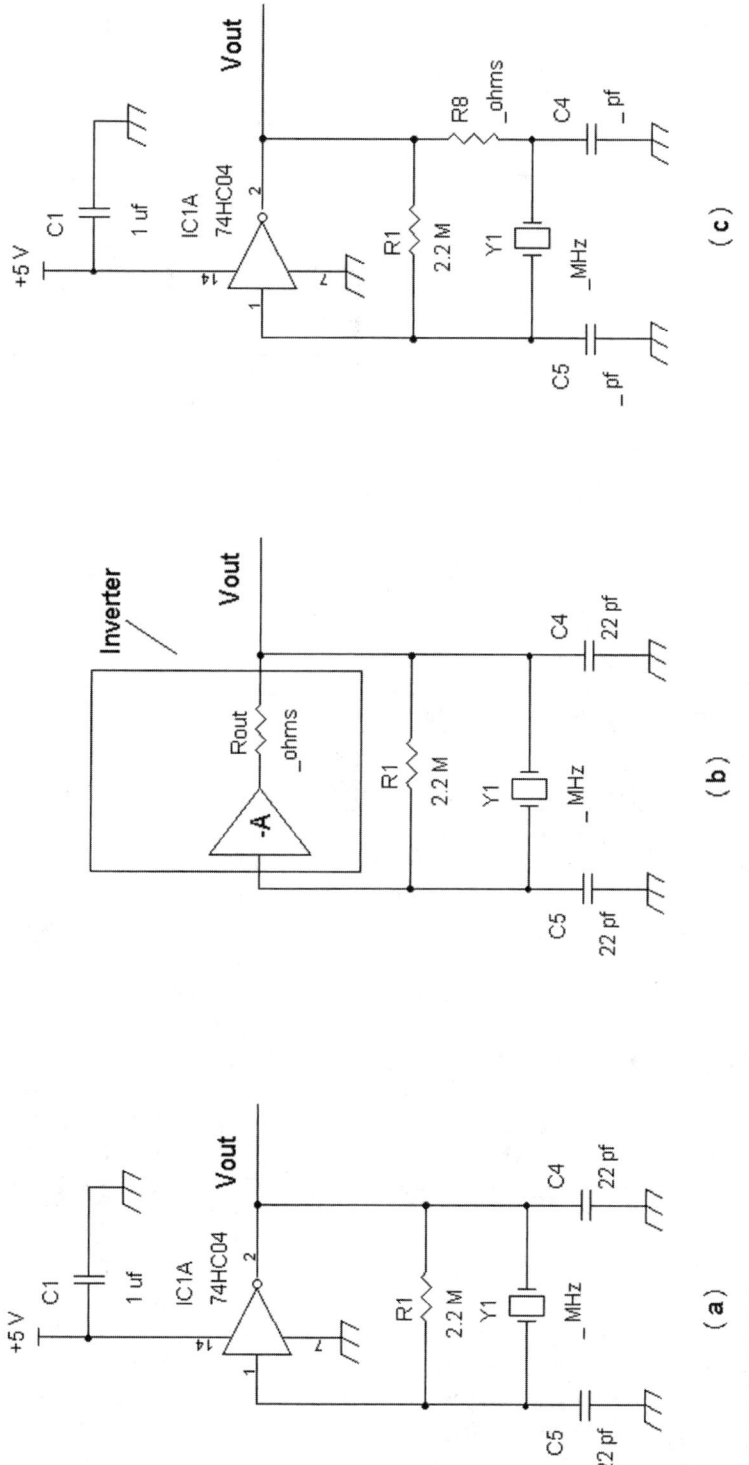

FIGURE 12-35 (a) A simple and not so reliable version of the one-gate oscillator, (b) a model of the oscillator, and (c) a more reliable design with series resistor R8.

315

can be set to the crystal's loading capacitance, which is specified generally between 11 pf and 39 pf, with common loading capacitances of 18 pf and 22 pf for C5. See Table 12-2 for suggested R8 and C4 values, types of inverter gates, and types of resonators Y1 that can include crystal or ceramic resonators.

TABLE 12-2   Suggested Values for Figure 12-35(c) Using Fundamental Frequency Crystals and Resonators Having Load Capacitance

Crystal Frequency	C4	C5*	R1	R8	Logic Family
32 kHz to 100 kHz	1000 pf	15 pf	4.7MΩ	4700Ω	CD4069, 74C04, 74HC04
1 MHz to 2 MHZ	150 pf	18 pf	2.2MΩ	2200Ω	CD4069, 74C04, 74HC04
2 MHz to 4 MHz	150 pf	18 pf	2.2MΩ	1000Ω	74HC04
4 MHz to 8 MHz	68 pf	18 pf	2.2MΩ	1000Ω	74HC04
8 MHz to 15 MHz	68 pf	18 pf	2.2MΩ	470Ω	74HC04, 74HCU04, 74AHC04
15 MHz to 30 MHz	47 pf	18 pf	2.2MΩ	470Ω	74AC04, 74AHC04, 74VHC04
Ceramic Resonators					
455 kHz to 503 kHz	330 pf†	100 pf†	2.2MΩ	2200Ω	74HC04, 74AHC04
3.0 MHz to 6.0 MHz	100 pf†	30 pf†	2.2MΩ	1000Ω	74HC04, 74AHC04
6.0 MHz to 13.0 MHz	47 pf	30 pf†	2.2MΩ	1000Ω	74HC04, 74AHC04

The nominal C5 value is the specified crystal load capacitance, which may be from 11 pf to 33 pf. See data sheet. In some cases where the load capacitance is in the 11-pf range, it may be preferable to use a slightly higher capacitance (e.g., 15 pf) for oscillator stability.

† Ceramic resonators, depending on the manufacturer, will have a different load capacitance value for C5. In some cases, C5 for ceramic resonators may be in the 15-pf to 33-pf range for 3 MHz and above.

Be Aware of Overtone Crystals

For most crystals in the ≤ 30 MHz frequency range, you can buy a fundamental frequency crystal. This means when you build any of the oscillator circuits as shown in Figure 12-35(a) or 12-35(c), the waveform will provide a frequency as marked on the crystal.

However, once you go above 23 MHz, you will find in many cases there is an overlap of fundamental frequency and third overtone frequency crystals. So, you will have to download the data sheet to be sure. Most crystals above 30 MHz will be third overtone versions.

For example, suppose we choose a 60-MHz crystal. It will most likely be a third overtone type, which means its fundamental frequency is 60 MHz/3 or 20 MHz. If you put a third overtone crystal into any of the previous circuits, it may oscillate at

20 MHz instead of the expected 60 MHz. To ensure the correct oscillation frequency, we have to add a band-pass filter at 60 MHz to the oscillator to eliminate the chance that the crystal will oscillate at a lower frequency. See Figure 12-36.

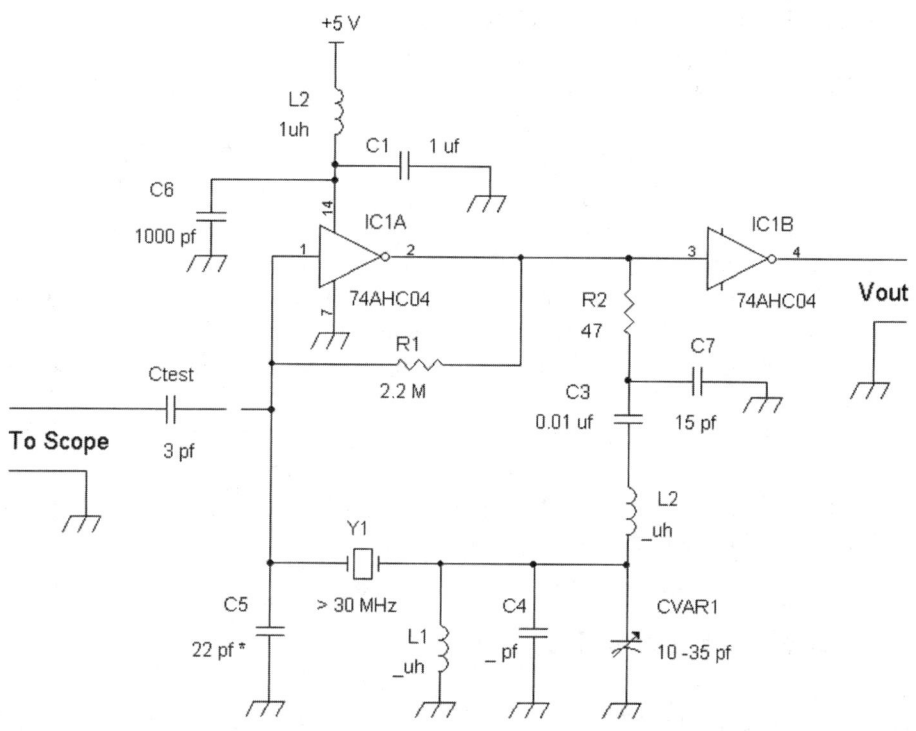

FIGURE 12-36 A third overtone crystal oscillator with a L1-C4-CVAR1 band-pass filter tuned to the crystal's frequency to avoid having a one-third oscillation frequency.

For higher frequencies above 30 MHz, the inverter gates should be of the 74AC04, 74AHC04, or 74VHC04 family. The 74HC or 4000 HC logic families top out at around 15 MHz for reliable oscillation. Although there is a slightly faster version called the 74HCU04, it is not nearly as fast as the AC, AHC, or VHC versions.

In Figure 12-36, there is a pre-filter R2 and C7 to prevent high-frequency parasitic oscillations and to prevent excessive output current flowing out of pin 2. Inductor L2 is typically 10 to 20 times the inductance of L1. Inductor L2 forms with L1 an inductive voltage divider circuit that has an equivalent ***Thevenin*** series inductance of ~ (L1||L2). Because L2 >> L1, L1||L2 ~ L1. Inductor L1 forms a high Q resonant low-pass filter circuit with (C4 + CVAR1), which provides 90 degrees of phase lag at the resonant frequency at:

$$f_r \sim 1/[2\pi\sqrt{L1(C4 + CVAR1)}]$$

Capacitor C3 provides DC blocking so that IC1A's output pin 2 does not short to ground in a DC manner via L2 and L1.

For example, if Y1 is a 57.3 MHz = f_r a third overtone crystal, then how do we pick some values for L1 and C4 with CVAR? One way is to pick (C4 + CVAR1) ~ 30 pf. This would mean C4 ~ 15 pf so that CVAR1 with a range of 5 pf to 35 pf will have enough range needed to tune with L1.

L1 = $1/[(C4 + CVAR1) \times (2\pi f_r)^2]$ = $1/[(30 \text{ pf}) \times (2\pi \, 57.3 \text{ MHz})^2]$ or L1 ~ 0.257 µH. We can have L1 = 0.22 µH and $10L1 \le L2 \le 20L1$, so if we choose L2 = 3.3 µH we are fine since the L2 inductance range is 2.2 µH to 4.4µH.

To adjust CVAR1, we need to temporarily add Ctest, a small capacitance value capacitor of 3 pf in series with C5 so that the ×10 oscilloscope's 11 pf to 22 pf probe capacitance does not load down the signal at C5. This way, we add at most 3 pf in parallel to C5. The oscillator is monitored with oscilloscope ×10 probes at the buffered output pin 4 (IC1A) and at Ctest as shown in Figure 12-36. CVAR1 is adjusted for maximum amplitude at Ctest. When adjusting CVAR1 is done, we can remove Ctest from the oscillator circuit.

An advantage of using logic inverter gates is that they can provide a buffered output (e.g., pin 4 in Figure 12-36) such that when the oscillator signal is driving a load that includes parasitic capacitance to ground (e.g., a board trace or cable capacitance), the frequency does not change.

Gain Bandwidth Product Revisited

There were examples on page 284 of setting the gain of an op amp based on gain bandwidth product (GBWP). In these examples we used the "best" case scenario calculations. For example, a TL082 has a 3 MHz GBWP, which means if you want a 100 kHz closed loop bandwidth (CLBW), the gain is set to 30. However, setting the closed loop gain to 30 (e.g., Figure 12-14 with [(R2 + R1)/R1] = 30) may not be repeatable because the TL082's 3 MHz GBWP is only typical. A better way is to set the gain lower by 3 fold such as a gain of 10 to get a typical 300 kHz CLBW, which will ensure a 100 kHz minimum bandwidth. The other solution is to set the gain to 30 but use a faster op amp by at least three fold such as a TLE082 that has a 10 MHz GBWP, which will again ensure at least a 100 kHz CLBW.

Summary

We have looked at a few circuits related to amateur radio, but there are obviously many more that can be explored. And there are whole books written just on ham radio circuits.

What's up ahead in the next chapter will be 555 timer circuits.

CHAPTER 13

Timer, CMOS, and Motor Drive Circuits

We will be looking into analog timing and oscillator circuits that use the NE555/LM555 (a.k.a., 555 circuit or chip) integrated circuit. Although this circuit was manufactured in the early 1970s, it still finds many uses in hobbyist circuits and projects. The 555 circuit is an 8-pin integrated circuit that can be configured into many different circuits. For example:

- A timing circuit to provide a fixed duration pulse when triggered by an external signal (e.g., closing a switch). The 555 can be configured as a *one-shot circuit* to generate pulses.
- 555 oscillator circuits.
- A simple CMOS (complementary metal oxide silicon) oscillator that includes a 555 pulse-width modulator circuit for driving a motor via a switch mode bridge amplifier.

Types of 555 Timer Chips

Before proceeding with some of the circuits, we should be aware that there are two types of 555 timer chips—bipolar transistor and complementary metal oxide silicon (CMOS).

The original design in the 1970s was done with bipolar transistors that had part numbers such as NE555, LM555, CA555, MC1455, and MC1555. This chip generally worked over a 5-volt to 15-volt power supply. One of its advantages is that the output signal can drive at 100 mA safely, and if pushed, up to 200 mA. Its disadvantage is that the standby current is in the 3-mA to 10-mA range for supply voltages at 5 volts and 15 volts, respectively. Top oscillation frequency of the bipolar 555 chip was about 500 kHz to 1 MHz.

By the 1980s the 555 integrated circuit was offered in CMOS, which dramatically reduced the standby current to less than 500 µA (200 µA typical at 5 volts supply). Typical part numbers were the TLC555, LMC555, and ICM7555 (or 7555 for a short-handed name). The CMOS version offered a higher oscillation frequency at about 2 MHz, which is twice the "speed" of the bipolar version. Later, an even faster CMOS version was available as the LMC555 that provides up to 3-MHz oscillation frequency. Although the ICM7555 guarantees at least a 500-kHz oscillation frequency, its data sheet chart shows oscillating frequencies of 2 MHz or 3 MHz.

The CMOS 555's power supply voltage range expanded to the low end starting at 2 volts while still having a maximum 15-volt rating. Output sourcing current is reduced to 10 mA, which means its drive current is only about 5 percent of the bipolar transistor version, LM555. However, it can sink or absorb about 100 mA at its output, which means if the load such as a relay's first terminal is connected to +5 volts, and the relay's second terminal is connected to the TLC555's output terminal, the relay can have up to 100 mA flowing through it.

NOTE: When connecting a relay in this manner, also connect a diode such as 1N4002 where the cathode is connected to the +5-volt supply and the anode is connected to the output pin of the TLC555 to suppress high voltage spikes from the relay coil.

Basic Modes of the 555 Timer Chip and Pin Outs

Figure 13-1 shows the 8-pin assignments for the 555 timer bipolar and CMOS chip, and also three basic operations.

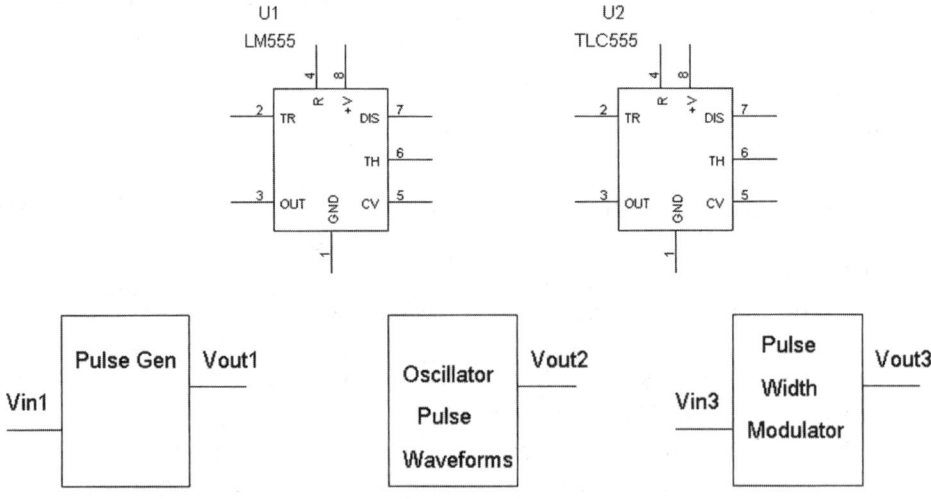

FIGURE 13-1   Pin outs for the bipolar LM555 and CMOS TLC555 integrated circuits and three basic modes of this timer chip.

As we can see in Figure 13-1, the bipolar (LM555) and CMOS (TLC555) versions have the same pin outs and they can in general be used interchangeably. In most cases the reset pin 4 is tied to the $+V$ pin 8 to enable the 555 circuit. To reset the output pin 3 to a low-logic level, the reset pin 4 must to be coupled to ground via a resistor (e.g., $\leq 1000\Omega$) or a wire.

The 555 timer chip can perform at least three basic functions as shown in Figure 13-1. This chapter will focus on it being a pulse generator (e.g., one-shot timing pulse) for an input pulse signal Vin with an output signal, Vout, an oscillator (Vout2), and a pulse-width modulator with an input modulating signal Vin3 with a pulse-width modulated signal at Vout3. Note that the pulse-width modulator may include a potentiometer for providing a modulating signal.

To make a timing pulse the 555 chip generally requires logic low (e.g., 0 volts) and logic high (e.g., $+V$ or the supply voltage) for Vin1 at pin 2 of the 555 chip to output a fixed duration pulse, Vout1 at pin 3. The output pulse starts when the input Vin1 goes from the high to low logic states such as $+5$ volts to 0 volts when the supply voltage $+V = +5$ volts. For example, you can make an LED or relay turn on for a fix duration when Vin2 is switched from $+5$ volts to 0 volts. When Vin1 goes from 0 volts to 5 volts, there is no change in Vout1. Only "negative" going transitions (e.g., $+5$ volts to 0 volts) will trigger an output signal having a fixed duration pulse at pin 3 (Vout1).

For the oscillator function, the 555 chip will output a pulse waveform at Vout2. The oscillation frequency ranges from less than 1 Hz to at least 500 kHz. The duty cycle of the output signal can be adjusted anywhere from a narrow negative going pulse to nearly a square-wave signal (~50 percent duty cycle).

The 555 chip can also be used to provide pulse-width modulated pulses. By pulse-width modulation we mean that the frequency of the output waveform does not change, but the output signal's duty cycle or pulse-width varies according to a modulation or control voltage input signal at Vin3.

The 555 Pulse Generator (a.k.a., One-Shot or Monostable Mode)

This 555 pulse generator circuit provides a fixed duration output pulse for an input signal that transitions from high to low logic levels (e.g., $+5$ v to 0 v). See Figure 13-2 where the 555 circuit is configured in "monostable" or pulse generator mode.

A "proper" input waveform for Vin is one that has narrow low logic voltage pulse-width that is shorter in duration than the output high logic state. In Figure 13-2(a) for Vin, the negative going pulse is shorter in duration than the output pulse's duration of t555. With the proper waveform, the output positive cycle pulse width is t555 ~ 1.1 R × C1 = τ.

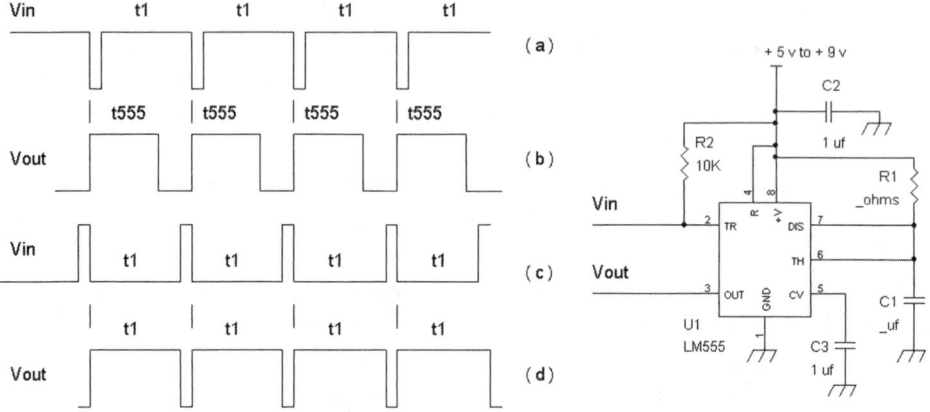

FIGURE 13-2 Input waveforms (a) and (c), and output waveforms (b) and (d) respectively with a schematic diagram of the monostable timing pulse generator.

However, if the input waveform has a longer duration low logic voltage, such as in Figure 13-2(c), where its duration t1 > t555, the output of the timing chip at pin 3 will result in a longer positive pulse of duration t1 of Figure 13-2(d) even though the pulse-width should really have a duration of t555 shown in Figure 13-2(b).

From a top overview of the 555 monostable pulse generator circuit, we see that the input signal into pin 2 (trigger input) requires a negative going transition to output a fixed duration pulse. This fixed duration pulse is logic high and has a time length of $\tau \sim 1.1\ R1 \times C1$, where τ is measured in seconds, R1 is measured in ohms, and C1 is measured in farads. For example, suppose we want to generate a 1 msec pulse (one thousandth of a second) or 0.001 second. We then have 0.001sec = 1.1 R1 × C1.

Suppose we first choose R1 = 1000Ω, then we can solve for C1 as:

C1 = 0.001 sec/(1.1 × R1) = 0.001/(1.1 × 1000Ω) → C1 = 9.09 × 10^{-7} farad

C1 = 0.909 × 10^{-6} farad

But 10^{-6} farad = 1 µf, so C1 = 0.909 × 1 µf, or C1 = 0.909 µf, which is a non-standard value.

If we can scale R1 by (1/1.1), R1 → 909Ω = 1000Ω/(1.1). This will get rid of the 1.1 factor in C1 = τ/(1.1 × R1). With R1 = 909Ω ~ 910Ω, then with **R1 = 910Ω** we have: C1 = 0.001sec/(1.1 × 910Ω) ~ 0.001 sec/(1000Ω) or **C1 = 1 µf**, a standard capacitance value.

One question to ask: Is there a safe range of resistance values for R1? For example, it probably would not be a good idea to make a 1 msec pulse generator with R1 = 0.91Ω and C1 = 1000µf because with R1 being less than one ohm, the circuit will have to draw several amps of current because the voltage at C1 varies between zero volts and one-third (e.g., ⅓) the supply voltage. For example, if +V = +5 volts and R1 = 0.91Ω, the voltage at C1 has to be in a range of 0 volts to +5 volts × ⅓ or 0 volts to 1.66 volts. Since R1 is connected to +V and C1, this means the voltage across R1 is (+5 volts − 1.66 volts) or 3.33 volts, or (+5 volt − 0 volts) or 5 volts. The current flowing through R1 would then be in the range of 3.33 volts/0.91Ω to 5 volts/0.91Ω or 3.66 amps to 5.49 amps. Most likely, the 555 circuit may be damaged or it may go into a current limiting mode. Either way, there would be no correct output signal or pulse at pin 3. Table 13-1 shows some suggested value ranges for R1 and C1 based on the output pulse duration.

TABLE 13-1 Suggested Values for R1 and C1

τ or pulse duration	R1	C1	555 version
10 secs to 1 μsec	10MΩ to 1KΩ	1 μf to 1000 pf	LM555
10 secs to 0.3 μsec	10MΩ to 1KΩ	1 μf to 330 pf	TLC555
10 secs to 0.3 μsec	10MΩ to 1KΩ	1 μf to 330 pf	LMC555
10 secs to 0.3 μsec	10MΩ to 1KΩ	1 μf to 330 pf	ICM7555

In Figure 13-2, capacitor C3 is a decoupling capacitor to ensure that the output pulse from pin 3 does not jitter due to noise coming through the power supply. If the power supply into pin 8 has ripple or other noises, C3 may have to be changed to a higher capacitance value such as an electrolytic capacitor, 33 μf at 25 volts, where the positive terminal of C3 is connected to pin 5 and the negative terminal of C3 is connected to ground. In general, the 555 timer chip should be powered by a regulated voltage to supply a "noise free" voltage to pin 8. However, this integrated circuit may be powered by batteries, which do not have ripple or noise. But be aware that a battery will slowly drop in voltage as it is supplying current to the 555 circuit.

The input level that triggers this one-shot circuit is at one-third the supply voltage, so to be on the safe side, the input signal should be between 50 percent to 100 percent of the supply voltage for a logic high voltage, and below one-sixth of the supply voltage for a low logic state. For example, if +V = 5 volts, then a "safe" logic high voltage is between 2.5 volts and 5 volts, and a "safe" logic low voltage is between 0 volts and +833 mV. The trip voltage is one-third of 5 volts or 1.66 volts, which means this voltage as an input voltage to pin 2 should be avoided or there will be "noise" or chatter at the output pin 3.

Table 13-2 shows some suggested low and high logic levels along with corresponding output levels.

TABLE 13-2 Input and Output Voltages for 555 Circuits Configured as a One-Shot or Monostable Circuit

Vs = Supply	Vin Low	Vin High	Vout Low	Vout High	555 Version
+5v to +15v	< 0.16Vs	≥0.50Vs	~ 0 volt	~ Vs	LM555
+2v to +15v	< 0.16Vs	≥0.50Vs	~ 0 volt	~ Vs	TLC555
+1.5v to +12v	< 0.16Vs	≥0.50Vs	~ 0 volt	~ Vs	LMC555
+3v to +15v	< 0.16Vs	≥0.50Vs	~ 0 volt	~ Vs	7555

Troubleshooting the 555 One-Shot Monostable Timer

1. With the circuit in Figure 13-2, find the DC conditions first using a voltmeter, or with an oscilloscope in the DC coupling mode set to 2 volts per division. Measure voltages with a voltmeter at pins 8 and 4 that should be the supply voltage, such as +5 volts (or some other positive supply voltage ≤ 15 volts). Confirm that the voltage at pin 1 that should be connected to ground is zero volts.

2. Now measure the DC voltage at pin 5; the control voltage pin should be 67 percent of the power supply voltage. For example, if pins 8 and 4 have +5 volts on them, then pin 5 should have 67% × 5 volts or 3.33 volts DC within 15 percent.

3. Connect a square wave signal generator that has a DC voltage swing from 0 volts to 5 volts to pin 2 of the 555 timer chip. To set the frequency of the signal generator, first determine the pulse width, pw = 1.1 R1 × C1. Set the signal generator's frequency to f_{gen} = 0.75/pw. For example, if the pulse width is pw = 0.001 sec or 1 millisecond, set the generators frequency f_{gen} = 0.75/0.001 or f_{gen} = 750Hz. Confirm that the output at pin 3 produces a positive pulse whose width is 1 msec at a 750 Hz rate.

4. Alternatively, if you do not have a signal generator and just want to see if the 555 chip is working, connect a voltmeter (DVM) to pin 3 and ground, where the positive test lead is at pin 3 and the negative test lead is connected to ground. Make sure pull-up resistor R2 in Figure 13-2 is installed. We are now going to make an 11-second timer. Let R1 = 1MΩ, and let C1 = 10 μf where the (+) terminal of the C1 is connected to pins 6 and 7. The DVM should be reading close to zero volts at pin 3. Now short pin 2 to ground temporarily for less than 1 second, then observe the DVM is at a logic high voltage > 3 volts DC for about 11 seconds. You can use a watch or a clock to measure the time. If this is OK within ± 30 percent, the circuit works. Be sure to measure both R1 and C1 if possible with a DVM (e.g., Extech MN26 or equivalent DVM). If you do not have a DVM, you can connect pin 3 to a series resistor (e.g., 470Ω to 4700Ω) to an LED (e.g., red, yellow, or green having about a 2-volt turn-on voltage). See Figure 13-3. The LED should be turned on for about 11 seconds. Then replace R1 and C1 with the original values. Again the output turn on time for an input signal from high to low logic states (e.g., input signal from +5 volts to 0 volts) is τ = 1.1 RC.

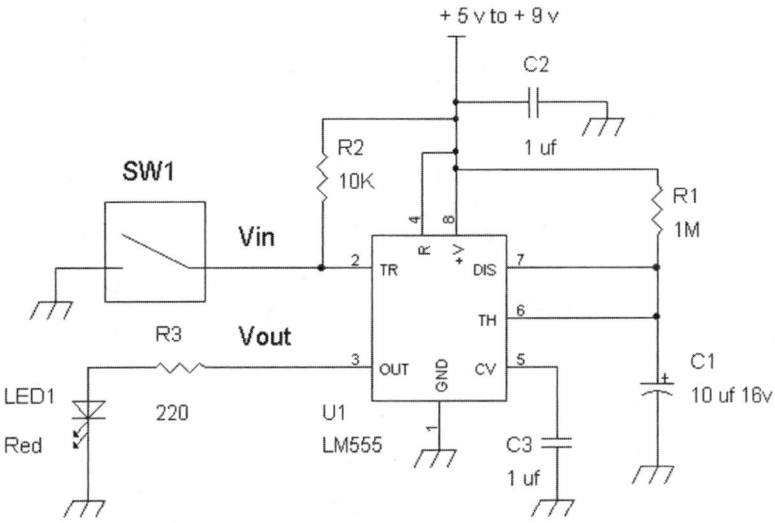

FIGURE 13-3   Switch SW1 is a momentary switch and it is closed circuit to ground for a split second, which should then turn on LED1 for about 11 seconds.

When You Want to AC Couple a Signal to Trigger a Pulse Output Signal

Figure 13-4(a) shows a simple DC restoration circuit with C3, CR1 and R3, and Figure 13-4(b) shows AC coupling methods to trigger pulses via a voltage divider circuit R2 and R3.

In both circuits, the input signal should have a peak to amplitude equal to the supply voltage of the 555 circuit. For example, if the supply voltage is 5 volts, the input signal Vin should be in the order of 4 volts to 5 volts peak to peak. However, it turns out that the input signal must have its *negative cycle duration* less than the 555's time output pulse duration of $\tau = 1.1R1 \times C1$, or the output pulse at pin 3 will be longer than the expected duration of $\tau = 1.1R1 \times C1$.

Figure 13-4(a) shows a negative peak DC restoration circuit where the negative peak of Vin is set to 0 volts via the DC restoration diode CR1. This means if Vin is a signal that has no DC offset or any DC offset voltage, the DC restoration circuit CR1 at its cathode will provide a signal "clamped" to 0 volts for the negative peak portion of Vin. This effect is useful when the incoming waveform (e.g., a pulse signal) with low or high duty cycle is used. DC restoration is often used to level shift a signal to a particular or well-defined negative cycle voltage. For example, if the following (Vin) incoming 5-volt peak to peak signal has a range of –2.5 volts to +2.5 volts, –5 volts to 0 volts, +5 volts to +10 volt, etc., the DC restoration circuit at the cathode of CR1

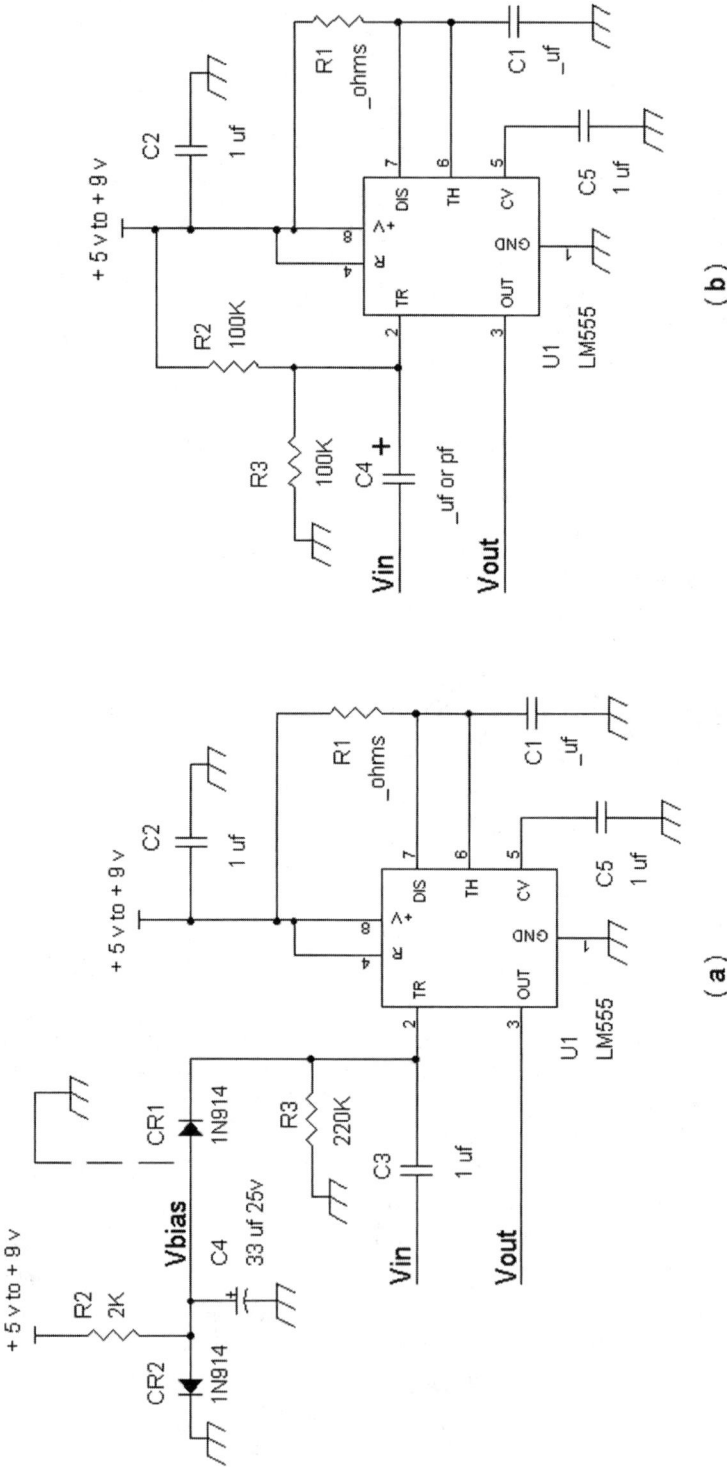

FIGURE 13-4 (a) AC coupling via a DC restoration circuit via C3, CR1, and R3 with Vbias ~ 0.7 volt; (b) AC coupling signals via a biasing network R2 and R3.

will provide a signal that ranges from 0 volts to +5 volts, a standard logic level range for a 5-volt system.

The circuit in Figure 13-4(b) allows AC coupling to the trigger pin 2 at 50 percent at the power supply voltage. Generally, C4 as 33 μf will work, but this large capacitance value can be a problem if Vin is a signal that has a longer negative cycle duration than τ = 1.1R1 × C1. For example, suppose we have τ = 1.1(910Ω) × 1μf = 1 msec. If Vin is a pulse whose negative cycle exceeds 1 msec, then the output signal at pin 3 will be longer than 1 msec. However, we can fix this by lowering the capacitance of C4 to about 220 pf to narrow the width of Vin. See Figure 13-5 for a comparison between having C4 = 33 μf and C4 = 220 pf.

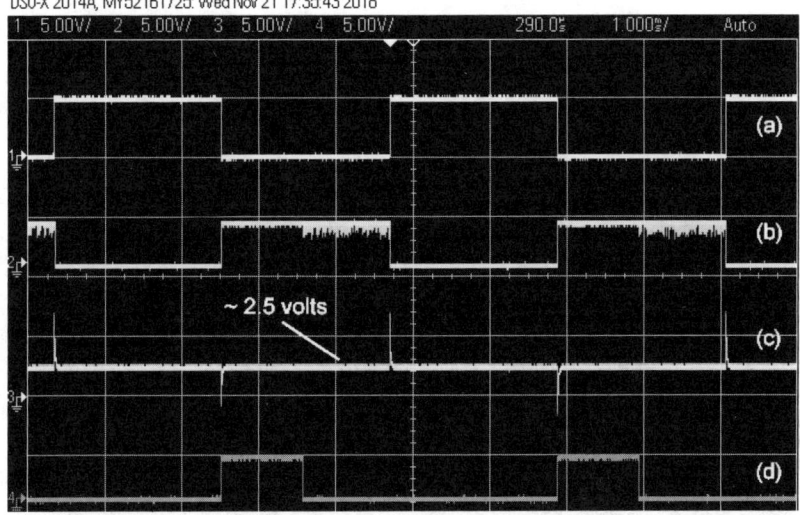

DSO-X 2014A, MY52161725: Wed Nov 21 17:35:43 2018

FIGURE 13-5   Comparing C4 = 33μf and C4 = 220 pf on (d) shows a more reliable output waveform, which conforms to τ = 1.1 R1 x C1 for the output pulse-width, than (b), which is longer than τ = 1.1 R1xC1.

With reference to the circuit in Figure 13-4(b) that uses AC coupling with C4, we see that with C4 = 33 μf, the waveform at pin 2 of the 555 chip is shown in Figure 13-5(a). This waveform is a square-wave signal, whose negative cycle duration is longer than τ = 1.1 R1 × C1 ~ 1 msec in this example. As a result of this longer duration negative cycle pulse the resulting output waveform is longer than τ = 1.1 R1 × C1 as shown in Figure 13-5(b). If we have instead C4 = 220 pf, then we see in Figure 13-5(c) that the negative going pulse is much narrower. The "flat line" in (c) is at 2.5 volts so that the negative going spike will easily go below the threshold voltage of 1.66 volts (e.g., 33 percent of the 5-volt supply) to trigger the correct duration pulse shown in Figure 13-5(d).

"Strange" Output Signals Observed via an Oscilloscope

We will now show two cases of pin 3 output signals in the monostable mode (one shot) where you can observe a high-frequency oscillating waveform from the 555 timer chip. See circuits in Figure 13-2 or Figure 13-4. The first example shows what happens when you forget to install the timing capacitor, or install the timing capacitor (e.g., C1) with an incorrect value where C1 ≤ 20 pf. See Figure 13-6, second trace from the top. The fix for this is rechecking your wiring and reading the capacitor's marking, then installing the correct value capacitor that is typically in a range of 1000pf to 10 μf.

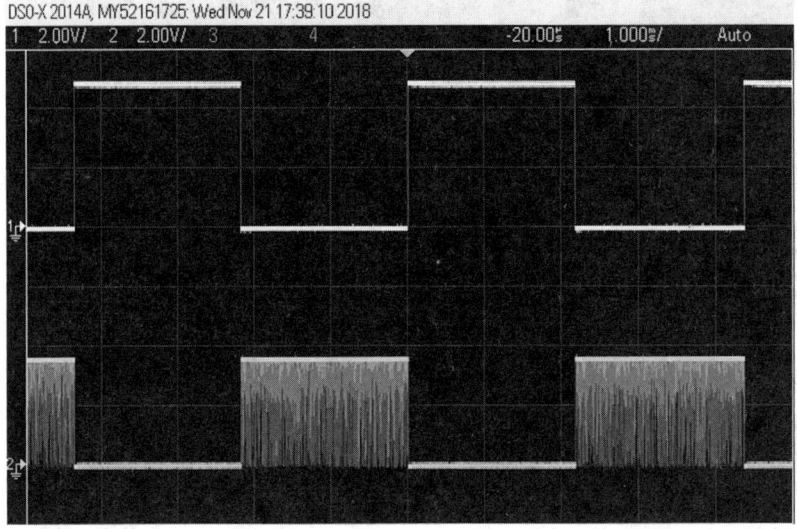

FIGURE 13-6 Completely gated oscillating waveform when C1 has insufficient capacitance. Top trace is the input signal, bottom trace is the output waveform.

If you also forget to wire a timing capacitor for C1, then you may see an output waveform on pin 3 as shown in Figure 13-6. Also, make sure the 555 pins 6 and 7 are connected.

The second error we will show is where the input signal's frequency and pulse-width can "force" the output signal at pin 3 to give out a slightly gated oscillating signal due to having a longer duration negative going pulse. See Figure 13-7, bottom trace. As a reminder, you can fix this problem by using the circuit in Figure 13-4(b) where C4 is a capacitor between 100 pf and 1000 pf to provide a narrowed negative going pulse into input pin 2 of the 555 chip.

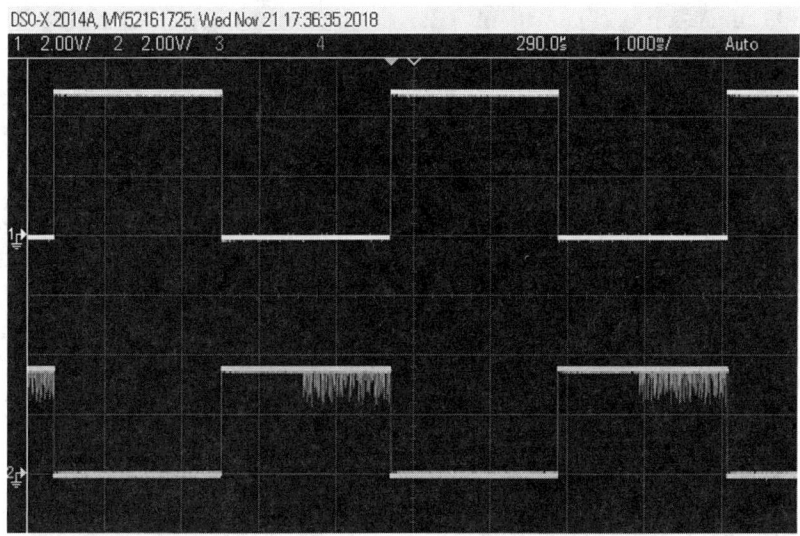

FIGURE 13-7 Top trace is the input signal and the bottom trace is the output waveform with noise.

Troubleshooting the 555 Oscillator (a.k.a. Astable Mode)

We now turn our attention to the 555 circuit configured as an oscillator with some troubleshooting methods. See Figure 13-8.

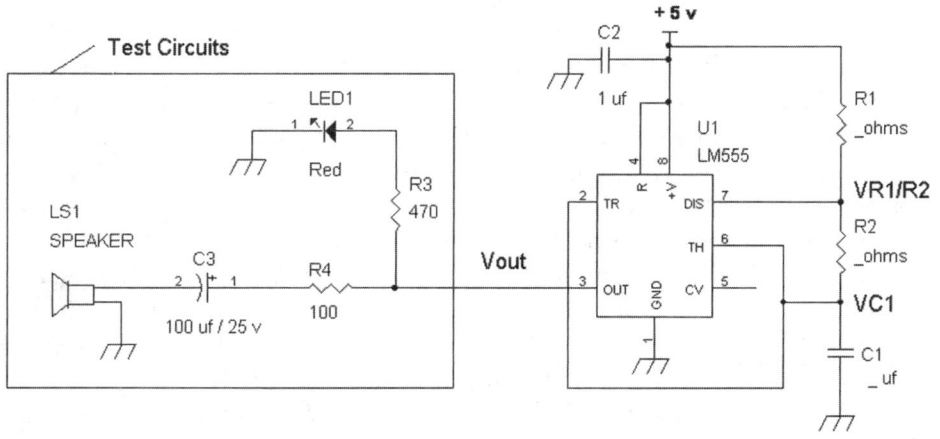

FIGURE 13-8 A 555 configured as an oscillator with test circuits including an LED and a speaker.

A 555 timer can oscillate at sub 1 Hz frequencies to at least 500 kHz. The frequency is determined by $f = 1.44/[(R1 + 2 \times R2)C1]$. For example, if C1 = 0.01 µf, R1 = 100kΩ, and R2 = 51kΩ, then $f = 1.44/[(100kΩ + 2 \times 51kΩ)(0.01 \times 10\text{-}6 \text{ F})]$, or f = 713 Hz. See Figure 13-8.

To confirm that the 555 circuit is oscillating within the audio range (e.g., 20 Hz to 15 kHz), you can use a loudspeaker (LS1) with an output coupling capacitor C3 and series current limiting resistor R4. The reason for the limiting resistor is to prevent the 555 chip from delivering excessively unsafe current that can damage its output circuit at pin 3.

Most speakers (LS1) have low voice coil impedance in the range of 4Ω to 8Ω. Thus, the speaker should not be connected directly to pin 3 of the 555 timer chip and ground for two reasons. One is to prevent loudspeaker damage, and the second is to avoid burning out the timer chip. A 555 circuit driving directly a load resistance of 4Ω to 8Ω is almost like shorting the output pin 3 to ground DC-wise. To prevent excessive DC currents flowing out of pin 3 of the 555, we need to add a series capacitor, C3, as shown in Figure 13-8. For low-impedance speakers, usually C3 ≥ 33 µf, with it typically being 100 µf to 470 µf.

If the 555 is set to lower frequencies below 1 Hz you can use an LED and observe the flashing rate. But you should also be able to hear the speaker "click" at the oscillation frequency such as 1 click per second if the frequency is 1 Hz. Finally, if you have an oscilloscope, you can probe VC1 and VR1/R2, which will have signals with an AC component such as an oscillating waveform.

If the 555's oscillation frequency is higher than the (human) audio hearing range, then you can divide the oscillation frequency to provide an audible tone. See Figure 13-9. Alternatively, you can connect/solder a "transmitting antenna" wire of a few inches to 15 inches to pin 3 and place this wire close to an AM radio to sense the oscillator. Generally, as you tune the AM radio, you will hear a reduction in hiss when the radio is tuned to a frequency related to the 555's oscillation frequency, or to a multiple (e.g., harmonic) of the oscillation frequency.

In Figure 13-9, the 555's output signal is connected to frequency divider U2A at its clock input pin 1, which provides square-wave signals at one-half (Q1A), one-quarter (Q1B), one-eighth (Q1C), and one-sixteenth (Q1D) of the 555's frequency. The resistor R3 as shown is connected to Q2D but can be connected to any of the signal outputs from Q1A to Q1D or from Q2A to Q2D. The output at Q1D is then fed to the second frequency divider clock input (pin 13) of U2B.

U2B continues the chain of frequency division to one–thirty-second (Q2A), one–sixty-fourth (Q2B), one–one hundred twenty eighth (Q2C), and finally one–two hundred and fifty sixth (Q2D). For example, if the 555 is set to 1 MHz, then the signal at Q2D will be 1 MHz /256 or 3.906 kHz, which would be an audible tone.

Of course, if you have an oscilloscope, you do not need to have U1A, U1B, and the associated speaker amplifier of TR1 with the speaker LS1. You can just probe the waveform at pin 3 of the 555 integrated circuit.

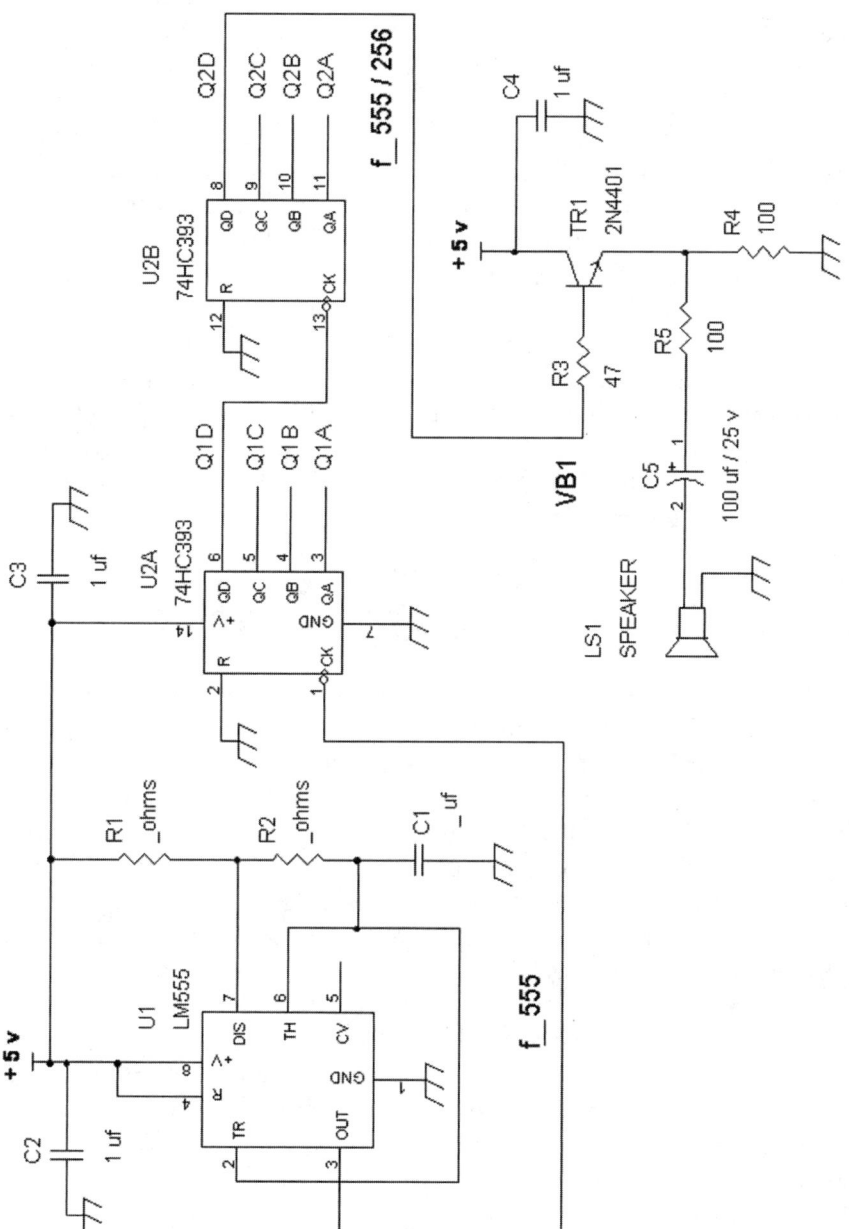

FIGURE 13-9 A 555 circuit with a frequency divider by 256 circuit using binary counters.

331

One More Example on Driving Speakers with the 555

You'll notice that the loudspeaker in Figures 13-8 and 13-9 uses a DC blocking capacitor. There is a good reason for this. Some circuits have the speaker driven by the 555 via a resistor to the pin 3 output. This will cause excessive DC current drain and also send a DC current into the loudspeaker, which can heat up its voice-coil unnecessarily. See Figure 13-10(a) and see the solution in 13-10(b).

When the series current limiting resistor R3 is in the order of a couple of hundred ohms, the loudspeaker can tolerate some DC current flowing into it. With 220Ω for R3 in Figure 13-10(a), the average DC voltage at pin 3 of the 555 is about 2 volts to 3 volts DC, depending on the values of R1 and R2, which determines the duty cycle of the output signal. With 2 to 3 volts DC at pin 3, there is then about 9 mA to 13 mA DC flowing into the loudspeaker LS1. This is power wasted. However, by using a coupling capacitor, C4, as shown in Figure 13-10(b), any DC current flowing into speaker LS1 is reduced to zero; and this then saves power because there is no 9 mA to 13 mA of DC current being wasted.

If the 220Ω resistor needs to be lowered to 47Ω so that the speaker produces a louder sound, then Figure 13-10(a)'s circuit will drain about $2.0v/47\Omega$ to $3v/47\Omega$ or 42 mA to 63 mA of wasted DC current. Having the DC blocking capacitor C3 in Figure 13-10(b) will then reduce the DC current to zero, while allowing the speaker LS1 to produce a louder sound.

Why Again an Output Coupling Capacitor Is Preferable

A 555 configured as an oscillator can be used to build a high-voltage converter. For example, with a 6-volt DC supply to the 555 circuit, you can make a greater than 60-volt DC source. This circuit uses a step-down 110-volt to 6-volt AC transformer in reverse to form a step-up from 6 volts AC to 110 volts AC. See Figures 13-11 and 13-12.

First, please **DO NOT BUILD Figure 13-11's circuit** because it will drain excessive DC current from the 555 output pin 3. In terms of DC current, the transformer's 6 V AC winding (T1), has close to zero DC resistance. The duty cycle of the negative pulse is R2/(R1 + 2R2), and in Figure 13-11 we see that R1 = $47K\Omega$ and R2 = $1K\Omega$, so the negative pulse duty cycle is $1K/(47K + 2 \times 1K) = 1K/(47K + 2K) = 1/49$ or about 2 percent. This then means that the positive pulse's duty cycle is then 100 percent minus 2 percent or about 98 percent. Because the 555 output pin 3 is a signal that is at least 98 percent duty cycle for the positive going pulse, this means that the average DC voltage is approximately 98 percent of the 555 power supply.

For example, if the supply voltage is 7.2 volts (six rechargeable 1.2-volt cells in series), the average DC voltage at pin 3 will be 7.2 volts $\times$ 98 percent or +7.05 volts DC. Because the 6 V AC winding of T1 is a DC short circuit to ground, the 555's pin

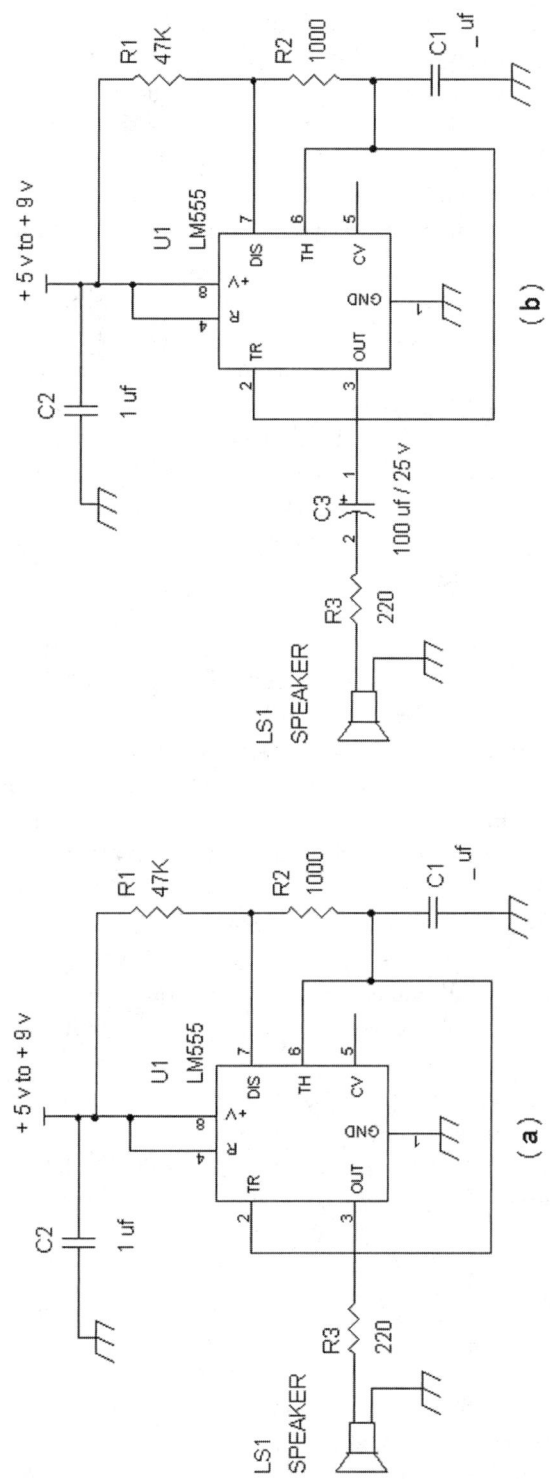

FIGURE 13-10 (a) The 555 is DC coupled to the speaker that causes a problem, and (b) a coupling capacitor, C3, blocks DC current.

333

3 will try to supply 7.05 volts to ground in terms of DC. This will then cause the 555 to go into current limiting, while **not** providing the correct output AC signal into T1. This is why we do not want to build the circuit in Figure 13-11, unless we fix it in terms of blocking DC currents from flowing via pin 3 of U1.

Again, **please do not build the circuit in Figure 13-11**, as it may cause damage to the 555 circuit. However, if we just add an AC coupling capacitor, C4, as shown in Figure 13-12, we avoid draining excessive DC current from output pin 3 of the 555 circuit.

In Figure 13-12, you can troubleshoot the circuit by using an oscilloscope and observing the waveforms, which have about a 2-kHz frequency. From the pin 3 of the 555 chip, we see a large duty cycle pulse in "wf1", waveform 1. It has DC levels from 0 volts to about the supply voltage minus about 1.2 volts. For instance, if the supply voltage is +6 volts the DC levels from wf1 is from 0 volts to (6v – 1.2v) = 4.8 volts. There will be some overshoot as shown in wf1 due to the inductive nature of the 6 V AC transformer winding. You can disconnect R3 temporarily and connect a DVM with 10MΩ input (e.g., a DVM that costs over $30 usually has a 10MΩ load resistance for the DC voltmeter setting) to D1's cathode. The voltage at C3 and the cathode of D1 should be > +100 volts DC when the supply voltage to the 555 chip is about 7.2 volts.

One "feature" to note in the waveform from pins 1 and 4 of T1 is that this voltage waveform is not symmetrical. See "wf2" in Figure 13-12, which is at the 120 V AC winding into D1's anode **when R3 is disconnected**. As shown, the positive peak is about 60 percent larger in magnitude than the negative peak. For example, if the positive peak is about +200 volts, the negative peak is about –125 volts. This means that you may get more voltage if you swap the leads of pins 1 and 4. When this circuit works properly and with R3 reconnected, the neon lamp may flicker or flash. This is normal since the filter capacitor, C3, provides very little current to the neon lamp, such that when the neon bulb lights up, it discharges C3's voltage somewhat and then D1 has to build back up C3's voltage to light the neon bulb. The neon has a firing voltage in the range of 95 volts or so. Below that, the neon bulb is not lit.

NOTE: Normally, in a power supply transformer such as in a 50 Hz or 60 Hz wall AC adapter, it does not matter which way you configure the leads because the AC voltage waveform is symmetrical. This means you get the same DC voltage after rectification and filtering regardless of the wiring orientation of the transformer. Swapping the leads here will give you the same DC voltage.

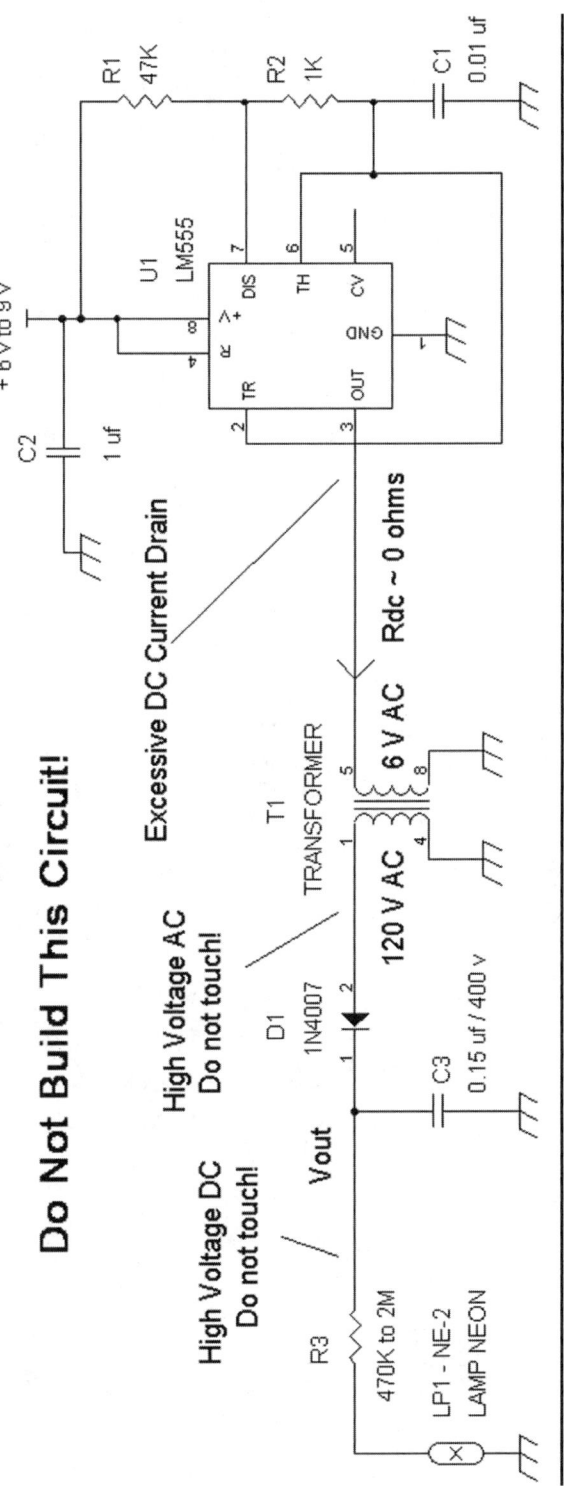

FIGURE 13-11 A high-voltage power supply that may draw excessive current.

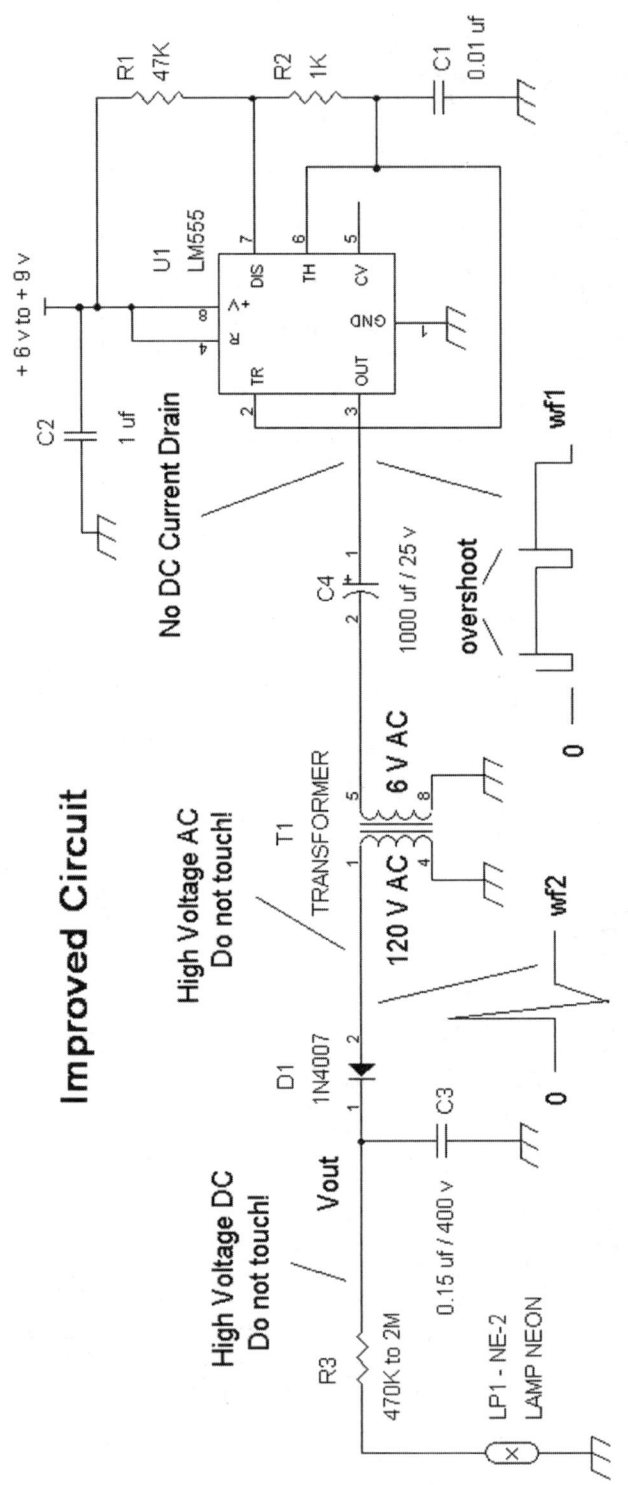

Improved Circuit

FIG 13-12 A high-voltage power supply with improved efficiency by using a DC blocking capacitor, C4. Note: wf1 < 15 volts peak to peak and wf2 > 100 volts peak to peak.

Note: T1 pins 1 and 4 may need to be swapped.

Average DC Level of Pulses

In Figures 13-11 and 13-12, the 555 timer outputs a high duty cycle pulse as shown in Figure 13-13(a), which produces a large average DC voltage, Vavg_hdc. In this example, the average level is almost the same as the logic high voltage, +V.

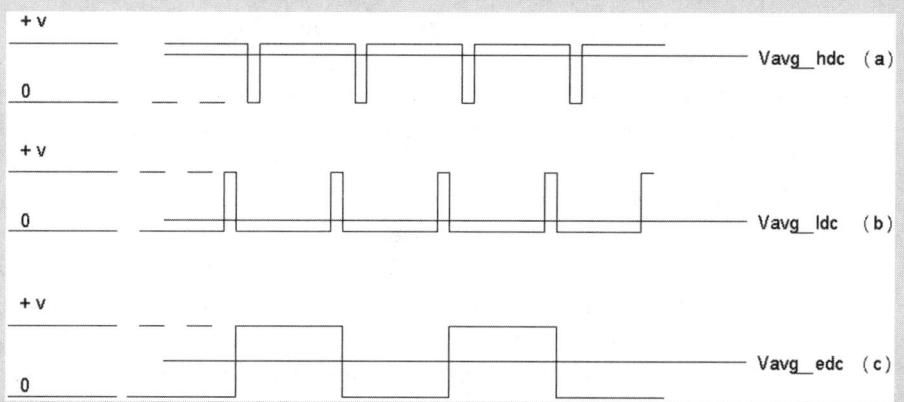

FIGURE 13-13   Respectively, average DC levels for (a) high, (b) low, and (c) even (50 percent) duty cycle pulse signals.

We see a high duty cycle waveform Figure 13-13(a) where the logic high level at +V stays on for a much longer time than when the waveform is at 0 volts. For pulses the average DC voltage is just the (duty cycle) × (+V), where duty cycle = $(t_{high\_level})/(t_{low\_level} + t_{high\_level})$, and where the frequency of the waveform is f = $1/(t_{low\_level} + t_{high\_level})$. This also means $(t_{low\_level} + t_{high\_level})$ = (1/f). For example, a 1000 Hz signal will have as $(t_{low\_level} + t_{high\_level})$ = (1/1000 Hz) = 1 millisecond. Note that millisecond = msec.

So, if t_{high_level} = 0.9 msec, then t_{low_level} = 0.1 msec since the sum of t_{low_level} and t_{high_level} must equal 1 msec.

The duty cycle then is $(t_{high\_level})/(t_{low\_level} + t_{high\_level})$ = 0.9 msec/(0.1 msec + 0.9 msec) or the duty cycle is 0.9/1.0 = 90%. Thus, the average DC voltage will be (90 percent) × (+V). For example, if +V is 6 volts, then the average DC voltage for this signal is 0.9 × 6 volts = 5.4 volts.

For Figure 13-13(b), a low duty cycle waveform, we can have as an example, t_{high_level} = 0.1 msec, and t_{low_level} = 0.9 msec. The duty cycle then is (0.1 msec)/(0.9 msec + 0.1 msec) = 0.1/1.0 or the duty cycle is 0.1, which is 10 percent.

In Figure 13-13(c) an even duty cycle waveform is where t_{low_level} = t_{high_level}; and the duty cycle is always 50 percent since $(t_{high\_level})/(t_{low\_level} + t_{high\_level})$ = $(t_{high\_level})/(t_{high\_level} + t_{high\_level})$ = ½ or 50.

These waveforms in Figure 13-13 will help us understand the next section on switch mode signals (pulses) in motor drive amplifiers.

Using a 555 to Drive Motors via Pulse-Width Modulation

We now turn our attention to using a 555 circuit, used as a monostable one-shot circuit to vary the pulse width that will control the direction and speed of a motor. By changing the pulse-width (of a 20 kHz high frequency pulse signal) the duty cycle varies, which in turn provides a different average DC voltage. See Figure 13-14.

This LM555 motor drive circuit was originally designed for an electronics class as an introduction to pulse-width modulation (PWM) signals. Instead of just using a DC power amplifier to drive the motor, a *switching Class D amplifier* is used because it has very high efficiency. The L239D chip sends out pulses of complementary duty cycle to the motor. For example, if V1 has an 80 percent duty cycle signal, then V2 has a 20 percent duty cycle pulse signal. Or if V1 has a 10 percent duty cycle pulse, then V2 has a 90 percent duty cycle signal. However, they can be equal when V1 has a 50 percent duty cycle pulse, which leads to a 50 percent duty signal for V2. The sum of the duty cycle of V1 and V2 must always total 100 percent.

Integrated circuit U2, L239D, is a bridge amplifier that receives pulse-width modulated input signals at pins 10 and 15. Because we want to be able to control the motor's direction (forward or reverse), the motor is connected to the bridge output terminals corresponding to pins 11 and 14. For example, if pin 14 has a DC voltage, V1, while pin 11 (V2) is near zero volts, we will measure a positive voltage across the motor from pin 14 to pin 11. However, if the DC voltage at pin 14 (V1) is near zero volts and there is a positive voltage V2 at pin 11, then there will be a negative voltage measured across pin 14 and pin 11. A negative voltage implies a reverse direction. Also, if the voltage is the same at pin 14 and pin 11, then there is no net voltage (or potential difference) across the motor, and it has stopped turning.

NOTE: Please use a small 12-volt DC motor.

Now let's look at Figure 13-14 starting with a relaxation oscillator U3A 74C14, which is a special logic inverter gate with Schmitt Trigger input characteristics. This type of input circuit allows for making an oscillator with a resistor (R8 + Pot 2) and capacitor C5. The oscillation frequency is $f \sim 1/[1.7(\text{R8} + \text{Pot 2})(\text{C5})]$. If you know the oscillation frequency, f, and have chosen a capacitor, then the resistance can be found as: $(\text{R8} + \text{Pot 2}) = 1/[1.7(f)(\text{C5})]$. If C5 = 0.0039 μf and f = 20 kHz, then (R8 + Pot 2) = $1/[1.7(20 \text{ kHz})(0.0039 \times 10^{-6} \text{ F})]$ or (R8 + Pot 2) = 7541Ω.

Note that the resistance and capacitance units in this equation is ohms for the resistance and farads for the capacitor; and that is why 0.0039 μf is expressed as 0.0039×10^{-6} F. Given that R8 = 1000Ω and Pot 2 can be adjusted from 0Ω to 10KΩ, there is sufficient range for providing a 20-kHz signal from U3A. That is, Pot 2 would be nominally adjusted to Pot 2 = 7541Ω – R8 = 7541Ω – 1000Ω = 6541Ω.

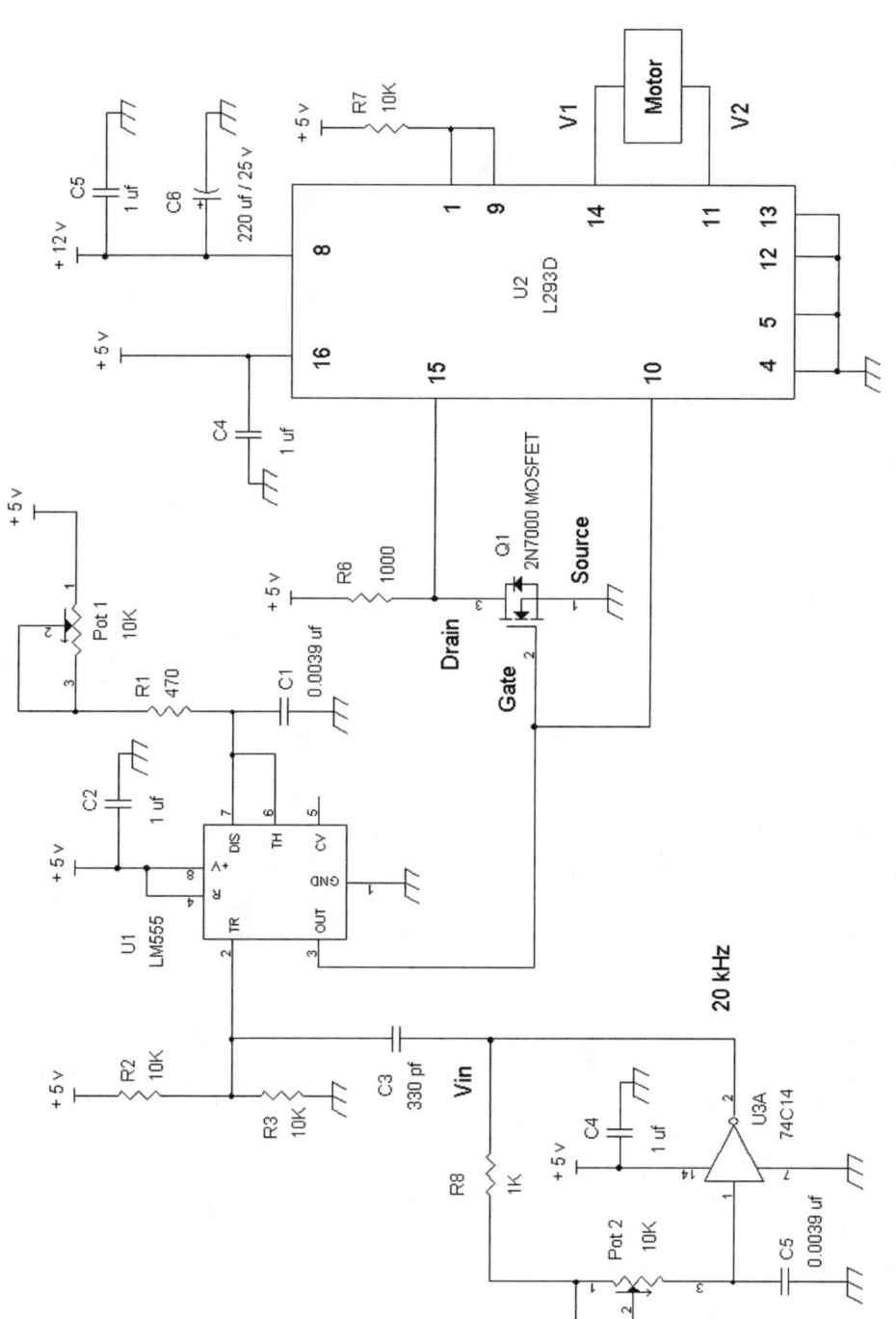

FIGURE 13-14 Motor drive circuit with Pot 1 for duty cycle adjustment to an H bridge amplifier.

339

But remember that C5 has a tolerance and the 74C14's oscillation frequency equation is approximate, so if you are out of adjustment range for Pot 2, you can increase R8 to 3300Ω, for example. To measure the frequency via pin 2 of U3A, you can use an oscilloscope, or a DVM (digital voltmeter) that includes measuring frequency. Most DVMs that cost over $30, such as the Extech MN26, will have this feature by setting the selector switch to Hz.

To provide a good range of the pulse-width such as from 10 percent duty cycle to 90 percent cycle from the 555 timer chip, the input signal needs a narrow pulse at pin 2 of U1. Capacitor C3 (330 pf) and resistors R2 and R3 form a pulse narrowing circuit that results in approximately a 3 μsec negative going trigger pulse. The pulse can be further narrowed to about 2 μsec if C3 → 220 pf. Pot 1 adjusts the pulse-width, and it is initially adjusted for about a 50 percent duty cycle to have the motor in the stopped position.

Since the motor drive amplifier U2, L239D, requires non-inverting and inverting 20-kHz pulsed signals, we provide the non-inverting pulsed signal from pin 3 of the 555 timer chip (U1) to pin 10 of U2. A MOSFET transistor Q1 logically inverts the pulsed signal from pin 3 U1. Q1's gate receives the non-inverting pulsed signal from pin 3 of U1, and with drain resistor R6, the drain terminal of Q1 provides a logically inverted pulsed signal to pin 15 of U2.

NOTE: Logically inverting a signal means low level goes to high level, and high level goes to low level. For example, a 0-volt low level logically inverted goes to +5 volts, and +5-volt logic level that is logically inverted goes to 0 volts. This is not the same as mathematically inverting a signal, which would mean +5 volts mathematically inverted is −5 volts, and mathematically +0 volt is inverted to −0 volt, but +0 volt = −0 volt mathematically.

Because the logic levels at pin 3 of an LM555 don't always swing to +5 volts with 5-volt supply, we can measure the duty cycle at the drain of Q1 with a DVM, or better yet with an analog volt ohm meter rated at 20,000 ohms per volt or more to ensure a reasonably high input resistance of the analog meter. That is, don't use a "cheap" 1000 ohms per volt analog meter for this measurement, which will load down the signal you are testing.

When you set Pot 1 for 50 percent duty cycle, the average voltage at the drain of Q1 should measure 50% × 5 volts = 2.5 volts. Then you can measure the voltage across pins 14 and 11 of U2, which should be close to zero volts. When you adjust Pot 1 in either direction, the duty cycle will change in a complementary manner at V1 and V2 (U2), and you should see a shift in DC voltage, which will start the motor turning. As the pulse's duty cycle at V1 becomes much greater than at V2, the motor's speed should increase. Originally, the circuit in Figure 13-14 used an external lab oscillator in place of U3A. The reason was to get the students to learn how to operate lab equipment. Because of this, there was not an inverter chip in the original design and that's why Q1 is used as a logic signal inverter. In the next circuit, we will see that since U3, 74C14, actually has six logic inverters (e.g., the 74C14 is known as a hex Schmitt Trigger inverter), we can substitute the Q1 circuit with another logic inverter in U3.

Also, to make this circuit a little more interesting, we can control the speed and direction of the motor turning with light sensors. See Figure 13-15. Here, we add two cadmium sulfide (CdS) light-dependent resistors, CdS1 and CdS2, which form a variable voltage divider or light-dependent potentiometer. The light sensors are physically spaced at least a few inches apart pointing up to the ceiling to receive equal light intensity. We then adjust Pot 1 for 50 percent duty cycle, which can be measured at U3D pin 8 for 2.5 volts DC. Then, if you cover or shade one of the light-dependent resistors more than the other, you should be able to control the duty cycle such that you can control the speed and direction of the rotation of the motor's shaft.

Also, note that R6 is used as a pull-up resistor at pin 3 of the LM555 to ensure that the logic high voltage goes all the way up to +5 volts. This resistor is added because the 74C14 logic inverter requires a high-input voltage to be confirmed as a logic high signal. Standard logic inverter chips, such as the 74C04 chips, only need about 3.0 volts for logic high, whereas the Schmitt Trigger inverter chips, 74C14, require about 3.75 volts for a logic high-input signal.

In Figures 13-14 or 13-15, waveforms from pin 2 U3A (74C14 oscillator), input pin 2 of the LM555 chip with pulse narrowing (via R2, R3, and C3), output pin 3 of the LM555, and the inverted signal at input pin 15 of the L293D motor drive amplifier are shown in Figures 13-16(a) to (d) respectively.

NOTE: On the L293D chip there was an error in the 1990 and 2002 datasheets that had the motor amplifier's power pin as pin 3 in their application notes, which is incorrect. Also the 1990 and 2002 datasheets show pin 3 in two places, which is in error. The subsequent L293D data sheets in 2004 and 2016 corrected the error and **the correct motor amplifier power pin is pin 8** (as shown in Figures 13-14 and 13-15). The motor drive amplifier pin 8 has 5 volts to 30 volts, and +12 volts is typical. Pin 16 needs to be +5 volts to match the LM555's supply voltage. *So the lesson here is to double check datasheets. Sometimes a datasheet may include typos.*

Also, the 20 kHz pulse frequency was chosen so that the motor could not respond to such a fast changing signal. Instead the motor's mechanical inertia and coil inductance filter out the 20 kHz AC signal but only responds to the average DC signal from the (high frequency) pulse waveform (e.g., due to the duty cycle).

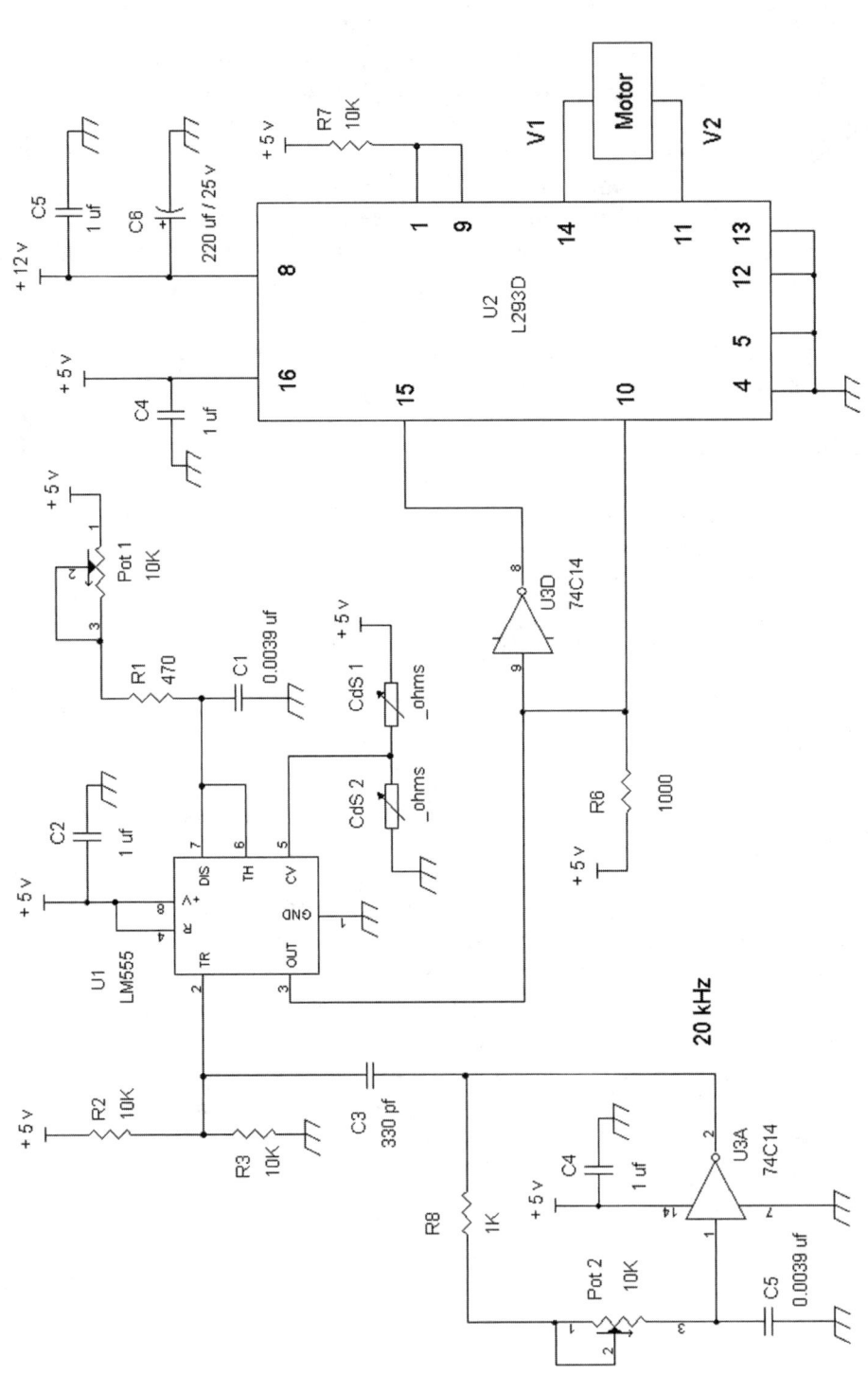

FIGURE 13-15 Motor drive with light sensors, CdS1 and CdS2.

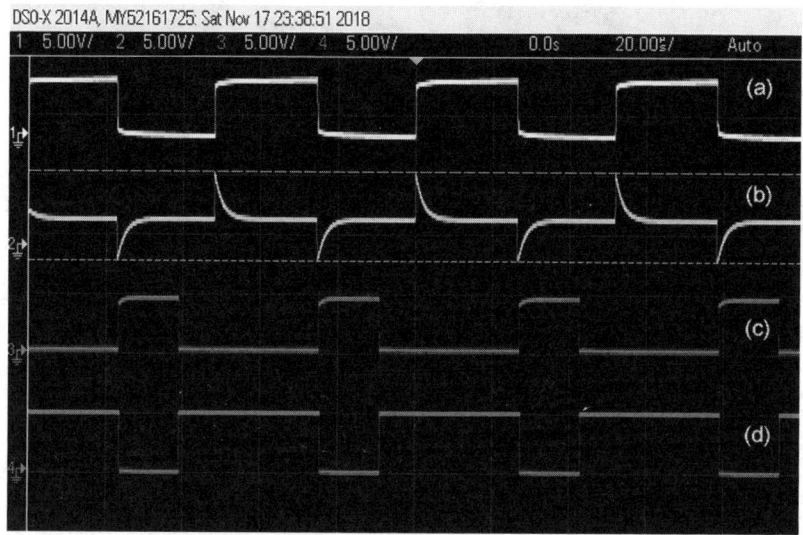

FIGURE 13-16 Waveforms at (a) pin 2 U3A, (b) pin 2 U1, (c) pin 3 U1, and (d) pin 15 U2 of the circuit from Figures 13-14 or 13-15.

We can see that the top trace of Figure 13-16(a) shows that the 74C14 CMOS oscillator does a pretty good job in providing approximately a 50 percent duty cycle waveform whose frequency is about 20 kHz. To narrow the negative cycle (logic low) portion of this waveform capacitor, C3 along with R2 and R3 form a high-pass filter as shown in the circuits with Figures 13-14 and 13-15. The narrowed negative going pulse in the (b) waveform is in the order of about 2 μsec, which is much less than the ~ 25 μsec negative cycle (low state) of the top trace waveform (a). In this example, the pulse width is set to about a 25 percent duty cycle as shown in the (c) waveform, and the inverted (logic) signal shown in the (d) waveform has a 75 percent duty cycle. Now let's look at how the duty cycle and average voltage works via Figures 13-17, 13-18, and 13-19.

In Figure 13-17(a), the oscillator's output is used as a reference signal. The waveform below it (b) shows the LM555's output signal. To drive the motor, voltages V1 and V2 are applied to it, and it is the difference in average voltage (V1avg – V2avg) that determines the rotation speed and direction of the motor's shaft. As we can see, the V1 waveform has about 75 percent duty cycle and the V2 waveform is at about 25 percent duty cycle.

Thus, the average voltages for a 12-volt supply (into pin 8 of the L293D chip) are about 12 volts × 0.75 = 9 volts = V1avg, and 12 volts × 0.25 = 3 volts = V2avg. We now have a voltage of (V1avg – V2avg) across the motor as shown in Figures 13-14 or 13-15, which is (9 – 3) volts, or +6 volts. When we change the pulse-width from the 555 chip to 50 percent duty cycle as shown in Figure 13-18(b), we see that the average voltages for V1 and V2 are the same. See Figure 13-18.

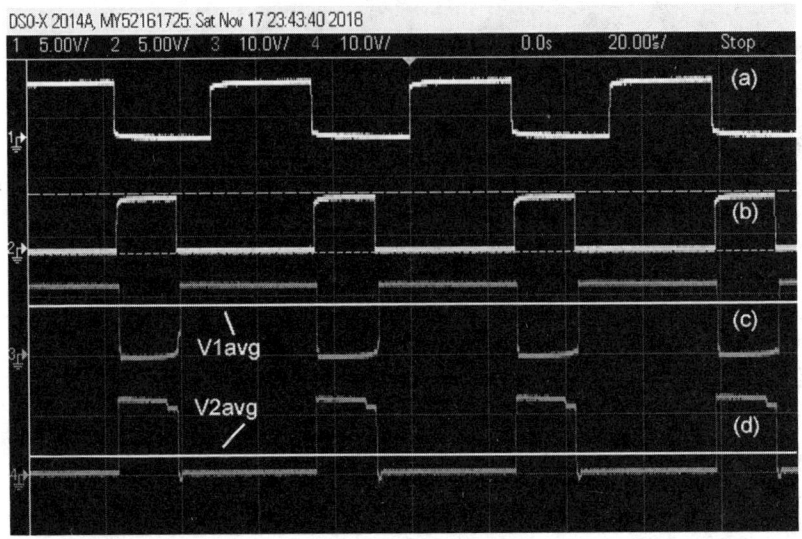

FIGURE 13-17   U3A pin 2 oscillator (a), U1 pin 3 output of LM555 (b), output at V1 pin 14 U2 (c) and its average voltage, and (d) output at V2 pin 11 U2 and its average voltage.

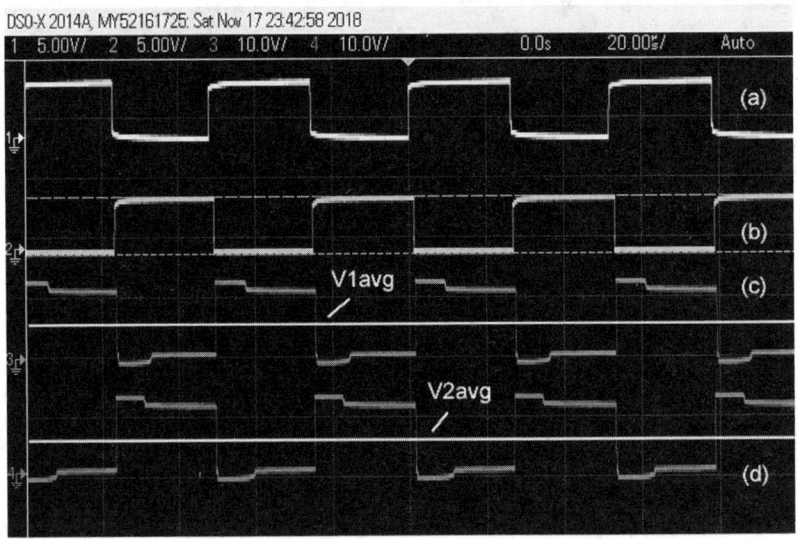

FIGURE 13-18   Top trace (a) is the oscillator signal at pin 2 of U3A, and increasing the 555's output waveform (b) to 50 percent duty cycle causes the motor drive waveforms (c) and (d) to also have 50 percent duty cycle.

By having the 555 timer's output waveform increased in pulse-width to 50 percent, voltages V1avg = V2avg. Thus, the voltage across the motor is (V1avg – V2avg) = 0 volts, and the rotation in the motor's shaft is no longer turning, and thus it is stopped.

When we then increase the LM555's pulse-width further as shown in Figure 13-19 (b), pulse-width at V1 has decreased, while V2's pulse-width has increased. See Figures 13-19 (c) and (d), respectively. V1's duty cycle is 25 percent so:

V1avg = 25% × 12 volts or 3 volts; and V2's duty cycle is now

75 percent with V2avg = 75% × 12 volts or 9 volts.

The voltage across the motor is: (V1avg – V2avg) = (3 – 9) volts, or –6 volts. With the minus sign (in –6 volts), this means the DC motor's shaft is rotating in the opposite direction.

A bridge amplifier like the L293D allows for controlling the speed and direction of the motor's drive shaft even though, if we note, this amplifier is running off positive power supplies. That is, it does not need a negative supply to run the motor in reverse.

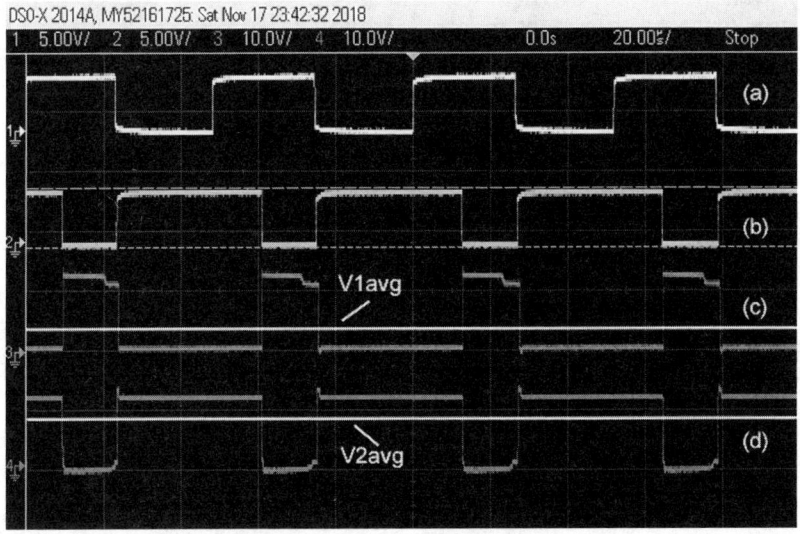

FIGURE 13-19   Top trace (a) is the oscillator signal at pin 2 of U3A, and increasing the 555's output waveform (b) to 75 percent duty cycle causes the motor drive V1 waveform (c) to have 25 percent duty cycle with V2's waveform (d) to have 75 percent duty cycle.

Note that the average voltages at V1 and V2 of U2 (L293D) or at pin 3 of U1 (LM555) can be measured with most DVMs in the range of 10 percent to 99 percent with good accuracy. The reason is because the oscillator frequency is at 20 kHz, where the DVMs tend to give an average voltage. If you suspect that a DVM is giving inaccurate readings for these measurements, you can measure the average voltage with an analog voltmeter. Again, try using an analog VOM that has at least 20,000 ohms per volt reading as to not load down the signals you are measuring.

Summary of Troubleshooting Techniques

- You can test timing circuits with LEDs to determine approximately how long the pulse is when the 555 as configured as a pulse generator (a.k.a., mono-stable mode).
- A loudspeaker or AM radio can be used to determine if the 555 circuit is oscillating. If the frequency is higher than what we can hear, you can connect the 555 circuit to a frequency divider (74HC393 or equivalent). Then couple a speaker amplifier to the output of the frequency divider.
- For a periodic pulse signal via the LM555, if you know the supply voltage and frequency, you can determine the pulse-width approximately by measuring the duty cycle indirectly by measuring the average voltage. You can measure the frequency of the periodic pulse signal with a DVM that has a frequency counter feature or use a frequency counter. For an approximate frequency measurement, an oscilloscope can be used.

CHAPTER **14**

Troubleshooting Other Circuits, Including Kits and Projects

So far, we have seen various ways to troubleshoot mostly basic circuits, such as operational amplifiers, 555 timer chips, LEDs, and transistors. For basic troubleshooting, there are always basic techniques to look at, such as checking the supply voltages, components' pin outs, and DC conditions. We also need to ensure that the power supply decoupling capacitors (e.g., 0.01 µf to 10 µf) have short leads and connected very close to the ICs' power and ground pins.

In this chapter we will look at circuits that may or may not work so well. These include electronic kits such as light-sensing circuits, light transceivers (transmit and receive circuits), temperature-sensing circuits, speaker amplifiers using transistors, and IC audio power amplifiers.

Also, we will explore circuits pertaining to power supplies, microphone preamplifiers, and oscillators.

Component Kits and Test Equipment

Before we start assembling boards/kits, there are at least a few items you should have on hand. The component kits, such as assorted capacitors, resistors, transistors, LEDs, etc., can usually be purchased on the web (e.g., www.amazon.com) for less than $15 per kit. Typically, the assorted parts, such as 300 assorted red, green, blue, and white LEDs, will be about $10. Having spare parts will allow you to repair, modify, or augment the electronic circuits you experiment with. Here are the components you should have:

- Ceramic capacitor kit from 10 pf to 1 µf or more.
- Quarter-watt resistor kit, either 5 percent, 2 percent, or 1 percent values from 1Ω to 1MΩ.

- Electrolytic capacitor kit of at least 16 volts rating, but 25 working volts or more is preferable for values from 1 µf to 470 µf, although 1 µf to at least 100 µf will work fine.
- Assorted NPN and PNP transistor kits such as 2N3904, 2N3906, PN2222, and PN2907, plus any power transistors.
- If you are into radio frequency circuits, you could purchase a kit of fixed value inductors from 1 µH to at least 1000 µH.
- A basic digital multimeter (DVM), or preferably a DVM that includes a capacitance meter and a frequency counter.
- Another measuring device you can purchase for < $50 is an inductance capacitance meter. You will be able to confirm a coil's inductance and capacitor's capacitance value. This will be very useful should you work with radio circuits, but also with some audio circuits where large value inductors (e.g., 1 Henry coils) are used.
- A good dual-power variable voltage power supply from 0 to ± 12 volts or more, with a maximum current output of at least 500 mA. You can purchase, for example, two independent adjustable voltage power supplies and with adjustable current limiting. Setting a current limit (e.g., to < 100 mA) is desirable to avoid destroying electronic components should there be a mistake in your wiring or parts installation.
- An alternative to power supplies in most cases is to use rechargeable batteries. To prevent damage to the batteries in case of an accidental short-circuit, wire a series 0.51Ω to 2.2Ω quarter-watt resistor with the battery. This way, if there is a short-circuit, the resistor will act like a fuse and burn out.
- A simple signal generator that provides sine waves and square waves. Sine waves are often used in audio and RF circuits, while troubleshooting some digital or logic circuits will require pulse waveforms such as square-wave signals.
- Finally, if you can afford an oscilloscope, get a two-channel version of at least 25 MHz bandwidth. For example, a new 100-MHz two-channel scope (Hantek) costs around $250 US or so in 2019 on Amazon.com. For good all-around troubleshooting, a 50-MHz or 100-MHz two-channel oscilloscope will work fine.

The reason for having extra parts is that some electronic kits may come with incorrect value resistors, capacitors, or other parts. Because assorted resistors, capacitors, inductors, transistors, and LEDs are generally under $10 each, it's a good investment to have spare parts handy. Not only can these parts be used in the electronic kit projects, but also for your own experimenting.

Of course, to build your electronic kits, you will also need a soldering pencil around 25 watts to 40 watts (e.g., a Weller SP23 or SP25 series with flat blade and conical point tips), or a soldering station that has temperature control (e.g., a Weller, Hakko, or Metcal solder station that generally costs > $100 US), solder such as 60/40 or 63/37 tin to lead composition of 0.020 inch to 0.031inch diameter with rosin flux, a pair of small long-nose pliers, a pair of small diagonal cutters, a wire stripper for various wire sizes, small screwdrivers for flat blade and Phillips screws or bolts, and wire from 18 AWG and 22 AWG, preferably some with single solid strand and stranded varieties.

LED and Sensor Kits

Our first electronics kit to build and test is a low power light switch that turns on in the presence of a dark room. Otherwise, with a normally lit room the switch is open circuited. This circuit uses a cadmium sulfide light-dependent resistor (LDR), which exhibits lower resistance when more light shines on it. In "total" darkness, the LDR will provide $\geq 1M\Omega$, while in a normally lit room the resistance will be $< 10k\Omega$. You can test your LDR using a DVM by setting the resistance to $200k\Omega$ (or $20k\Omega$). See Figure 14-1, which shows examples of two LDRs measured with DVMs. As shown on the left side where the LDR's front side is turned toward the back wall with less light, the left side DVM shows about $6.3k\Omega$, whereas the LDR on the right (with its front side turned toward the light source) receives more light and measures at $2.8k\Omega$. Also note that the LDR on the right side shows a serpentine or zigzag pattern, which is the front side of the LDR where light is sensed.

NOTE: The LDR's two leads can be connected either way because they are not polarized.

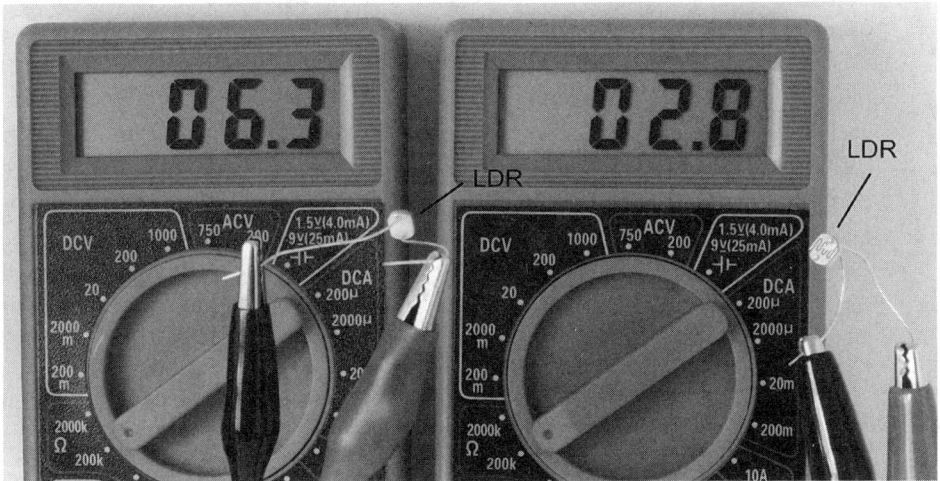

FIGURE 14-1 LDR's resistance measurements. On the left side, the LDR is facing less light and measuring 6.3kΩ. On the right the LDR is sensing more light with a lower 2.8kΩ resistance; also note the serpentine pattern on the LDR's front/face side. Again, the LDR is a non-polarized device, and their two leads are to be treated like the two leads of a standard resistor.

Now let's take a look at a DIY light sensor kit with a cadmium sulfide LDR. See Figure 14-2.

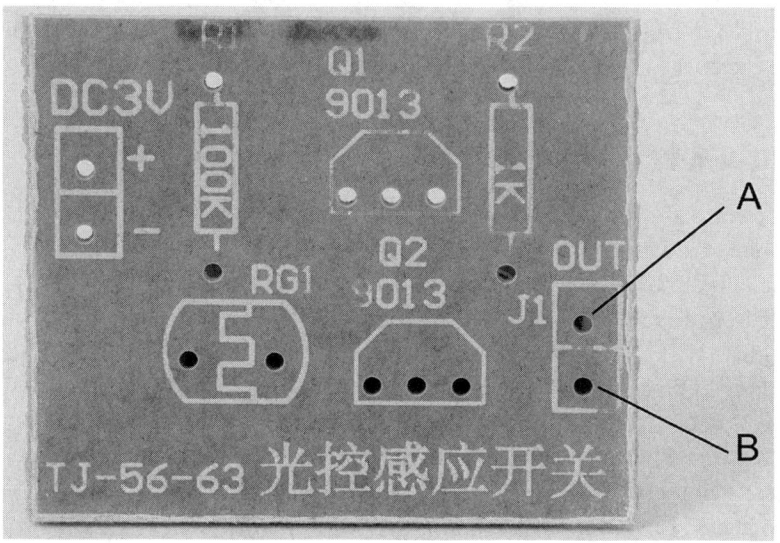

FIGURE 14-2   The light sensor kit's unloaded board with "A" and "B" associated with the circuit shown in Figure 14-3.

As a reference, you should take a picture of the blank printed circuit board first. This way, you have a record of where each resistor, capacitor, transistor, etc. goes. Once you load and solder in the parts, the silkscreen reference values or reference designations may be obscured. By having a picture of the blank board, you can quickly check whether you have loaded the parts in the correct locations. For example, if you loaded R2 with some value *other* than 1kΩ, you can refer to Figure 14-2 and confirm that R2 should be a 1kΩ. Now let's look at the schematic in Figure 14-3.

This DIY light sensor project includes the transistors Q1 and Q2; light-dependent resistor, RG1; and resistors R1 (100kΩ) and R2 (1kΩ). Referring to the original circuit in Figure 14-3, we see that the connections "A" and "B" are used for turning on a circuit with the 3-volt supply (BT1). In this case, two AA cells in series provide the 3 volts. When light strikes the light-dependent resistor, RG1, it provides a sufficiently low resistance to provide base current, IB1, to Q1. Q1's base current is amplified by the current gain (β_1) of Q1, which is > 100, such that the voltage across the collector and emitter of Q1 is then [3 volts − (β_1 IB1)R2], and wherein the lowest collector voltage is 0 volts (Note that R2 = 1kΩ.). That is, in reality, the quantity [3 volts − (β_1 IB1)1kΩ] is ≥ 0 volts, but [3 volts − (β_1 IB1)1kΩ] is ≤ 3 volts (or BT1).

So, when the sensor is a room that is at least normally lit (e.g., normal room light to "blazing" sunlight), RG1 has low resistance and turns on Q1 as a switch such that Q1's collector-to-emitter voltage ~ 0 volts. Because Q1's collector is connected to Q2's base, this means that Q2's base-to-emitter voltage also ~ 0 volts, which means

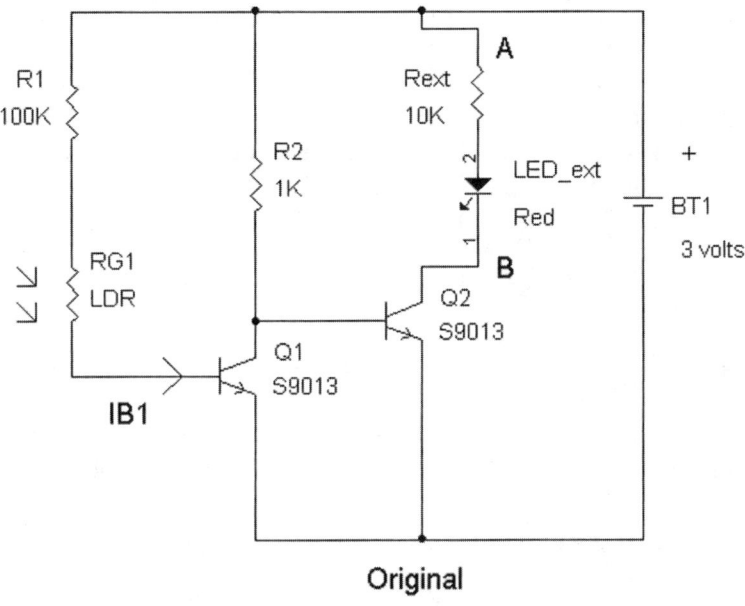

Original

FIGURE 14-3 Schematic of the original circuit, which corresponds to Figure 14-2 with "A" and "B" terminals related to J1 OUT.

Q2 acts like an open circuit. Therefore Q2's collector does not draw any current and the external light-emitting diode, LED_ext, and series resistor R_ext have no current flow and LED_ext is thus not lit. However, when the light sensor circuit receives no light (e.g., light sensor module is placed in a dark room), Q1's base current drops sufficiently low such that Q1's collector current is low enough that Q1 is essentially an open circuit from Q1's collector and emitter terminals. This means R2 (1kΩ) acts like a base-driving resistor to Q2's base, which then turns on Q2 as a switch to turn on LED_ext. Thus, with sufficient base current from R2, the Q2's collector voltage is close to 0 volts with respect to Q2's emitter or with respect to the (−) terminal of BT1.

To troubleshoot this circuit, you can use a digital voltmeter (DVM) and connect the negative (black) test lead to the (−) terminal to BT1, and then with the DVM's red lead measure voltages at the base and collector terminals of Q1 and then at the base and collector terminals of Q2. When in a lit room, Q1's base voltage should be ~ 0.6 volts ± 20 percent, and Q1's collector voltage and Q2's base voltage should be less than 0.6 volts but no lower than 0 volts (e.g., no negative voltage reading at the collector of Q1). If the LED is not lit, you can test the sensor circuit by covering the light-dependent resistor, RG1, with your hand or place an opaque material over the RG1 that blocks out the light. If the light-dependent resistor is very sensitive (e.g., able to provide a low resistance even when there is very little light shining on it) or the Q1's current gain is very high, you may have a harder time getting the external LED to light up in the dark.

One solution is to solder a resistor about 47kΩ (Rext_2) across the base and emitter of Q1, which should desensitize the current gain of Q1. See the #1 modified circuit with Rext_2 having a 47kΩ value in Figure 14-4. That is, RG1 now has to have a high enough resistance with R1 (100kΩ) to form a voltage divider with this extra Q1 base-to-emitter resistor Rext_2 (47kΩ) such that the voltage across Q1's base-emitter voltage is $VBE_{Q1} < 0.6$ volt.

For the condition that $VBE_{Q1} < 0.6$ volt, the following equation holds:

$$VBE_{Q1} = 3 \text{ volts } [47k\Omega/(47k\Omega + R1 + RG1)]$$

or

$$VBE_{Q1} = 3 \text{ volts } [47k\Omega/(47k\Omega + 100k\Omega + RG1)]$$

We want $VBE_{Q1} < 0.6$ volt so that Q1 is off and Q2 is turned on. Suppose in a dark room, RG1 = 200kΩ, then $VBE_{Q1} = 3$ volts $[47k\Omega/(47k\Omega + 100k\Omega + \mathbf{200k\Omega})]$ which leads to:

$$VBE_{Q1} = 3 \text{ volts } (47k\Omega/347k\Omega) \text{ or } VBE_{Q1} = 0.4 \text{ volt}$$

which is sufficient to turn off Q1 (e.g., collector of Q1 is an open circuit with the Q1's emitter) and allow R1 to act as a base-driving resistor to the base of Q2. In general, the LDR is capable of having a dark (e.g., no light) resistance at least 1MΩ, which will provide an even lower than 0.4-volt VBEQ1.

We can now show a slight modification of the sensor circuit in Figure 14-4, the #2 modified circuit where we use a variable resistor (POT1_ext) across Q1's base-emitter junction.

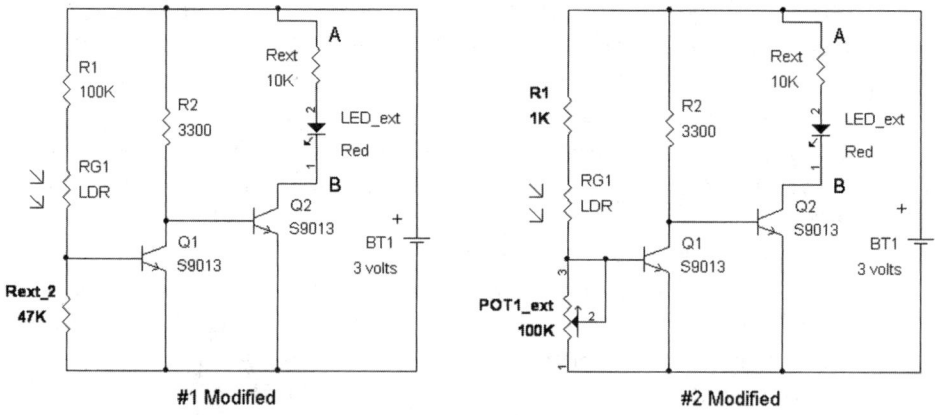

FIGURE 14-4 #1 modified circuit with Rext_2 = 47kΩ; and #2 modified circuit with variable/adjustable resistor POT1_ext.

In Figure 14-4 with the #2 modified circuit, you can choose POT1_ext with resistance values of 20kΩ to 100kΩ. As shown, POT_ext = 100kΩ, and resistor R1 is changed from 100kΩ to 1kΩ because RG1 can vary (with light intensity) from about 1MΩ to 100Ω. So R1 = 1kΩ is used as a current limiting resistor such that if RG1 = 100Ω, RG1's current is limited less than 3 mA. Resistor R2 is increased from 1kΩ to 3.3kΩ for increasing the sensitivity of the circuit if needed. However, the POT1 allows for a wide range of adjustments for light and dark turn-on and turn-off thresholds. Typical set values for POT1 are in the 1kΩ to 10kΩ range.

NOTE: Not every light-dependent resistor (LDR) RG1 has the same resistance per light intensity. You can generally find the data or specification sheet on the web for a specific LDR.

Figure 14-5 shows the two circuits including the modified version with a 100kΩ variable resistor (POT1 in the schematic) denoted as a potentiometer labeled as "104" on the right side.

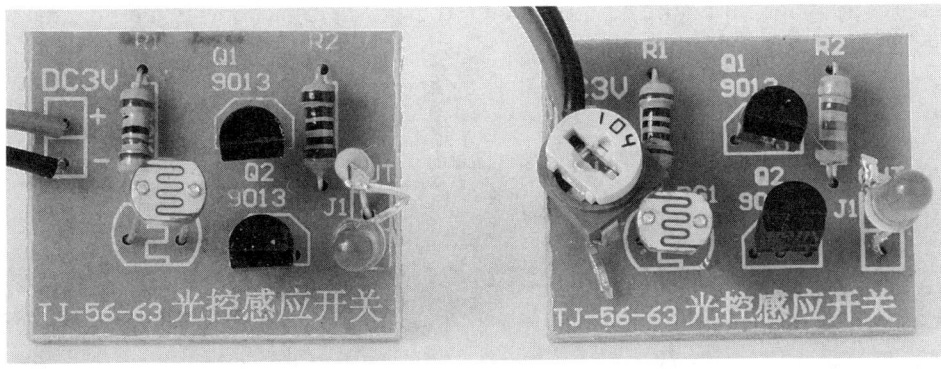

FIGURE 14-5   The LDR sensing circuits, original and modified, with external LED indicator lamp. Notice RG1, the LDR device (with its serpentine pattern), is to the left side of Q2.

It should be noted that either LDR circuit in Figure 14-5 can have the following:

- The transistors Q1 and Q2 may be substituted with virtually any general-purpose NPN transistor with a pin out of emitter-base collector. This means a 2N3904, 2N4401, 2N4124, or MPS2222 type can be used. As long as the current gain, β, is greater than 30, the circuits will work.
- You can replace the RG1, the light-dependent resistor, with other types or sizes as long as the resistance ranges from about 100kΩ to 100Ω, from dark to bright light shining into it.
- Although the LDR circuit shown works with a 3-volt supply, it can work at higher power supply voltages up to 15 volts or more. For the modified version with POT1, make sure the series resistor is scaled accordingly for 1kΩ/3 volt

supply. For example, if you use a 12-volt supply, R1 = (12 volt /3 volt) × 1kΩ
or R1 → 4kΩ (e.g., either 3900Ω or 4300Ω is close enough to 4kΩ).

A Quick Detour with the LM386 Audio Power Amplifier IC

We will now look into a very popular audio power amplifier that may be used to drive
a wide range of loudspeakers from the very small ones used in laptop computers to
those used in "book end" speakers, for example. The commonly used **LM386N-1** can
deliver up to about half a watt with a 9-volt supply. Typically, this amplifier runs off
batteries or a regulated power supply between 4 volts and 12 volts. It can work off a
raw supply, but caution must be observed to ensure the absolute maximum 15-volt
rating is not exceeded. If higher voltages are required, you can order the **LM386N-4**
part number that has a 22-volt absolute maximum supply rating.

When using batteries or regulated power supplies ≤ 12 volts, a circuit such as the
one shown in Figure 14-6 can be used. Note all capacitors should be rated at least
25 volts.

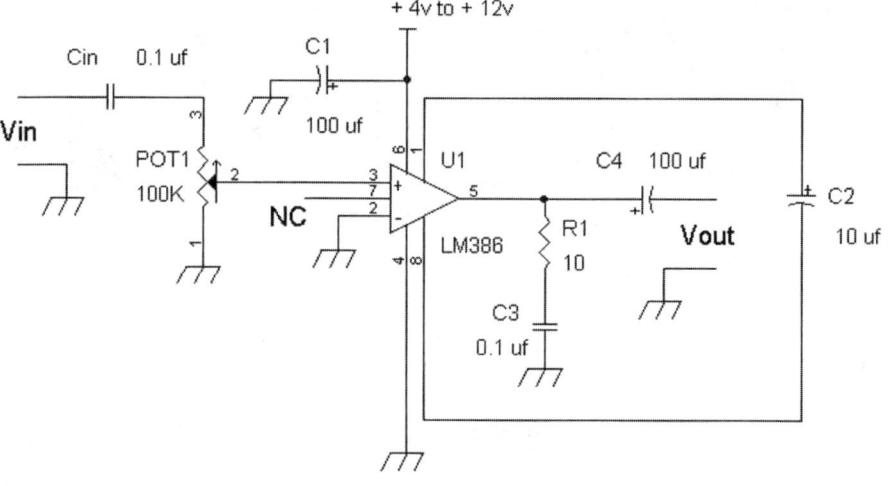

FIGURE 14-6   An example LM386 audio amplifier with battery or regulated supply with
input gain control POT1. The maximum gain is 200 with C2, and 20 with C2 removed.

For battery or regulated power supply operation, pin 7 is left as a no-connect,
NC, where you can optionally add an extra power supply decoupling capacitor that
will be shown in Figure 14-7. One of the most important aspects of this amplifier is
the series resistor-capacitor "snubbing" network R1 and C3. This snubbing network
provides a resistive load at high frequencies so that the amplifier is free of oscillations should the output Vout be connected to long wires or to a loudspeaker system
that represents an "unstable" or reactive load. Without R1 and C3 as configured, it is
possible that the amplifier can oscillate at high frequencies. With a speaker con-

nected, you can probe Vout with an oscilloscope to confirm no high-frequency signals are present. In general, R1 has a value between 4.7Ω and 10Ω and C3 has a range from 0.047 μf to 0.22 μf (e.g., ceramic or film capacitors ≥ 50 volts).

One of the worst implementations of the snubbing network is to set R1 = 0 Ω and have Vout connected directly to C3, which will increase the chance for oscillation. Please make sure R1 is in the 4.7Ω to 10Ω range in series with C3 to ensure stable operation.

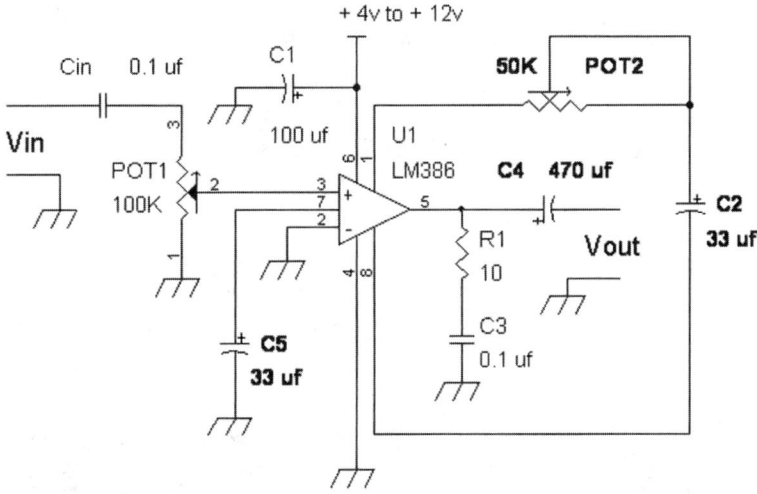

FIGURE 14-7   A more generalized LM386 audio amplifier circuit with decoupling capacitor C5 and feedback gain control POT2 to set the maximum gain in the range of 20 to 200.

If you are using a 6-volt or 9-volt DC wall charger, chances are that it will provide DC with some ripple (e.g., hum) due to its simple rectifier and single capacitor circuit. If you are unsure how high a DC voltage the wall charger will provide, use an **LM386N-4** part number and keep the wall charger's voltage rating ≤ 12 volts. The reason is that a wall charger's voltage under light current load can provide up to 40 percent more voltage. For example, a 12-volt DC wall charger can give up to 17 volts for its open-circuit voltage. Since the **LM386N-1** is rated at 15 volts, it can be damaged by a 12-volt DC wall charger that can give out 17 volts when lightly loaded. For safety reasons, a 9-volt or 6-volt version may be a better choice to avoid damage to the **LM386N-1**.

The LM386 can "ignore" the hum from the power supply, providing C5 is connected to pin 7 as shown in Figure 14-7. The larger the capacitance value for C5, the lower the hum will appear at Vout. Typically, C5 = 33 μf will work fine, but larger values such as 100 μf to 470 μf will reduce hum even further.

POT2 shown as 50kΩ can actually be a value in the range of 10kΩ to 20kΩ for finer or easier control of the voltage gain. In some cases, if the maximum gain is too high at 200, which can pick up background noise in Vin, then you can lower the maximum gain to about 20 via turning/setting POT2 to 50kΩ, its maximum resistance value. See Figure 14-7.

Alternatively, you can remove C2 (33 μf) to lower the voltage gain to about 20, which would disconnect POT2 from the LM386.

Again, note that the series resistor-capacitor snubbing network, R1 and C3, is essential for ensuring that the amplifier does not oscillate at high frequencies when Vout is connected to a loudspeaker. R1 is required and should be from 4.7Ω to 10Ω, and if R1 = 0Ω, then the LM386 amplifier will likely cause oscillations at Vout.

We now will look at a light transmitter and receiver system that uses an LM386.

Photonics: A Light Transceiver System

Our next DIY kits involve a modulated light source (LED) for wirelessly transmitting audio signals to a photo-detector with playback on a loudspeaker. See Figure 14-8. The transmitter and receiver kit boards are both labeled ICSK054A on the bottom sides; and the transmitter board has provisions for two LEDs and a TO-92 transistor, while the receiver board has an 8-pin IC.

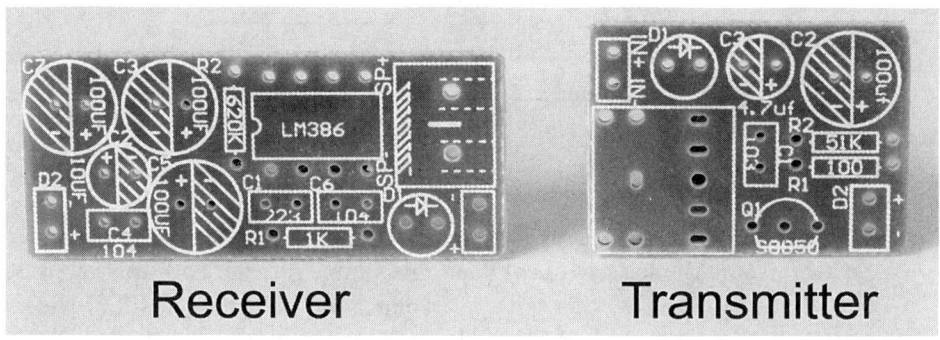

Receiver   **Transmitter**

FIGURE 14-8   Light receiver board with an LM386 8-pin IC, and transmitter board with a Q1 transistor, S8050 (NPN with 1.5-amp maximum collector current).

Now let's take a look at the transmitter's schematic in Figure 14-9.

In Figure 14-9, the red LED, D1, works not only as a visible light emitter, but also to let you know that the LEDs, D1 and D2, are installed correctly. If D1 does not light up, you need to check the pin outs of D1 and D2. Chances are either or both D1 and D2 have been installed reversed. Resistor R1 acts as the LEDs' current-limiting resistor, but we can use it for monitoring the LEDs' current. By measuring the voltage across R1, VR1, as shown, the LED currents will be VR1/100Ω. For example, if VR1 = 2.7 volts, then the LED current flow through D1 and D2 will be 2.7v/100Ω or 27 mA. *Also, just remember that the longer lead of an LED is the anode or (+) terminal.* Q1's DC emitter voltage is just the sum of the forward voltages across a red

LED and an IR (Infrared) LED, which is about (1.8 volts +1.3 volts), respectively, or about 3.1 volts.

Now let's take a look at the original receiver circuit as shown in Figure 14-10.

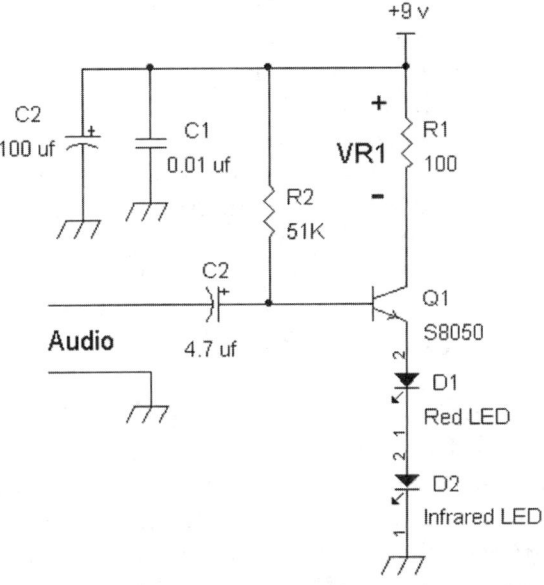

FIGURE 14-9   An emitter follower amplifier, Q1, driving two LEDs in series with an audio signal modulation (e.g., via a headphone output signal from a phone or player) via Q1's base. With +9 volts supply and a random sample of Q1 (S8050), VR1 ~ +2.7 volts.

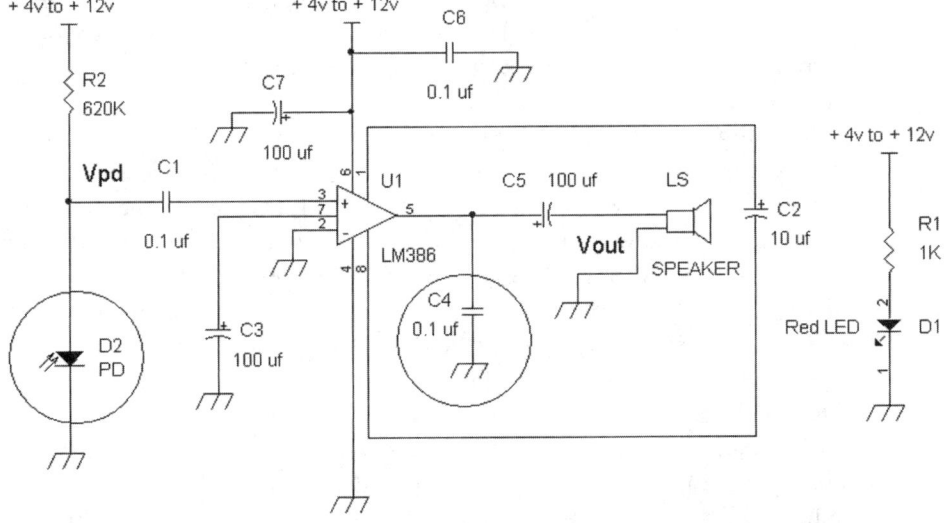

FIGURE 14-10   Original photodiode receiver design with circled problem areas.

The original design shown in Figure 14-10 needs to be modified for reliable operation. The first problem is C4 is connected directly to the output pin 5 of the LM386 amplifier, which can lead to instability or oscillations in the output signal Vout. A series resistor will be needed with C4 to correct this problem.

The second problem shown is that D2, the photodiode, is biased incorrectly from anode to cathode via R2 in forward bias mode, when in fact D2 should be biased in reverse bias mode (e.g., positive voltage at the cathode with respect to the anode). With a positive bias voltage via R2 into the anode of D2, the photodiode is conducting DC current set by R2. The DC voltage at the anode of D2 is about $+0.7$ volts. When a photodiode is biased incorrectly in the forward bias mode, the photodiode's signal current is "short-circuited" back into the photodiode's internal low resistance, which results in a very small signal at node Vpd. We can fix this problem by simply reversing the anode and cathode leads of photodiode D2 (see Figure 14-11).

If you happen to build the circuit as shown in Figure 14-10 (without any circuit correction to D2), you will still receive signals from Figure 14-9's transmitter modulated light signals, but the speaker's (LS) volume will be low.

Figure 14-11 shows the circuit corrections, which are shown inside the rectangular outlines. As shown, we see now that the photodiode, D2, is reversed in connection such that a positive voltage is provided to the cathode. Depending on the light shining on the photodiode, the DC voltage at D2 is [Vsupply $- (I_{pd} \times R2)$].

For example, if the supply voltage is 6 volts and the photodiode current is 1 μA with R2 = 620kΩ, the D2 cathode voltage is 6 v $-$ 1μA $\times$ 620kΩ = (6 volts $-$ 0.62 volt), or $+5.38$ volts at D2's cathode. Resistor R2 also sets the receiver gain since it is a load resistor for the photodiode current received by D2. For lower gain, R2 can be set to about 10kΩ, and for maximum gain R2 can be 620kΩ, the original value. Note that the input resistance of the LM386 is already 50kΩ, which is in parallel with R2 AC signal-wise.

Rext2 and Cext1 are optional if the power supply is from a wall charger whose DC voltage includes ripple or hum. Otherwise, you can remove Cext1, and set Rext2 = 0Ω as a wire.

You will get the furthest range between the transmitter and receiver with the component values shown in Figure 14-9 and Figure 14-11. The modified photodiode receiver in Figure 14-11 is very sensitive and will pick up the flickering from room lights and produce hum or buzzing noise. If you turn off the lights, the buzzing noise will be reduced dramatically. Also, if the photodiode receiver is playing too loud for comfort, you can reduce the audio gain tenfold by removing C2, or alternatively reduce the sensitivity even further by having R2 = 4700Ω.

The correct pin out for the photodiode D2 is shown in Figure 14-12, which includes a photo of the transmitter and receiver with loudspeaker. Again, the transmitter's input jack will accept line-level audio signals from a phone, digital player, or radio. Signals from a microphone are too small in amplitude and will require a preamplifier with a gain typically between 30 (electret microphone) and 300 (dynamic microphone) depending on the type of microphone.

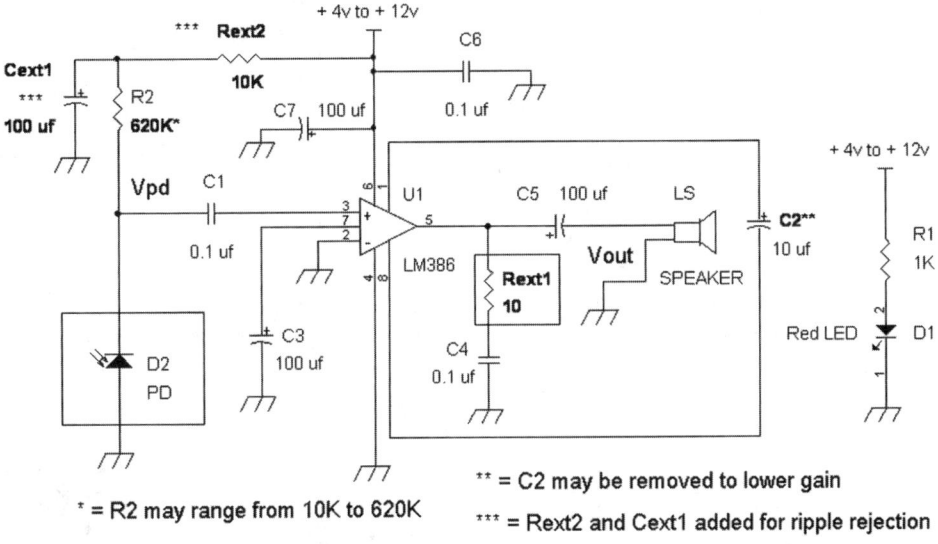

FIGURE 14-11   Modified photodiode receiver circuit to avoid oscillations by adding Rext1 and to ensure D2 (photodiode) is in the correct biasing mode. Corrections are pointed out by the rectangles surrounding D2 and Rext1.

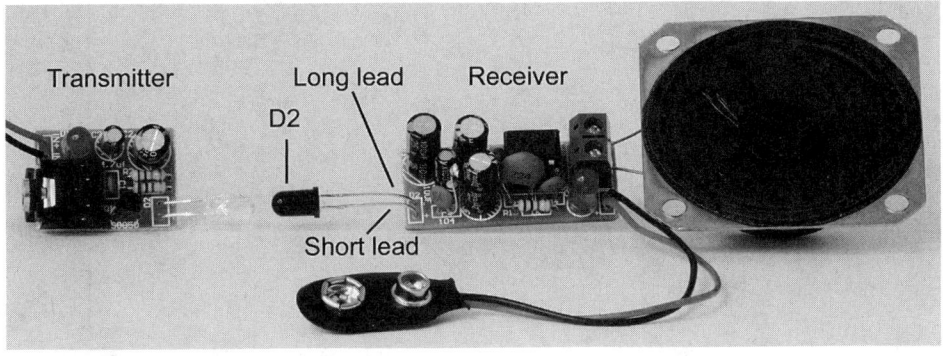

FIGURE 14-12   Photodiode D2's correct pin outs as soldered to the receiver board.

See Figure 14-13, which shows examples of microphone preamplifiers that amplify signals from microphones to about the same level that comes from a digital player or radio.

For a dynamic microphone, usually a (voltage) gain of about 50 dB or 300 is required; that is shown in Figure 14-13 (a) where the feedback resistors R1 and R2 set the gain at (1 + R1/R2) or the gain is (1 + 100K/330) ~ 304. A dynamic microphone usually wants to be loaded with ≥ 1kΩ resistance, which is R4 ∥ R5 = 11kΩ due to having two 22kΩ resistors in parallel. Note that R3 to C4 is an AC short-circuit to ground due to its large capacitance (100 μf). To remove noise from the power supply line, +V, R3 and C4 form a low-pass filter at about 5 Hz. This means

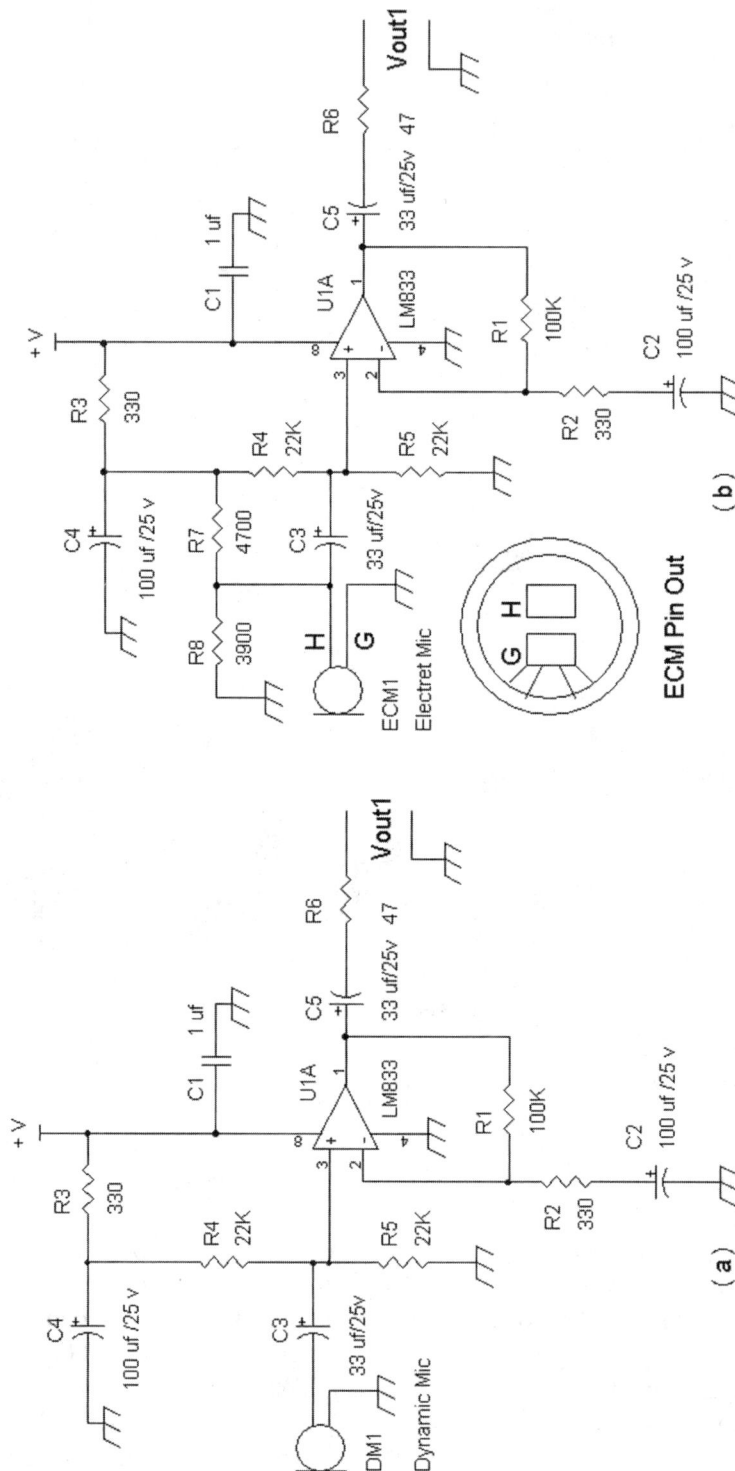

FIGURE 14-13 (a) Preamplifier circuits for dynamic microphones and (b) for electret condenser microphones (ECM). The power supply may range from 7.2 volts to 15 volts for +V.

if +V has some 50 Hz or 60 Hz ripple voltage, it will be attenuated by about tenfold or more at the plus terminal of C4. Having a low-noise DC bias voltage is essential for microphone amplifiers. If more filtering is required, C4 should increase to 1000 μf, if necessary, to provide low noise at Vout1.

Figure 14-13 (b) shows an electret condenser microphone preamplifier. Electret microphone capsules that are commonly used in DIY kits have an identifying ground lead as shown in the ECM Pin Out diagram. Most electret microphone capsules have a ground or shield lead, G, and a hot or + lead denoted by H. When you view the electret microphone's bottom side, you will see that the canister or metal body is connected to the ground terminal, G, via traces to the outer ring or case. See the illustration that is shaded in gray. Since the output from an electret microphone is typically 10× more than the dynamic microphone, the preamplifier's gain is set to about 31 via R1 = 10kΩ and R2 = 330Ω. The gain is then (1 + R1/R2) = (1 + 10K/330) ~ 31. To bias the electret microphone, it requires an equivalent 1000Ω to 2200Ω load resistor with a bias voltage from 3 volts to 9 volts max. Generally, a 3-volt bias is common.

For Figure 14-13 (b), the voltage at C4's plus terminal is still about +V due to a very small low current flowing through R4 and R5 that incurs a small voltage drop across R3.

If +V = 9 volts, then the voltage at the junction of R8 and R7 is:

9 volts × [R8/(R7 + R8)] or 9 volts × [3900/(4700 + 3900)] = 4.08 volts. The equivalent load bias resistor to ECM1 is R7 || R8 or 3900Ω || 4700Ω = 2131Ω. Resistors R4 and R5 are 22kΩ that present an extra 11kΩ AC signal load to the microphone. However, we can make R4 = R5 = 220kΩ to provide a negligible loading to the electret microphone, but should be aware that the DC bias voltage will take about 10 seconds to settle down or stabilize due to the time constant with C3, which is:

$$\tau = C3 \times [R4 \, || \, R5] \text{ with } C3 = 33 \text{ μf and due to } R4 \, || \, R5 = 220k\Omega \, || \, 220k\Omega \text{ or}$$
$$R4 \, || \, R5 = 110k\Omega \text{ so } \tau = 33 \text{ μf} \times 110k\Omega = 3.6 \text{ seconds.}$$

Normally it takes about three time constants for the circuit to settle, which is then 3 τ = 3 (3.6 sec) = 10.8 seconds.

One other thing to notice is that the values for R7 and R8 were chosen to be unequal with resistors R8 < R7 to have less than half the supply voltage at C3's negative terminal. This is done so that electrolytic capacitor C3 is DC biased correctly. The reason is so that the DC voltage at the negative terminal of C3 is guaranteed to be less than the DC voltage at the positive terminal of C3 where the resistors R4 and R5 are equal to provide about 50 percent of +V.

Thermal Sensing Circuit via Thermistor (Temperature-Dependent Resistor)

This DIY project uses a thermistor that looks almost like a ceramic capacitor. See Figure 14-14. However, it is a resistor that changes resistance values with tempera-

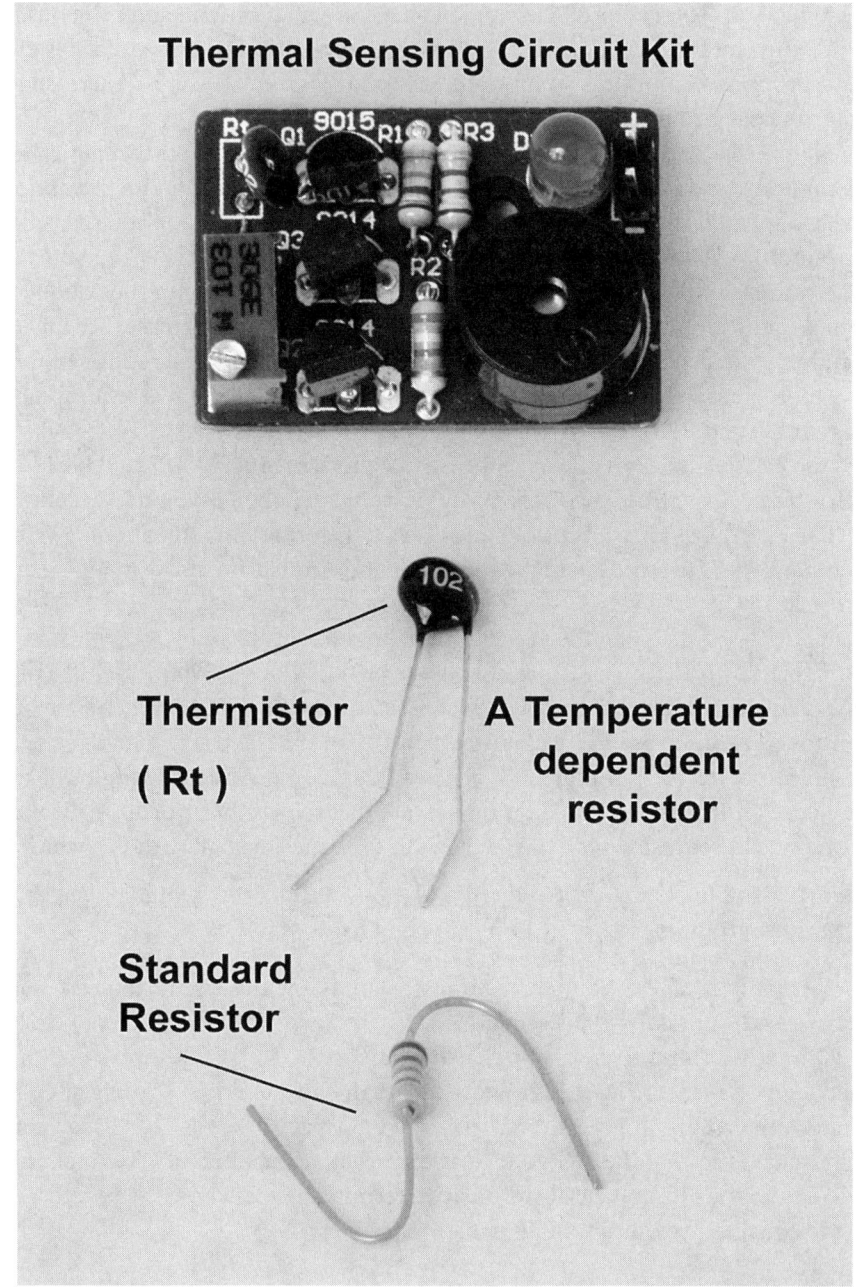

FIGURE 14-14   Sensing circuit (top), thermistor (middle), and resistor (bottom).

ture. There are two types of thermistors, one with a negative temperature coefficient and the other with a positive temperature coefficient. The negative temperature coefficient (NTC) version has reduced resistance when heated up, and increased resistance when cooled. And the positive temperature coefficient (PTC) thermistor has increased resistance when heated, and decreased resistance when cooled. For this project, we will be using the negative temperature coefficient version that has lower resistance when the temperature is raised.

The DIY circuit also includes a multiple-turn potentiometer. See Figure 14-15.

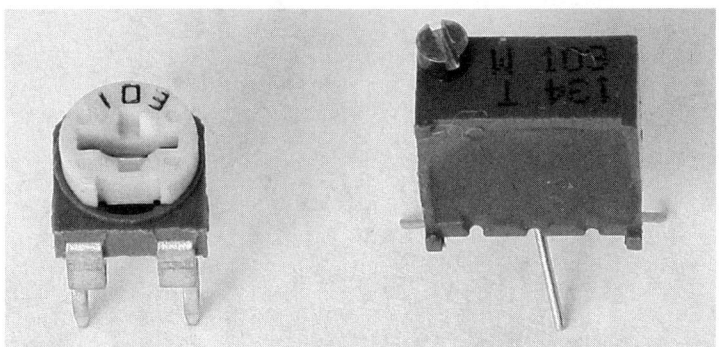

FIGURE 14-15   A single-turn potentiometer on the left and a multiple-turn version on the right. The single turn potentiometer allows setting the mid-resistance point by observing the slit's reference position above the "103" for 10kΩ, whereas the multi-turn potentiometer has no such reference position on its adjustment.

Now let's take a look at the original DIY circuit in Figure 14-16.

Let's start with the Themistor & Resistor Graph first, which shows for an ideal resistor, R, whose resistance does not change with temperature. However, a negative temperature coefficient (NTC) thermistor, Rt, in the graph shows higher resistance at cold temperatures and lower resistances at hot temperatures. This graph is an approximation of an NTC thermistor, part number TTC05102 made by TKS. In this example let's start out at room temperature, 25 degrees C, Rt = 1000Ω. When we look at the graph, this means at a very cold –20 degrees C, the resistance is tenfold greater or Rt = 10 × 1000Ω = 10kΩ, and at a very hot +80 degrees C, the resistance is at 10 percent of the room temperature resistance or Rt = 0.1 × 1000Ω = 100Ω. A way to test a thermistor is to connect it to an ohm meter (DVM) and observe that the resistance will change when you heat the thermistor by just holding it with a finger and thumb. For example, if you take the 1000Ω thermistor (e.g., marked 102) from the DIY kit before soldering it to the board, measure it first. Set the DVM to the 2000-ohm setting and observe the resistance before placing your hand on it. As you raise the thermistor's temperature with your finger and thumb, the thermistor's resistance should drop slightly.

The circuit in Figure 14-16 has a voltage divider formed by Rt and variable resistor RP1. RP1 is set first to light up the LED, and then RP1 is adjusted again until the

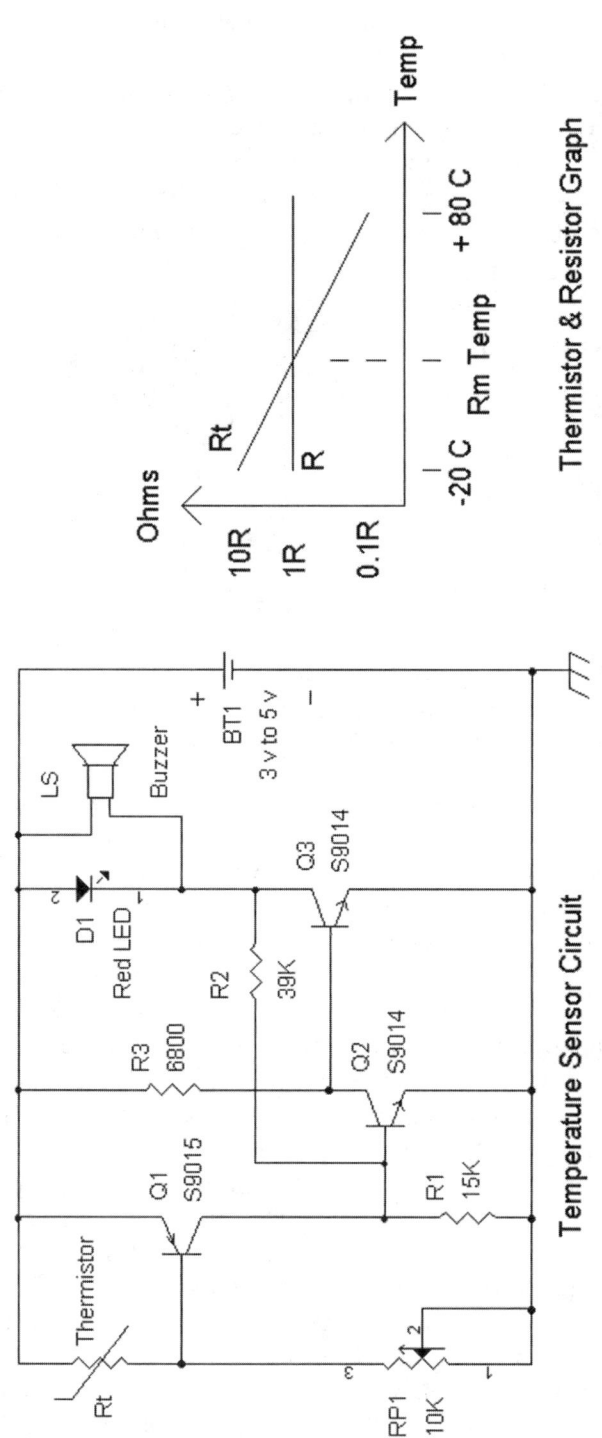

Thermistor & Resistor Graph

Temperature Sensor Circuit

FIGURE 14-16 Temperature sensor circuit with a graph of an example of the thermistor's resistance, Rt, and an ideal resistor, R, plotted in terms of temperature versus resistance. The graph shows a negative temperature coefficient (NTC) characteristic for Rt, the thermistor.

LED just turns off. When the LED just turns off this means Q3 is turned off because the base-emitter voltage at Q3 is close to 0 volts due to Q2 being turned on. Q2 acts as an inverting logic gate with the base as the input and collector via load resistor R3 as the output. With Q2 turned on, this means there is sufficient base current drive via Q1's collector so that the collector-to-emitter voltage of Q2 is close to zero volts. With Q1's collector supplying current to the base of Q2, this means that the voltage divider with RP1 and Rt forms about 0.6 volt at the emitter-base junction of PNP transistor Q1. **When thermistor Rt warms up,** its resistance drops and the voltage across Q1's emitter-base junction voltage drops below the 0.6-volt turn-on voltage for Q1, which means there is no collector current flowing into Q2's base. When this happens Q2 is at cut-off, and there is no Q2 collector current, which means R3 now supplies base current to Q3 and provides about 0.6 volt at the base-emitter junction of Q3. With Q3 turned on, Q3's collector current sinks current and turns on the LED. The speaker/buzzer, LS, is connected across the LED, which provides about 1.8 volts via the LED to LS that is supposed to turn it on for providing a noticeable alerting sound. **When the thermistor Rt cools down,** its resistance increases sufficiently to provide an emitter-to-base voltage to Q1 such that Q1's collector current is provided to Q2, which then turns on Q2's collector-to-emitter junction and provides, via R3, a logic low voltage to the base-emitter voltage of Q3 so that the LED and buzzer are turned off due to Q3 being at cut-off.

NOTE: An intuitive way of figuring out Figure 14-16 is to "pretend" that at low temperatures, Rt = open circuit, or Rt can be thought of as removed from the circuit; and at high temperatures, Rt = 0Ω or Rt is a short-circuit. Often, you can quickly analyze a circuit by looking at the two extremes of a particular parameter (e.g., resistance). In this case it is Rt's resistance range. When Rt is an open circuit = infinite ohms, Q1 will turn on via pulling base current via RP1, and Q1 will supply collector current to Q2's base. And when Rt = 0Ω, this means the emitter-base junction of Q1 is shorted out to zero volts. This puts Q1 in the cut-off mode, and there will be no collector current from Q1 flowing into Q2's base, which causes Q2 to be in cut-off as if Q2's collector was disconnected to R3 and to the base of Q3. With Q2's collector disconnected, we then have R3 via the supply voltage providing base current to Q3, which then turns on the LED and speaker/buzzer.

So, does the circuit in Figure 14-16 looks like a workable design? Yes and no. There are a few problems. One is that if the adjustable resistor RP1 is accidentally set to zero ohms or close to zero ohms such as 10Ω, you may end up with burnt-out parts for Q1, Q2, and RP1. Also, the other problem is that the LED's maximum current may be exceeded via Q3's collector current because there is no reliable current-limiting mechanism. Q3's collector current is set by the base drive resistor R3. For example, if we have BT1 = 5 volts, then the base current into Q3 is about (5 volts – VBE_{Q3})/6800Ω = (5 V – 0.7 V)/6800Ω or 632 μA.

Q3's collector current is $\beta \times 632\ \mu A$. If we have Q3's current gain $\beta = 100$ then Q3's collector current is $100 \times 632\ \mu A$ or 63 mA, which is excessive for driving LEDs (e.g., LEDs generally operate at 1 mA to 20 mA). Third, the buzzer really requires about 3 volts or more to produce sufficient volume. The LED, D1, supplies only about 2 volts to the speaker/buzzer, LS, which sounded very soft. See Figure 14-17.

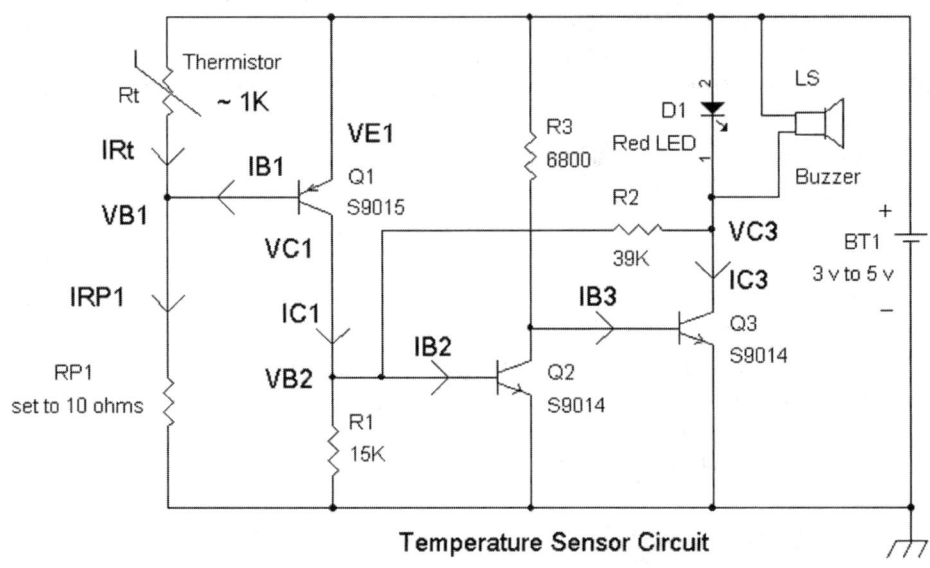

FIGURE 14-17 When RP1 is adjusted to 10Ω, Q1, Q2 and RP1 itself may be damaged due to excessive current or power dissipation that leads to burnt-out components.

Now let's take a look at Figure 14-18, which substitutes a diode and resistor to show the excessive currents and power dissipation.

In Figure 14-18 if the supply voltage is 5 volts, then the voltage across RP1 set to 10Ω is VRP1 = (5 volts – 0.7 volt) = 4.3 volts. The 0.7 volt is due to the turn-on voltage across the emitter-base junction diode (Q1 of Figure 14-17) connection as being represented by DEBQ1. Most variable resistors or potentiometers have a 250 mW or 500 mW rating. With 4.3 volts across RP1 = 10Ω, the power dissipated into RP1 = $[(4.3)^2/10] = [18.49/10]$ watts or 1.849 watts that can burn out RP1 that is set to 10Ω. The current through DEBQ1 is VRP1/10Ω or 4.3 volts/10Ω = 430 mA. The maximum (base or collector) current rating for both S9015 and S9014 is about 100 mA. This means Q1 may burn out. And when we look at the base current into Q2, it is similar at (5 volts – VBE_{Q2})/RECQ1 ~ (4.3 volts/10Ω) ~ 430 mA, where VBE_{Q2} ~ 0.7 volt. Therefore, Q2 is running at excessive current and it too may burn out.

To ensure a safe and reliable circuit that reduces any chance of burning out RP1, Q1, and Q2, we need to add current limiting resistors.

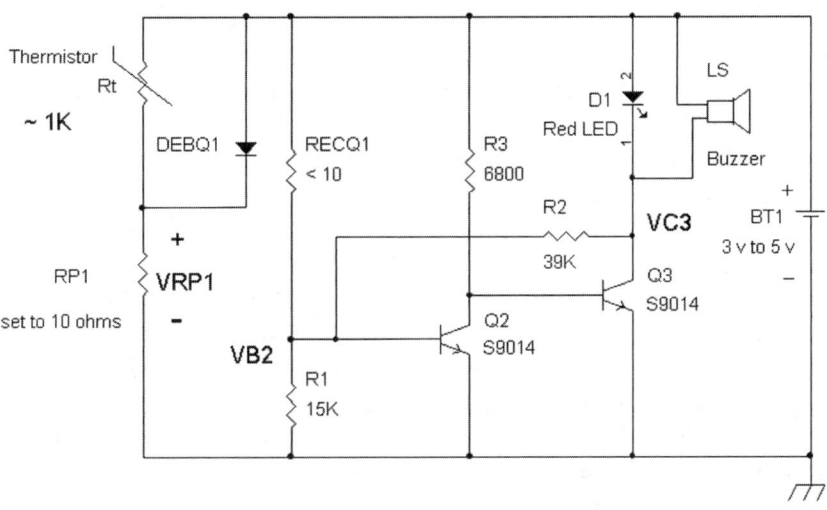

FIGURE 14-18   Excessive current via diode DEBQ1 (equivalent diode representation of the emitter-base junction of Q1) and resistor RECQ1 (an example emitter-to-collector resistance of Q1) when RP1 is set to 10Ω.

See Figure 14-19 that shows one such example solution as highlighted in rectangles around resistors Rext1, Rext2, and Rext3. Resistors Rext1 and Rext2 protect RP1, Q1 and Q2, while Rext3 protects D1 (Red LED) from over-current. When D1,

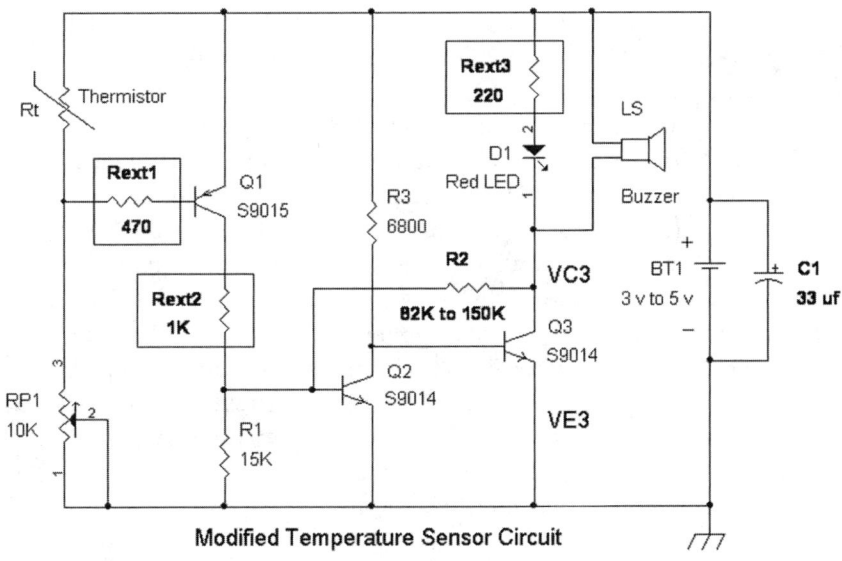

FIGURE 14-19   Current-limiting resistors Rext1, Rext2, and Rext3, which prevent damage to D1, the red LED, and transistors Q1 and Q2. C1 is added to reduce noise in the supply line.

the LED, is turned on, via Q3 being turned on, and with Q3 acting like a switch such that the collector is effectively grounded, the LED current is limited by $(BT1 - V_{LED})/Rext3$. This leads to the LED current as: (5 volts − 1.8 volts)/220Ω or (3.2 volts/220Ω) = 14.5 mA = LED current. See the circuit in Figure 14-19 for a reference to the text on this page.

Again, Rext3 also allows Q3 to saturate and act like a switch when R3 supplies current to its base terminal. With Q3 in saturation, the collector voltage, VC3, is nearly the emitter voltage, VE3, which is the ground voltage.

It should be noted that the resistor values for Rext1, Rext2, and Rext3 can be other values typically in the range of 220Ω to 1000Ω. For example, you can make all of these resistors 220Ω. What to be cautious of is to not make Rext1 or Rext2 too large of a value such as 470kΩ, which will render the circuit inoperable. For example, if Rext2 = 470kΩ, there is no way that Q2 will turn on when Q1's collector is turned on to +5 volts. This is because the voltage at the base of Q2 will be +5volts [R1/(R1 + Rext2)] = +5 volts [15K/(15K + 470K)] = 0.154 volt, which is less than the required + 0.6 volt to turn on Q2's base-emitter junction.

Resistor R2 is a positive feedback resistor from Q2's input base to Q3's output collector. Recall that each grounded common emitter amplifier has an inverting property in terms of phase from input to output. When you cascade two inverting amplifiers or two inverted logic gates, the result is an in-phase output signal from the collector of Q3 referenced to an input signal to the base of Q2. R2 serves the purpose of slightly biasing the base of Q1 when Q3 is off. There is some slight leakage current via D1 and Rext3 into R2 at the collector of Q3 to the base of Q1. R1 forms a voltage divider circuit with R2. If R2 is too low in value, then the circuit can latch up because R2 will provide a sufficiently high logic voltage to the base of Q2 such that Q2 is turned on and Q2's collector will output a logic low. This will cause Q3 to turn off all the time because there is no way now to lower the voltage at Q2's base to a logic low since Q1's collector current can only add or increase voltage to the input base of Q2. Typically, you can measure the base voltage with a DVM and select a value for R2 such that it is in the 0.10-volt to 0.4-volt range when Q3 is turned off. When the supply voltage is increased, R2 should be increased as well (e.g., such as proportionally). For example, in Figure 14-19, the 82kΩ value was chosen for a 4-volt supply. If we increase to 5 volts, then R2 = [5 volts/4volts] × 82kΩ or R2 = 102.5kΩ ~ 100kΩ.

However, the real reason for R2 is to provide a *hysteresis* effect on the temperature sensor. By having hysteresis, this means that the circuit acts like a thermostat in the way the LED/buzzer is turned on and off due to temperature. That is, the trip points for turn on and turn off are at different temperatures so as to avoid chatter or flickering in the LED. The turn on will be at one temperature, but the turn off will be at another temperature many degrees lower. For example, if the sensor is set to light the LED at +27 degrees C, it will not turn off until the temperature is lower, such as at +22 degrees C.

Finally, on this sensor, there are other ways to add the current-limiting resistors. For example (but not shown here), in pins 1 and 2 of RP1 that are connected to ground, we can break the ground connection and place a 470Ω series resistor connected to ground and to pins 1 and 2 of RP1. These modifications (Rext1, Rext2, and Rext3) as shown in Figure 14-19 can be done easily on the printed circuit board (e.g., without cutting printed circuit board traces). We only need to lift the base and collector leads of Q1 and solder one lead of Rext1 and Rext2 to the traces, while soldering the other ends of Rext1 and Rext2 to the base and collector leads of Q1. The same approach was made to the D1 LED, where either cathode or anode lead can be lifted with one lead of Rext3 soldered to the printed circuit board, and the other lead of Rext3 soldered to the LED. See Figure 14-20 for a close-up photo.

FIGURE 14-20 The modified version on the left side and the original circuit on the right side. The large round device is the speaker/buzzer.

In Figure 14-20 (left side), it was found that separating the thermistor, Rt, from Q1 worked better in terms of testing it. For example, if you heat both the thermistor (Rt) and Q1, the sensing effect slightly cancels out each other. Also, a decoupling capacitor is added across the supply line because when the speaker sounds off, some noise is added to the supply line. A 33-µf decoupling capacitor (C1) reduced this noise. See Figure 14-20 (left side) for the added C1.

The speaker/buzzer, LS, is a piezo speaker that has infinite DC resistance, but provides a tone similar to the one from a smoke alarm when activated with a DC voltage across it. Note: The speaker buzzer is polarity sensitive and must be installed correctly (see Figure 14-21).

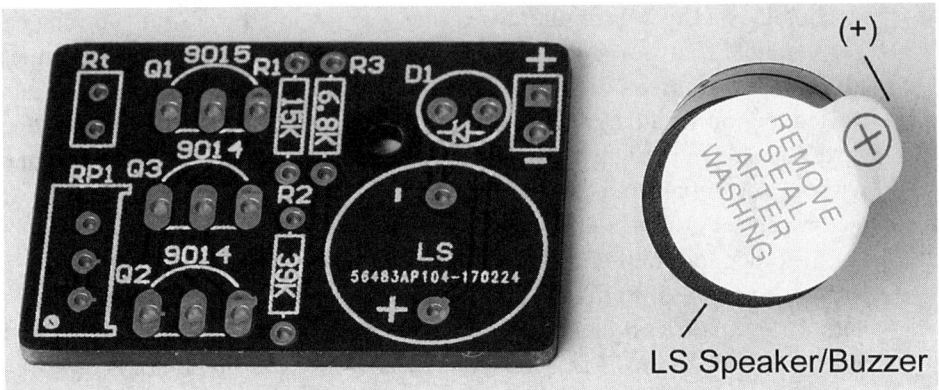

FIGURE 14-21 The speaker/buzzer in this DIY circuit is polarized and must be installed according to LS (–) and (+) pin outs in the board's silkscreen marking. The LS speaker/buzzer is shown on the right side with the identified positive (+) terminal.

If the speaker/buzzer, LS, is installed backwards, it will not produce the alarm tone. This device is really not a speaker in the usual sense, but instead a tone buzzer that requires DC to operate it.

A Circuit Using an Electrolytic Capacitor Incorrectly

Here's a strange audio amplifier circuit at first glance that has a reverse biased electrolytic capacitor to include a leaky internal resistive load to a transistor's collector terminal. This amplifier was used with a single IC (e.g., MK484 or TA7642) AM radio kit. See Figure 14-22. Adjusting the volume is done with potentiometer, P1.

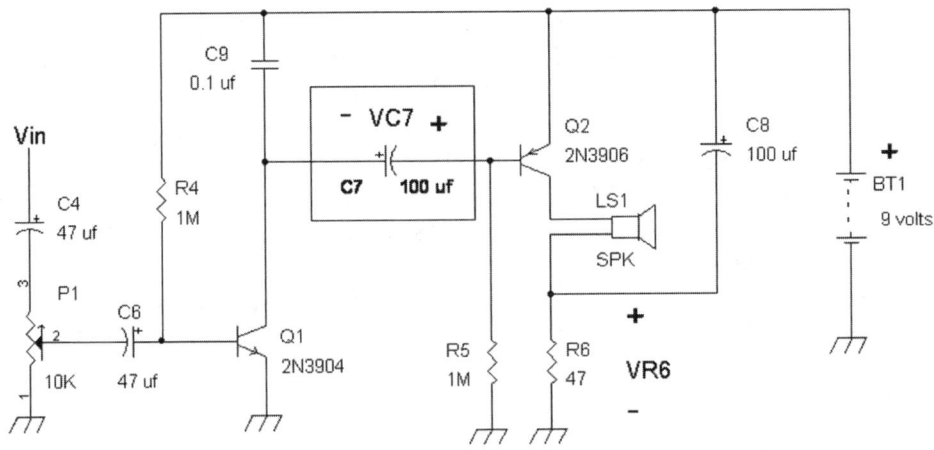

FIGURE 14-22 An audio amplifier with no apparent DC path for Q1's collector because of capacitors C9 and C7 apparently blocking DC current flow.

This circuit actually works with a polarized electrolytic capacitor C7 (highlighted within the rectangle) because C7 is incorrectly reverse biased, and C7 will then include a leaky internal resistor in parallel with it. Reverse biasing C7 is strongly not recommended because it can be unsafe. If not for the collector of Q1 to limit the current flow of C7, this circuit can cause C7 to be damaged. As configured, the voltage across the 47Ω resistor, $R6 = VR6 \sim 7$ volts DC. The 7 volts is actually bad because there is almost 1 watt of power dissipation across R6, a quarter-watt resistor. Recall that power, $P_{R6} = [V^2/R] = 7^2/47\Omega = (49/47)$ watt ~ 1 watt. Worse yet, the collector current of $Q2 = ICQ2$ is also the current flowing into R6, which is 7 volts/47Ω $\sim$ or $IC_{Q2} = 149$ mA. Q2, a 2N3906, has a maximum collector current rating of 200 mA. Normally you try to run your transistors at < 50 percent max collector current for reliability reasons.

And if we look at Q2's power dissipation, $P_{Q2} = VEC_{Q2} \times IC_{Q2}$. To find VEC_{Q2}, we need to find Q2's emitter voltage, which is $BT1 = 9$ volts, and also Q2's collector voltage, $VC_{Q2} = $ collector current $IC_{Q2} \times$ (resistance of LS1 + R6). If LS1 is an 8Ω speaker, then $VC_{Q2} = 149$ mA $\times (8\Omega + 47\Omega) = 8.19$ volts. $VEC_{Q2} = VE_{Q2} - VC_{Q2} = 9$ volts $- 8.19$ volts or $VEC_{Q2} = 0.81$ volt and thus, $P_{Q2} = VEC_{Q2} \times IC_{Q2}$ or $P_{Q2} = 0.81$ volt $\times 149$ mA $= 120$ mW $= P_{Q2}$. However, Q2's maximum power dissipation is about 200 mW, so again we are running the transistor hot.

If we take into consideration battery life from BT1, a 9-volt battery with about 200 mAH rating, Q2 is draining in the order of 149 mA. This means about 1 hour of usage before the battery is exhausted. So, although Figure 14-22's circuit works as an audio speaker amplifier, it will drain the battery too quickly, not to also mention that it's dangerous to have electrolytic capacitor C7 put in backwards (i.e., unsafe when C7 is incorrectly biased or reversed biased).

Now let's look at Figure 14-23. It shows an equivalent circuit, including an internal resistor to C7.

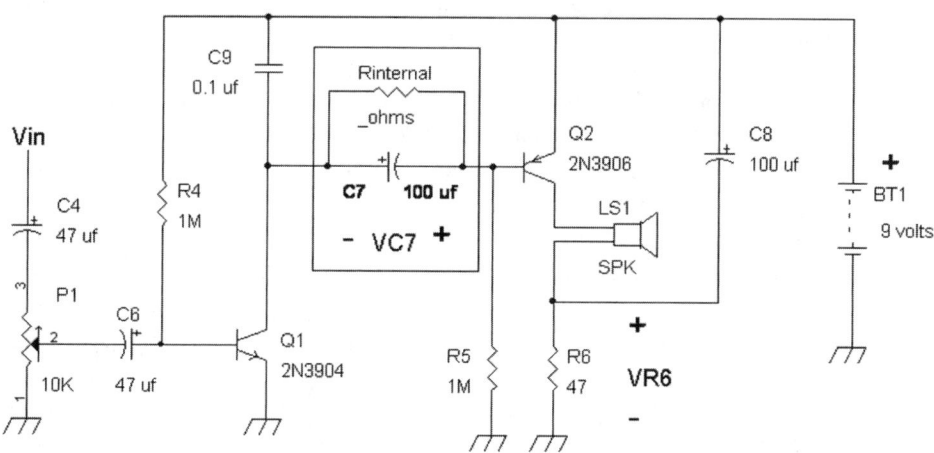

FIGURE 14-23 A reverse biased electrolytic capacitor C7 can be approximated with an internal resistor, Rinternal, as shown here. Potentiometer P1 adjusts for volume.

With C7's equivalent internal or built-in resistor, Rinternal, in parallel with C7, we can now see that from Q2's base, DC current flows to Q1's collector. In essence, Rinternal is now an equivalent load resistor for Q1. Various standard 100-µf electrolytic capacitors were tried in this circuit and VR6 measured around 7 volts. However, when three 33-µf capacitors were paralleled (to equal ~ 100 µf) and placed into the circuit as C7, VR6 dropped to 1.5 volts. *This means that Q2's collector current cannot be reliably set by the reverse biased leakage resistance, Rinternal.*

Also keep in mind that there are many different types of electrolytic capacitors, including low leakage ones that will vary Q2's collector current. For example, if C7 is a very low leakage electrolytic capacitor, even when reverse biased, this means that Q1's and Q2's collector currents may be "starved" and thus the amplifier will deliver insufficient signal current to the loudspeaker. We can do an initial fix to this circuit via Figure 14-24.

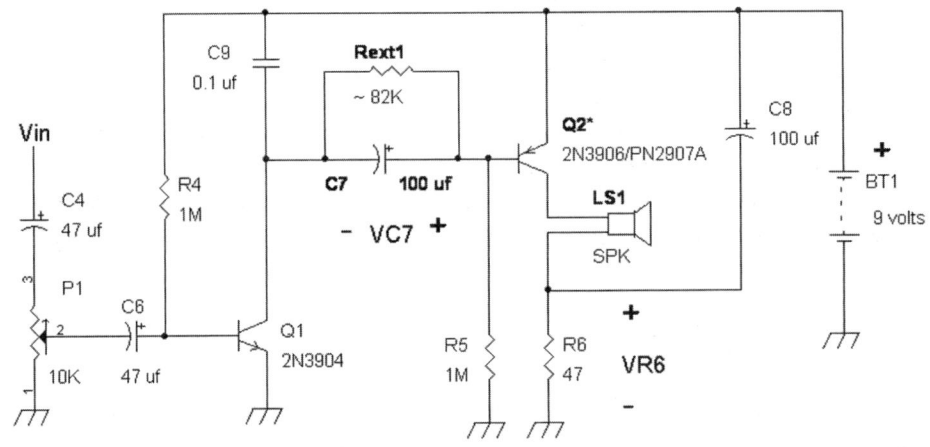

FIGURE 14-24   C7 is installed correctly and external biasing resistor Rext1 is added as shown. Potentiometer P1 adjusts for volume.

When an electrolytic capacitor (in Figure 14-24) is connected correctly polarity-wise, the internal leakage resistance goes up to near infinite ohms. So, the electrolytic capacitor, biased properly, truly blocks DC current. So, if C7 is connected correctly and without an external resistor in parallel with it, Q1 will have no load resistor and its collector current goes to zero.

Also notice that Q2 can be upgraded in terms of maximum collector current and power dissipation by choosing it as a PN2907A, which has 600 mA collector current and 600 mW power ratings as compared to the 200 mA and 200 mW ratings of the 2N3906.

In Figure 14-24 the external biasing resistor Rext1 in parallel with the correctly installed C7 (positive terminal of C7 connected to Q2's base), we provide a load resistor for Q1's collector, and we can set Q2's collector current. For example, if Rext1 = 82kΩ for a particular set of Q1 and Q2 transistors, then the voltage across R6, which is VR6 = 2.01 volts. This means Q2's collector current is (2.01 volts/R6) = 2.01 volts/47Ω or 42.7 mA, which gave sufficient speaker volume with the IC radio using the MK484 or TA7642. Given the much lower current from 149 mA originally, at least the battery will last almost four times longer. In another example, when Rext1 → 56kΩ, VR6 → 2.85 volts or Q2's collector current → 60.6 mA (2.85 volt/47Ω = 60.6 mA). The drawback to this circuit is that **Rext1 in Figure 14-24 needs to be selected accordingly** because Q1 and Q2 have a wide range of current gain (β) variation (e.g., 2:1 of β variation) that will affect Q2's final DC collector current.

However, we can modify this circuit to be insensitive to the transistors' current gain β with the negative feedback amplifier circuit shown in Figure 14-25.

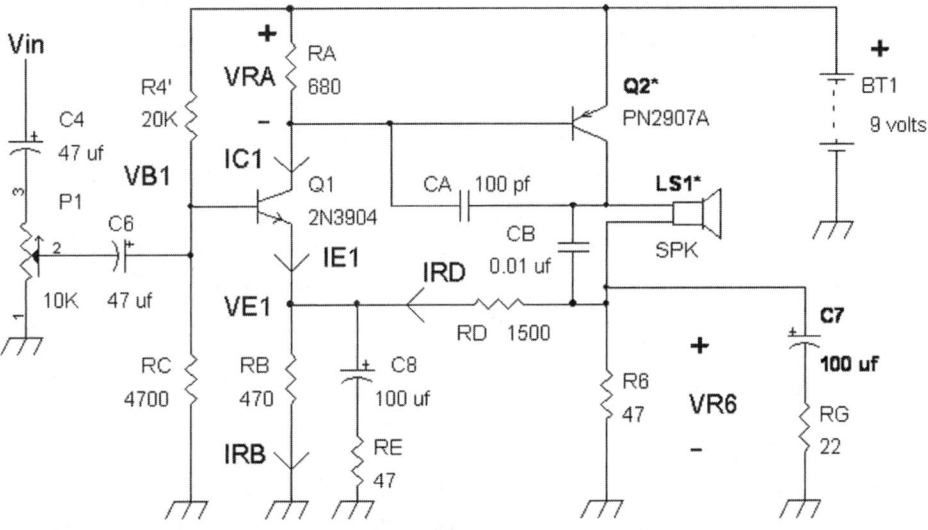

FIGURE 14-25   A feedback amplifier circuit for a more reliable Q2 DC collector current setting. Potentiometer P1 is used for adjusting volume. R6 (47Ω) is rated at 1 watt.

Let's look at Figure 14-25 first in terms of DC conditions. We treat the capacitors as open circuits or infinite DC resistance devices (e.g., you can imagine all capacitors to be removed for DC analysis). This is a current sensing amplifier with R6 as the sense resistor. Whatever the voltage across R6 is determines Q2's collector current, which in turn determines the drive current into the loudspeaker LS1. To provide adequate loudness from the speaker, we generally need something in the range of 10 mW to 50 mW. For example, some of the earliest simple two or three transistor radios in the 1960s typically had about 50 mW of audio power, which played loudly with its 2.25-inch speaker. For adequate playing volume, having at least 10 mW will suffice with a reasonably efficient speaker such as a 2.25-inch to 4-inch 8Ω speaker.

For the analysis **with C7 or RG removed**, the power into the speaker from this amplifier is about 0.5 $(IC_{Q2})^2 \times R_{speaker}$, where IC_{Q2} = Q2's collector DC current and $R_{speaker}$ is the LS1 loudspeaker's impedance that is typically 8Ω. For example, if we set

VR6 = 2.5 volts, then IC_{Q2} = 2.5 volts/47Ω or IC_{Q2} = 53 mA. The audio power output is then P = 0.5 $(IC_{Q2})^2 \times R_{speaker}$ or 0.5 $(0.053 Amp)^2 \times 8Ω$ = 22.6 mW. As we can see, just having a higher-impedance speaker proportionally increases the power via P = 0.5 $(IC_{Q2})^2 \times R_{speaker}$. For example, given the same conditions of 53 mA, if we use a 16Ω speaker instead of the 8Ω version, then the power to the speaker is increased twofold to 2 $\times$ 22.6 mW or 45.2 mW. And if we choose a 45Ω speaker, then the power output will be 0.5 $(0.053 Amp)^2 \times 45Ω$ = 127 mW.

Figure 14-25 shows a two-transistor common emitter amplifier for both stages Q1 and Q2. It actually looks sort of like an op amp with the output at Q2's collector, the non-inverting input at the base of Q1, and the inverting input at the emitter of Q1. The **DC feedback elements** are resistors RD and RB. To understand how this circuit works, we need to look at the DC currents flowing through Q1's emitter, IE1, with current flowing through RD via IRD, and current flowing through RB, IRB. We will need to balance all the currents accordingly so that Q1's emitter current plus the current from RD via VR6 equals the current flowing through RB. Put in another way we can say that IRD + IE1 = IRB. Also, in order to make this circuit work properly in terms of output current swing into the loudspeaker (SKR), the voltage VR6 across R6 must be greater than Q1's emitter voltage referenced to ground.

Q1's base voltage, VB1, is set by the voltage divider circuit R4' and RC, which is: 9 volts $\times$ [RC/(R4' + RC)] = 9 volts $\times$ [4700/(20,000 + 4700)] or 1.71 volts. We can approximate that Q1's base-to-emitter voltage VBE_{Q1} = 0.7 volts, so Q1's emitter voltage is Q1's base voltage minus Q1's base-to-emitter voltage. Q1's emitter voltage = (1.71 volts – 0.7 volt) or Q1's emitter voltage **VE1 ~ 1 volt**. The current (IRB) flowing through RB = 470Ω, then Q1's emitter voltage is divided by 470Ω or **IRB = 2.12 mA**.

Generally, we can ignore Q2's base current adding to Q1's collector current. And make the approximation that the collector current of Q1 related to the voltage across RA, which is the VEB turn-on voltage of Q2 of about 0.7 volt. Q1's collector current is VEB/RA = 0.7 volt/RA or Q1's collector current = 0.7 volt/ 680Ω or IC1 ~ 1 mA. Since the collector and emitter currents are generally the same, IC1 = IE1 for large current gain β for Q1, and we also have **IE1 = 1 mA**.

This means Q1's emitter current **IE1 ~ 1 mA** since the collector current and emitter currents are approximately equal when the transistor is operating as an amplifier.

If we want the voltage across R6, VR6 set to 2.5 volts, then the current flowing through RD is: IRD = (VR6 – Q1's emitter voltage)/RD or (2.5 volts – 1 volt)/RD or **IRD = 1.5 volts/RD**. Now let's put everything together.

Since **IE1 + IRD = IRB** (and equivalently IRD = IRB – IE1), and with **IE1 = 1 mA, IRD = 1.5 volts/RD and IRB = 2.12 mA** we have:

1 mA + 1.5 volts/RD = 2.12 mA

Or put in another way by subtracting 1 mA from both sides:

1.5 volts/RD = 1.12 mA, which leads to RD = 1.5 volts /1.12 mA = 1.34kΩ, or RD = 1340Ω. Again please refer to Figure 14-25.

With RD → 1.5kΩ, VR6 will be a little bit higher than 2.5 volts as originally set. We can find VR6 via the following, which you may skip because of the equations.

The current flow through RD is (VR6 − VE1)/RD = IRD. But from the previous equation we know that IRD = IRB − IE1 and that IRB = 2.12 mA and IE1 = 1 mA. So, this leads to IRD = 2.12 mA − 1 mA or IRD = 1.12 mA. Now we can find VR6 by (VR6 − VE1)/RD = IRD = 1.12 mA. Since VE1 ~ 1 volt and RD → 1.5kΩ we have: (VR6 − 1 volt)/1.5kΩ = 1.12 mA or equivalently:

(VR6 − 1 volt) = 1.12 mA × 1.5kΩ or (VR6 − 1 volt) = 1.68 volts, which leads to: VR6 = 1.68 volts + 1 volt or **VR6 = 2.68 volts**.

Although this analysis is a bit long-winded, it accurately predicts what VR6 will be. But when making an approximation, it's just easier to build the circuit and measure VR6 when there is no audio input signal present.

To get an approximation of the AC gain, G, from the base of Q1 to VR6, it is with C8 taken as an AC short-circuit at typical audio frequency from 100 Hz and up, and with RD = 1500Ω, RB = 470Ω and RE = 47Ω and (RB ∥ RE) = (470Ω ∥ 47Ω) ~ 42Ω:

$$G = \{1 + [RD/(RB ∥ RE)]\}$$

For this example, G = [1 + (1500Ω/42Ω)] ~ 36.

Finally, to ensure that this amplifier does not oscillate, capacitors CA and CB are added. The loudspeaker LS1 can be modeled as a resistor in series with an inductor. The inductive part of the LS1 comes from its voice coil winding. Capacitor CB is connected in parallel with the speaker to act like an AC short-circuit at higher frequencies and thus "short" out the inductive portion of the loudspeaker. Capacitor CB also provides an AC path at high frequencies to the current sense resistor R6 that is grounded. With Q2's collector coupled to R6 via CB at higher frequencies, capacitor CA is used as a compensation capacitor that ensures the amplifier works without causing oscillation or peaking high-frequency response. When swept from 100 Hz to 100 kHz at the input terminal, Vin, the output at VR6 responded flat to about 50 kHz before rolling off at frequencies above 50 kHz. Capacitor CB actually may have a capacitance range from 0.01 µf to 0.1 µf.

Identifying and Fixing "Bad" Circuit Designs

When I was starting out in electronics, there were many projects from hobbyist magazines and books. Most of these worked, but a few got my head scratching. Some circuits would not work, and I went for help to my nearby TV repair shop.

Here's one example of a variable DC power supply shown in Figure 14-26.

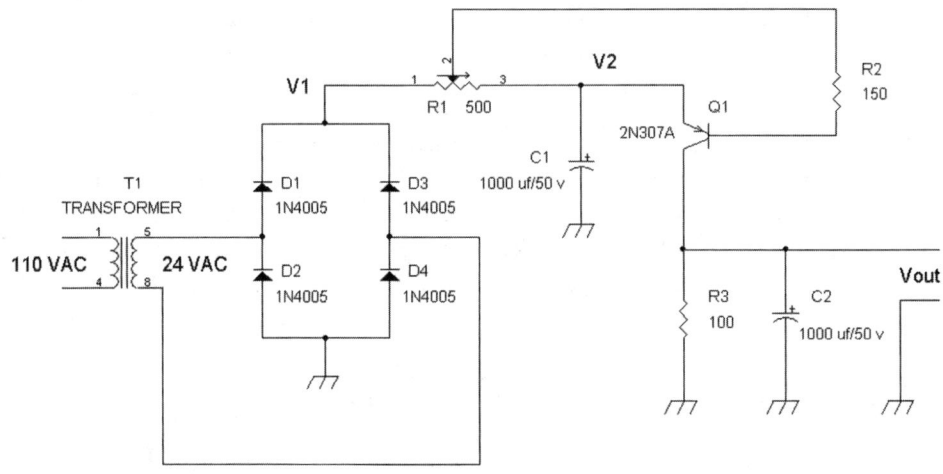

FIGURE 14-26   A DC power supply circuit from a hobbyist book from the 1960s. This circuit does not work well. (The original schematic labeled R1 as a potentiometer.)

The circuit uses a 24-volt AC secondary winding transformer, T1, that delivers 24 volts RMS AC to a full-wave bridge rectifier from diodes D1, D2, D3, and D4. At the cathodes of D1 and D3, there is a pulsating DC voltage that peaks out at about 34 volts since the peak voltage of a sine wave is the RMS voltage $\times$ 1.414, which is 24 volts $\times$ 1.414 = 34 volts peak. This full-wave rectified pulsating DC voltage is low-pass filtered via potentiometer, R1 (500Ω), which is connected to the positive terminal of filter capacitor C1 (1000 µf).

Power transistor Q1's emitter is also connected to C1 so there is smoother DC voltage there at V2 than at V1 where diodes D1 and D3 cathodes are connected. The slider of R1 is connected back to the cathodes of D1 and D3 at V1, which provides a "somewhat" lower voltage, which has more "ripple" at V1 than at V2.

In this circuit we see that the PNP power transistor Q1, 2N307A (an old germanium TO-3 power transistor) works mostly as a current source via its collector connection to C2. Load resistor R3, 100Ω, should be rated at 10 watts or more.

The problem with the circuit in Figure 14-26 is that Q1's base voltage has to be lower than its emitter voltage to start turning on. This would require that current is flowing through R1 from pins 1 to 3. If there is not enough current draw, the circuit will not output any voltage at Vout. When this circuit was built, there was very little control over the voltage output via R1. It just did not work well. And even if it did, the source resistance for this circuit would be 100Ω from R3 when Q1 is acting like a current source. This means Vout will drop in half if it is loaded to a 100Ω resistor.

A better approach is to use the same rectifiers and filter capacitors and modify the circuit to Figure 14-27. The transformer, T1, is now connected directly to C1 to provide a 34-volt DC voltage that has some ripple. This voltage at the positive terminal of C1 is coupled to a Zener diode circuit R4 and ZD1 to provide a regulated voltage into variable resistor R1. The voltage at the anode of ZD1 is filtered further via

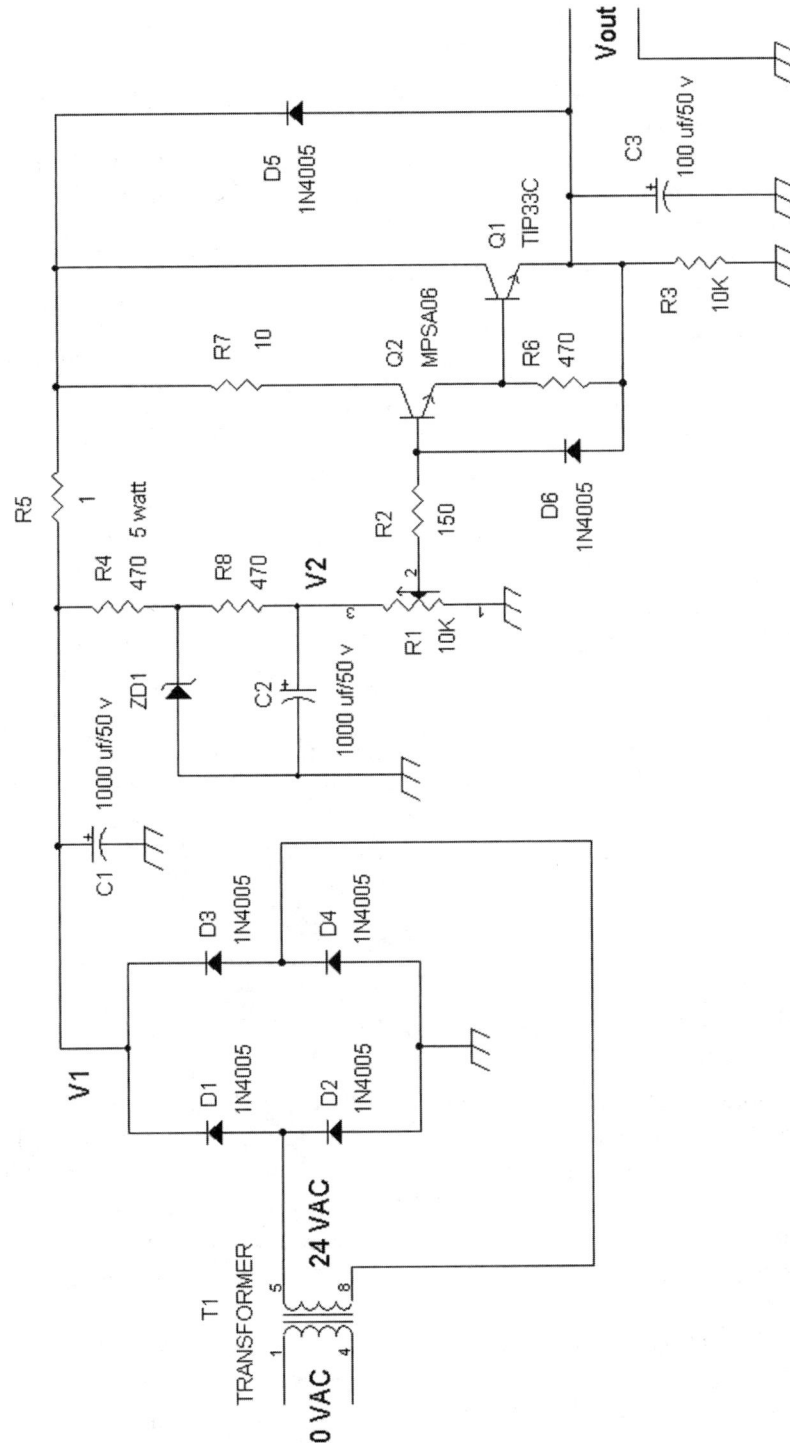

FIGURE 14-27 Fixing the DC power supply's design, with improvements via emitter followers Q2 (MPSA06) and Q1 (TIP33C). Note: In keeping with the original schematic (Figure 14-26), R1 is designated as a potentiometer.

377

R8 and C2, which provides an almost ripple-free DC voltage to potentiometer R1 that is changed to 10kΩ instead of 500Ω in Figure 14-26. The slider of R1 provides a variable DC voltage to the base of emitter follower transistor Q2. The output of Q2 is its emitter, which is connected to the base of the output transistor Q1, which is also an emitter follower circuit. Any ripple voltage at the collectors of Q1 or Q2 is rejected and attenuated because emitter follower circuits prevent noise at their collector terminals from leaking into the emitter terminals.

The transistors Q2 and Q1 are protected from over-current via current-limiting resistors R5 and R7, which are half-watt resistors. Should excessive current arise at Vout, these resistors can act like a fuse to save the transistors from burning out.

The "fixed" design now has capacitor C1 directly connected to the full-wave bridge rectifier circuit (D1 to D4) to provide a peak DC voltage of 24 volts × 1.414 ~ 34 volts DC. By using double emitter follower amplifiers, Q2 and Q1, and a well-filtered DC voltage at V2, this variable voltage power supply provides essentially ripple-free DC voltages at Vout. Q1's emitter follower amplifier provides a low source resistance at Vout, which provides a "stiffer" voltage source than the collector output of the 2N307A transistor in Figure 14-26.

The DC voltage is regulated (e.g., if the power line (mains) voltage changes, so does Vout), as shown in Figure 14-27, via Zener diode ZD1 connected across C2. Typically, you want the Zener diode's voltage to be lower than the raw 34 volts at V1. For example, you can try something like a 24-volt, 5-watt Zener diode (e.g., 1N5359B). Make sure R4 is at least a 2-watt resistor when you add the Zener diode. Of course, other Zener diode voltages can be used instead of 24 volts, such as 20 volts or 22 volts. Catch diodes D5 and D6 are added to prevent reverse bias breakdown of the transistors Q1 and Q2 when the power is turned off.

NOTE: Q1, TIP33C, should be on a heat sink insulated from ground.

An Example of the Missing Ground Connection

In our next circuit, a 3.00-MHz crystal oscillator, strange signals were measured on the oscilloscope. See Figures 14-28 (a), (b), and (c). The original circuit was built on a printed circuit board. All connections for the circuit appeared to be correct and resemble the circuit shown in Figure 14-28(c). However, output signal Vout in Figures 14-28 (a) and (b) showed incorrect DC voltages and waveforms. Vout should output a 3.00-MHz pulsed waveform close to 50 percent duty cycle from 0 volts to +5 volts. Let's see what was found.

In Figure 14-28(a), it is not known yet that pin 7 of IC1A (74HC04) is not really connected to ground. By using a DVM or scope, the average DC voltage is checked at pins 1, 2, 3, and 4. If the 74HC04 logic inverters are working properly, we should measure about 2.5 volts average on all these pins because the oscillation waveform

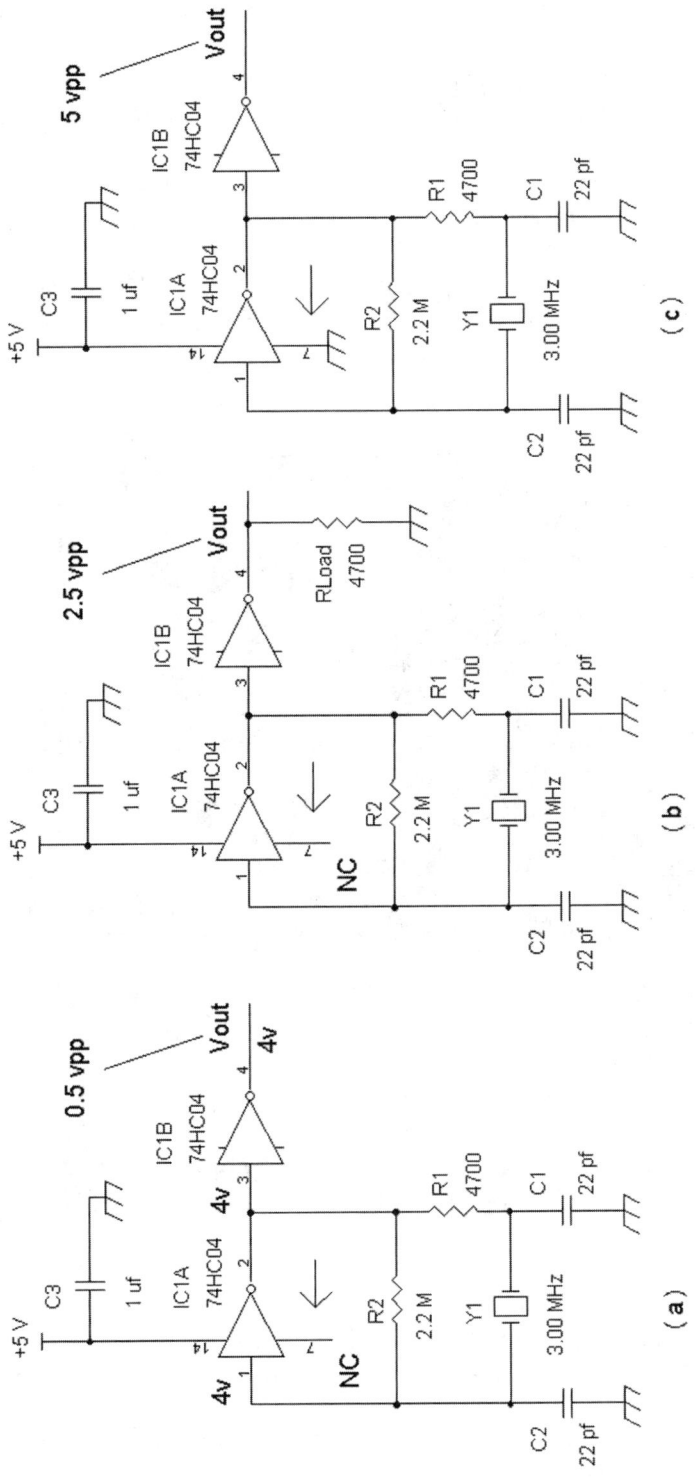

FIGURE 14-28 (a) No connection on ground pin 7; (b) no connection on pin 7 with test load resistor, RLoad; and (c) with ground connection re-established to provide proper output signal.

will be close to a 50 percent duty cycle pulse that goes from 0 volts to 5 volts. So, the average voltage would be 2.5 volts. Instead, as shown in Figure 14-28(a), we see that the average DC level is about 4 volts on pins 1, 2, 3, and 4. This would not make sense if the input to IC1B at pin 3 is 4 volts, a logic high voltage, then by "definition" the inverter gate's output at pin 4 should be close to 0 volts. But this is not the case; output pin 4 is also about 4 volts. So, there is something wrong here. When we look closely for an oscillation signal, we see a small amplitude waveform shown in Figure 14-29. There are again problems with this waveform, as it is around 0.5 volts peak to peak and at the wrong frequency, 2.5 kHz, which is way off the crystal's frequency of 3.00 MHz. Furthermore, the waveform does not resemble a 50 percent duty cycle pulse (square wave). Instead, it is a narrower duty cycle pulse.

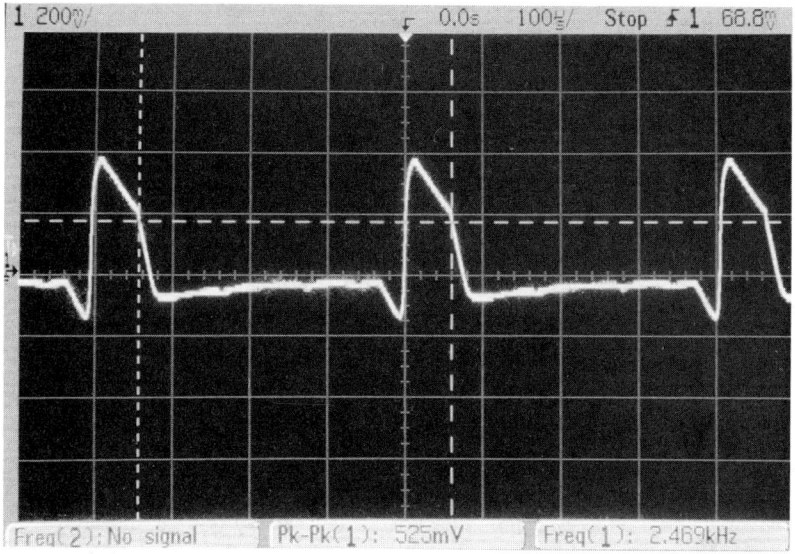

FIGURE 14-29   Vout of the circuit in Figure 14-28(a), an oscillating waveform at about 2.5 kHz, which is the incorrect frequency since the crystal is at 3.00 MHz.

The next thing to try is to see what happens if the output is loaded with 4700Ω as shown in Figure 14-28(b). We would expect no change in the waveform. But instead we see a larger amplitude waveform shown in Figure 14-30. The amplitude has increased from 0.5 volt peaked (with no load resistor) to 2.5 volts peak to peak. So, loading the output pin 4 unexpectedly increased signal level. This is indeed strange.

Using an ohm meter, it was found that the 74HC04's ground pin 7 was not tied to ground. After re-soldering the ground pin, the output signal looked normal as shown in Figure 14-31, and as depicted in the schematic in Figure 14-28(c).

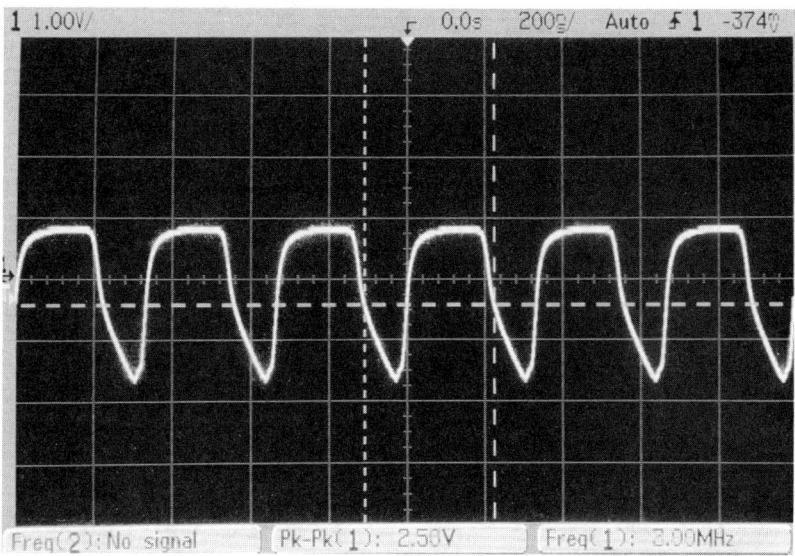

FIGURE 14-30 With Vout connected to an external 4700Ω load resistor, the frequency is now correct at 3.00 MHz, but the waveform shape and amplitude are incorrect.

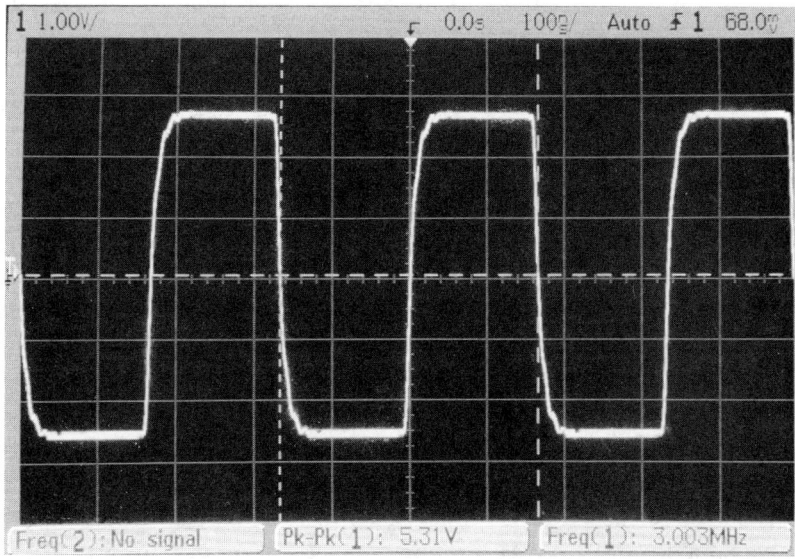

FIGURE 14-31 After connecting pin 7 of the 74HC04 chip to ground, Vout provided a 5-volt peak to peak signal at 3.00 MHz with the correct logic levels.

So, there are a couple of lessons here to take. One is to use an ohm meter to confirm the ground pin of the IC is really connected to ground. The other is that if you just scoped the waveforms and the logic levels are incorrect, check the ground pin for continuity to ground.

Ferrite Beads to Tame Parasitic Oscillations

When working with high-frequency oscillators, parasitic oscillations can occur. See Figure 14-32(a) and (b). Figure 14-32(b) provides a solution to stop parasitic oscillations via a base series resistor or using a series base ferrite bead.

The Colpitts oscillator uses a tuned circuit including inductor L2, variable capacitor VC_osc, and capacitive voltage divider circuit C4 and C5. Crystal Y1 works as a very high Q series resonant circuit. The variable capacitor is adjusted such that with L2 and C4 and C5, the circuit oscillates at the crystal frequency (e.g., 26.59 MHz). If there is not sufficient loop gain to sustain an oscillation, the loop gain can be increased. This can be achieved by having the collector current increased via lowering emitter bias resistor R3 (e.g., from 1000Ω to 560Ω), or the resistance of R4 may be increased twofold (e.g., to 68kΩ). Another way to increase loop gain is by decreasing C5's capacitance (e.g., from 680 pf to 330 pf) to provide less attenuation via the capacitive C4/C5 voltage divider circuit. Also, C4's capacitance may be increased from 33 pf to 39 pf while C5 can be decreased to 470 pf. Variable capacitor VC_osc will be readjusted to provide the (sustained) oscillation signal.

If the crystal oscillator's frequency is changed to be ≥ 50 MHz, the inductor, L2, may need to be wound with thick gauge wire to ensure high Q at these very high frequencies. For example, L2 may be wound with 18 to 14 AWG (American Wire Gauge) wire.

Essentially, positive feedback is performed by coupling the signal at VC (collector) via the voltage divider C4 and C5 to the emitter terminal via crystal Y1 that acts as almost a short-circuit at the crystal's stated frequency but a high impedance at other frequencies. The crystal then predominates in terms of setting the frequency and not L2, VC_osc, C4, and C5. Tuning VC_osc will vary the oscillator frequency extremely little, usually by less than 0.1 percent. By coupling the collector terminal's signal into the emitter via C4, transistor Q1 is a grounded base amplifier that provides no signal inversion from emitter (as the input) to the collector (as an output terminal). Q1's base is AC ground via the 0.01-μf capacitor C3. Note that if C3 is not low impedance at the oscillation frequency, the oscillator will not start up, and no signal will be at Vout. For example, if C3 is → 10 pf (due to soldering in a wrong value capacitor), the oscillator will not work (e.g., Vout → 0 volts AC) because at 26.59 MHz, a 10-pf capacitor has about 599Ω of impedance, which is too high an impedance value to effectively AC ground Q1's base.

A capacitor with capacitance, C, has an impedance magnitude in ohms, $|Z_c| = 1/(2\pi f_{crystal} C)$, where C is measured in farads, and $f_{crystal}$ is the crystal's frequency in Hz. For example, if $f_{crystal}$ is 26.59 MHz and C = 0.01 μf, then $|Z_c| = 1/(2\pi f_{crystal} C) = 1/(2\pi \times 26.59 \times 10^6 \times 0.01 \times 10^{-6})$ then we have in ohms: $|Z_c| = 1/(2\pi \times 26.59 \times 0.01) = 1/(1.67) = 0.599Ω$. An easier estimation for finding the impedance magnitude of 0.01 μf = 10,000 pf is if we know from the previous paragraph that 10 pf at the same frequency, 26.59 MHz, has 599Ω for $|Z_c|$, then we

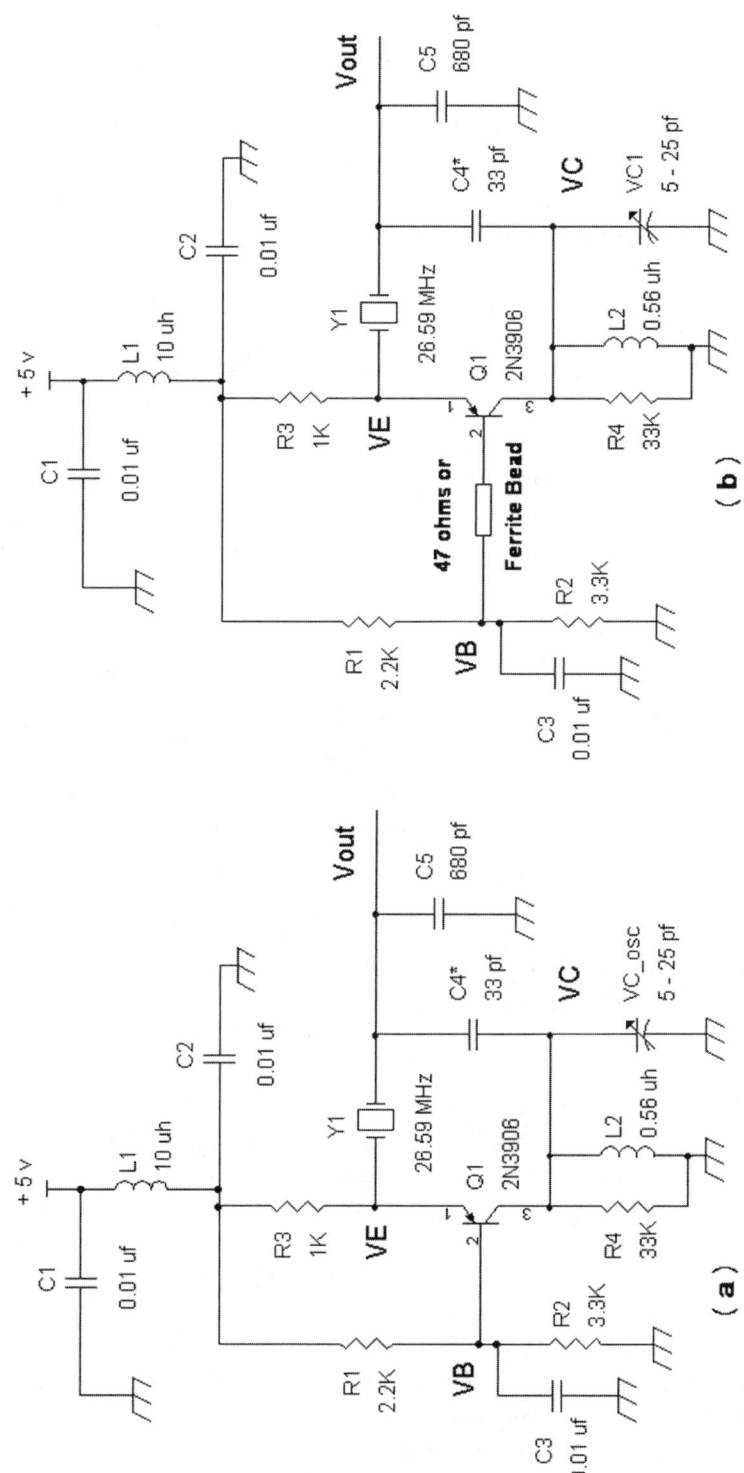

FIGURE 14-32 (a) A Colpitts crystal oscillator that may not oscillate at the crystal frequency due to parasitic oscillations; (b) a Q1 base series 47Ω or ferrite bead to remove the parasitic oscillation.

383

know that 0.01 μf equals 10,000 pf has to be 1000× lower in impedance magnitude of the 10-pf capacitor. The reason is the 10,000 pf is a 1000× multiple of 10 pf. Thus, the impedance drops from 599Ω to 599Ω/1000 = 0.599Ω. So, for all practical purposes, the 0.01-μf capacitor C3 is indeed like an AC short-circuit given its 0.599Ω impedance magnitude at the 26.59 MHz oscillation frequency.

The output signal is generally not taken from the collector terminal because any stray capacitances can detune the circuit easily. And if the collector terminal is loading into a medium to low resistance load (e.g., 1000Ω), then the oscillator can stop working due to (too much of a) reduced loop gain. Providing the output signal via the capacitive voltage divider at Vout generally gives the "cleanest" waveform. Capacitive voltage dividers C4 and C5 provide a lower impedance source, which can then drive other circuits. Emitter terminal signal VE can provide an output signal but it is a bit more distorted than the signal at Vout.

Figure 14-33 shows various implementations of adding series resistors and ferrite beads to a transistor to stop parasitic oscillations. This can be especially important if you use ultra-high-frequency (oscillator) transistors for Q1, such as the 2SA1161, a 3.5 GHz PNP transistor.

FIGURE 14-33   The left transistor has a solder series resistor, the center transistor has one ferrite bead slipped into its base lead, and the one on the right shows two ferrite beads installed.

One important item to note when using ferrite beads as shown in this photo, is to confirm with an ohm meter that the ferrite beads you slip into the transistor's base lead are indeed non-conductive. There are ferrite beads that can be slightly conductive, which will partially short out the base, collector, and emitter. Use the 2MΩ setting on your ohm meter to confirm there is infinite resistance.

Generally, the series base resistor can be on a printed circuit board, but if there is no space or if the board inadvertently omitted the series base resistor, you can solder one in as shown on the left side of Figure 14-33. The actual resistance value depends on the frequency you are operating at. Normally, you can start with about 22Ω, but it can go as high as 220Ω. Alternatively, you can slip one or two ferrite beads into the base lead. Having two beads stacked in series is sometimes needed if the parasitic oscillation exists with one bead. If you have a longer ferrite bead, you can use that in place of two beads. Now let's look at Figure 14-34, an output signal waveform from Figure 14-32(a) with some higher-frequency parasitic oscillation where the oscillator transistor (e.g., 2N3906) does not have a series resistor or ferrite bead.

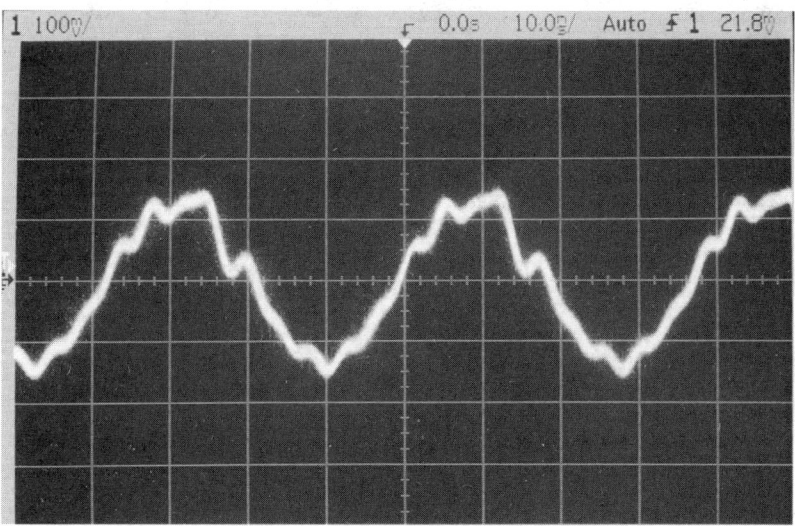

FIGURE 14-34   A "jagged" waveform at Vout due to a high-frequency parasitic oscillation signal riding on top of the lower frequency desired oscillator signal.

Now let's see what happens when a single ferrite bead is inserted into the base lead of the same transistor (2N3906) to stop the parasitic oscillation. Figure 14-35 shows a much cleaner output waveform at Vout from Figure 14-28(b).

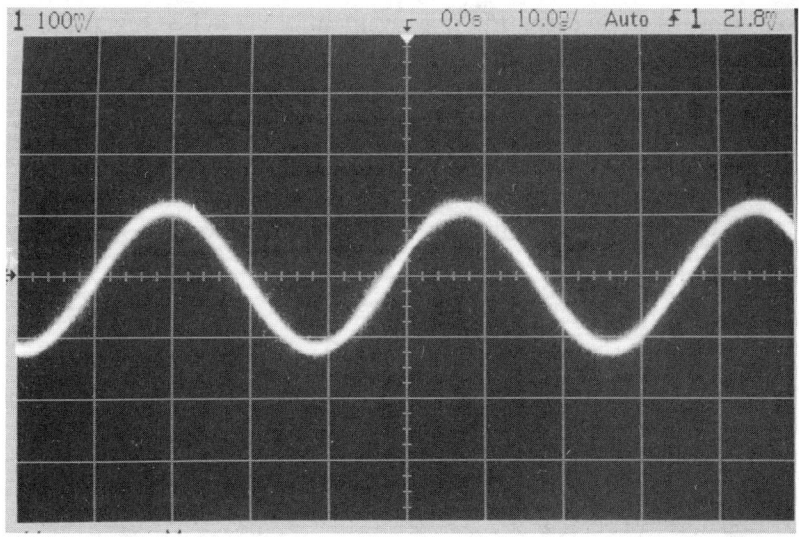

FIGURE 14-35   A single ferrite bead inserted into the oscillator transistor's base lead results in a much cleaner Vout waveform.

Similar results were achieved using a 47Ω series base resistor in the oscillator transistor's base lead. See Figure 14-36.

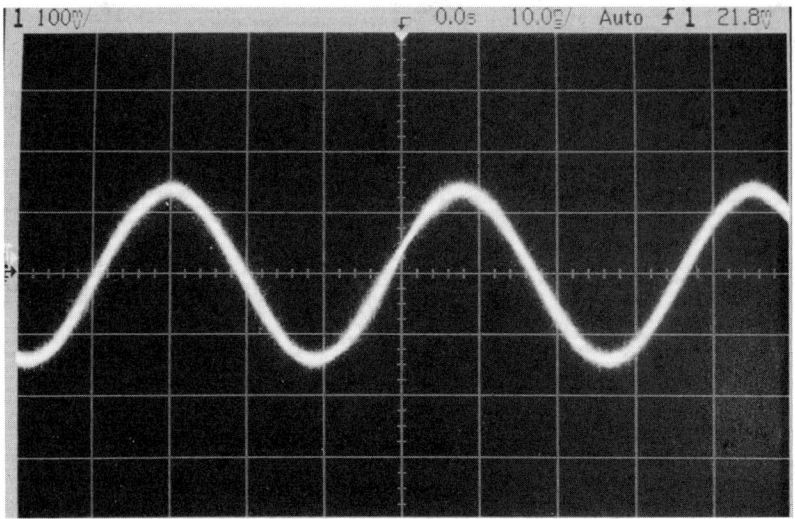

FIGURE 14-36   A series base resistor (e.g., 47Ω) stops the parasitic oscillation to deliver a cleaner sinusoidal waveform.

Summary

We showed in this chapter how to analyze not-so-well-designed circuits and then improved them. In some cases, the improvement or fix is just reversing the leads (e.g., the photodiode (D2PD) in Figure 14-10 or the capacitor (C7 = 100 μf) in Figure 14-23). Another example showed increasing the reliability of an audio power amplifier by adding a correct snubbing network at the output terminal (Figure 14-11, by adding series resistor Rext1 = 10Ω with C4 = 0.1 μf) to avoid inducing parasitic oscillations. Thus, troubleshooting sometimes requires that we look at the circuit design carefully because the original circuit may not have worked so well in the first place.

CHAPTER 15

More Tips and Final Thoughts

We have looked at many types of analog circuits on the individual component level with resistors, capacitors, diodes, LEDs, transistors, and ICs.

However, when we troubleshoot circuits, sometimes we have to work from a schematic diagram that is not clearly drawn. In this chapter, we will present an example of how certain harder-to-read schematics are actually drawn today (2019) and how to fix them.

If you are stuck trying to figure out how to troubleshoot particular circuits such as high-fidelity amplifiers, radio circuits, etc., sometimes it is easier to download the service manual or schematics of similar circuits via the web and compare them with yours. For example, if you are troubleshooting a stereo power amplifier, you can go to websites such as:

- **https://www.hifiengine.com/** Or you can pay a fee for schematics via SAMS Photofact.
- **https://www.samswebsite.com/** If you have the manufacturer's model number or chassis number, then you may find used copies via eBay. For example, if you have a Lafayette LR-4000 stereo receiver, you will find that that the SAMS Photofact booklet is "MHF-58," for which you can do a search on the web.

The manufacturers' service manual often includes DC bias point voltages, and some AC signal waveforms along with setup procedures.

There are other websites that post their schematics and service manuals and these can be used as a reference for troubleshooting. What you are doing is looking for products that should already work, and their circuits will generally reveal good engineering designs, but not always. The reason for looking at various commercial products' schematics is to get a feel for how these circuits are designed well enough for repeatability.

I will also share with you how one "retro" circuit is driving analog meters the "wrong way" and the right way. Also, from many years ago, I will show how a circuit was fixed without fancy test equipment by looking for a similar schematic diagram.

Before I give my final thoughts, we will cover some basic problems due to bad connections on IC sockets and how to choose a better one for reliable operation.

Deciphering Schematics with Too Many Connection Flags

We will now look at a schematic example that is scattered and can be rearranged to show a clearer idea of what the circuit does. See Figure 15-1.

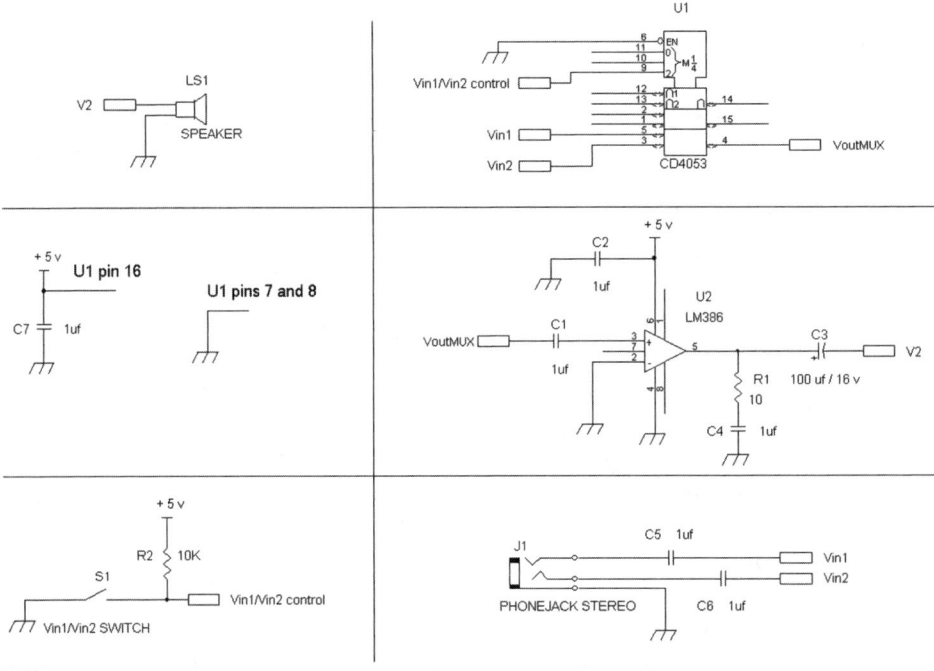

FIGURE 15-1 A "disjointed" schematic using net flags such as V2 from C3 connected to LS1.

We can start rearranging the schematic as shown in Figure 15-2 by connecting all the net flags and having the input on the left side and the output at the right side.

We can further improve the schematic by replacing U1 CD4053 with a better symbol.

Because U1 in Figure 15-2 looks more like a "black box," we should find a schematic symbol that shows its operation. See Figure 15-3 that shows U1 having three analog switches.

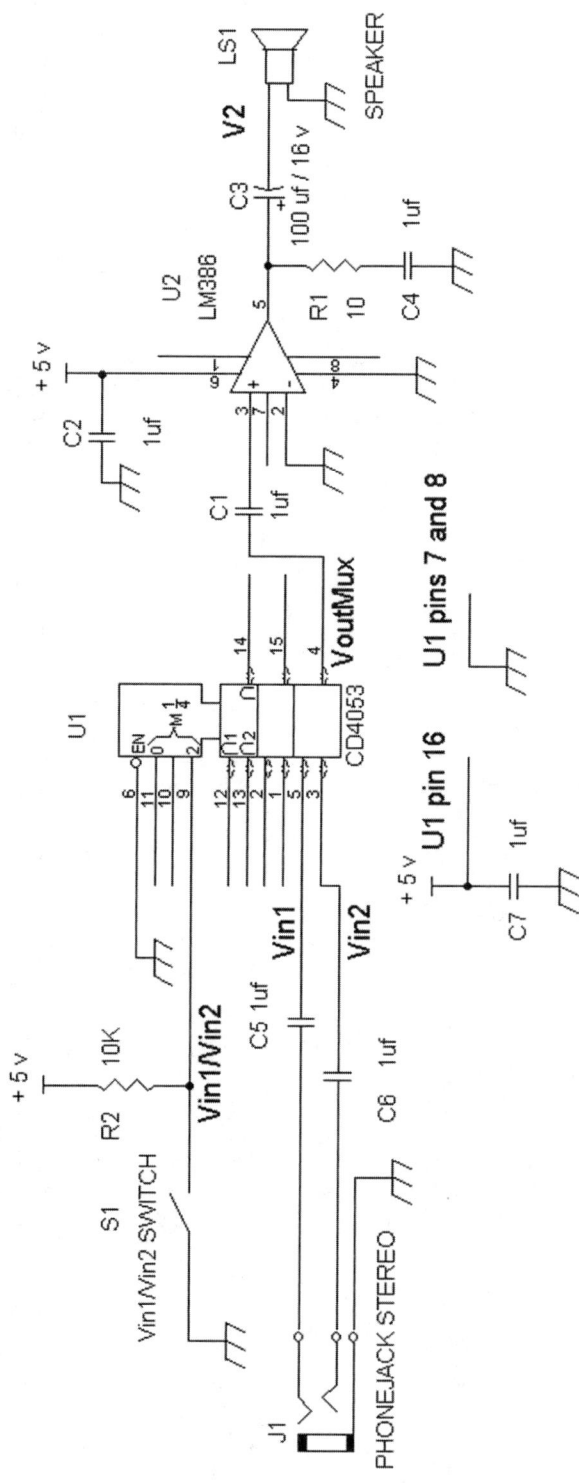

FIGURE 15-2 A rearranged schematic that is starting to make more sense, but does it work?

389

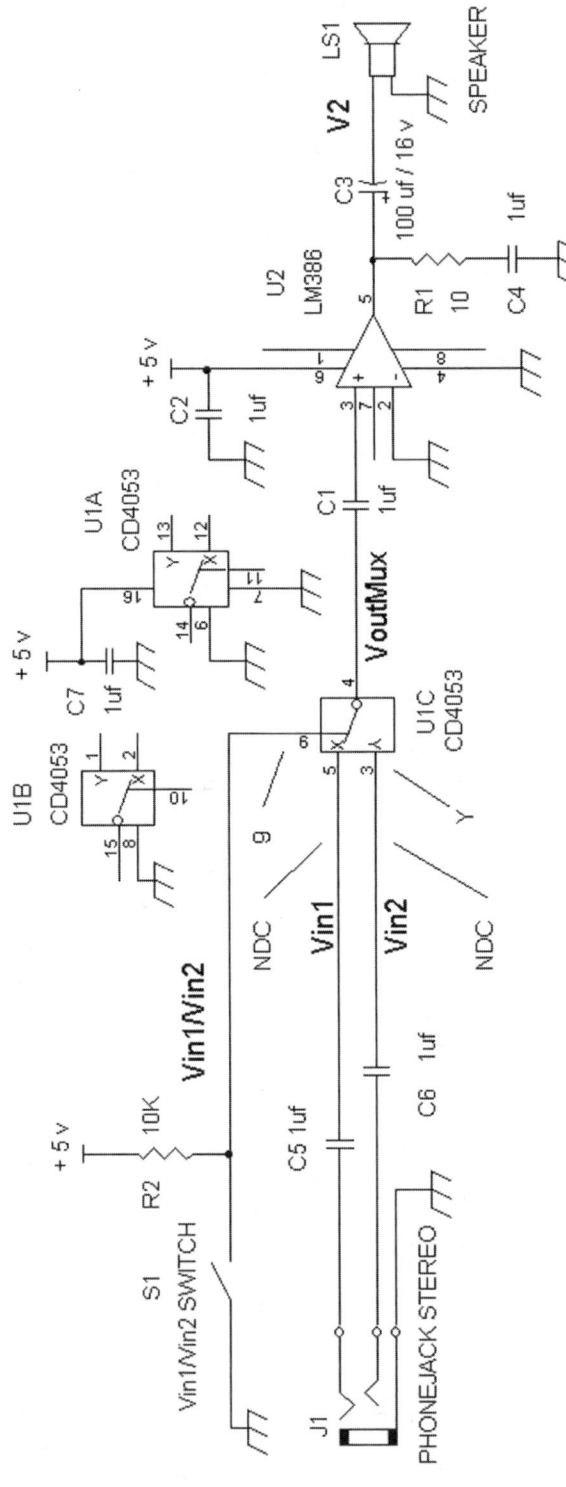

FIGURE 15-3 The CD4053 is revealed to be an X-Y two-channel analog switch that passes either an input signal from Vin1 or Vin2 via selector switch S1. Amplifier U2 LM386 then amplifies the selected input voltage for playing back on loudspeaker LS1.

390

From Figure 15-3 we see that the CD4053 switch is working off a single +5-volt supply due to its –VEE pin 7 being connected to ground instead of a negative supply. This means that the X and Y inputs at pins 5 and 3 should be biased to a DC voltage typically at one-half the DC supply voltage (e.g., 2.5 volts) for maximum AC signal swing. As shown in Figure 15-3, Vin1 and Vin2 are floating DC-wise and have no DC (labeled NDC) bias voltage.

As we can see by rearranging the schematic circuits from Figures 15-1 and 15-2 and inserting a more detailed circuit version of the CD4053, DC bias problems are found. We can now fix the circuit as shown in Figure 15-4 by adding a 2.5-volt biasing circuit.

In Figure 15-4, R5 and R6 form a voltage divider circuit that provides 2.5 volts into filter capacitor C8's positive terminal. Capacitor C8's capacitance is high so as to provide an AC short circuit to ground. This way, any AC voltage at R3 and R4 cannot cause crosstalk to each other. Without C8, some signals from Vin1 will "leak" over to Vin2 via R3 and R4. But with C8 as an AC ground we provide good isolation between the input signals Vin1 and Vin2. Capacitor C8 also reduces noise from the +5-volt power supply so that the output signal to the speaker is relatively noise free (e.g., no hum or no extraneous power supply noise).

Also note that there is a signal coupling capacitor C1 into U2 (LM386) pin 2 that does not seem to have a DC bias resistor. It turns out that the LM386 has an internal 50kΩ resistor from each of its input terminals (pin 3 and pin 2) to its ground pin 4.

When switch S1 is open circuit the Vin1/Vin2 control signal is logic high via +5 volts from R2 to control pin 9 of the CD4053, which selects the "Y" channel or Vin2. To select Vin1 or the "X" channel, S1 is closed, connecting R2 to ground. This (then) sends a logic low signal to control pin 9 of the CD4053.

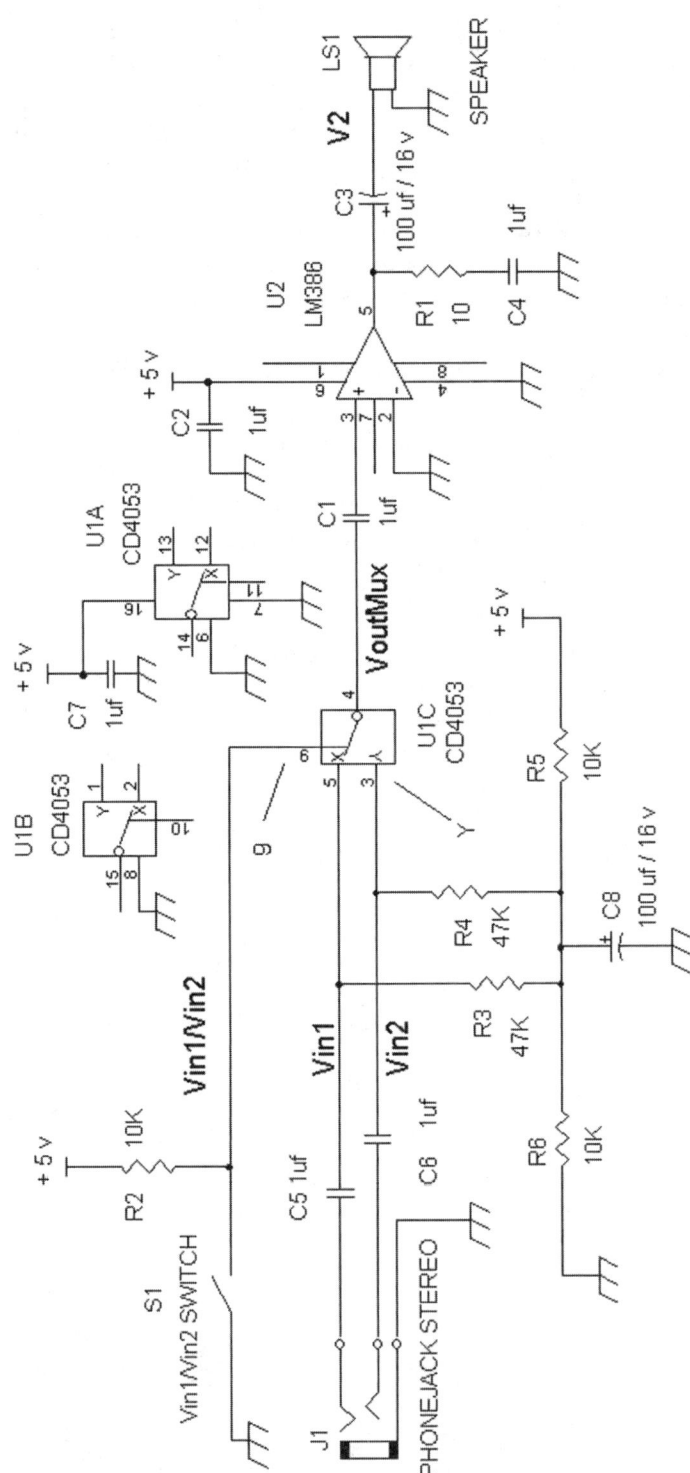

FIGURE 15-4 Adding DC bias circuit R5, R6, and C8 provides 2.5 volts via R3 and R4 to the CD4053's input pins 5 and 3 for Vin1 and Vin2.

Troubleshooting with Minimal Test Equipment

The next two sections will show methods that do not necessarily require any test equipment. Instead, we will use deductive reasoning or examples of circuits that are already commercially manufactured. The examples shown will be from my own experience before I worked in a TV/radio repair shop and also before I decided to study electronics formally. As a hobbyist back then, all I had were example circuits from other equipment.

Analog Meter Driving Circuits for AC Signals

This will be a story about learning how to drive analog meters correctly. See Figure 15-5.

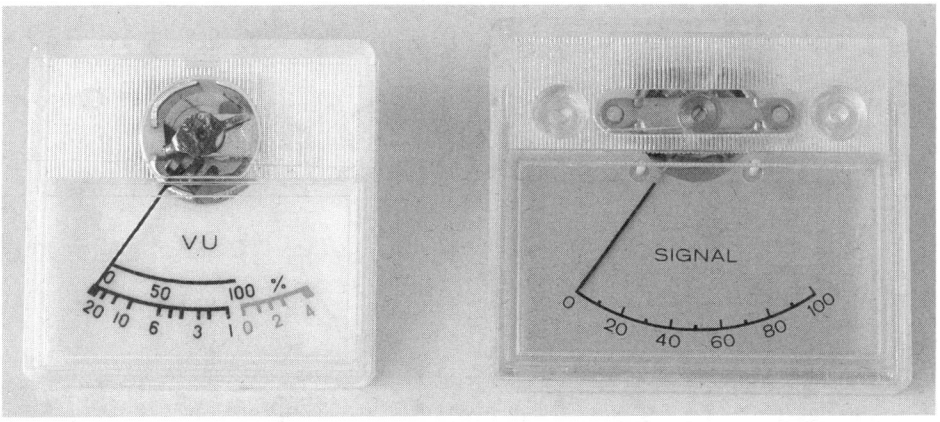

FIGURE 15-5   An analog "VU" meter (Volume Unit audio meter) and a signal strength meter.

A circuit that works for sure to convert AC signals into DC currents is a full-wave rectifier. See Figure 15-6(a), which uses an AC coupling capacitor C1, where typically R1_meter ≥ 2000Ω to prevent burning out the analog meter, M1.

With C1 the full-wave rectifier works well, and this circuit is used in many commercially made audio mixers such as the Shure M67. But what if you do not have four diodes for full-wave rectification as shown in Figure 15-6(a) and decide to use a half-wave rectifier circuit in Figure 15-6(b) instead? What you will find out is that the input capacitor C1 will block any DC current once it is charged up and the meter M1 will return to zero and will not operate as expected.

NOTE: Meter M1 is usually rated at 50 µA to 1000 µA full scale and R1_meter will be in the range of 2000Ω to 20kΩ. For example, the VU meter in Figure 15-5 (left

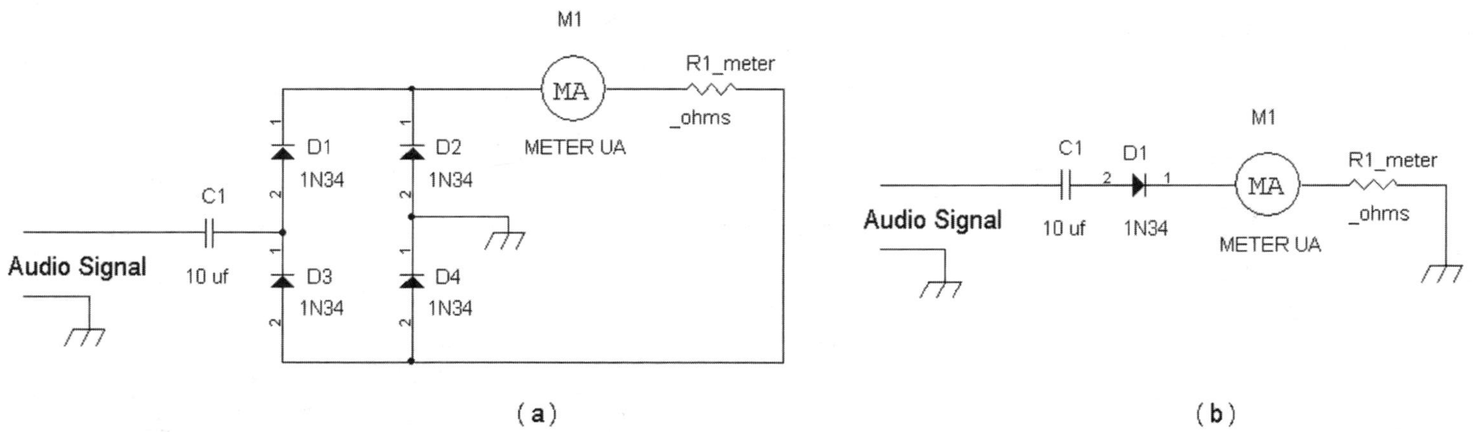

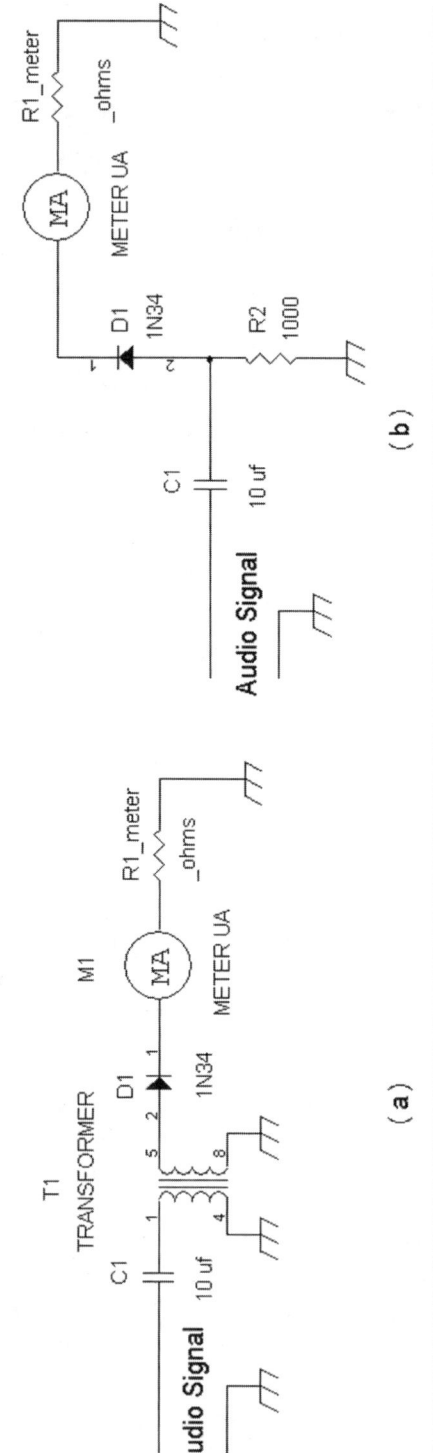

FIGURE 15-6   Full rectifier circuit that works (a) and one that does not work because of C1 (b).

FIGURE 15-7   (a) Meter circuit now works by using a transformer; (b) an alternative using R2.

side) has a full-scale reading at about 150 µA. Meter M1 has a 600Ω DC resistance, whereas the signal strength meter on the right side of Figure 15-5 has a 1200Ω DC resistance. Series resistor R1_meter prevents excessive current into meter M1.

CAUTION: Connecting a voltage source directly to meters usually results in burning them out. This is why a series resistor, R1_meter, is used to protect the meter, M1.

One idea we can "borrow" from power supply circuits is to use a transformer, which for sure produces a pulsating DC current with a half-wave rectifier. See Figure 15-7(a).

At the time the solution was in fact using an audio transformer such as a 10kΩ to 2kΩ primary to secondary impedance ratio transformer. In Figure 15-7(a), transformer T1's primary winding at pins 1 and 4 allows for currents to flow in both directions through C1. Put in other words, AC currents are allowed to flow through C1 and T1's primary winding. An AC signal voltage is then transferred to the secondary winding at pins 5 and 8, such that D1 rectifies the AC voltage at the secondary winding and provides a pulsating DC voltage to the M1 that then displays an average DC current set in part by series resistor R1_meter (e.g., 10kΩ).

In Figure 15-7(b) resistor R2 provides a DC path to ground and also allows currents to flow in both directions through C1. By simply adding R2, the meter circuit now works. This is a simpler solution given that resistors are more common than transformers.

Another circuit we can borrow from power supplies is the voltage doubler circuit. See Figure 15-8.

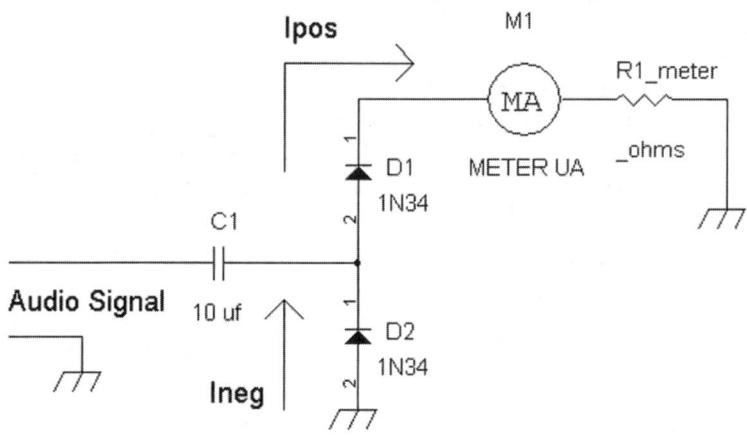

FIGURE 15-8 A voltage doubler circuit with additional diode D2 gives yet another circuit solution.

Diode D2 may be looked upon as a DC restoration circuit with C1, which means it level shifts the audio AC signal such that negative peak of the AC signal is "clamped" to about 0 volts or ground level. Diode D2 also provides a path for C1 to allow for signal flow in the negative direction (Ineg), while D1 allows C1 to flow currents in the positive direction (Ipos). At D1's anode there is already a pulsating DC signal, and D1 ensures only positive going currents, Ipos, are sent to meter M1.

NOTE: The diodes used in most analog meter circuits are small signal germanium diodes such as 1N34, 1N60, 1N270, etc., to minimize forward diode drop losses (e.g., ≤ 200 mV). You can use small signal Schottky diode replacements such as BAT42, BAT43, BAT85, and BAT86 with ~ 300 mV turn-on voltage. Using standard diodes such as 1N914 or 1N4148 will have turn-on voltages in the 500 mV to 600 mV ranges, which means the meter circuit will require very large AC voltages greater than 5 volts peak to peak to have reasonable measurement accuracy. To avoid audio distortion caused by the meter circuit's nonlinear load, a buffer amplifier can drive the meter circuit to isolate the main audio output signal path. Or instead, a near 0 Ω output resistance amplifier can drive a meter circuit and provide an audio output.

If you are wondering why the meter circuits are not directly connected to an op amp with positive and negative power supplies, then you have to consider start-up conditions and catastrophic failure modes. When an op amp is first powered on, there may be a large DC offset transient that may damage the meter. Having a DC blocking series capacitor (C1) ensures protection against long- or short-term DC voltages into the meter. Also, if a directly coupled amplifier such as an op amp fails, it may deliver a large sustained DC current into the meter that can cause damage.

Troubleshooting an Older Push Pull Audio Amplifier in a 1950s Transistor Radio

This is a story where the problem was solved over many years. The circuit was put aside because after replacing virtually every resistor and capacitor, there was still a problem. Since I did not have any more than an analog VOM and no schematics, I was not able to troubleshoot the problem. Also, I was just starting out in learning electronics at a time way before attending high school. See Figure 15-9.

The problem was the radio had a low volume and it sounded somewhat distorted.

JUST A REMINDER: I was just starting out in repairing radios like this and had no formal electronics background. And I only knew some basics like Ohm's law but had no idea how to apply them because I had not taken algebra yet.

So how would one go about troubleshooting without much electronics resources? For example, I just had an analog VOM and no signal generator. At first, I "shotgunned" it by replacing every resistor, capacitor, and transistor, but this did not help. It did not improve anything.

FIGURE 15-9   A General Electric model/chassis number P-716A six-transistor radio (year 1958).

The way to signal trace the circuit is to hack another working transistor radio, which I had, to make it an audio signal tracer or speaker amplifier. I can then trace the audio signal at the General Electric P-716A's volume control and subsequent audio stages. The audio via the makeshift audio signal tracer showed that the radio appeared to be fine up to the first audio transistor, but not at the speaker output.

So, I also traced out a partial schematic of the radio. Note that back in those days, there were no silkscreen reference designations for the electronic components. Also, these electronic parts such as resistors and capacitors were loaded with uneven spacing and at different non-standard angles (e.g., at 30 degrees, 45 degrees, etc.). Today's printed circuit boards are usually loaded with parts at a reference zero degrees, such as East-West, or 90 degrees, such as North-South. Because the boards had a single layer of copper traces, components often had longer spreads to jump over traces beneath them. For example, the standard 400 mil (0.4 inch) spacing for quarter watt resistors may be extended to anywhere from 500 mils to 700 mils and mounted at an angle (referenced to either edge of the board). The type of component arrangement makes it hard to identify if all the components are put in their proper locations. See Figure 15-10. Note all audio transistors, Q1, Q2, and Q3 are germanium transistors.

If you are familiar with transistor amplifiers, you can spot the problem right away. But what if you lack the knowledge and need help because you think that the circuit in Figure 15-10 looks "reasonable" as a working audio amplifier? One way is to find a similar radio with a schematic diagram. See Figure 15-11.

The Arleigh radio has a similar audio amplifier circuit so we can trace its circuit by taking it apart, or if there is a schematic available, we use that instead. Fortunately, there is a schematic diagram printed inside its back cover. See Figure 15-12.

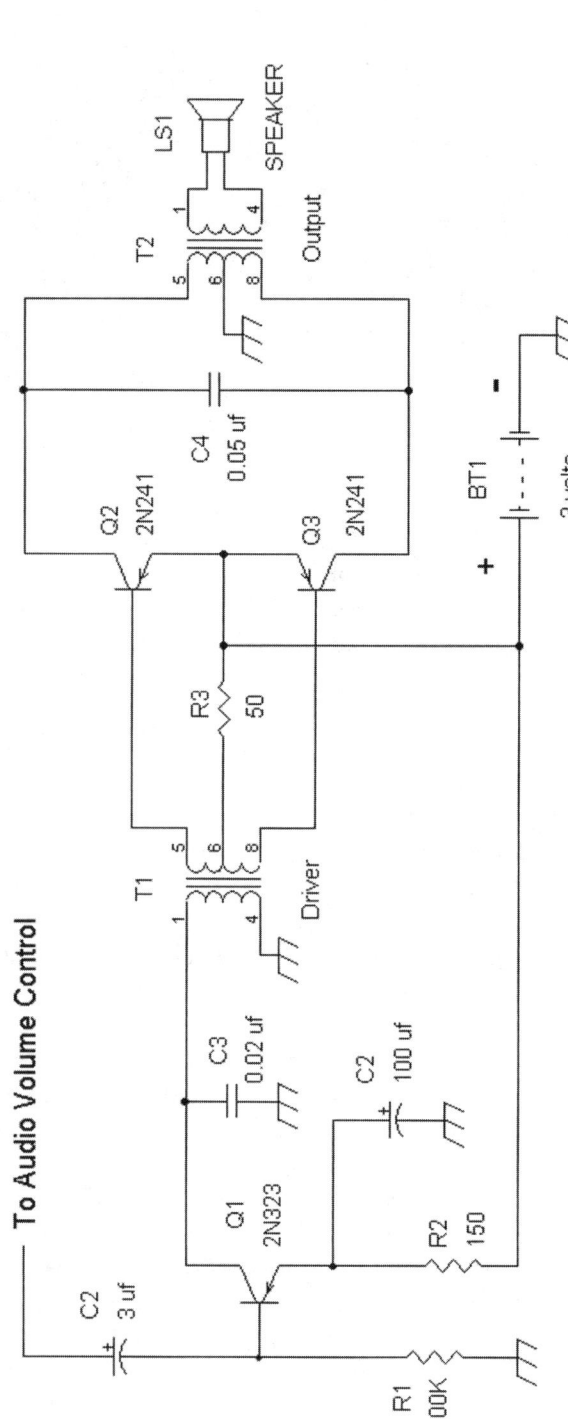

FIGURE 15-10 A partial traced-out circuit diagram of the GE transistor radio model P-715A, which uses two AA batteries to supply 3 volts.

FIGURE 15-11 An "Arleigh" six-transistor radio that has a similar audio amplifier as the GE radio.

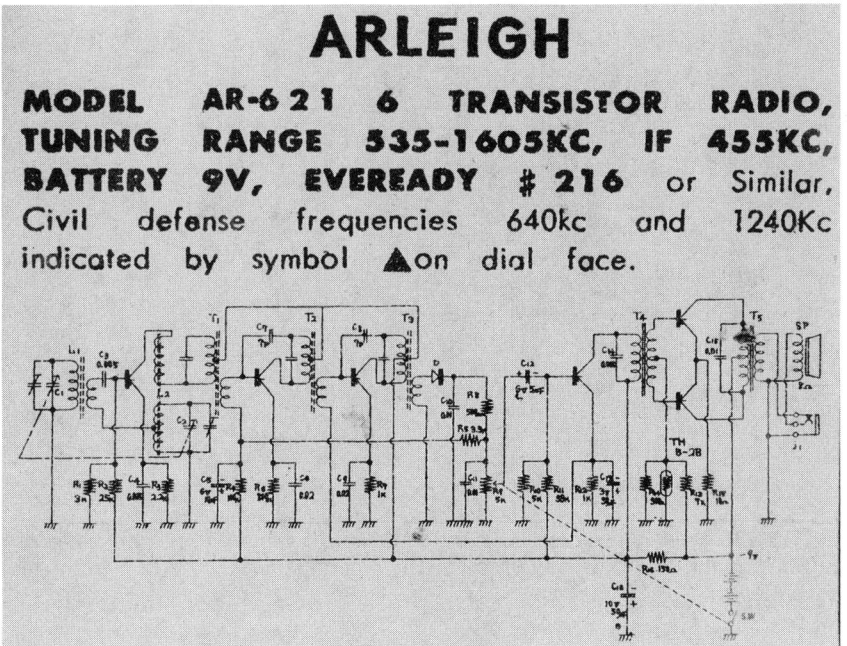

FIGURE 15-12 Arleigh radio's complete schematic.

NOTE: There are biasing resistors (R13 and R14, which are not clearly seen here) connected to center tap of transformer T4 in Figure 15-12.

There is also a thermistor (temperature dependent resistor), TH B-2B, which is connected in parallel with R14.

For a closer look, see Figure 15-13, where extra notes have been added:

- The audio driver transistor is labeled as Q4, and the audio output transistors are labeled as Q5 and Q6.
- Resistors R13, R14, and R15 are identified. Resistor R15 = 10Ω, so essentially the emitters of output transistors Q5 and Q6 are connected to ground, or more importantly, they are connected to the plus terminal of the 9-volt battery.

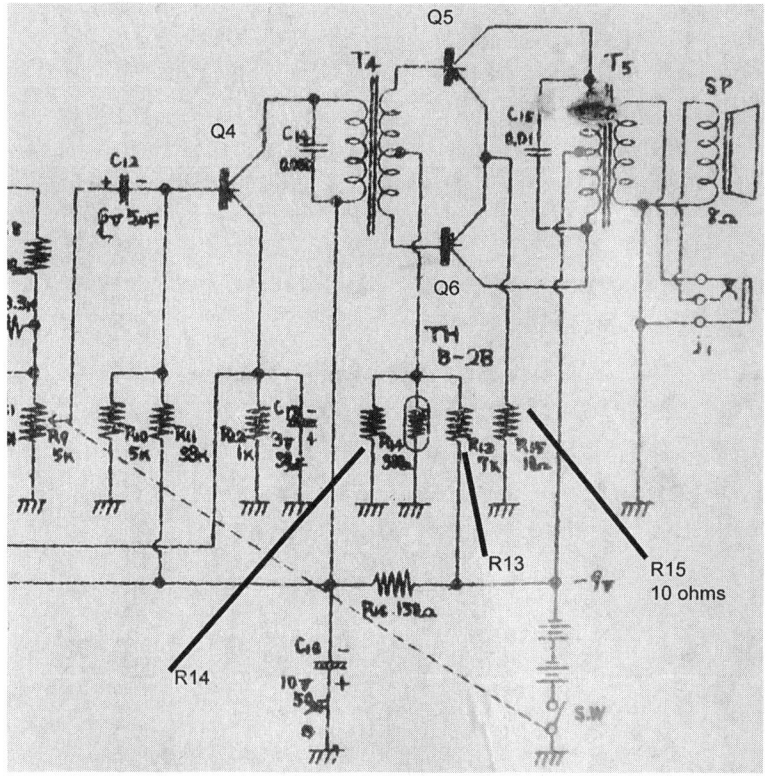

FIGURE 15-13 Biasing resistors R13 and R14 are connected to the center tap of driver stage transformer T4 of the Arleigh radio. The ground is connected to the plus terminal of the battery.

If we just look at the connections to the 9-volt battery, its positive terminal is connected to ground via the on/off switch, S.W., and the battery's negative terminal is connected to the lower side of R13 of Figure 15-13. The emitters Q5 and Q6 via R15 are essentially connected ground since R15 is 10Ω. This means the emitters of Q5 and Q6 are essentially connected to the positive terminal of the battery via S.W., connected to ground and the positive terminal of the battery. The lower side of resistor R14 is connected to ground, which means the lower side of R14 is also connected to the positive terminal of the 9-volt battery via S.W. In essence, R14 is connected across the center tap of T4 and "indirectly" connected to the emitters of the output transistors Q5 and Q6 (via the 10Ω resistor R15 in Figure 15-3).

We also have R13 at 7kΩ with one side connected to the negative terminal of the battery and its other side connected to the center tap of T4 and R14. If we look carefully, R13 with R14 and TH B-2B form a bias network for turning on Q5 and Q6. Thermistor TH B-2B is connected in parallel with R14.

This bias voltage across the emitter and base of Q5 and Q6 will be in the order of about 0.25 volts, which is the turn-on voltage of germanium transistors.

So, the mystery problem we had was that there was a part missing!

Now let's look at the GE radio's schematic again; this time we will add the missing biasing resistor. See Figure 15-14.

The R4 in the GE radio is sort of like the R13 (7kΩ) in the Arleigh radio. And R3 in the GE radio is like R14 and TH B-2B in the Arleigh radio. So, where do we start? Since the Arleigh radio uses 9 volts instead of the 3 volts in the GE P-715A, we can scale R13 down to about one-third of its value or R4. So R4 ~ R13/3 or 7kΩ/3 ~ 2333Ω ~ **2200Ω = R4**.

With R4 ~ 2200 ohms the voltage across **R3 (50Ω)** in Figure 15-14 is: 3 volts × [R3/(R3 + R4)] or 3 volts × [50/2250] ~ 0.067 volt. We will get an improvement in sound with R4 = 2200Ω, but it is still not good enough. If we reduce R4 to 1000Ω, the voltage across R3 will be 3 volts × [50/1050] ~ 0.143 volt. This will be a dramatic improvement and could work out OK because germanium transistors have a turn-on voltage range between 0.10 volt and 0.3 volt.

Over 20 years later when I finally found the SAMS Photofact schematic for the GE P-715A radio, it clearly showed that the R4 value was 680Ω, which would yield across R3 a voltage of 3 × [50/(50 + 680)] ~ 0.206 volt.

So, yes, if I had found the schematic back then via SAMS Photofact, I would have solved the problem immediately. However, there will be times when the circuit diagram is not available and with limited resources (and sometimes limited electronics knowledge), we have to find other ways to troubleshoot the problem.

NOTE: Had I looked at the GE radio's printed circuit board very carefully, I would have found two pads or two small holes that were there for a missing part.

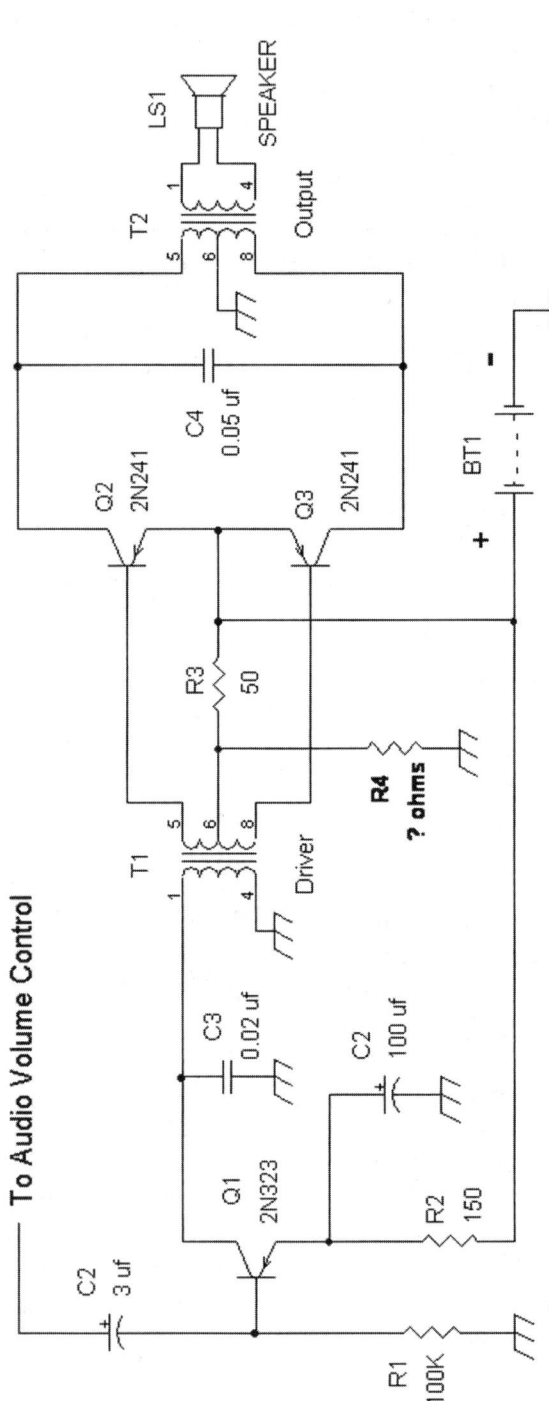

FIGURE 15-14 GE radio schematic with added resistor R4 (in bold fonts) connected to the center tap of T1 and to ground or equivalently to the negative battery terminal.

Thus, one way is to learn from an example of a known working circuit. These days you can find example circuits by various manufacturers' products via websites such as:

- http://www.nostalgiaair.org/Resources/ for old antique radios and television circuits
- http://bama.edebris.com/manuals/ for almost everything related to test equipment, HiFi products, etc.
- https://www.vintage-radio.info/heathkit for Heathkit schematics and manuals
- https://www.vintageshifi.com/repertoire-pdf/Dynaco.php for Dynaco HiFi stereo products (preamps, power amps, and tuners)

Also, you may find specific schematics posted on the web by using Google. For example, if you need to know product or model numbers of certain products, you can look through archives of older electronics or stereo magazines such as: https://www.americanradiohistory.com/High-Fidelity-Magazine.htm for High Fidelity Magazine or https://www.americanradiohistory.com/Audio-Magazine.htm for Audio Magazine. Once you have the manufacturer's model number, you can look up a service manual or schematic from https://www.hifiengine.com/.

Reducing Noise on the Power Supply Bus with Multiple Circuits

When a large system having multiple circuits is supplied with a common power supply voltage, each circuit can generate noise back into the common power supply bus or line. See Figure 15-15. In the ideal world VCC = VCC1 = VCC2 = VCC3, but this does not really happen.

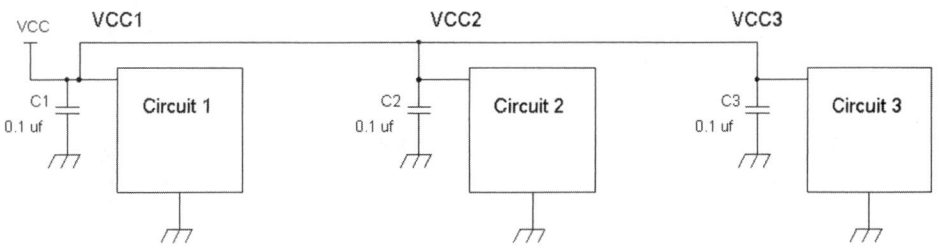

FIGURE 15-15   A common VCC voltage supplying three circuits where the drain from each circuit can generate its own noise at its supply node, VCC1, VCC2, and VCC3.

While in the ideal world, VCC supplying all three circuits would have no noise, each circuit can contribute to its own noise due to AC signals running in Circuit 1, Circuit 2, and or Circuit 3. For example, if Circuit 1 is an oscillator circuit, and Circuit 2 is a microprocessor circuit, and Circuit 3 is a memory circuit, each circuit will be turned on and off in such a way to cause transient noise spikes that may be at different times. These noise voltage transients along the power supply lines mean that VCC1 ≠ VCC2 ≠ VCC3 at different times. If the noise transient spikes are large enough, they can interfere with proper signal processing for each circuit. One solution is to isolate the noise transient spikes and improve power supply decoupling for each circuit. See Figure 15-16 where the circuits are isolated with inductors L1 and L2.

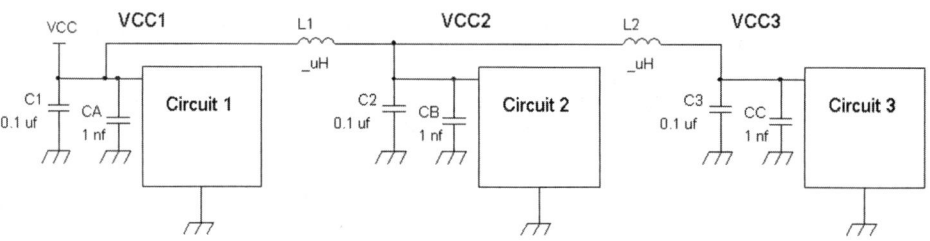

FIGURE 15-16 Circuits 1 to 3 have series inductors forming a pi filter (π filter) with C2/CB, L2 and C3/CC and another with C1/CA, L1 and C2/CB.

A pi filter refers to an analogous mechanical construction having two "pillars" supporting a horizontal bar that is denoted by π. The capacitors form the pillars and an inductor forms the bar.

In Figure 15-16, typical inductance values for L1 and L2 range from 1 μH to 100 μH. Note that added capacitors CA, CB, and CC are lower in capacitance at 1000 pf (1 nf) for better decoupling at higher frequencies such as ≥ 50 MHz. If the circuits are running in the GHz range, the decouple capacitors have to be reduced in value accordingly, such as CA, CB, and CB = 220 pf for power supply decoupling. The reason is that at higher frequencies virtually every capacitor includes a (significant) series inductor. That series inductor decreases with smaller capacitance values, which allows the capacitor to effectively remove high-frequency noise at the power bus. For example, if your circuit is running at 100 MHz, using a 0.1 μf to 1 μf capacitor does almost nothing in terms of acting like an AC short circuit at that frequency. The reason is because 0.1 μf or more has a relatively large series inductor that negates a low impedance across the capacitor at 100 MHz. However, a 1000 pf capacitor has a much smaller series inductance and will be effective in acting like an AC short circuit at 100 MHz.

Figure 15-17 shows another example of isolating noise from the power bus (e.g., VCC).

In Figure 15-17, if there is noise at VCC1, VCC2, and or VCC3, each noise component will be filtered or attenuated at VCC via inductors L1, L2, and L3. Any noise at VCC then gets filtered again via L1/C1/CA, L2/C2/CB, and L3/C3/CC. This arrangement requires more components to be added but can be very effective in ensuring power supply noise from one circuit (e.g., Circuit 1 or Circuit 2 or Circuit 3) does not "contaminate" the supply voltages of the other circuits.

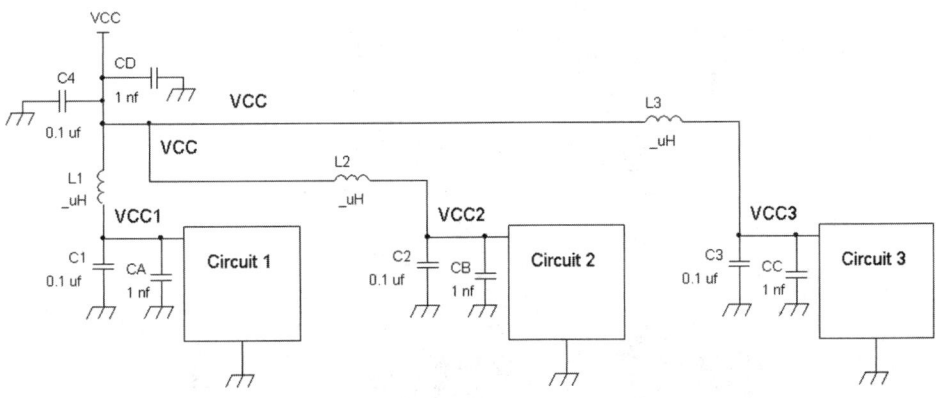

FIGURE 15-17   The common supply voltage VCC is spread out via inductors L1, L2, and L3.

In general, for Figure 15-16 or Figure 15-17, each of Circuit 1, Circuit 2, Circuit 3, and so on can include a single stage amplifier such as an IF (intermediate frequency) amplifier, RF (radio frequency) amplifier, and or logic gate/circuit. The inductors L1, L2 . . . may be replaced with ferrite bead inductors. If the current draw of each circuit is not too high, such as < 10 mA, the inductors may be substituted with a 10Ω to 22Ω resistor, but the isolation characteristics in terms of power supply noises between the circuits may not be as high when compared to using inductors (e.g., L1, L2, and L3).

Bad Connections from Some IC Sockets

Whenever you purchase IC sockets, especially the ones that are "all plastic," there may be a chance that they may not make good contact or electrical continuity with the IC's pins. And if the integrated circuit itself has slightly oxidized pins, some sockets may cause an open circuit between the pins of the IC and the IC socket terminals. See Figure 15-18.

An IC is installed on a "fully" plastic socket in Figure 15-18(a). These types of sockets are generally fine but can have continuity problems. To be sure, use an ohm meter to measure each pin of the IC with its corresponding IC terminal lead. Figure 15-18(a) shows an example of how to measure pin 1 of the 8-pin IC, but you will have to go through every pin for a continuity check via an ohm meter.

Next, Figure 15-18(b) shows the unloaded plastic socket. Sometimes if you have already soldered this type of socket and found a continuity problem, one way to resolve the issue to remove the IC from the socket (e.g., via a small flat blade screwdriver or with an IC pulling tool) and then reinsert the IC back into the socket. By doing this, you may have "cleaned" the connections.

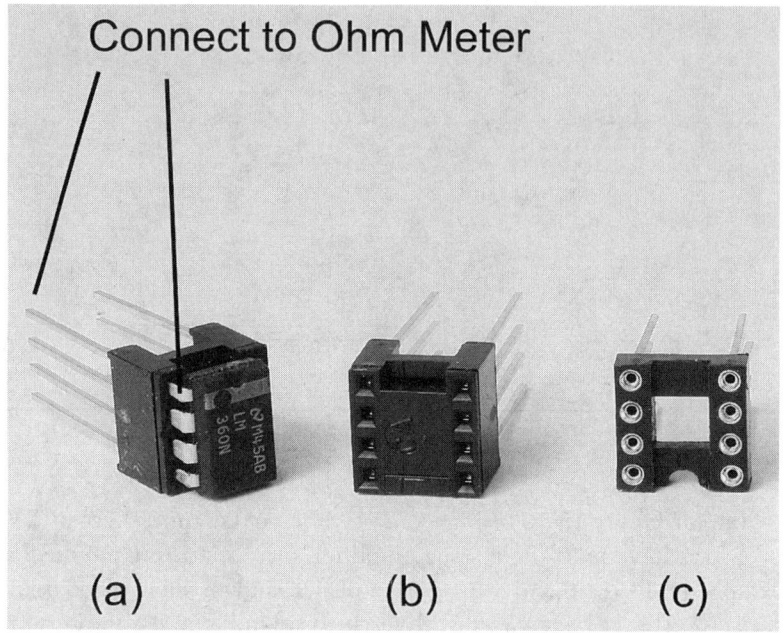

FIGURE 15-18   Full plastic sockets shown in (a) and (b), with a socket with round pins for better reliability shown in (c).

If you want higher reliability, you can use a socket with round pins as shown in Figure 15-18(c). The round "wells" allow self-cleaning of the IC's pins by wiping against the edges of the IC pins to ensure better connections.

Summary and Final Thoughts

We've seen that troubleshooting circuits includes the following:

- A knowledge base of components and how to use them properly. This includes operating them below their maximum voltages/currents/power ratings (e.g., capacitors, transistors, resistors, logic circuits/gates, etc.) and connecting them with the correct polarity (e.g., electrolytic capacitors, LEDs, transistors, ICs, etc.).
- Using your volt meter to confirm the DC voltage operating points in your circuits and power supplies are at the correct voltage.
- Using your ohm meter to measure resistors with the circuit power shut off and with at least one lead of the resistor disconnected.
- You can buy DVMs to measure capacitors, which is very important since many capacitors have the same size but very different capacitances (e.g., 10 pf to 1 μf). Thus, it can be easy to install the wrong valued capacitors.

Also, if you are working with surface mount capacitors, they may not be marked, so a capacitance meter comes in handy.

- Most DVMs will also allow you to check NPN and PNP transistors and diodes. This is very useful since the pin outs to transistors are not the same such as a 2N3904 (emitter, base, collector) versus a BC550 (collector, base, emitter), or an ultra-high-frequency transistor 2SA1161 (base, emitter, collector).

- Look up specifications for your parts and download a "library" of data sheets. This includes transistors, ICs, and diodes. Keep the data sheets handy for reference. For example, if we look for low–drop-out (LDO) voltage regulators, then we will see that there can be two different types. There is one with a typical drop-out voltage ~ 0.5 volt (e.g., adjustable voltage LM2941 or fixed voltage LM2940) that will have a PNP output transistor's collector connected to the output of the regulator. However, there is another one with higher drop-out voltage of ~ 1.0 volt to 1.2 volts (e.g., adjustable voltage LM1117), where the emitter of the NPN output transistor is connected to the output terminal. Note that the LM317 with a Darlington emitter follower output circuit is a standard adjustable voltage regulator, and it has a typical drop-out voltage of about 1.8 volts. Studying data sheets is a must if you want to ensure the circuits are using the correct electronic components.

- If you are using power supplies, always turn the settings to 0 volts for each output before connecting to your circuit because the supply may have been already set to an excessively high voltage that will damage your transistors or ICs. Set the voltage first, confirm with a DVM, confirm the polarity, and then connect the power supply leads to your circuit. If the power supply has a current limit feature, set it at about 100 mA unless you know your circuit drains more than that, which means you can increase the current limit.

- When using positive and negative supply voltages, you can need to connect a common lead such as the (–) lead of the positive voltage source to the (+) lead of the negative supply. Then this common lead will be used as the ground connection to your circuit.

- Have a kit of spare parts such as resistors, capacitors, transistors, diodes, etc. Sometimes it is just easier to replace a component to see if the circuit is fixed. For example, in discrete audio power amplifiers, you start with replacing the output transistors since they are usually the first ones to blow out. Also, in some cases with RF circuits, having spare crystals is a must. Crystals can stop working, or some of them could have been damaged due to excessive AC signal voltages across them. For example, some two-way radios have crystal oscillators where the crystals fail in about 10 or 20 years.

We have presented troubleshooting techniques for various circuits, including some not exactly well-designed circuits that were posted on the web or published. In many cases, troubleshooting is about detective work. This includes gathering data via specification sheets, schematics of similar circuits, using your basic test equipment such as a DVM or signal generator, and confirming the correct operation of the electronic components (e.g., an op amp used as an amplifier should have approximately 0 volts across its (+) and (–) input terminals). By process of elimination you can pinpoint the problem. For example, you can start probing for an output signal and work back to the input circuits.

As with any skill, troubleshooting circuits requires experience. The more circuits you are able to fix, the better you will be able to utilize your institutional knowledge for solving future problems. So even if you are not repairing a circuit at the moment, a good way to build up your electronics knowledge is to just read and study lots of schematics of manufactured products. These are readily available on the web these days.

If you are designing circuits or putting together sub-circuits, then whenever possible build a breadboard before you commit to a printed circuit board design. The reason for this is that not all circuits will "talk" to others flawlessly. By building a prototype first, you get to see just how well each circuit behaves under non-ideal conditions, such as with noisy power supplies or with a range of supply voltages. For example, does the circuit still work under worst-case conditions such as a specified low supply voltage or under maximum voltage input to the circuit?

Another item to look out for is to avoid the trap of KISS (Keep It Simple, Stupid) by trying to design your first circuit with minimum components count. You should look at the big picture and try to get the design to work first. Then, if possible, see if the parts count can be reduced without sacrificing performance and reliability. Some of my colleagues went by the KISS philosophy and found that their designs were lacking in performance and reliability. This caused them to cut traces and/or add piggy-back boards to their printed circuit boards. If they had just added a few more components in the first place, all this would have been avoided.

If you are designing circuits based on application notes, it's best to also breadboard these circuits. By doing this you will find out what works in terms of laying out the components and of the best wiring routes that will ensure a less noisy circuit. For example, some signal traces may have to be shielded via one or more ground planes on the board. Also, you may find that not all power supply decoupling capacitors work equally. For example, some ceramic capacitors may be too lossy or have poor capacitance tolerance due to working DC voltage limit and temperature. Again, for example, a 1-µf ceramic capacitor rated at 90 percent of its DC working voltage may only deliver about 50 percent of its specified capacitance such as 0.47 µf. In some circuits, such as active filters using op amps, 2 percent tolerance film capacitors and 1 percent metal film capacitors may be a better choice of components than using 10 percent ceramic capacitors and 5 percent resistors.

On another note, there are some DC-to-DC converter circuits that will input +5 volts from your USB port and provide one or more output voltages, such as

+12 volts, +12 volts and −12 volts, +3.3 volts, etc. Generally, these DC-to-DC converters will have high-frequency ripple noise at their outputs, and may radiate high-frequency noise via associated inductors. You may need to follow each output with a linear regulator (e.g., low–drop-out [LDO] adjustable voltage regulator), or add some series inductor-capacitor low-pass filters. For example, add a 1000-µH series inductor with a 1-µf shunt capacitor, such as L3 and C3 shown in Figure 15-17. Definitely if you are working with preamplifier circuits, make sure you do not directly connect a switching supply's voltage to a microphone or photo diode bias circuit. For an electret microphone bias circuit, add a low-pass filter such as a 220Ω resistor in series with the switching supply that is followed by a 470-µf shunt electrolytic capacitor and be mindful of polarity (e.g., if you are using a positive supply, the negative lead of the electrolytic capacitor is grounded). If you are biasing a photodiode such as supplying a negative voltage to the anode lead, then you can connect a 2200Ω resistor between the negative switching supply and the anode of the photodiode. Then connect a 1-µf to 10-µf ceramic capacitor to the photodiode's anode lead and ground the other lead of the ceramic capacitor to form a low-pass filter (that will reduce power supply noise).

So, these are my final thoughts, and good luck with troubleshooting circuits!

(Also following this chapter, be sure to check out Appendix A on test equipment.) See below for a preview on test equipment. Figure 15-19 shows a tester.

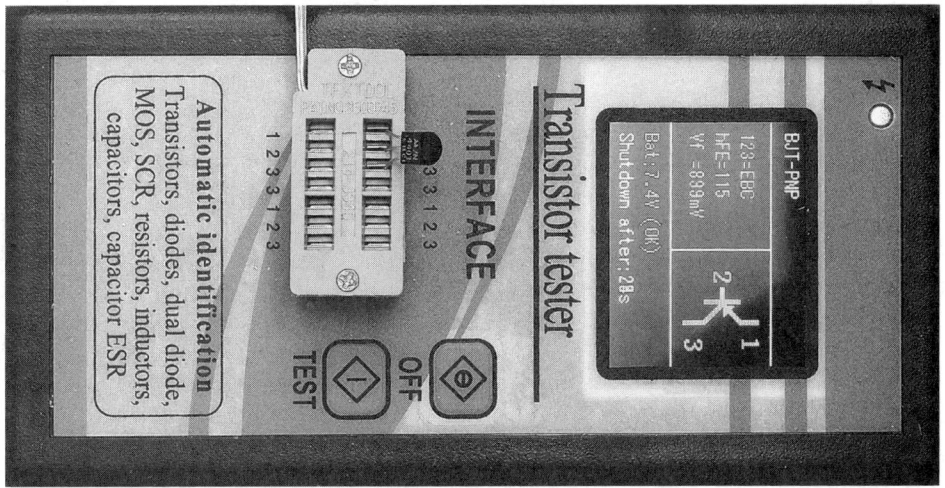

FIGURE 15-19 A component tester that not only tests various devices (transistors, FETs, diodes, capacitors, inductors, and resistors), but it also identifies the pin outs and polarity of the device. As shown above, the device is a PNP transistor with EBC pin out.

Choosing Test Equipment

Lab Power Supplies (Adjustable)

When purchasing a power supply we have to pay attention to whether it is a switching type or linear version. Switching power (switch mode) supplies are lighter than linear ones by a factor of at least two to one. For example, a 0- to 30-volt 10-amp switching supply weighs less than 4 pounds whereas a linear 0- to 30-volt 5-amp supply weighs about 12 pounds. See Figure A-1.

Both supplies can be adjusted for an output voltage of 0 to 30 volts, and allow for setting the current limit for safety. To adjust for current limiting, initially turn the current limit knob fully counterclockwise, which will set the current to zero. Then turn the voltage knob about halfway. Temporarily connect a wire across the negative and positive binding posts. You can now slowly turn up (e.g., clockwise) the current limit knob to your desired maximum current limit such as 100 mA. Remove the wire across the negative and positive binding posts.

The supplies show three terminals, negative, ground, and positive, which allows the supply to be a floating voltage source (e.g., think of a battery) not connected to ground when only the negative and positive terminals are used. To have a positive voltage source with respect to ground, connect a wire from the negative terminal to the ground terminal. And to provide a negative voltage source with respect to ground, connect a wire from the positive terminal to the ground terminal.

One note of caution, the switching power supply shown in Figure A-1 (left side) has slightly smaller diameter binding posts, which do not appear to accept standard banana plugs from the United States. But the linear supply in Figure A-1 (right side) has standard binding posts and there were no problems using standard banana plugs including double banana plug connectors.

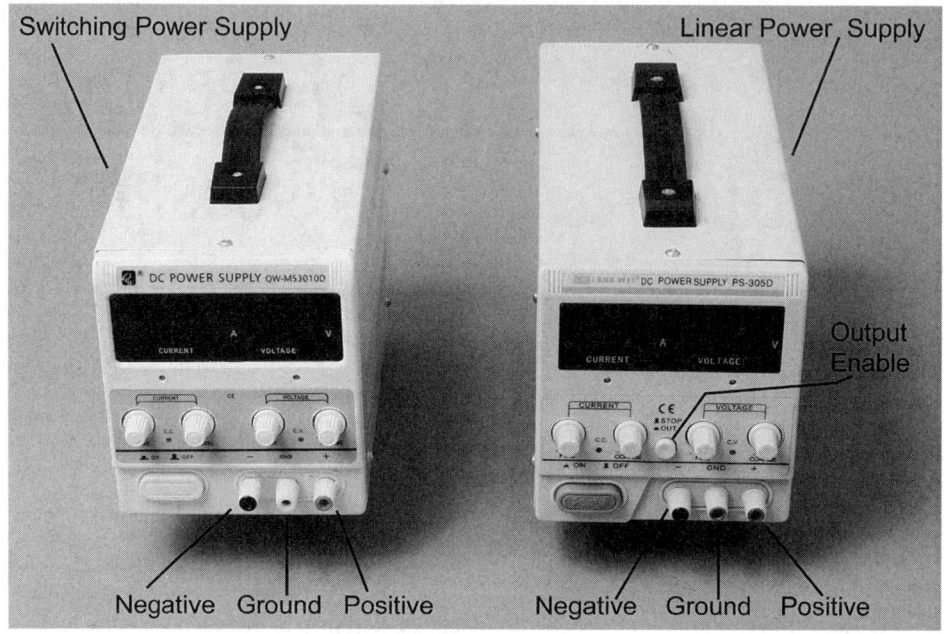

FIGURE A-1   On left side, a 10-amp switching supply such as the QW QW-MS3010D; and on the right side a 5amp linear version such as the LW Long Wei PS-305D, Lavota BPS-305, or Dr. Meter PS-305DM. Both supplies are adjustable from 0 to 30 volts.

Although both supplies in Figure A-1 look and cost (approximately $55 US) about the same, they are not. For almost all purposes, it's better to use the *linear* supply that measured with very low ripple/noise output, typically < 3 millivolts peak to peak, and a CV ripple rating of ≤ 1 mV RMS (root mean square).

However, the *switching* power supply is not desirable for working with analog circuits. The output ripple noise can be very high in the order of a few hundred millivolts to about a volt peak to peak. If we measure the noise from the switching supply of Figure A-1, we will see high amounts of noise. See Figure A-2, which shows the noise/ripple voltage from the QW model QW-MS3010D switching power supply.

As we can see in Figure A-2, the ripple noise looks like a high-frequency carrier (measured around 60 kHz, the switching frequency) that is amplitude modulated by a low-frequency signal with about an 8-millisecond period that translates to 120 Hz. Although the peak-to-peak voltage is high at over 1 volt, the RMS (root mean square) noise voltage will be much lower. For radio frequency work, a switching power supply is not recommended due to its excessive noise.

To identify a switching power supply, look for the "CV" ripple rating. If it is in the 10 mV RMS range it is likely to be a switching supply. In contrast a linear supply will typically have ≤ 1 mV RMS of CV noise or ripple. Also if the shipping weight is about half the weight of a comparable or even half amperage linear power supply, most likely again it's a switch mode power supply. For example, in Figure A-1 the

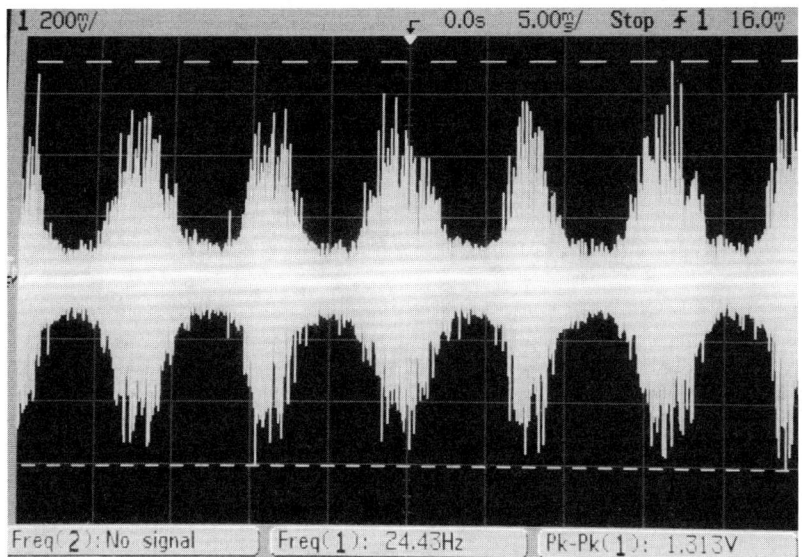

FIGURE A-2   Ripple from the switching power supply set to 12 volts DC from Figure A-1, where the ripple voltage is at least 1 volt peak to peak given the 200 mV/div scale.

weight of the switch mode (0–30 volts, 10 amps) is about 5 pounds whereas the linear version (0–30 volts, 5 amps) weighs about 12 pounds. One key word to look out for is "linear" when buying a power supply.

For general lab work, it's better to have a dual or triple linear power supply. You can typically purchase new ones at about $150 to $200 US. However, it is possible to order a used or refurbished analog/linear power supply such as an older Instek GW GPC-3020 triple power supply (two independent 0–30 volts at 2 amps, one fixed 5-volt floating supply at 3 amps, and it weighs 17 pounds). See Figure A-3.

Most double or triple supplies will be linear types and will weigh in the 15- to 30-pound range. For example, you can purchase a new linear dual supply with two adjustable 0- to 30-volt supplies or a linear triple supply with 0– 30 volts and a fixed voltage supply.

For purchasing a linear supply, look for the telltale clues: > 15 pound weight, low CV ripple noise rating ≤ 1 mV RMS, and the word "linear" in its description.

Today many linear supplies include fans to reduce their heatsink size, which can be noisy sound-wise. However, you can find a new supply with a giant heatsink and no fan, such as the triple supply Dr. Meter HY3005-3, dual 0– 30 volts at 5 amps with a fixed 5-volt, 3-amp output for about $200 US.

Also, you can also get an older supply without a fan as the Leader LPS-152, which delivers 0–25 volts at 1 amp and 0–6 volts at 5 amps refurbished for under $200 US, but make sure there is a warranty just in case. Other older, quieter linear supplies include the ones made by Lambda and Hewlett Packard (HP).

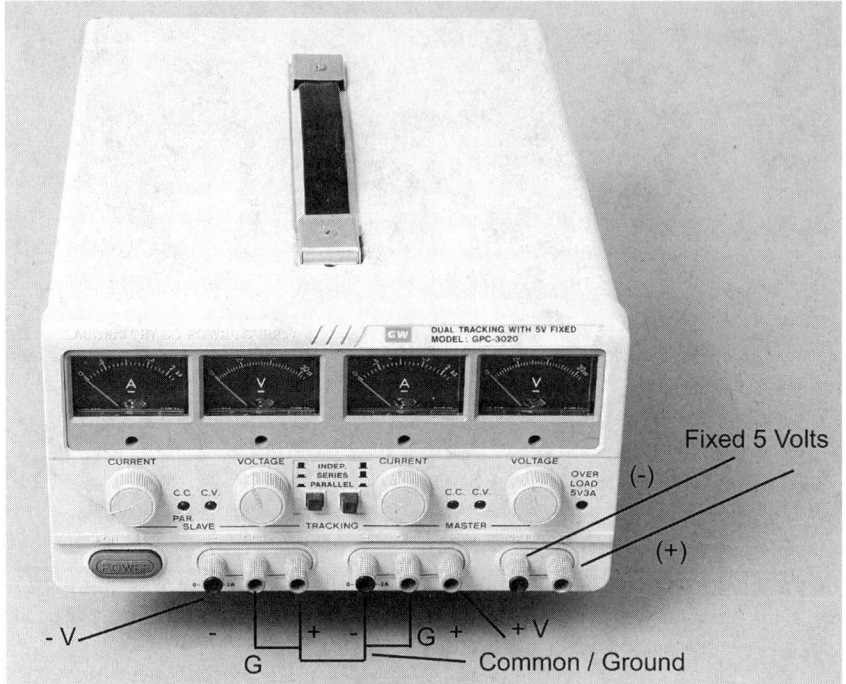

FIGURE A-3 A linear triple power supply bought used for less than $100 US. Example connection shown to provide +V and −V referenced to ground and a floating 5 volt source that can be configured as a +5-volt supply when its (−) terminal is connected to a ground terminal, G.

Signal Generators

You can purchase inexpensive general purpose signal generators at relatively low prices below $100 US. If you want higher performance, there are excellent signal generators in the $250 to $500 range. See Figure A-4 where these generators use Direct Digital Synthesis (DDS) technology to deliver almost any type of waveform with very stable and accurate frequency.

Both generators are crystal-controlled and provide two independent signals with the frequency accuracy of a crystal oscillator (e.g., < 0.01 percent frequency tolerance).

The JPS-6600 sometimes goes by other names such as KKmoon, Koolertron, or Dominty. You can buy it in versions of 25 MHz to 60 MHz with increasing prices. This generator works well in delivering two independent channels of signals, or they can be "tied or synchronized" together such as providing I and Q signals (0 and 90 degree signals) of the same frequency. Sine wave harmonic distortion is about 0.50 percent, so it may not be suitable for evaluating high-fidelity amplifiers. But you can add a low-pass filter (e.g., three-pole filter) to lower the distortion.

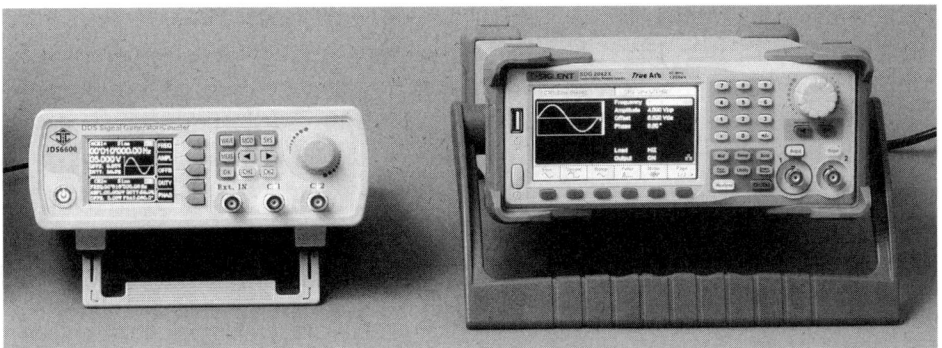

FIGURE A-4 On the left side, JDS6600, a relatively inexpensive two-channel DDS signal generator (approximately 100 USD), and on the right the Siglent SDG2042X, a very high-quality, two-channel generator at about 500 USD. Note that the JSD6600 also includes a precision frequency counter via its Ext. In BNC connector so that you can measure the frequency of a signal.

For very high performance the Siglent SDG2042 works very well and its sine wave distortion is typically < 0.05 percent.

Both generators will provide various signals such as sine, square, triangle, pulsed, and arbitrary waveforms. In addition they will generate modulated signals including AM (amplitude modulation) and FM (frequency modulation), and other modulated waveforms.

One note to pay attention to: If the output resistance/impedance/Z is set to 50Ω, then this means that the amplitude you set will be twice the amplitude when loading into a high-impedance load such as ≥ 100kΩ, and will be the correct voltage when the generator is loading into 50Ω. For example, if you set the amplitude for 4 volts peak to peak, the generator will deliver 4 volts peak to peak into a 50Ω load, but will give 8 volts peak-to-peak open circuit. This has been the standard for most generators made by HP (Agilent, Keysight), Leader, Siglent, and others.

For a lower-cost generator, such as the JDS6600 in Figure A-4 or the higher performance FY6800 (not shown), it follows a nonstandard convention. If you set the output Z (resistance) to 50Ω and set the amplitude to 5 volts peak to peak, the output voltage will be 2.5 volts peak to peak into 50Ω, and 5 volts peak to peak into an open circuit.

These generators have many features so you will need to read their instruction manuals.

Oscilloscopes

Probably one of the most useful troubleshooting instruments you can purchase in addition to a digital voltmeter (DVM) is an oscilloscope. It measures voltage as a function of time. This is useful, for example, in tracing sine wave signals at various inputs and outputs of an amplifier. It for instance can tell you where the test signal "disappears" in a circuit. There are two types of oscilloscopes—analog and digital.

The analog oscilloscope is very easy to operate compared to the menu-driven digital scope. For troubleshooting analog circuits, the analog scope will more than suffice. However, only a few manufacturers are selling them new.

The classic analog dual trace Tektronix 465 and 475 are still among the best to use for troubleshooting analog circuits. These Tektronix scopes include a Channel 1 vertical output signal via a BNC connector that drives 50Ω input test equipment such as an RF spectrum analyzer. This way you can probe the circuit on Channel 1 with a high impedance probe (e.g., 10 Meg Ω) while observing the frequency spectrum via a spectrum analyzer of the (time domain) signal displayed on the oscilloscope. Note that if you connect the 50Ω input of the spectrum analyzer straight to parts of your circuit, you will most likely load down the signal or cause an oscillator circuit to stop oscillating.

Digital oscilloscopes in 2019 are relatively inexpensive compared to those sold in the 1990s (e.g., > $2,000 US). A good lab version costs about the same as a high-end DVM (approximately $250 US). See Figure A-5 with three 2-channel versions.

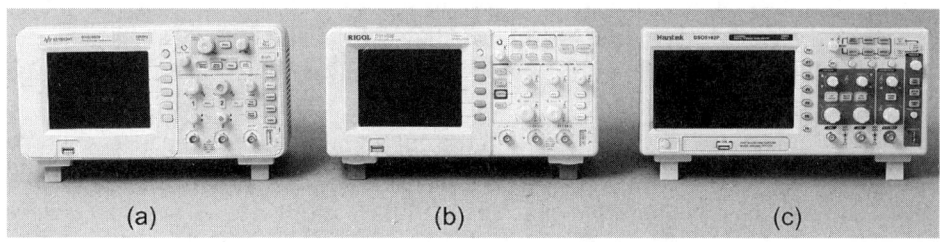

FIGURE A-5   Three digital oscilloscopes: (a) Keysight DSO1052B 50 MHz (approximately $300); (b) Rigol DS1102E 100 MHz (approximately $300); (c) Hantek DSO5102P 100 MHz (approximately $260).

Two 4-channel digital oscilloscopes that can be very handy for troubleshooting analog and digital circuits are shown in Figure A-6. These have 480×800 displays and ≥ 12M memory points (pts).

Some of the main features to look for in a digital storage oscilloscope (DSO) are:

- A 100 MHz bandwidth is good enough for most purposes. If you are troubleshooting RF circuits, you may need ≥ 200 MHz bandwidth. For the beginning hobbyist, a 50 MHz oscilloscope will work very well. Note all oscilloscopes will still respond to at least another 20 percent higher than its rated bandwidth, but the waveform displayed at these frequencies will not be accurate amplitude-wise due to a roll-off in high-frequency response. For example, a 50 MHz oscilloscope can usually still display a 60 MHz or 100 MHz signal, but with some attenuation.

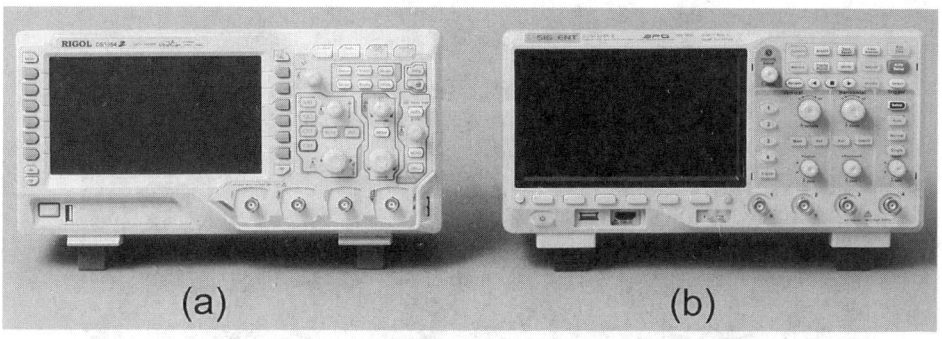

FIGURE A-6 Two 4-channel digital oscilloscopes: (a) Rigol DS1054 50 MHz (approximately $375); (b) Siglent SDS 1204X-E 200 MHz (approximately $760).

- Display resolution. The smaller display oscilloscopes, such as the Keysight DSO1052B and older Rigol DS1102E, have a 240×320 pixel display, which reproduces waveforms in a stepped manner. The Hantek DSO5102P, Rigol DS1054, and Siglent SDS 1204X-E have 480×800 pixels that will display a smoother, more analog-like waveform.
- Number of points stored, which can be as low as 16 thousand and as high as > 14 million. For critical work, go for at least 100 thousand points. Preferably, ≥ 1 million points will work very well.

Examples of Display Resolution and Number of Memory Points

Let's take a look at Figure A-7, which shows imperfect sine wave reproduction due to display limitations and limited memory (Keysight has 32K points and Hantek has 40K points). *Note: pts = points.*

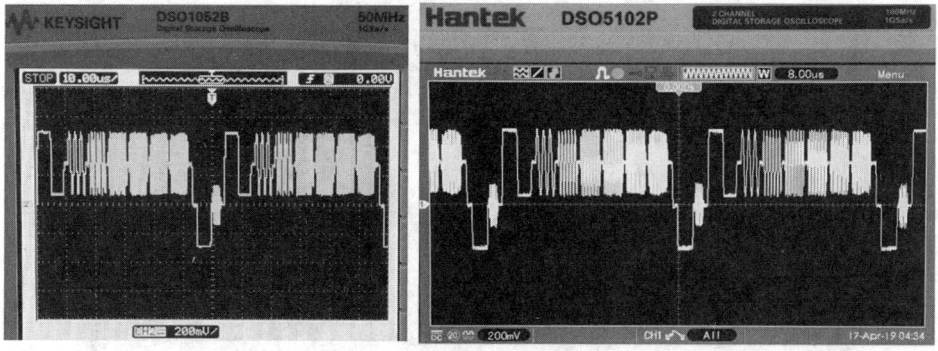

FIGURE A-7 Keysight with 240x320 pixels (32K pts); Hantek with 480x800 pixels (40K pts).

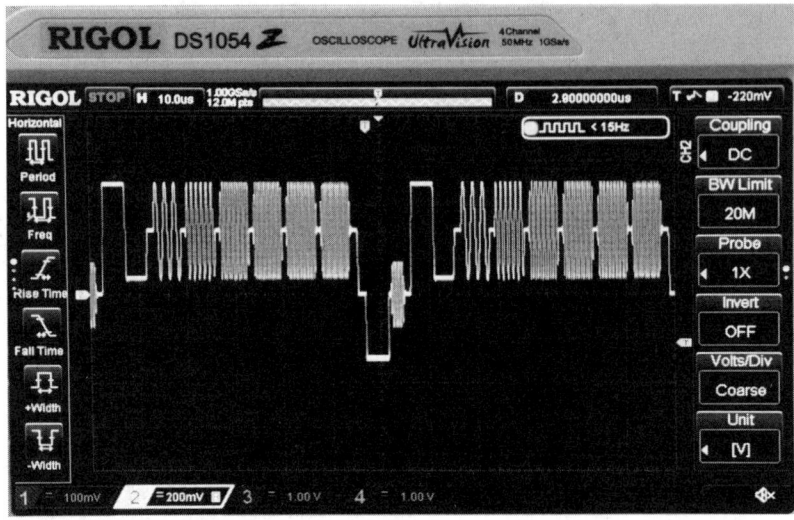

FIGURE A-8   Rigol 480x800-pixel display but with 12M points gives a better looking waveform.

As we can see in Figure A-8, the number of pixels and memory depth (e.g., number of memory points) really makes a difference in displaying a waveform with fidelity. We can also look at an analog oscilloscope's display of this waveform in Figure A-9.

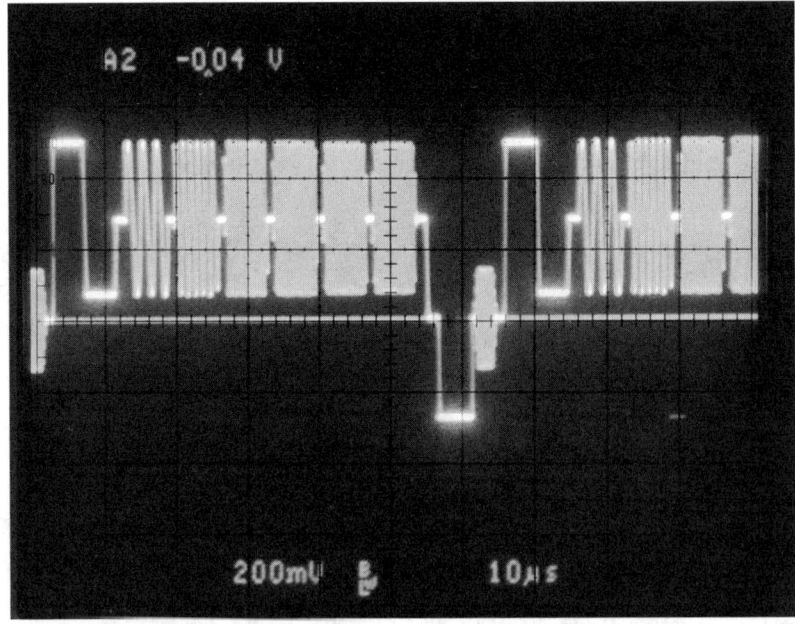

FIGURE A-9   A Tektronix analog oscilloscope shows a really clean display of the test signal. Note the lack of "jaggies" when compared to Figures A-7 and A-8.

NOTE: The test signal is a video multiburst signal starting with one cycle of squarewave, then followed with sinewave packets at increasing frequencies of 0.5 MHz, 1 MHz, 2 MHz, 3 MHz, 3.58 MHz, and 4 MHz.

When a digital oscilloscope has less than 100K memory points, it will generally display the waveform with aliasing problems due to insufficient sampling points of the waveform.

If we take the same test signal and slow down the horizontal sweep from 10 μsec/div (or 8 μsec/div) to 200 μsec/div we see the oscilloscope with the most memory points reproduces the waveform more correctly. See Figure A-10 where the waveforms are stored at 200 μsec/div.

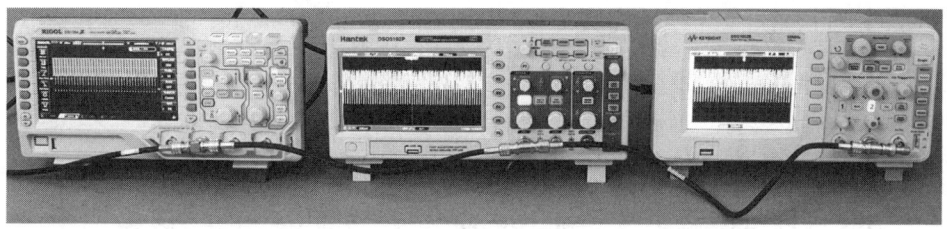

FIGURE A-10   From left to right: Rigol (12M points), Hantek (40K points), and Rigol (32K points), and all three oscilloscopes are set to the "Stop" mode to store the waveform. Only the Rigol (on the left side) with 12 million memory points seems to display the test signal properly at 200 μsec/div. The other two have random gaps in the waveform that are caused by sub-Nyquist sampling.

We can zoom in or magnify the stored waveform by setting the horizontal sweep to 20 μsec/div. See Figure A-11.

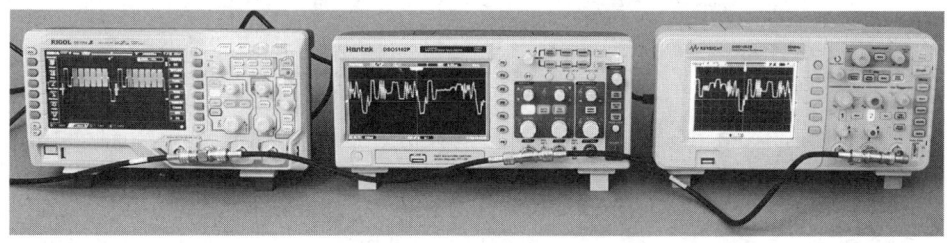

FIGURE A-11   Zooming into the waveform with 20 μsec/div, we see that only the Rigol on the left side reproduces the test waveform correctly. The Hantek is in the center.

As shown in the Rigol oscilloscope, the waveform is reproduced accurately with each sinewave packet increasing in frequency. See Figure A-11 above. However, we see with the 40K point Hantek, the test signal looks distorted in that we do not see sinewave packets increasing in frequency and the waveform is replaced with a pulse or a lower-frequency triangle wave.

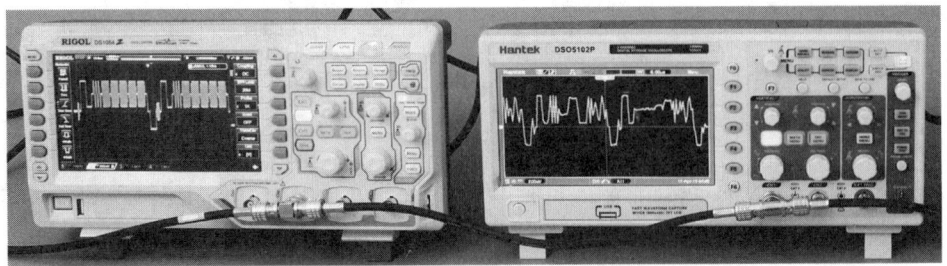

FIGURE A-12   A closer comparison between the $375 12M point 50 MHz Rigol DS 1054Z (on the left) and the 100 MHz $260 40K point Hantek DSO5102P (on the right).

In Figure A-12, the Rigol oscilloscope is the most expensive compared to the Hantek and Keysight. So this comparison is not taking into account the cost.

Fortunately, Hantek and Keysight make oscilloscopes with more memory points, such as the 100 MHz, 1M pt Hantek DS5102B ($359), or the 70 MHz, 1M pt Keysight DSOX1102A ($563) with higher-resolution display (compared to the 50 MHz Keysight DS1052).

Note that all digital oscilloscopes also include a set of built-in measurement functions such as an FFT spectrum analyzer, peak-to-peak voltage, frequency, pulse width, etc. Thus, it is important to read through the operating manual.

Oscilloscope Probes

You will need to use oscilloscope probes to monitor/measure various nodes of your circuit. In general, you should use a 10× probe for lowest capacitance loading (e.g., 10 pf to 20 pf, depending on the make and bandwidth).

NOTE: Sometimes the probes are named x1 or 1x and x10 or 10x.

Most fixed 10× probes have a bandwidth specification of 100 MHz or more, and they have typically a tip-to-ground lead capacitance of 10 pf to 15 pf. The load resistance will be 10MΩ when connected to an oscilloscope's BNC connector. A 10× probe actually attenuates the signal you are probing by 10-fold. For instance, if you are monitoring a 10-volt signal with a 10× probe, the oscilloscope will receive only 1 volt at its BNC input connector. Although there is attenuation, you have the advantage of essentially not loading the signal with your 10× probe. This is particularly important when troubleshooting high-frequency or high-impedance circuits.

In Figure A-5, fixed 10× probes are standard accessories for Keysight oscilloscopes.

When you buy a new oscilloscope today such as one from Hantek, Rigol, or Siglent, most likely they will supply you with switchable 1×/10× probes. See Figure A-13.

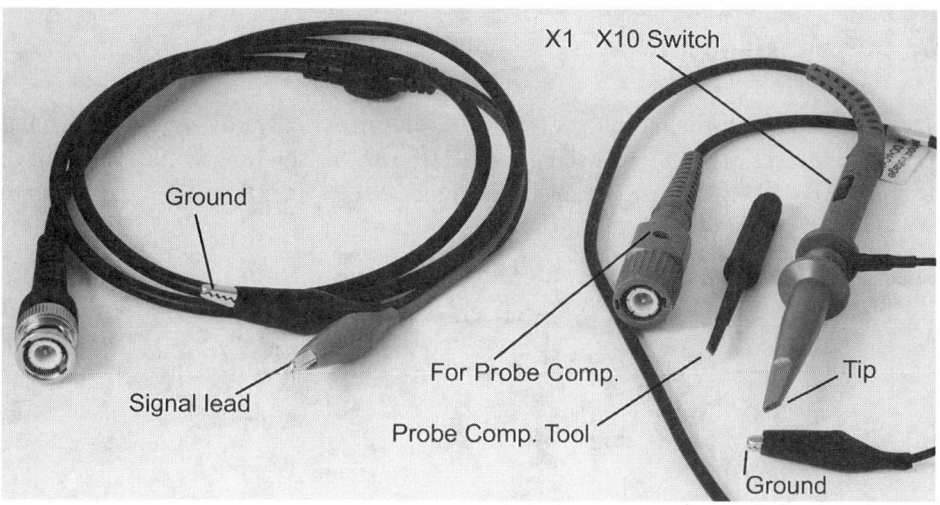

FIGURE A-13   A coaxial cable on the left, and a X1/X10 probe on the right with a probe compensation tool.

NOTE: The cable or probe's ground lead connects to the circuit's ground.

The coaxial cable in Figure A-13 is an inexpensive way to connect to your oscilloscope, is the same as a X1 probe with about 100 pf capacitance, and will have a 1MΩ load resistance via the oscilloscope's BNC input. Generally, we should avoid using the coaxial cable when working with high-frequency or high-impedance circuits. You can use them but just remember that the cables may load down your signal. For example, measuring with a coaxial cable (or with a X1 probe) may make your circuit appear worse in high-frequency response than it really is.

The switchable X1 and X10 probe acts pretty much like a coaxial cable when switched to the X1 setting (approximately 80 pf to 150 pf capacitance and 1MΩ resistance), while in the X10 mode it has an 11 pf to 20 pf load capacitance and 10MΩ load resistance.

Also shown in Figure A-13 is a small flat blade tool to adjust for probe compensation (comp.), which is a procedure for only 10× probes. The ×10 probe is connected to a squarewave calibration signal supplied in the front panel of the oscilloscope. Via a hole at the BNC connector (or at the probe), you adjust with the tool until the oscilloscope displays a "perfect" squarewave without rounding in the corners or without overshoot. Each probe you use must go through the probe compensation adjustment; otherwise, the frequency response you are measuring with an uncompensated probe will result in a non-flat frequency response that will cause measurement errors.

An Inexpensive Lab

If you are on smaller budget (< $100), you can actually buy an analog signal generator for less than $10, an oscilloscope in the $14 to $66 range, and a DC wall charger/power supply in the $10 range. See Figure A-14.

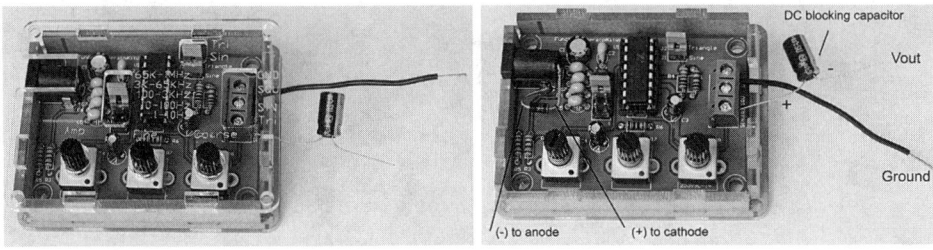

FIGURE A-14   A < $10 generator kit. Generator with supply protection diode.

You need to assemble this analog signal generator kit with a soldering pencil and wire cutters. Also, you need a power supply in the 10 volts to 18 DC range with the ground/sleeve portion labeled as the minus terminal (–) and the tip of the power connector (2.1 mm) marked as the positive terminal (+). A protection diode was added just in case the power supply is connected with wrong polarity. Also, a DC blocking capacitor (e.g., 100 µf 25 volts) was added because the output terminals have large DC offset voltages > 1 volt DC.

If you need a 1-channel oscilloscope that will work with low-frequency projects such as audio amplifier or circuits at < 50 kHz, there is an inexpensive $14 to $20 oscilloscope that requires a 9-volt DC power supply. Also, you can buy an oscilloscope with a rechargeable battery (via USB connector) for about $40. These oscilloscopes are not recommended for serious troubleshooting but can come in handy out in the field when you have limited space. See Figure A-15, which shows two oscilloscopes or o'scopes.

There are also USB oscilloscopes with two channels available from about $66 and up (e.g., the 20 MHz, 1M pt Hantek 6022BE), but you need a computer to display the waveforms. See Figure A-16. While a USB o'scope can be useful for troubleshooting out in the field, it may not have the high analog bandwidth or enough memory to compete with a lab oscilloscope such as one with a 200 MHz bandwidth and 14M points. Also, it is generally easier to use a lab scope with its knobs rather than using a USB oscilloscope with a computer's cursor.

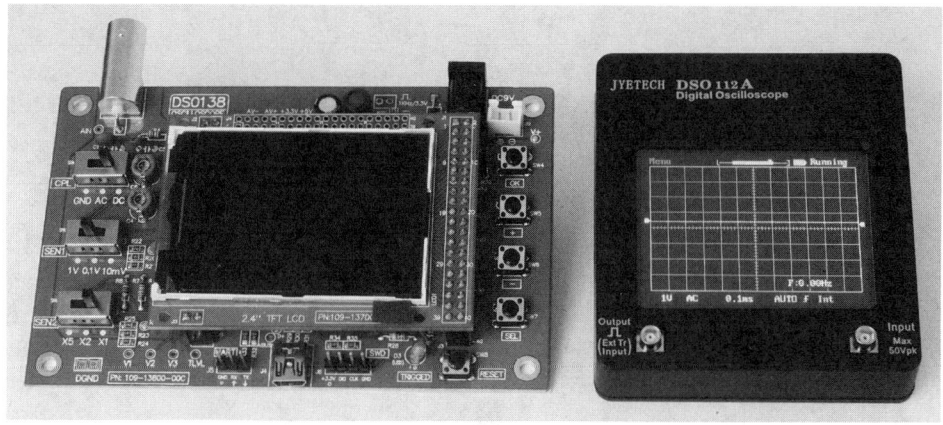

FIGURE A-15   A $14 o'scope on the left, and a $40 rechargeable battery version on the right.

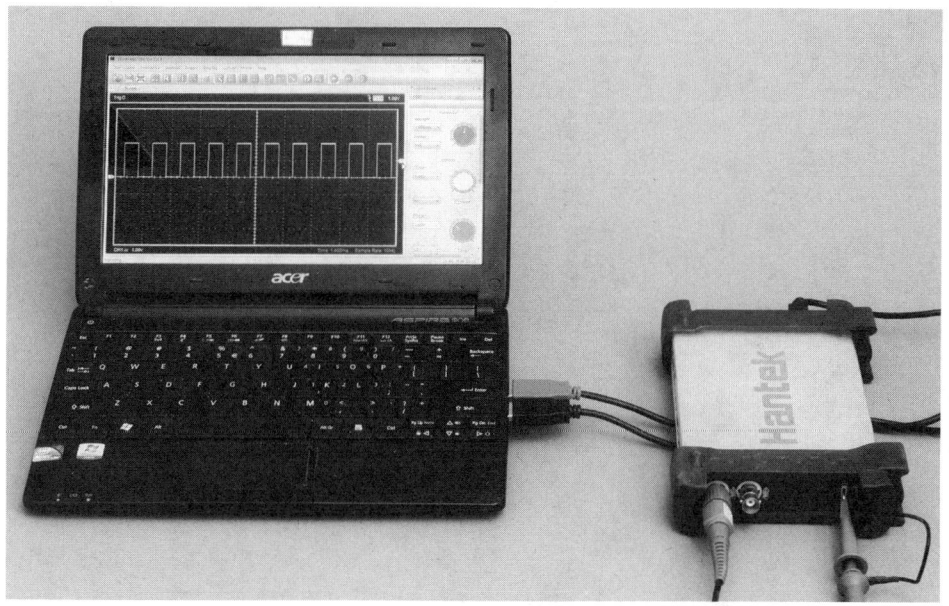

FIGURE A-16   A Hantek 2-channel USB oscilloscope with a 10-inch laptop computer.

APPENDIX B
Online Learning Resources

Below are some links from YouTube on using test equipment and learning electronics.

- Dave L. Jones at http://www.eevblog.com/ has lots of cool videos on test equipment, including discussions of electronic circuits

- Jeri Ellsworth has many videos on science and electronics at http://www.youtube.com/user/JeriEllsworthJabber

- The Kahn Academy on YouTube has many great videos on math and science at http://www.youtube.com/user/khanacademy

- For the advanced hobbyist, try the MIT OCW site for their televised classes at http://ocw.mit.edu/courses/find-by-department/

- Element14 electronics and technology videos are available on YouTube at http://www.youtube.com/user/element14

- The Amp Hour Podcasts showcase many technology related interviews by Chris Gammell and Dave Jones at https://theamphour.com/

APPENDIX C

Components and
Parts Suppliers

General Electronic Components

Transistors, FETs, Diodes, LEDs, Photodiodes, and ICs

- Mouser Electronics at www.mouser.com
- Digi-Key Corporation at www.digikey.com
- Anchor-Electronics at www.anchor-electronics.com
- Jameco Electronics at www.jameco.com
- Amazon at www.amazon.com; almost types of electronic parts and kits as well
- Joe Knows Electronics at http://www.joeknowselectronics.com; other types of electronic parts as well
- Frys Electronics at www.frys.com
- AliExpress at http://www.aliexpress.com

Low-Noise Transistors and JFETs, Including Matched Pairs

- Linear Integrated Systems, Inc. at www.linearsystems.com

Passive Components, Resistors, Capacitors, Fixed Valued Inductors, Transformers, Tools, Soldering Irons, Breadboards, and Solder

- Mouser Electronics at www.mouser.com
- Digi-Key Corporation at www.digikey.com
- Anchor-Electronics at www.anchor-electronics.com
- Jameco Electronics at www.jameco.com
- Frys Electronics at www.frys.com
- AliExpress at http://www.aliexpress.com

Kit Parts for Transistors, Diodes, Capacitors, Resistors, LEDs, and More

- Amazon at www.amazon.com
- eBay at www.ebay.com
- Mouser at www.mouser.com
- Digi-Key Corporation at www.digikey.com
- AliExpress at http://www.aliexpress.com

Ham Radio Parts

Crystals, Inductors, Capacitors, Transistors, RF Transistors, Transformers, and ICs

- Mouser Electronics at www.mouser.com
- Digi-Key Corporation at www.digikey.com
- Anchor-Electronics at www.anchor-electronics.com
- Jameco Electronics at www.jameco.com

Oscillator Coils, IF Transformers, and Audio Transformers

- Mouser Electronics at www.mouser.com for Xicon transformers and coils
- Centerpointe Electronics at www.cpcares.com for Xicon transformers and coils. For IF transformers, search under Xicon I.F. Transformer.

Antenna Coils

- Mike's Electronic Parts (Hard to Find Electronics Parts) at https://www.mikeselectronicparts.com for ferrite antenna coils, variable capacitors, and crystal earphones.
- AliExpress at http://www.aliexpress.com search for "diy AM/FM magnet coil," or ferrite rod or bar (use insulated magnet wire to wind your coil).
- eBay. Search under "ferrite bar" or "ferrite rod". These blank rods or bars can be wound with wire such as Litz wire to the desired inductance. Typically, the length is at least 2 inches or at least 50 mm.

Variable Capacitors

- Mike's Electronic Parts (Hard to Find Electronics Parts) at https://www.mikeselectronicparts.com for variable capacitors, antenna coils, and crystal earphones.

- eBay. Search under the terms "polyvaricon", "variable capacitor", "tuning capacitor", or "crystal radio parts". Note that some of the variable capacitors on eBay are timer capacitors (e.g., < 100 pf).
- AliExpress at http://www.aliexpress.com search for "polyvaricon" or "radio variable capacitor" and some come in two or four sections.
- Amazon at www.amazon.com

Science Kits, Cool Things, and Everything Else

- Evil Mad Scientist Laboratories, http://www.evilmadscientist.com/; components such as resistors, LEDs, displays, and also many DIY (do-it-yourself) kits including a discrete transistor version of the 555 timer and 741 op amp by Eric Schlaepfer.
- Elenco, http://www.elenco.com/product/educational; all sorts of educational kits.
- Amazon at www.amazon.com
- Jameco Electronics at www.jameco.com
- AliExpress at http://www.aliexpress.com search for "DIY kits" or "electronic kits"

Index

Page numbers followed by *f* and *t* refer to figures and tables, respectively.

0₁ ₁₄
√ √